The Whole Building Handbook

The Whole Building Handbook

How to Design Healthy, Efficient and Sustainable Buildings

Varis Bokalders and Maria Block

Earthscan works with RIBA Publishing, part of the Royal Institute of British Architects, to promote best practice and quality professional guidance on sustainable architecture

publishing for a sustainable future
London • Sterling, VA

First published by Earthscan in the UK and USA in 2010

Copyright © Varis Bokalders and Maria Block, 2010

All rights reserved

ISBN: 978-1-84407-833-2 hardback
978-1-84407-523-2 paperback

Translated by Susanne Kredentser and Miles Goldstick

Typeset by Domex e-Data, India
Cover design by Rogue Four Design www.roguefour.co.uk

For a full list of publications please contact:
Earthscan
Dunstan House
14a St Cross St
London, EC1N 8XA, UK
Tel: +44 (0)20 7841 1930
Fax: +44 (0)20 7242 1474
Email: earthinfo@earthscan.co.uk
Web: www.earthscan.co.uk

22883 Quicksilver Drive, Sterling, VA 20166-2012, USA

Earthscan publishes in association with the International Institute for Environment and Development

A catalogue record for this book is available from the British Library

Library of Congress Cataloging-in-Publication Data
Bokalders, Varis, 1944-
  [Byggekologi. English]
  The whole building handbook : how to design healthy, efficient, and sustainable buildings / Varis Bokalders and Maria Block. -- 1st ed.
      p. cm.
  Originally published: Byggekologi. Stockholm : Svensk byggtjèanst, 2004.
  Includes bibliographical references and index.
  ISBN 978-1-84407-833-2 (hardback) -- ISBN 978-1-84407-523-2 (pbk.) 1. Sustainable architecture. 2. Sustainable buildings--Design and construction. I. Block, Maria. II. Title.
  NA2542.36.B6513 2009
  720'.47--dc22
                                                              2009002968

This book was printed in the United Kingdom by Butler Tanner & Dennis, an ISO 14001 certified company. The paper used is certified by the Forest Stewardship Council (FSC), and the inks are vegetable based.

Contents

Introduction ix

1 Healthy Buildings 1

1.1 Materials and Construction Methods 5
1.1.0 *Selection of materials* 6
1.1.1 *Criteria for selection of materials* 9
1.1.2 *Knowledge of materials* 14
1.1.3 *Assessment of materials* 28
1.1.4 *Choosing a construction method* 101

1.2 Services 111
1.2.0 *Interior climate* 112
1.2.1 *Ventilation* 117
1.2.2 *Electrical services* 144
1.2.3 *Plumbing* 155
1.2.4 *Heating systems* 162

1.3 Construction 171
1.3.0 *Build correctly from the start* 172
1.3.1 *Damp* 174
1.3.2 *Radon* 178
1.3.3 *Sound and noise* 182
1.3.4 *Ease of cleaning* 187

1.4 Implementation 193
1.4.0 *Environmental management* 194
1.4.1 *Planning and procurement* 198
1.4.2 *Economics* 211
1.4.3 *The construction site* 215
1.4.4 *Recycling construction materials* 219

2 Conservation   225

2.1	**Heating and Cooling**	**232**
2.1.0	*Heating efficiency*	233
2.1.1	*Insulation*	248
2.1.2	*Windows*	254
2.1.3	*Heat recovery*	264
2.1.4	*Architecture*	269

2.2	**Efficient Use of Electricity**	**284**
2.2.0	*Use of electricity*	285
2.2.1	*Appliances*	288
2.2.2	*Lighting*	296
2.2.3	*Electrical devices*	306
2.2.4	*Getting things done without electricity*	309

2.3	**Clean Water**	**317**
2.3.0	*Water use*	318
2.3.1	*Water-saving technology*	321
2.3.2	*Hot water*	327
2.3.3	*Water supply*	331
2.3.4	*Water purification*	341

2.4	**Waste**	**345**
2.4.0	*Waste from human activity*	346
2.4.1	*Waste sorting*	352
2.4.2	*Composting*	358
2.4.3	*Recycling*	362
2.4.4	*Ecological design*	368

3 Ecocycles — 371

3.1 Renewable Heat — 376
3.1.0 *Biomass, solar energy and accumulation tanks* — 377
3.1.1 *Bioenergy* — 383
3.1.2 *Solar heating* — 398
3.1.3 *Heat pumps* — 404
3.1.4 *Cooling buildings* — 411

3.2 Renewable Electricity — 416
3.2.0 *Production of electricity in a sustainable society* — 417
3.2.1 *Combined heat and power with biomass* — 423
3.2.2 *Hydropower* — 427
3.2.3 *Wind and wave power* — 431
3.2.4 *Solar cells* — 436

3.3 Sewage — 447
3.3.0 *Sewage in ecological cycles* — 448
3.3.1 *Sewage separation at the source* — 451
3.3.2 *Technological methods of purification* — 461
3.3.3 *Natural methods of purification* — 466
3.3.4 *Nutrient recycling* — 471

3.4 Vegetation and Cultivation — 479
3.4.0 *Permaculture* — 480
3.4.1 *Vegetation structures* — 484
3.4.2 *Vegetation on and in buildings* — 490
3.4.3 *Gardens* — 496
3.4.4 *Ecological agriculture and forestry* — 503

4 Place 511

4.1 Adaptation to Natural Surroundings 515
4.1.0 *Local conditions* 516
4.1.1 *Geology and topography* 519
4.1.2 *Hydrology* 524
4.1.3 *Flora and fauna* 530
4.1.4 *Adaptation to climate* 534

4.2 The Social Fabric 546
4.2.0 *The sustainable municipality* 547
4.2.1 *Traffic* 553
4.2.2 *The holistic town* 575
4.2.3 *Town–country* 589
4.2.4 *Cultural values* 603

4.3 Existing Buildings 611
4.3.0 *The use phase* 612
4.3.1 *Operation and management* 615
4.3.2 *Energy conservation* 618
4.3.3 *Decontamination* 625
4.3.4 *Rebuilding* 629

4.4 People 643
4.4.0 *People's needs* 644
4.4.1 *Comfort* 648
4.4.2 *Room for everyone* 652
4.4.3 *Participation* 655
4.4.4 *Beauty* 660

Bibliography 671
Index 675

Introduction

There is overwhelming scientific evidence that environmental problems, including climate change, are a serious global threat, demanding an urgent global response. Many researchers and analysts warn that we have only a few decades in which to achieve sustainable development, to prevent catastrophic environmental changes. The design and methods of construction of our built environment – our homes, workplaces and cities – have an enormous impact on both the global environment and locally on the communities that inhabit them. Creating a sustainable built environment is therefore a crucial part of the transformation needed to achieve true sustainability.

Environmental Changes and Impacts

Human activities and the technologies we use can cause many problems. Our extensive use of fossil fuels and hazardous chemicals pollute the atmosphere, water and soil – the essential commodities for our survival. Burning of fossil fuels discharges carbon oxides (CO_x), sulphur oxides (SO_x) and nitrogen oxides (NO_x), which affect the climate and the ozone layer, and contribute to acidification of soil and water. Many current human activities, including the burning of fossil fuels and massive deforestation, cause increased CO_2 and other 'greenhouse gas' levels in the atmosphere. This creates the enhanced greenhouse effect, heating up the average global temperature and causing climate change. Calculations show that a temperature increase of 1.6–6°C is possible, which would lead to a 15–100cm rise in ocean water levels. There will be a shift in climatic zones over a large part of the Earth. Extreme weather conditions such as storms, floods and drought will become more common. The social and economic impacts of climate change could be huge. In 2006, Sir Nicholas Stern, former Head of the British Government Economic Service and Adviser to the Government on the economics of climate change and development, wrote a report on the economics of climate change. He reported that stabilization of CO_2 levels requires that annual emissions be brought down to at least 80 per cent below current levels. The cost of action was estimated to be limited to around 1 per cent of global GDP each year if adequate action were to be taken immediately. Since then, Stern has revised this up to 2 per cent of GDP because global warming is happening faster than previously predicted. If no action is taken, the cost could be between 5 and 20 per cent of GDP each year, now and forever. And it will be difficult or impossible to reverse the changes.

Population growth and our overexploitation of resources also cause massive problems. It is currently believed that oil reserves will last for 30 years, natural gas reserves for 70 years and coal reserves for 250 to 300 years. Fresh water is scarce and unequally distributed, groundwater reserves are being used up, and water is being polluted so that it is undrinkable, meaning that water management will very soon be one of the world's greatest problems. Arable land is a limited resource: fertile topsoil is being lost because of erosion and agricultural land is being destroyed by salinization, water logging and the building of cities on arable land. The seas and oceans are heavily over-fished; without a decrease of current fishing, many fish stocks will disappear. Considerable deforestation is taking place in many parts of the world, especially in the tropical rainforests. Biological and genetic depletion is increasing both on land and in the water. Cheap fossil fuels have made possible a growth in population that is without historical precedent. According to predictions, the global population will increase from 6.7 billion in 2007 to 9 billion in 2050. How large a population can our Earth support?

Sustainable Development

The 1987 Brundtland Commission's report, *Our Common Future*, at the request of the United Nations, established an ethical principle that should be self-evident: 'We must satisfy our generation's needs without destroying the opportunities for future generations to satisfy their needs.' It was the Brundtland Commission that launched the concept of 'sustainable development'. The task the United Nations gave to the countries of the world was to merge technology, economics and sustainable development with a new lifestyle based on equity. It is thus a question of ecological, economic and social sustainability. Human survival and human welfare may depend on our success in transforming principles of sustainable development into a global ethic: 'thinking globally and acting locally'.

Conventional economic practices take no account of vital natural assets that have no monetary price, such as clean air, clean water, nature, etc. A country's wealth and well-being is typically judged by its GNP. A re-evaluation of the GNP concept, where negative environmental effects are included as minus amounts in a country's welfare – a green GNP – would be a better measure of national economic development.

In Balance with Nature

Planning a sustainable society requires a holistic approach in which we learn from and cooperate with nature. Our planet and its ecosystems is a complex whole where plants, animals, people and micro-organisms all form an integral part: everything is connected to everything else, nothing disappears, everything must go somewhere. We can take from this some fundamental principles for environmental sustainability, including: renewable resources must be managed in a sustainable way; non-renewable resources must be recycled; air, water and soil need to be kept clean and biodiversity has to be maintained.

Lifestyle

Transition to sustainable technology and renewable energy sources is not enough to achieve sustainable development. We also have to change our lifestyle. There is a huge difference in lifestyle and resource use between poor and rich countries, and between poor and rich people within countries. If all of the people in the world had a lifestyle similar to the average person in the European Union, four planet Earths would be needed to satisfy everyone's energy and resource needs.

Western lifestyles result in consumption of energy and resources in three main sectors: transport, food and housing. To achieve a sustainable society we have to change how we travel, eat and live. This includes using less petrol, eating less meat, living in energy-efficient buildings and changing our focus from quantity to quality, from material consumption to non-material well-being.

Sustainable Building

A very large proportion of the energy used in the world, and the greenhouse gases that are released from this energy use, is connected to the building sector. It is clear that no move towards sustainable development can go ahead without radical changes in architecture, construction and spatial planning. We are now seeing a huge drive to conserve energy, increase efficiency and create zero-carbon buildings, all of which are vital in reducing the environmental impacts of buildings. But building sustainably must also take a broader approach, including the whole impact of a building – on the environment, people's health and social well-being – throughout its whole lifetime. In order to build truly sustainable buildings and cities, architects and planners need to think holistically and have a comprehensive grounding in all aspects of sustainable building.

This Book

A Holistic Perspective and the Tree Structure

This book provides the knowledge required to understand what is involved in planning and building sustainably. Our thesis is that sustainable planning and building requires a holistic perspective based on a comprehensive and integrated approach, and an understanding of the different parts that are important for the whole. Our tree, presented on the inside cover of the book, represents the whole, and the branches and canopy of leaves represent the interconnected components. The tree is a tool that not only illustrates the holistic perspective but also the structure of this book.

Professionals working with sustainable planning and building must understand the whole, be able to break down the whole into its parts and learn about the full range of components. Once knowledge is acquired about the various components, they can be used to construct new wholes. The tree structure can be used as a checklist in this process.

The book is based around four fundamental aspects, which form the four main chapters, or branches, of the book:

1 Healthy Buildings: Constructing healthy buildings is achieved through choosing materials that are suitable from the perspective of health and the environment. Services should provide a healthy and comfortable interior climate. Technical workmanship should avoid problems with moisture, radon and noise, as well as facilitating cleaning and maintenance. Environmental goals guide the entire planning and building process.

2 Conservation and Efficiency: Conserving resources is achieved by making buildings that use resources efficiently, where heating needs and electricity use are minimized and water-saving technologies are used. The amount of waste is reduced, and waste is separated into different categories, to be composted, recycled or reused.

3 Ecocycles: Closing ecological cycles is achieved by producing heat and electricity using renewable energy. Sewage systems are designed so that nutrients can be recycled. To ensure that organic material from waste and sewage is returned to arable land, vegetation and cultivation must be integrated with settlements.

4 Place: Adaptation to local conditions means that a site must be studied with respect to nature, climate and community structure, as well as human activities. In order to achieve harmony with nature and for people, the conditions of the site must be used as the point of departure for planning. Existing development is made use of and environmentally adapted.

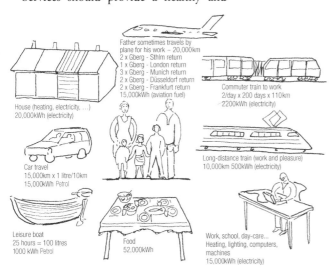

A normal Swedish family's energy use. Architect Hans Eek, who works in Gothenburg and lives in Alingsås, has analysed his family's energy consumption pattern. The energy use can be roughly divided into three: how we transport ourselves, what kind of houses we live and work in, and what we eat.

INTRODUCTION xi

The Natural Step Framework's definition of sustainability includes four system conditions (scientific principles) that must be met in order to have a sustainable society.

To become a sustainable society we must ...

1. eliminate our contribution to the progressive build-up of substances extracted from the Earth's crust – for example, we must not take heavy metals and fossil fuels out of the earth and systematically distribute them throughout the world in a way that causes damage.

2. eliminate our contribution to the progressive build-up of chemicals and compounds produced by society – for example, we must not produce stable organic toxins (chemicals) like dioxins, PCBs, and DDT and systematically distribute them throughout the world in a manner that causes damage.

3. eliminate our contribution to the progressive physical degradation and destruction of nature and natural processes – for example, we must not systematically extract more from nature than the rate of re-growth allows. With regard to forests, plants, fish and animals, we should live on the interest (additional natural growth) and not on the capital (deforestation).

4. eliminate our contribution to conditions that undermine people's capacity to meet their basic human needs – for example, as the Earth is a planet with limits, and the population is growing we should conserve resources and strive for a just division. The more people there are who squander resources, the larger the number of people forced to live in poverty.

Varied International Knowledge and Experiences of Sustainable Planning and Building

Interest in sustainable planning and building has grown with rising oil prices. The increased understanding of climate change and environmental problems in general, and corresponding knowledge about how to plan and build in a sustainable manner, have improved over the last 30 years. However, this expansion of knowledge has varied in different parts of the world and is often poorly communicated between different countries and regions. As environmental problems become increasingly apparent, we believe that the knowledge developed in different regions needs to be combined and utilized.

In this book we present approaches and experiences from around the world with a focus on those from Scandinavia, our home region, in which we have worked for 35 and 15 years respectively. We have compiled information from a wide range of literature and research reports, many of which are unavailable in English. During this work we have developed a structure that shows the complexity of the issue and we have organized this book by that structure in order to guide the reader through the different parts of our holistic approach. Our aim has been to bring together the necessary knowledge to enable others to plan and build sustainable buildings for a sustainable future.

Healthy Buildings

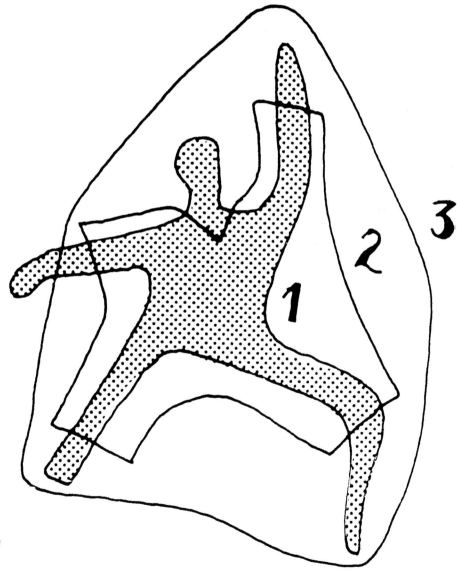

A building influences the well-being of those who spend time in it. A building can be called the third layer of skin, and clothes are the second.

Source: *De siste syke hus* (The Last Sick Buildings), Björn Berge, 1988

Building Healthy Buildings

Building healthy buildings deals primarily with the interior environment, i.e. how to construct buildings that people feel good in, but also how to do so without adverse effects on construction workers and the environment. In some climates people spend more than 90 per cent of their time indoors. This means, of course, that people are highly influenced by the buildings they live and work in. This concept can be compared with the kind of clothes we wear and how their characteristics greatly influence us and our well-being.

Healthy Construction Materials –
Choosing Materials From Environmental and Health Viewpoints

Construction materials are described here according to their chemical content and their influence on the environment from a life-cycle perspective. A general assessment is given of materials that have the least impact on environment and health.

Supply –
Choosing Services for Ventilation, Electricity and Water

Which supply options should be chosen to get a comfortable interior climate with pleasant warmth and low electromagnetic fields, and how can a building be protected from water damage? How should ventilation, electricity and pipework systems be designed?

Design –
Choosing Good Design for Resource Conservation and Well-being

Even with healthy materials and good supply systems, things can go wrong if the design isn't right. How can problems with moisture, radon and noise be avoided? How can building design achieve ease of cleaning and maintenance?

Implementation –
How the Chosen Environmental Aims are Achieved

Building sustainably requires new knowledge, better coordination and environmental management so that consideration for the environment is not lost along the way. New routines are needed at construction sites, together with a different attitude to construction waste.

Questionnaire for surveying peoples' experience of the interior climate.

Source: Örebro Länslasarett (County Hospital), Kjell Andersson et al

Current Problems

During the last three months have you had any of the problems/symptoms below?
(Answer each question even if you haven't had any of the problems/symptoms!)

	Yes, often (every week) (1)	Yes, occasionally (2)	No, never (3)	If yes: Do you think it has to do with your working environment? Yes (1)	No (2)
5–6 Fatigue	☐	☐	☐	☐	☐
7–8 Heavy-headed	☐	☐	☐	☐	☐
9–10 Headache	☐	☐	☐	☐	☐
11–12 Nausea/dizziness	☐	☐	☐	☐	☐
13–14 Difficulty concentrating	☐	☐	☐	☐	☐
15–16 Itchiness, burning, eye irritation	☐	☐	☐	☐	☐
17–18 Irritated, stuffed up or running nose	☐	☐	☐	☐	☐
19–20 Hoarseness, dry throat	☐	☐	☐	☐	☐
21–22 Cough	☐	☐	☐	☐	☐
23–24 Dry or reddened facial skin	☐	☐	☐	☐	☐
25–26 Flaking/itchiness in the scalp/ears	☐	☐	☐	☐	☐
27–28 Dry, itchy, red skin on the hands	☐	☐	☐	☐	☐
29–30 Other..	☐	☐	☐	☐	☐

Environmental Impact of Building Materials

Choosing the right material from an ecological perspective is not always easy since there are many factors to take into consideration. Important aspects are how materials influence the interior climate (and so the people who spend time there); how materials impact the outdoor environment in general; and how plants and animals are influenced by the discharges caused by the production and disposal of such materials. Furthermore, there is also the issue of the working environment for all those who work producing materials, and undertaking construction and demolition.

Sick Buildings

There has been debate about sick building syndrome (SBS) since the 1980s. The number of sick buildings has increased with time. It is important to try and understand the causes of sick building syndrome so that they can be avoided in future. About 10–30 per cent of modern buildings may cause SBS. Those who spend time in these buildings show one or more of the following symptoms: eyes/nose/throat irritation, throat infections, sinus infections, a feeling of dryness in mucous membranes, dry lips and skin, itchy face and scalp, skin redness, eczema, fatigue, lack of concentration, nausea, headaches and allergy problems.

It is not possible to confirm that a building is sick by using technical measurements. Peoples' experiences and problems are used as the basis for judgements. At Örebro County Hospital (Kjell Andersson, et al) a suitable interview model has been developed to determine whether a building can be regarded as sick. First, people fill in a specially formulated questionnaire about their experiences of the interior climate. Then the answers are presented graphically in 'roses', one for environmental factors and one for problems and symptoms experienced.

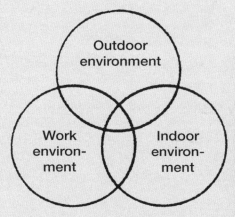

When choosing materials from an environmental point of view, it is important to weigh up a material's impact on the outdoor environment together with its impact on the interior environment (the health of users), as well as on the working environment.

Working Environment			
During the last three months have you **been bothered** by one or more of the following factors at **your workplace**? (Answer each question even if you haven't felt bothered!)	Yes, often (every week) (1)	Yes, occasionally (2)	No, never (3)
28 Draught	☐	☐	☐
29 Room temperature too high	☐	☐	☐
30 Varying room temperature	☐	☐	☐
31 Room temperature too low	☐	☐	☐
32 Stuffy ("bad") air	☐	☐	☐
33 Dry air	☐	☐	☐
34 Unpleasant odour	☐	☐	☐
35 Static electricity so that you get shocks easily	☐	☐	☐
36 Second-hand tobacco smoke	☐	☐	☐
37 Noise	☐	☐	☐
38 Light that is too weak or causes glare and/or reflection	☐	☐	☐
39 Dust and dirt	☐	☐	☐

Causes of Sick Buildings

Much is known about why buildings make people sick or cause difficulties:

- poor materials (release of chemical emissions, fibres or allergens);
- poor ventilation (relative humidity, temperature, odours);
- moisture problems (poor construction, carelessness, insufficient drying times, etc.);
- electric and magnetic fields (wiring, appliances, metal, etc.);
- problems with the ground (radon, moisture);
- disturbance from the surrounding environment (traffic, noise, air pollution, etc.);
- inappropriate use (use of chemicals, more showering than a waterproof area can handle, smoking);
- insufficient cleaning and maintenance;
- lack of comfort;
- lack of architectural attractiveness.

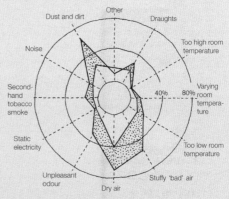

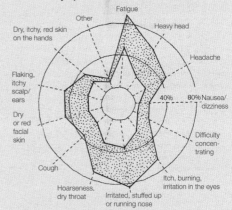

The results from an interview survey of a suspected sick building are compiled in 'rose' figures.

Source: Örebro County Hospital, Kjell Andersson et al

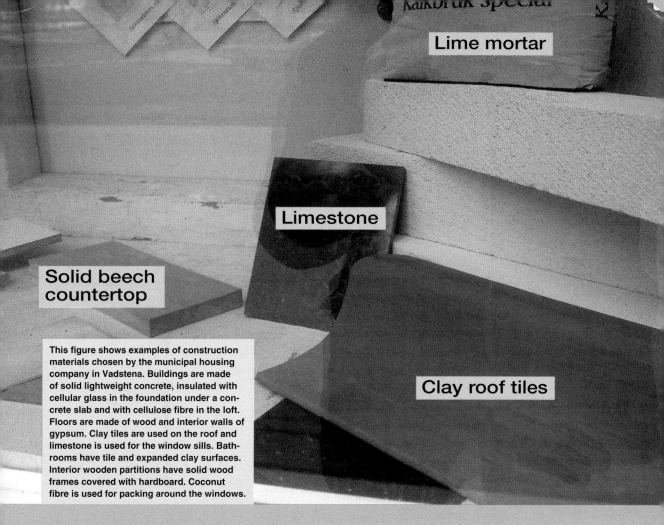

This figure shows examples of construction materials chosen by the municipal housing company in Vadstena. Buildings are made of solid lightweight concrete, insulated with cellular glass in the foundation under a concrete slab and with cellulose fibre in the loft. Floors are made of wood and interior walls of gypsum. Clay tiles are used on the roof and limestone is used for the window sills. Bathrooms have tile and expanded clay surfaces. Interior wooden partitions have solid wood frames covered with hardboard. Coconut fibre is used for packing around the windows.

1.1 Materials and Construction Methods

It is difficult to choose the right material since a lot of knowledge is required. For example, knowledge of chemistry is needed since chemicals influence health and the ecosystem. Furthermore, information is needed about resource consumption and the environmental impact of materials in production and use. The environmental impact of a material from cradle to grave can be measured using life-cycle analysis (LCA) and reported in the form of an environmental profile. Some analysts study materials from cradle to cradle, as nothing should be wasted in a sustainable society. The next problem is to consider all the facts and make a judgement. It is a matter of understanding how a material influences the interior environment and the health of users, how management of the material influences the health of construction workers, as well as how use of the material impacts the ecosystem.

1.1.0 Selection of Materials

In order to evaluate a material it is necessary to know what it contains, how it is made, the quantity of resources used, the emissions released, how much and what kind of energy is used in production and transport, and what the resulting residues are. This means that access to product information documentation (environmental declarations) is required. This can be difficult to obtain and is not always complete. Similar materials may have different contents and different production processes. In order to make a good environmental choice, it is not enough to know the impact of a material type, the impact of the specific product must be to hand.

Construction Material Specifications

The specification of a construction material should include the following:

- material name, short description and area of use;
- producer and supplier and their environmental policies and environmental management systems;
- declaration of ingredients, noting any recognized hazardous substances and environmental labelling.

1 Input materials – resource consumption: Raw materials and additives and in what amounts. Emissions to air, water and land. Creation of hazardous waste. Are recycled materials included? The sum total of energy used (raw and recycled materials). Source and method of transporting raw materials/ingredients. Type of energy source used (renewable or non-renewable).

2 Production – production process: Type of energy source, energy consumption, energy quality, emissions to air, water and land. Are residual products created, and if so are they used in other production? Is hazardous waste created?

3 Distribution – finished construction material: Production location/country, method of transport, forms of distribution, type of packaging, does the producer take back the packaging material?

4 Construction phase – building process: The need for equipment and machines. Document the need for consumable supplies for the building process. Emissions to air, water and land. Are hazardous wastes created? Are there any customized materials? Are leftover materials taken back?

5 Use phase – operation, maintenance: Operation, energy source, materials necessary to maintain function and characteristics during the use phase (e.g. cleaning compounds and lubrication oil). Maintenance (surface treatment, filters, parts that wear out, etc.). Life.

6 Demolition – dismantling: Ease of dismantling. Does the product require special measures for protection of health and the environment?

7 Residual products – reuse, recycling: Can the residual products be reused, the material recycled or energy extracted from it? Emissions during combustion?

8 Waste products – dumping: Emissions to air, water and land during dumping. Should the waste be managed according to hazardous waste regulations?

9 Inner environment – interior environment: Ingredients contain substances

Declaration of Ingredients

Material	Proportion, weight/per cent	Classification	Einecs-no	CAS-no
Butane [<0.1 % 1,3-butadiene (203-450-8)]	1–15	F+;R12	203-448-7	106-97-8
Dimethylether	1–15	F+;R12	204-065-8	115-10-6
Propane	1–15	F+;R12	200-827-9	74-98-6
Metylene difenyl diisocyanate polymer	30–60	Xn;R20 Xi;R36/37/38 R42/43	–	9016-87-9
Chlorine-paraffin C14-17	10–25	–	287-477-0	85535-85-9

Source: Part of a product information sheet. The declaration of ingredients is for a joint filler from Würth.

hazardous to health that have CAS (Chemical Abstracts Service) numbers. Requirements for storage and carrying out work in order to avoid negative effects on the interior environment. Emissions and odours from the building itself. Requirements for the surrounding materials.

The following threshold values are used for substance declarations:

- in general 2 per cent by weight;
- 1 per cent by weight if the product contains substances that are health hazards, corrosive, irritating, allergenic or carcinogenic, or a mutagenic toxin, and the same applies to substances hazardous to the environment;
- 0.1 per cent by weight for very poisonous, carcinogenic substances, mutagens or reproductive toxins.

Comparisons During Operation

The environmental impact of cleaning and maintaining a material must also be evaluated. A floor, for example, can have a greater environmental impact over its life from cleaning methods than from its production, transport, etc. There are no maintenance-free materials.

Product Information

It is best is to obtain a product environmental declaration from the supplier. Some countries are creating their own product databases, where products are checked against a list of the worst chemicals. In Sweden a number of large owners and developers have agreed on a standard and a system called *Byggvarubedömningen* (Building Supply Assessment) (www.byggvarubedomningen.se). The Building Supply Assessment database includes environmental assessments for the most used products and goods in the property business. Folksam insurance company and the private company SundaHus Miljödata from SundaHus in Linköping AB are other examples of databases. The databases list whether or not a building product contains dangerous substances of a particular class and amount. To a large degree they use the European Union's REACH regulations and the Swedish Chemicals Agency recommendations. Large construction companies may also have their own product databases, but the problem with these is that they all have different bases for evaluation, so it is difficult to understand how the evaluations were arrived at, and access fees are expensive. The long-term ambition should be to have a national product database for the construction sector that includes product environmental declarations.

Environmental Labelling

Environmental labelling is a system that guarantees that a product or service fulfils certain environmental standards. An environmental label does not certify that a product is good for the environment, but it often causes less environmental stress than similar products that don't fulfil the standards.

The currently established environmental label, the EU flower, so far only applies to a small number of products. Criteria have been determined for indoor paint and varnish (see Paint and Surface Treatments), hard flooring, textiles and mattresses. Some other national systems have been in existence for over ten years and are widely used.

The fair trade symbol is an international symbol indicating that the company behind the product respects human rights and ensures a good working environment for employees. Workers and growers receive a fair wage for the work they do. Such companies oppose child labour. Organic growing is encouraged, and democracy and the right to organize are promoted.

Most labels do not cover building materials and so few are labelled. The Swan label is the most important environmental label in Scandinavia, used on houses, furniture and hotels, among other things. Nature Plus is an international environmental organization whose aim is the development of a culture of sustainability within the building sector. To this aim the association has developed a label to enable future-oriented building products to reach a far stronger and sustainable market position. Öko Test in Germany is a consumer magazine that tests different products, among them building materials. They have a good reputation and producers use the Öko Test label if they are best in test. IBO (Österreichisches Institut für Baubiologie und Bauökologi) is a well-trusted label in Austria. The R-symbol in Germany shows how much of a product comes from renewable, mineral and fossil sources. The Forest Stewardship Council (FSC), which is the environmental labelling organization for forests, has developed a set of regulations for timber. The FSC logo identifies products that contain wood from well-managed forests certified in accordance with the rules of the Forest Stewardship Council. Some Swedish labelling symbols in use.

The KRAV symbol

The Swan label

The Bra Miljöval (Good Environmental Choice) label

The EU flower

The Fair Trade symbol

The Nature Plus mark

The Öko Test label

The IBO label

The R-symbol

The Forest Stewardship Council symbol

1.1.1 Criteria for Selection of Materials

There are primarily two things to think of when choosing materials. How does the material affect health and the ecosystem, and how does it affect resource use and environmental damage? Regarding health, it is the emissions and chemical ingredients that are decisive. Regarding resource use, environmental profiles produced using life-cycle analysis can be examined.

Health Aspects

The idea behind methods used to determine health aspects is that the material should not contain dangerous chemicals, and above all should not release these to the indoor environment.

Chemicals in the Building Industry

Global use of chemical substances has increased at an explosive rate. As we use innumerable goods, chemicals are spread through society and the environment: by waste incineration, leaking rubbish dumps, leakage from products, intentional spreading (pesticides) and releases during production.

It is important to rid the building industry of as many unpleasant chemicals as possible. Chemicals can present risks in many ways – for example, the persistent bioaccumulative and toxic (PBT) chemicals, substances that are long-lived, accumulate in the body and are poisonous. Well-known environmental poisons such as DDT and polychlorinated biphenyl (PCB) are in this group, as well as brominated flame retardants. Chemicals that are very long-lived and accumulate in the body to a high degree are called vPvB (very persistent and very bioaccumulative). Substances that are carcinogenic, mutagenic or reprotoxic are called CMR substances. Long-lived organic environmental poisons are referred to as persistent organic pollutants, or POP substances. The latter group of substances can cause nerve damage, behaviour disturbances, cardiovascular sicknesses, kidney damage, hormone disruption, brittle bones, cancer, foetal damage, reduced fertility and sterility.

REACH – The EU's Regulatory Framework for Chemicals

REACH stands for 'Registration, Evaluation, Authorisation and Restriction of Chemicals'. The EU has developed a regulatory framework for controlling the use of chemicals. All chemicals, new and old, that are manufactured or imported in amounts over 1 tonne require registration. Registration means that the industry is forced to produce certain basic data. The industry itself is required to evaluate and make public the risks of the registered chemicals. After an evaluation, the authorities determine if a substance can be distributed, is totally banned or is authorized.

The registration requirement in article 7.1 of REACH applies to substances in goods that meet all of the following conditions:

- the substance is intended to be released under normal and reasonably predictable conditions, and
- the total amount of the substance exceeds one tonne per producer or importer per year, and
- the substance is already registered for the specific use.

The amount of the substance that should be registered is the total amount found in all goods produced or imported per company.

The Swedish Chemicals Inspectorate's 'PRIO' (Priority Setting Guide) database (in both English and Swedish), can help in preparation for compliance with REACH. Searching in the PRIO database can therefore give an indication of which substances may be subject to the REACH approval process.

Emissions

Construction materials release a number of substances. Limits are only set for some of them, and there is limited knowledge of the combined effects of various substances. Therefore, a simplified method measures total emissions (TVOC, total volatile organic compounds). However, TVOC measurements are no longer of interest, since even materials known experientially to be healthy give off emissions, such as newly baked bread or wood. Now the emissions for individual substances are measured instead. Emissions from most materials decrease over time.

The Swedish Chemicals Inspectorate's PRIO Database

The legal basis for the PRIO database is in the Environmental Code. In this context, the most important parts are knowledge of the contents, the precautionary principle and the substitution principle. These regulations are generally applied to producers, importers, vendors and users of chemical products and goods.

PRIO is made up of several parts, including a database with more than 4000 chemical substances. PRIO divides substances into two priority levels: phased-out and priority risk-reduction substances. Substances in the phased-out category have such serious properties that they should not be found in society, regardless of how they're used. They largely reflect the criteria that are used as a basis for authorization in REACH, the European chemical regulations.

Priority risk-reduction substances have properties that should get special attention. The measures required for handling a substance depends on which group it is in. Priority risk-reduction substances must always be evaluated according to their particular use. The PRIO database also includes practical tips, for example how to make an inventory of chemicals and how to estimate hazards and carry out a risk analysis.

Global use of chemicals in 1959 (left) and 2000 (right). How such chemicals influence health is largely unknown. Effects depend on how much enters the body over what period of time and at what rate.

Source: The Swedish Society for Nature Conservation

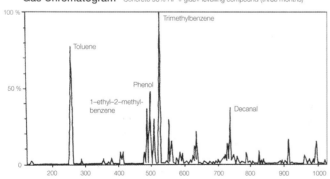

Building materials contain different chemical substances and release a portion of these to the interior air.

Source: Kjell Andersson, Örebro County Hospital

Fifteen of the Worst Chemicals

Biocides (some preservatives and fungicides)	Glue, paint, caulking compounds, wetroom silicone sealant
Brominated flame retardants	Cellular plastic, plastic in electronics, textiles
Bisphenol A – diglycidyl ether	Epoxy, polycarbonate
Freons (HCFC, HFC)	Air-conditioning units, refrigerators and freezers
Phthalates	PVC flooring, PVC-covered sheet metal, paint, glue, caulking compounds, cables, plastic coated fabric
Isocyanates	Polyurethane, glue, caulking compounds, caulking foam
Isothiazolinones	Glue, paint
Chlorinated paraffins	Softeners in metal roofing sheet paint, softeners in plastic
4-chloro-m-cresol	Preservatives in glue
Wood rosin	Paint, linoleum
Methylethylketoxine	Paint, caulking compounds
Nonylphenols, nonylfenol ethoxylate (alkylphenol) nonylphenol etoxylates (alkylphenols)	Paint, varnish, glue, caulking compounds, primers
Organic solvents	Paint, caulking compounds
PCB	Caulking and floor compounds, sealed glazing unit

Seven Environmentally Hazardous Metals

Arsenic, arsenic compounds	Wood protection agents and caulking compounds
Lead, lead compounds	Metal roofing sheets, window frames, pipes, stabilizers in hard PVC, light bulbs, and heating, water and sanitary fittings
Cadmium, cadmium compounds	Surface treatments, stabilizers, batteries
Copper, copper compounds	Water pipes, metal roofing, wood preservatives, tanks
Chrome, chrome compounds	Surface treatments, pigments, and heating, water and sanitation fittings
Mercury, mercury compounds	Thermometers, fluorescent lights, batteries
Organic tin compounds	Caulking compounds, stabilizers in plastic, PVC-covered metal sheets

Environmental Profiles

There are several systems used to describe a material's, a construction's, or a whole building's resource consumption and environmental impact (sometimes called ecological effects or ecological footprint). Creating an environmental profile is a big task since so many facts need to be gathered. Life-cycle analysis is used, where every phase of a material's life cycle is studied. The amount of energy and raw materials used, as well as releases to air, water and land are investigated. Databases are drawn up to collect relevant information in one place. It is important to use the same method for different materials in order to make a fair comparison. The phases studied are production of the material, making of the product, transport, building, use and demolition.

Life-Cycle Analysis (LCA)

Life-cycle analysis is a method used to examine a material's external environmental impact. The amount of environmental disturbance caused in the form of energy consumption and releases to land, air and water is calculated. A problem with life-cycle analysis is that it takes much time, and the result can be influenced more by the production technique in question than by the qualities of the material itself. In any case, life-cycle analysis is a good tool for improving a production process.

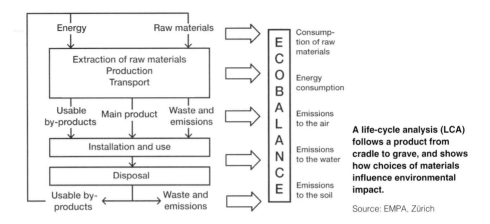

A life-cycle analysis (LCA) follows a product from cradle to grave, and shows how choices of materials influence environmental impact.

Source: EMPA, Zürich

The MIPS Model

The MIPS (Material Intensity per Sequence) model, which originated in Germany, involves examining resource consumption, i.e. how much organic and inorganic material, water, air and land is required in the manufacturing process for each material, e.g. insulation.

In order to insulate a section of wall to a certain U-value, a certain weight of insulation material is required. With the MIPS model, a comparison can be made of the resource consumption of different materials in order to achieve the same function.

Natureplus

Natureplus is an international organization based in Germany that certifies building materials and household products according to strict health, ecological, functional and lifetime criteria. Life-cycle analysis is also used. At least 85 per cent of a product must be made up of renewable raw materials, and any minerals used must be in abundant supply. There are strict rules for synthetic materials and hazardous chemical substances. It is estimated that only 20 per cent of products in certain categories are approved for the Natureplus label.

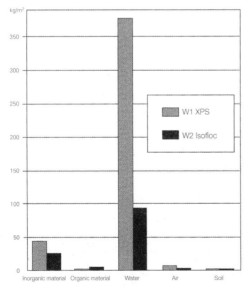

The MIPS model shows use of materials for a certain functional unit. Shown here is a comparison between two insulation materials in a wall with a U-value of 0.4W/m²K: cellular plastic (XPS = extruded cellular plastic) and cellulose fibre (Isofloc).

Source: *Das Wuppertal Haus* (The Wuppertal Building), Friedrich Schmiedt-Bleek

The NaturePlus logo (see www.natureplus.org).

The Danish Environmental Profile Model

The Danish Building Research Institute has a database called BEAT 2000, which contains life-cycle analyses (LCAs) of various building materials. The environmental profiles present resource and energy consumption as well as environmental impact in the form of greenhouse effect, acidification and nitrogen load (CO_x, SO_x and NO_x). In some cases values are given for waste management and toxicity to humans.

The TWIN Model

In the TWIN model in The Netherlands, both health and environmental impacts are investigated, and both qualitative and quantitative assessments are made. The qualities of a material are presented in two vertical bars, one for health impacts and one for environmental impacts.

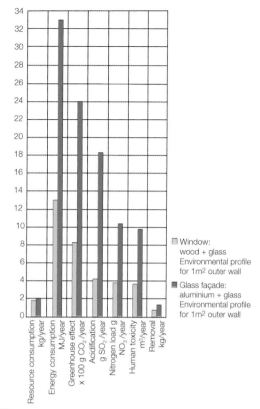

BEAT 2000
The environmental profile for a glass façade shows that large and double-pane façades are very resource-intensive, partly due to the use of aluminium sections.

Source: *Arkitektur og miljø – form konstruktion materialer – og miljøpåvirkning* (Architecture and Environmental Construction Materials – and Environmental Impact), Rob Marsh, Michael Lauring, Ebbe Holleris Petersen, Denmark, 2000

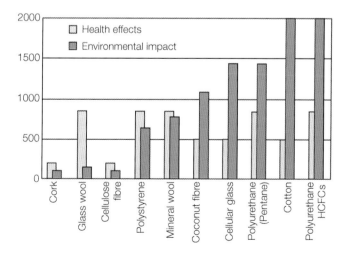

TWIN
The dark columns show environmental impact (0–3500) and this value (calculated according to a mathematical model) takes into account use of raw materials, pollution caused, waste, energy consumption, life, etc. The light columns show the value for how healthy a product is (0–3500). Health aspects take into consideration physical and chemical impact, biological effects, ergonomics and safety.

Source: TWIN-model, Handboek Milieu Classificaties Bouwproducten, NIBE Nederlands Instituut voor Bouwbiologie en Ecologie bv, dr. ir. Michiel Haas, 1998

1.1.2 Knowledge of Materials

It is valuable from an environmental perspective to gain knowledge of all types of materials and how they function in various environments, both individually and together. They can be divided into: organic materials such as wood and cellulose fibre; mineral materials such as stone, concrete, glass and lime; metals; and synthetic materials such as plastics.

Organic Materials

The term 'renewable materials' is sometimes mentioned, meaning organic materials that grow and are created with the help of solar energy and photosynthesis. The basic principle is that as long as use doesn't exceed growth, these materials are preferable from an environmental perspective. These are primarily wood and wood products, but also other types of natural fibres, from both plants and animals.

Wood

From an environmental perspective, wood is one of the best materials in almost all situations. It can be used in nearly all parts of a building, from construction and insulation to claddings, roofs and interior furnishings. The main rule is that wood should not be exposed to the ground and moisture, and should this happen, it should be allowed to dry quickly. Timber quality depends on how it is grown, when and how a tree is felled, and how the drying and management takes place. There is a loss of quality when a single species is cultivated intensively. It is preferable to avoid impregnated timber in favour of more environmentally safe alternatives such as pine heartwood repeatedly treated with oil. There are methods that use heating and pressure treatment of beech, ash and pine that preserve wood to the same degree as conventional impregnation. Imported timber should be FSC- or PEFC-labelled, which is a guarantee that the tree comes from environmentally adapted conditions. Mankind has a thousand years' experience of living in wooden buildings and knows that wood is a healthy material. Initially, some kinds of wood, such as pine, release a lot of terpenes, and time therefore needs to be allowed for the emissions to diminish. From an environmental perspective, increased use of local deciduous trees can increase biological diversity of ground flora, insects and birds. The most common type of timber, and that most often used for building, is softwood, derived from conifers (usually pine and spruce). Deciduous timber (hardwood) is used mostly for floors, interior furnishings and furniture, but alternatives are possible.

Organic Fibre Materials

Paper and pulp are made from wood and used in building in the form of wallpaper and insulation. Cellulose fibre insulation is made from either waste paper or new paper pulp. Substances are added to counteract micro-organisms, small animals and fire.

Fibres that are in principle good from an environmental perspective include straw, flax, coconut, sisal, jute, hemp, cotton, as well as peat, sheep's wool and cork. Possible uses include insulation and matting. However, aspects that should also be taken into consideration when choosing a material are the environmental impacts of transport, cultivation and production methods (such as spraying pesticides on cotton). Straw bales can be used to make buildings, straw can be used as a roofing material, and is used in mixtures for building with clay. Flax is used to make insulation sheets, as is hemp and sheep's wool. Coconut is a good packing material. Jute and sisal are used for matting and wallpaper. Cork is a good choice for sound impact insulation in both flooring and insulation.

Alder is a tree whose timber can be used for flooring, furnishings and carpentry.

USES FOR DIFFERENT KINDS OF TIMBER

Roofing, boarded roofing, wood roofing, shingles:

Heartwood of oak, pine and larch, as well as aspen can be used untreated. There should be ventilation from underneath. Tar extends its life but at the same time increases the fire hazard.

Exterior boarding:

The same timbers as for roofing, as well as close-grained spruce.

Windows, outside doors:

Heartwood of oak, pine and larch as well as close-grained spruce.

Construction:

Pine, spruce, larch as well as hardwood such as aspen for studs and beams if relative humidity during construction is kept low. For visible columns and beams, some layers in glue-laminated wood; and for beams with low beam height, deciduous trees such as ash, oak, elm, birch or aspen can be interesting alternatives.

Flooring, hard:

Oak, beech, ash, elm, whitebeam, juniper and sometimes maple.

Flooring, medium hard:

Larch, cherry, maple, mountain ash and sometimes birch.

Flooring, soft:

Spruce, pine, aspen, alder, linden.

Interior panelling:

All types of wood can be used as available, if the relative humidity in the room is constant. Hornbeam is the least suitable, and caution should be used with, among others, beech and cherry if moisture levels are variable.

Ceiling:

The same as for interior panelling. Heavier types of wood such as oak require good anchoring. Most untreated woods yellow over time. Aspen remains light and is suitable both untreated and painted. Spruce is lighter than pine.

Mouldings:

The same as for interior panelling. Care should be taken to obtain dry wood with straight grain, preferably quarter-sawn.

Carving:

Linden and alder are easy to work with. Oak and birch are harder and more durable.

Worktops:

Oak, elm and other types of wood if kept away from water. Note however the risk of staining and corrosion with oak.

Saunas:

Aspen, possibly alder for benches, floors and panelling. Can go rather grey where water is poured over them. Spruce for panelling and flooring if this scent is desirable. Pine releases resin.

Wetrooms:

Alder, oak, elm, pine, larch, possibly spruce and aspen. Untreated wood goes grey on contact with water when not dried quickly. On other surfaces and with good ventilation, more types of wood can be used. Note the risk of staining and corrosion with oak.

A well-restored Åland-style windmill (post mill) at Ortala works in Uppland, Sweden.

Wood fasteners:

For strong and tough types of wood, the same wood can be used as that to be joined. Juniper, lilac, mountain ash and blackthorn, among others, provide extra strong bonding.

Windmills, post mills:

The post was made of oak so that it could withstand weather and wind and be strong, the building of pine, the blades of spruce (since it is flexible). The mill wheel was made of birch with cogs of whitebeam.

Minerals

Minerals are considered to be non-renewable; however, many of them exist in such quantities that they can be used in construction without significantly affecting the environment. What does affect the environment is primarily energy consumption during production and transport. Stone quarries and gravel pits do, of course, scar the countryside.

Glass

Glass is made of quartz sand, soda or potash, and limestone. It's melting temperature is about 1400–1550°C. There are many different types of glass. Old glass can be recycled into new glass. Production of glass is energy-intensive. Mechanical fastening should be used so that environmentally hazardous caulking compounds are avoided. Besides being used for windows and glass partitions, glass can be used as thicker material for bearing components and in hardened foam form as insulating filler material.

Rock

Rock is most often good from an environmental perspective provided that it comes from a nearby quarry. Care should be taken to avoid rock that requires lengthy transport, or those few types of rocks that are radioactive, as well as dust from rock preparation (long-term inhalation can cause silicosis). Crushed aggregate for concrete (macadam) is the most frequently used material in construction. Natural gravel should be used sparingly as it is in short supply in many places. Instead, concrete, brick and other crushed rock can be used as a fill material.

Bricks

Bricks are made from clay. Clay is dried, preheated and fired at 800–1100°C for about three hours. Some types of high-fired bricks are fired at 1100–1200°C. Brick fired at sufficiently high temperatures so that it sinters completely (fireproof brick) is used mostly in chimneys and fireplaces. Brick with a varying degree of high-firing characteristics is available and it is best to use brick which has been fired at as low a temperature as possible. There are solid bricks and hollow bricks. Additives such as sand, sawdust or finely ground brick can be added to minimize shrinkage during firing. When sawdust is added, it burns off and leaves spaces that have an insulating effect. Brick is, in principle, maintenance-free. A new product is honeycomb brick. Walls made with this are no thicker than 2–3mm. It is made of especially fine-pored clay using thin cellulose fibres as porosity builders. With honeycomb brick, insulation values similar to lightweight concrete can be achieved. Honeycomb brick has a coefficient of thermal conductivity of $\lambda = 0.12 W/mK$. Porous hollow bricks have $\lambda = 0.20 W/mK$. Brick can be used as construction material, lining, and floor and ground covering (pavers). There are also special hollow brick block units that can be used in ceilings or walls as well as in floor structures. Brick surfaces can be treated with linseed oil (floors), can be polished, whitewashed or painted (walls).

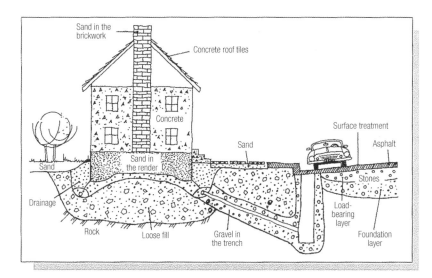

Natural gravel or crushed aggregate for concrete (macadam) are the most frequently used materials in construction. As natural gravel is in short supply in many places, construction that economizes on gravel can be used or crushed concrete can be used.

Source: *Bygningsmaterialenes Økologi* (The Ecology of Building Materials), Björn Berge, 1992

Bricks can be reused after the mortar (weak) has been removed. If brick is to be reused, quality control checks for compressive strength, frost resistance and contamination should be carried out.

Concrete

Concrete is composed of cement, water, aggregate (usually gravel >8mm or macadam) and various forms of admixtures. Currently, radon levels in aggregate are checked. There are many admixtures for concrete such as components to accelerate or decelerate hardening, stabilizers, plasticizers, air-entraining agents, water-reducing admixtures and solvents. Qualities to take into consideration in concrete production are: thermal qualities and thermal storage, sound

A drystone wall chapel from the Middle Ages in Ireland.

Kingohusene: a brick residential area by architect Jørn Utson, near Helsingør, Denmark.

insulation, frost degradation, reinforcement corrosion, as well as moisture properties. Most solvents are melamine-based and contain formaldehyde. The addition of admixtures to concrete should take place in a closed system if the materials would otherwise pose unnecessary risks in the work environment. In general it can be said that the contents of admixtures should be examined before use to avoid substances which are harmful to the environment.

Concrete is heavy, stores heat well, and has good sound-reduction properties. It is fire-resistant, insulates poorly against the cold, prevents moisture diffusion and does not absorb moisture well. It takes a long time for concrete to dry after casting, which means that building with organic material on concrete has to wait until the moisture content falls below a critical value. Wet concrete is often a cause of sick house syndrome. Finished concrete binds admixtures and does not release them into the air. Formaldehyde emissions can, however, occur during the first weeks after casting.

Expanded Clay Aggregate

Stephen Hayde invented and patented the method of making a structural grade lightweight aggregate early in the 20th century. Expanded clay aggregate is made of lime-deficient clay that is heated to about 1150°C in a rotary kiln. The clay expands and small air-filled cells are formed in the Haydite balls. At the same time the surface of the ball becomes so hot that it sinters, in other words it fuses and the ball becomes covered with a dense ceramic shell. The end product is a strong and durable lightweight aggregate used in numerous types of construction throughout the world. Hayde named the product after himself. Although the original patents have long since expired, the term Haydite is used by several companies in marketing their expanded shale lightweight aggregate.

Binding Agents

Many construction materials contain binding agents that have a significant effect on the environment. It is not possible to entirely avoid binding agents in construction, but their use can be minimized and those which cause the least environmental damage can be used. Cement production is very energy-intensive. Lime is mined in open pits and has to be burned, which consumes a lot of energy. Gypsum (used in plaster) is widely used in buildings, especially indoors. Clay is labour-intensive but an environmentally friendly alternative. Asphalt is a residual product of the oil industry. Adhesives, which are also binding agents, are described in the Adhesive Compounds section.

Asphalt

Asphalt (bitumen) is a by-product from the refining of oil. The products of (catalytic) cracking, which contain appreciable amounts of polycyclic aromatic hydrocarbons (PAHs), are suspected of being carcinogenic. On the other hand bitumen that is 'pure' and distilled in a safe manner is not considered to contain unhealthy components. Research has shown it is only the PAH fumes and aerosols that may be released when asphalt is heated that are injurious to health. Asphalt is used to impregnate fibreboard, as an adhesive in placing cellular glass panels, and as a coating on cellar walls, floors, bathroom walls and ceilings, in order to prevent moisture penetration.

Cement

Portland cement dominates the market and consists of 64 per cent quicklime (CaO), 20 per cent silicic acid $Si(OH)_4$, 5 per cent feldspar-bearing clay (Al_2O_3), 2.5 per cent iron oxide-bearing clay (Fe_2O_3), and 8.5 per cent other material such as calcium sulphate, blast furnace slag, fly ash, oil shale, limestone flour, pumice powder, bentonite, etc.

These raw materials are ground down and burned at a temperature of about 1500°C. Cement has the property that, when mixed with water, it hardens into a rock-hard mass.

Cement production is very energy-intensive and generates emissions of primarily CO_x, SO_x, and NO_x as well as small amounts of mercury. The cement industry has therefore developed methods to minimize emissions. Cement can contain small amounts of chromium, about 0.01 per cent, as it is naturally present in limestone. In order for cement mortar to have a long shelf-life and harden quickly, chemical additives are used. Epoxy resin is an additive to be especially avoided. It contains the carcinogenic substance epichlorhydrin. Furthermore, epoxy resin is allergenic, an eye irritant, and injurious to mucous membranes. People with a chrome allergy should avoid direct skin contact with cement. Globally, cement production represents 7 per cent of energy requirements, a figure that could be reduced to 3 per cent solely through improved processes.

Gypsum

Gypsum has good sound-insulating and heat-storage properties and also provides good fire protection. Gypsum is especially suitable for interior use for plastering, stucco, in the subfloor and in plasterboard.

There are two main sources of gypsum: natural gypsum that is mined and power station (known as REA) gypsum that is a solid waste from industrial desulphurization of flue gases. A third source is phosphogypsum, a by-product of the phosphate fertilizer refining process. REA gypsum has properties similar to natural gypsum. Phosphogypsum, however, can contain dangerous heavy metals and naturally occurring radioactive materials, and should therefore be checked before use in construction.

Natural gypsum is made from gypsum rock (anhydrite) which is composed of calcium sulphates. Commercial quantities of gypsum rock are found in Germany, England, Canada and the US. It is fired at 200°C, at which point water is forced out and a gypsum powder (half hydrate) is obtained. When the gypsum powder is mixed with water, the mixture hardens and gypsum rock (dihydrate) is reconstituted. This process is used in the production of plasterboard and gypsum plaster.

Lime

Lime is extracted by burning limestone, which for the most part is mined in open pits. When limestone is heated to 900°C, it is reduced to quicklime (CaO). If water is added to quicklime – a process that evolvos much heat – it is transformed to slaked lime ($CaOH_2$).

Slaked lime mixed with water is used to whitewash buildings and barns. It then combines with CO_2 in air and turns back to solid limestone. The classic plaster made from only sand and lime is excellent as it both absorbs and releases moisture and has disinfecting qualities as well. Lime is used as a binding agent in mortar, plaster and in sandlime brick.

Cast houses were built using a historical technique where pieces of slag, crushed brick or rock were cast together with lime mortar in a mould. The lime mortar used was a mixture of two parts unslaked lime and one part sand. Later people discovered that lime mortar could be thinned down with more sand in order to conserve lime. In lime-mortar houses, the mortar is composed of lime and sand in a 1:4 ratio and in sand-mortar houses, only 10 per cent lime is used.

Clay

Clay is found almost everywhere in the ground and consists mainly of aluminium silicates. It has a long history as a construction material. When people talk about 'earth' houses, it does not refer to pure clay but a clay and sand mixture (e.g. in a 1:2 ratio). Such clay is used as wall material, mortar and as plaster. It can also be called clay concrete as the clay functions as a binding agent and the sand functions as an aggregate. Sometimes other materials are mixed into clay, partly in order to increase its insulating properties (e.g. straw, chopped straw, sawdust, or expanded clay – Haydite), partly to improve its strength (twigs and branches, flax waste or animal hair), and partly to increase the material's plasticity (e.g. cattle manure or horse urine). Building with clay is labour-intensive and should be done during the summer when it is dry and warm so that the construction dries out well. It is important to protect the construction from running water, which necessitates large roof projections and a capillary break. 'Earth' houses have a good interior climate: they are cool in summer and hold heat well in winter. Clay has a very good capacity to buffer moisture, so that clay buildings enjoy stable and comfortable humidity conditions.

There are several techniques for building with clay. The method chosen depends partly upon the character of the clay and partly upon local tradition. Rammed earth houses (pisé) have load-bearing walls of thin clay compressed into formwork. When one part of the wall is compressed, the formwork is moved so that the next part can be built. In cob houses, weight-bearing walls are built without moulds, using thick clay that often has a little straw and some small stones added. Adobe houses are made by producing unburned brick in moulds, or

Professor Gernot Minkes' own house, built of clay, in Kassel, Germany.

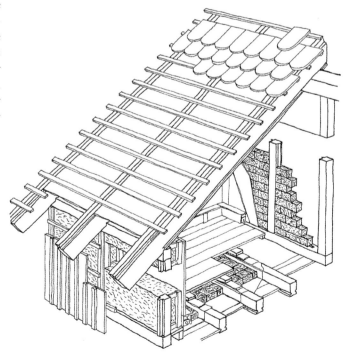

Constructing buildings of clay is common in many cultures around the world. Clay can store both heat and moisture and is an inexpensive, environmentally friendly material.

Source: Drawing from Klaus Schillberg, Heinz Knieriemen, *Naturbaustoff Lehm*, 1993

Adobe wall in Hus Torkel, Kovikshamn (Bohuslän), Sweden. As the wall faces a bathroom, the bottom two rows are fired brick in case of water leakage.

'loaves', which are dried in the sun. The sun-dried clay brick is laid with clay mortar to build walls and roofs. The wattle-and-daub method of house construction consists of a load-bearing wooden framework interlaced with a wickerwork of poles which is covered with clay on both the inside and outside. A clay mix with a high straw content is called 'light clay'. Expanded clay or other insulating material which is mixed with clay sludge composed of thick clay is known as 'isolating clay'. Adobe that is compressed by machine (compressed earth blocks) produces stronger blocks. In Germany efforts are being made to rationalize clay construction. It is possible to buy clay plaster and clay mortar in sacks, clay bricks by the pallet, and sheet material made of clay reinforced with reeds.

Clay mortar is weak and soft, which means that it can expand and shrink without cracking, and it is easy to dismantle the construction and knock off the mortar. That is why clay mortar is always used in tiled stoves. Clay mortar is also used to build walls of sun-dried clay bricks and to build cordwood or stovewood buildings.

Clay plaster is not only used in woven wattle-and-daub construction. Clay plaster is used both inside and outside to plaster buildings made of adobe or rammed earth, and wattle-and-daub was commonly used in the past inside log houses to make them more wind- and watertight and to provide a smoother surface for wallpapering. Interior wattle-and-daub can also be used in concrete buildings to improve the interior climate.

Adobe wall painted with casein paint, Hus Torkel, Kovikshamn (Bohuslän), Sweden.

Clay construction techniques have experienced a renaissance in the West, primarily among self-builders. Usually processed clay, mixed with straw or expanded clay pellets to increase its insulation properties, is used.

Metals

Metals should be used sparingly. Mining creates scars in the countryside and environmentally dangerous wastes. Metallurgical processes put stress on the environment. Alloys and surface treatments result in long-term environmental consequences that are difficult to determine. Since the discharge of metals into the environment increases concurrently with society's use of metals, the level of metals in the soil and sediments also increases. It is important to reuse metals, especially aluminium, which requires a great amount of energy to produce (32,000–71,000kWh/tonne). If aluminium is reused, the energy required is only 5 per cent of that required to produce new aluminium. The best metals from an environmental perspective are iron and steel, reused aluminium and in certain cases stainless steel. If metals are used outside, they are subject to corrosion and must be protected, otherwise valuable material is wasted. It is important to try to minimize metal use or if possible to use other construction materials, such as laminated timber, ceramic materials and concrete hybrids (see also Roofing).

Steel consists mainly of iron together with small amounts of carbon, manganese, phosphorus and sulphur. Steel is the metal that requires the least energy to produce but is subject to rust and must be protected with a surface coating. Galvanized steel contains about 5 per cent zinc. Galvanizing steel can produce nickel, chromium, cyanide and fluoride discharges. A problem with steel as a load-bearing material is that it must be covered, for example, with several layers of gypsum, or painted with fire-resistant paint. Fire-resistant paints are not environmentally friendly. Continuous steel structures as well as welded mesh reinforcement and metal studwork should be earthed in order to reduce the health effects of electromagnetic fields.

Copper, zinc and chromium are listed in the Swedish Chemicals Agency PRIO database of priority risk-reduction substances (of which reduction of use is recommended) as they are highly toxic to aquatic organisms and can result in harmful long-term effects to the aquatic environment. For copper, this is primarily in the ion form. Chromium is also hazardous if breathed in and highly toxic when exposed to the skin and when ingested. Nickel can cause allergies and is suspected of being a carcinogen. Brass is a copper alloy, where copper is alloyed with zinc. Other alloy substances can be found. The most common however is brass, which is made up of 60 per cent copper and 35 per cent zinc. Lead, chromium and nickel may also be used.

Aluminium is fairly corrosion-resistant but its production is very energy-intensive and produces poisonous fluorine pollution. Large amounts of environmentally deleterious red sludge is created. Aluminium is made from cryolite and bauxite (a clay-containing aluminium). Resmelting of aluminium scrap metal is very common today and some aluminium products are produced from reused material. Only aluminium components composed primarily of reused aluminium should be used. Aluminium use is growing faster than any other metal used in our industrial society.

The Natural Step Foundation has developed a consensus document about metals. In summary, they concluded that all use of metals involves a strain on the environment. Metals have varying degrees of danger which can be determined from the metal's FCF (future contamination factor). This compares the natural level of metals in the Earth's crust with the amount of metal that has been extracted from deeper layers.

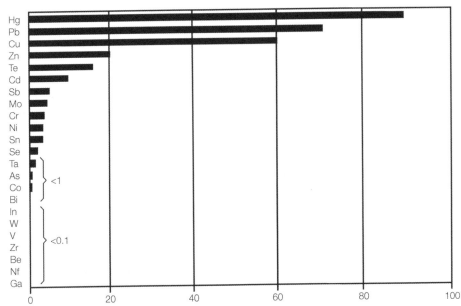

The future contamination factor (FCF) of metals. Certain metals are worse than others, such as heavy metals, e.g. copper and zinc.

Source: The Natural Step's Consensus Document on Metals, 1993

Synthetic Materials

Synthetic materials do not exist naturally but are human-made. Plastics are the most used synthetic materials.

Plastics

Plastics are materials that are produced by linking monomers together into long chains or nets, which results in polymers. Monomers are produced in the (catalytic) cracking of crude oil and natural gas. About 4 per cent of the total crude oil and natural gas production in the world is used for plastic production. Polymers are mixed with admixtures to achieve certain properties. Plastics with brominated flame retardants should be avoided.

Admixtures (additives) are made of a number of substances in order to provide plastic with different properties. Examples include fillers, colour pigments, hardeners, catalysts, accelerators, inhibitors, antioxidants (ageing protection), softeners, ultraviolet (UV) or heat stabilizers, flame retardants, anti-static additives and blowing agents. Plastic properties are determined primarily by the arrangement and chemical structure of the constituent polymer, but admixtures can influence these properties. Admixtures are one of the greatest problems with plastics, partly because they can be stable, xenobiotic substances and partly because they can be health hazards. Furthermore they complicate reuse of plastics. In fact it isn't sufficient to sort plastic according to plastic type. Knowing which admixtures the various plastics contain is also important. During plastics production, solvents, lubricants and mould-release agents are used in the handling process.

Thermoplastics, which are mouldable, are based on polymers that are either linear or forked. Heating makes it possible for these polymer molecules to move in relation to each other, which means that the plastic softens and can be moulded. With cooling, the molecules are again bound closely to each other and the plastic hardens.

In principle, thermoplastics can be repeatedly melted and reformed (theoretically) and are therefore well suited to recycling. However, today's plastics are difficult to sort and present-day plastic material can be reused a maximum of 4–5 times.

In thermoset plastics, the polymers are bound together in large, three-dimensional meshes. The grade of the mesh, close or coarse, determines the possibility and degree of deformation. Cross-linking takes place during hardening which is done using either heat or chemicals. Because of their mesh structure, thermoset plastics do not soften but break into pieces when temperatures exceed a certain level. Thermoset plastics are therefore not suitable for recycling but can only be used for energy extraction through incineration or consumption.

Some Common Plastics

The most environmentally friendly plastics are those with simple carbon-hydrogen compounds such as polyethylene, polypropylene and polyolefine. LDPE (low-density polyethylene) has no additives and is often used to package foods. HDPE (high-density polyethylene) has additives, is not decomposable, but can be incinerated. Polyolefines consist of mixtures of polyethylene and polypropylene.

Polystyrene (PS) and polyester (PET) consist of carbon, hydrogen and oxygen. These are not as good from an environmental perspective because aromatic hydrocarbons (POMs and PAHs) are produced during incineration and should not be allowed to accrue systemically. Controlled incineration should thus be required for such plastics. Their production is complicated, energy-intensive and environmentally dangerous. Benzol, which is produced from petroleum, is both carcinogenic and affects the immune system. Styrene is produced from benzol and is hormone-disturbing. Xylene and styrene are given off during production as well as by the finished material, especially during the first two to three months. Production thus takes place in a closed system.

Plastics to be avoided are, for example, epoxy and bakelite, which are thermoset plastics that cannot be recycled. Epoxy is highly allergenic in its unhardened form. Polyurethane is also a thermoset plastic. When it is burned, isocyanates are formed which are toxic and can provoke allergies and asthma. ABS plastic (acrylonitrile-butadiene-styrene) should also be avoided. PVC (polyvinyl chloride) consists of carbon, hydrogen and chlorine atoms and is, from an environmental perspective, one of the most controversial plastics. Chlorine causes environmental problems during its production, incineration and breakdown. During its production, large amounts of mercury are used, albeit in a closed process. During incineration, hydrochloric acid and dioxins (toxic, long-lasting organochloride compounds) can arise. Chlorine and high temperatures create a highly reactive environment. PVC contains admixtures such as softeners, stabilizers and antioxidants. A common softener is a phthalate which comprises 15–35 per cent of the plastic. Admixtures migrate and leak to the surface.

Plastics of the Future

Plastics can be one of the great materials of the future if they are produced from renewable resources, contain admixtures that do not disturb the environment, and can be recycled many times. Under these conditions, plastic is an excellent, resource-saving material that can replace more environmentally dangerous materials such as metals. Plastics can be produced from corn and potato starch, cellulose, charcoal or vegetable oils. These kinds of plastics are available on the market and function as well or even better than plastics made from crude oil.

Abbreviations for Plastic and Rubber Materials

Thermoplastic, thermoplastic elastomer		Rubber/hard plastic	
ABS	acrylonitrile-butadiene-styrene	AEM	ethylene acrylic rubber
EVA	ethylene vinyl acetate	BR	butadiene rubber
PA	polyamide (nylon)	CR	chloroprene rubber
PBT	polybutylene terephthalate	ECO	epichlorohydrin rubber (ethylene oxide copolymer)
PC	polycarbonate	EP	epoxy resin, also ethylene-propylene (hard plastic)
PE	polyethylene	EPDM	ethylene propylene diene monomer rubber
PET	polyethylene terephthalate (polyester)	FPM	fluorocarbon rubber
PMMA	polymethyl methacrylate (acrylic resin)	HNBR	hydrogenated nitrile rubber (acrylonitrile-butadiene rubber)
POM	polyoxymethylene (acetal)	IR	isoprene rubber (synthetic natural rubber)
PP	polypropylene	IIR	butyl rubber
PPS	polyphenylene sulphide	NBR	nitrile rubber
PS	polystyrene	NR	natural rubber
PTFE	polytetrafluoroethylene	Q	silicone rubber
PUR	polyurethane	SBR	styrene butadiene rubber
PVC	polyvinyl chloride	UP	unsaturated polyester resin (hard plastic)
SBS	styrene butadiene styrene		

Rubber

Rubber is a polymer that through a vulcanization process has become elastic with the help of chemically reactive additives, e.g. zinc oxide. The sulphur vulcanization process involves cross-linking between molecular chains so that a network is created. Rubber can be deformed and after the pressure is released it can, in effect, regain its original shape. Sometimes rubber material has a woven fabric backing. The raw material for synthetic rubber is petroleum. Natural rubber comes from rubber trees, which grow in tropical climates. The primary energy required to manufacture natural rubber is about half as great as for the manufacture of synthetic rubber. The work involved in transportation is great for both kinds. From an environmental perspective, natural rubber-based products are generally preferred to synthetic rubber. Natural rubber products are often mixed with synthetic rubber and there is synthetic rubber that has the same structure as natural rubber. For all rubber products, in order to evaluate how environmentally friendly a specific product is, it is important to investigate the additives used.

Nanomaterials

Nanotechnology deals with producing and using structures as small as a nanometre (a millionth of a mm or about five atoms wide). At the nanoscale, there is no clear limit between physics, chemistry, biology and materials science, which results in totally new phenomena and functions. In many cases, an attempt is made to imitate nature. Animals such as butterflies, beetles and peacocks use nanotechnology. The shimmering colour on their wings, shells and feathers is achieved with the help of special nanostructures of the substance chitin.

All branches of industry are expected to be influenced by nanotechnology.

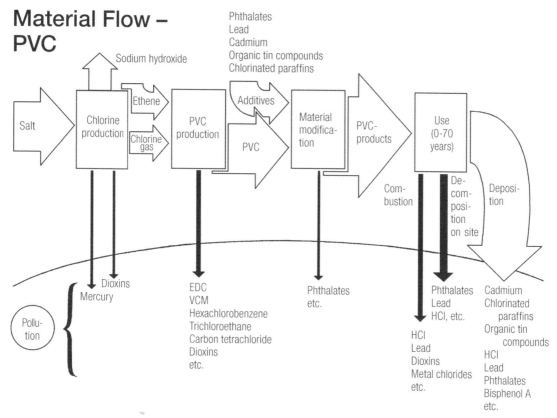

According to Greenpeace, PVC should be avoided for the sake of the environment. In response to criticism, the PVC industry has removed softeners and thermal stabilizers. The production process has been improved. Despite this, PVC is a chlorinous material (57 per cent), and hazardous substances can thus be formed during waste management and in the event of a fire. PVC consists of 43 per cent material derived from fossil fuels.

Source: *PVC+kretslopp=sant?*, Greenpeace, 1993

Areas affected in the context of building construction are colouring of glass and rubber, and production of paint colouring and cement. Also, nanostructured surfaces that do not get dirty can be, for example, self-cleaning windows and textiles.

The fears of nanotechnology are regarding that nanoparticles are so small that they can get into people via the skin and lungs. If particles 'only' micrometre in size are breathed in they are caught in the mucus on the walls of the trachea and carried away. Nanoparticles however penetrate deeper into the lungs. They can reach all the way into the alveoli, where the vitally important oxygenation of blood takes place, and stay there. Research has shown that exposure to carbon nanotubes causes substantial inflammation in the lungs of mice. The symptoms are similar to those caused by asbestos fibres.

1.1.3 Assessment of Materials

All evaluations are subjective since it isn't possible to compare 'apples and pears' in a mathematically correct way. In this book, the following methods are used. First, chemical contents are examined to see which materials contain environmentally hazardous chemicals. Then, environmental profiles of materials with similar functions are compared, e.g. ceiling and floor materials. Finally, materials are divided into three categories: recommended, accepted and to be avoided.

Our Choices

We summarize how environmental institutes in Europe evaluate building materials, and how trustworthy authors on the subject view building materials. The results are presented in tables divided into material categories.

The following institutes and books were sources of guidance:

Germany: Holger König, Wege zum Gesunden Bauen, Ökobuch, 1997.

Germany: Thomas Smitz-Günter, Living Spaces. Ecological Building and Design, Ökotest, Könemann, 1999.

Norway: Björn Berge, Ecology of Building Materials, second edition, Architectural Press, 2009.

Austria: Österreichisches Institut für Baubiologi und Ökologi (IBO), several publications.

The Netherlands: Michiel Haas, Nederlands Instituut voor Bouwbiologie en Ecologie (NIBE), several publications.

The Netherlands: Stuurgroep Experimenten Volkhuisvestin, Handbook of Sustainable Building, James & James, 1995.

Switzerland: Bosco Büeler, Genossenschaft Information Baubiologie (GIBB), BauBioRatgeber, 2001.

Sweden: Birger Wärn, Tyréns Ingenjörsbyrå, *Guide för materialval* (Guide for Choosing Materials), database, 2003.

Sweden: *Sunda Hus* (Healthy Buildings), database of products in the construction sector (2009), which uses the Swedish Chemicals Agency guidelines.

Sweden: ABKs (AB Kristianstadsbyggen) *Miljö och Kvalitetskrav* (Environmental and Quality Requirements), Kristianstad 1998.

Sweden: Göran Stålbom, *Miljöhandbok* (Environmental Handbook), published by VVS-information, 1999.

Sweden: Teknikhandboken 2009, VVS Företagen.

All the institutes and experts have not evaluated every category of material, and they all carried out their evaluations in slightly different ways.

Ground Cover

Ground cover around a building affects the environment in several ways. Not only does production of material require energy, some ground cover contains dangerous chemicals. Furthermore, ground cover affects surface-water management, depending on how porous it is. It also affects the microclimate, depending on how much solar radiation it absorbs and how much water evaporates from it. Does the cover withstand car traffic? How is the cover affected by rain, frost and ground frost? Another factor is maintenance, for example how gravel paths might be weeded.

Asphalt (bitumen) is a residual product from oil refining. Asphalt that allows water to percolate through it, so making it easier to deal locally with runoff, can be ordered from most asphalt plants. This material can be used for car parks, and on top of drainage layers. Asphalt can become clogged over time when dirt collects in it and must then be cleaned using a high-pressure spray.

Concrete for ground cover is available in many forms, sizes and colours. Concrete slabs are available with a smooth or patterned surface. A typical concrete slab is 450mm square and 40mm thick or 350 × 350 × 50 or 70mm. Avoid using blocks for paving in order to conserve material. The foundation can be made with a 35mm-deep bed of sand that has to be firmed and levelled. Cracks should be carefully filled with fine grain sand or dry mortar consisting of fine grain sand and cement in a proportion of 6:1 by volume. Brush the sand or mortar into the joints and then brush away the excess. Sprinkle water over the concrete to rinse off the block or slabs and moisten the joints so the cement can harden.

Grass can be used both as a lawn and as a meadow. The more often grass has to be cut, the more energy is used. Grass surfaces have a good capacity to let rain through. Grass surfaces are sensitive to wear, especially in the spring when the ground may be frozen even though the surface is soft.

Use of reinforced grass makes it possible to lay an almost invisible road with a grass surface. There is often a need for driveable surfaces used only by emergency vehicles. A reinforced grass car park can be less conspicuous. After laying the reinforcement, spaces are filled in with soil and sown with grass. Paving stones vary in appearance and can be made of different materials. The most common materials are cement or plastic.

Brick used on the ground and ceramic tiles used for paving are weather-resistant and non-slip and are available in a large variety of colours. A high firing temperature and ground tiles of slate clay increase frost resistance. For small areas that receive little traffic, a 30–100mm-deep sand bed is laid. More heavily trafficked surfaces require a 150–200mm-deep layer of macadam under the sand. The sand has to be compacted and levelled. The same laying procedure is used as for concrete.

Stone paving in garden paths and on terraces can be laid almost anywhere, on the ground, sand or cement. Natural stone varies in thickness and it is so natural that sand beds (about 50mm) are adjusted as the stones are laid. The foundation must be stable so that the finished path will be smooth and level. If desired, the stones can be stabilized in mortar (with a cement/sand ratio of 1:3) and joints can be filled with a sealant containing cement/fine sand in the ratio 1:6.

Loose inorganic materials, such as fine gravel or crushed brick of different colours, or larger gravel that is used for roads, can be used. Gravel covers must be regularly maintained by replenishing with new gravel. Pit-run gravel is a non-renewable resource. It is important that more and more recycled fill material be used as gravel. Gravel on a sloping surface is easily washed away by rain. Remove all vegetation and dig the area to a depth of about 50mm. The ground should slope somewhat in one

Sidewalk pavement of ceramic tiles in Ängelholm.

direction to facilitate drainage. One way to avoid weeds is to put plastic or a dense geotextile under the whole area. To avoid the use of weedkillers, weeds can be removed manually or burned away using hot steam (special equipment for steam spraying is available).

Loose organic paving materials such as bark and wood chips are used for ground cover on paths, around trees and in garden beds. Wood chips and bark are the most suitable materials if a medium-hard surface is desired. New material must be added after about four years. Since pressure-impregnated wood should be avoided, stone, brick or cement can be used as edging materials.

Wood on the ground rots, so it is best to raise wood above the ground in the form of decking. It is important for timber to be able to dry out properly after it has been saturated by rain. Slow-growing heartwood of oak, larch, pine or elm should be chosen. Blocks of oak that are scorched underneath and set in sand are relatively durable.

Granite sidewalk stone.

Pressure-impregnated timber should be avoided for environmental reasons but regular oil treatment extends the life of wood. There is currently timber available that has been pressure-impregnated with linseed oil. Another method to make wood resistant to moisture is to expose it to high pressure and heat. This process changes the cellular structure of the wood and gives it other properties.

Geoproducts

Geoproducts are woven or non-woven textiles, nets, mats or membranes used in the ground for separating, reinforcing, draining, filtering or protecting from groundwater and erosion. Geoproducts are produced all over Europe. Their service life is estimated to be more than 60 years.

Drainage liners can be made of polypropylene or polyethylene. Other materials

Brick ground cover at the Gärtnerhof eco-village in Austria.

which may be used consist of a core of polyamide threads surrounded by a polyester cover. Drainage liners can be used instead of graded gravel in pipeline trenches around building foundations, in drainage ditches and under road embankments.

Georeinforcement (geonet, erosion mats) can be made of synthetic or natural materials. The synthetic varieties are usually made of polyethylene, polypropylene or polyester. Other synthetic materials which may be included are polyamide, polyvinyl alcohol and aramid fibre. Coconut fibre is the natural material which may be used. Georeinforcement is used to stabilize and reinforce slopes and road embankments, as well as to strengthen underground and load-bearing layers in the construction of roads and foundations. It is also used to prevent erosion and to support the establishment of plants.

Geomembranes are made of polyethylene, polypropylene and/or polyester. They may contain a UV stabilizer (e.g. 2 per cent carbon black). One variation of the geomembrane is made of bentonite (clay material which swells upon contact with water) encapsulated between two layers of geotextile. Geomembranes of butyl rubber contain other polymers as well as EPDM rubber, carbon black, filler, processing chemicals, antioxidants and a vulcanizing system. There are geomembranes made of PVC and others that contain an aluminium foil layer.

The chemical content of synthetic rubber geomembrane should be carefully scrutinized before acceptance. Geomembrane is used as a weatherproofing layer in the ground, in some instances to separate runoff from contaminants (near landfills) and also during the construction of ponds. They can also be used as radon protection in building foundations as they are impermeable to both water and gas.

Geotextiles (geocloth) are made of polypropylene with a UV stabilizer as well as of polypropylene mixed with polyester or polyethylene. Geotextiles are primarily intended to separate materials. They can also function as a porous membrane (both separating and filtering). Geotextiles are also used to protect geomembranes.

Ground cover fabrics are used in cultivation to maintain ground moisture, reduce radical temperature swings, and prevent weed growth, thereby providing good conditions for the establishment and growth of plants. There are synthetic ground cover fabrics as well as biodegradable ground cover fabrics made from coconut fibre with a double, light-screening, foil layer of biodegradable plastic.

Structural Materials

In a sustainable building, it is important to find environmentally friendly alternatives

Ground Covers and Geoproducts: Environmental Evaluation

Recommended	Acceptable	To be avoided
Gravel, sand	Concrete slabs*	Asphalt***
Grass	Geoproducts, plastic materials	Pressure-impregnated wood
Geoproducts, natural materials	Grass reinforcement**	PVC
Organic spreading materials	Ceramic slabs	
Stone, domestic	Paving brick	
Wood, e.g. blocks	Stone, imported	

Notes: * Reinforced concrete slabs are not acceptable. Concrete slabs with recycled aggregate are recommended.
** Plastic grass reinforcement is acceptable only if it is recycled PP or PE plastic.
*** Asphalt cannot be avoided but the area of surfaces to be asphalted can be reduced.

to reinforced concrete and metals. There are structural materials available that are also thermal insulators, for example, lightweight concrete, expanded clay, sandlime brick, foamed concrete, cellular glass and wood-wool cement boards. Structural constructions in brick or sandlime brick in combination with infill walls are also acceptable from an environmental perspective. The most environmentally friendly construction materials are wood and clay. Solid wood buildings up to about 12 floors can be economic to construct.

Concrete (see Minerals in 1.1.2 Knowledge of Materials) can be poured on location or supplied as prefabricated components or slabs. Most concrete constructions are reinforced. Because of its great resistance to pressure and moisture, concrete is an important material in the building of foundations. Concrete has a high compressive strength and with steel reinforcement or non-metal reinforcement such as fibreglass, concrete is getting tensile strength and is therefore used for columns, beams and floor structures.

Foamed concrete (see Insulation) is used in solid constructions both as a support and an insulating material.

Glass (see Claddings)

Cellular glass blocks are made from loose cellular glass (85 per cent) and cement (15 per cent). They are used as load-bearing and insulating building blocks. They keep their shape, withstand frost, mould and fire, and do not contain hazardous additives. These blocks are used in a similar way to lightweight aggregate blocks.

Clay materials (see Binding Agents) have been used in construction for thousands of years, with a variety of techniques. People have used untreated clay, worked clay and clay with straw or other materials mixed into it. Traditionally, buildings have been constructed of earth or turf blocks which have been dug directly from the ground, as in Icelandic turf houses.

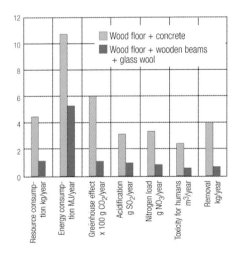

If the environmental load for various construction materials is compared, wood is always preferable to reinforced concrete. The diagram compares environmental profiles (per 1m²) of a wooden floor on concrete beams to a wooden floor on insulated wooden beams.

Source: *Arkitektur og miljø – form konstruktion materialer og miljøpåvirkning,* Rob Marsh, Michael Lauring, Ebbe Holleris Petersen, 2000

Clay-stone wall built with a load-bearing wood construction.

Source: Hus Torkel, Kovikshamn, Bohuslän

An example of how a concrete building foundation can be reinforced with fibreglass instead of metal.

Source: Hus Torkel, Kovikshamn, Bohuslän

Metals (see 1.1.2 Knowledge of Materials). It is not possible to build a modern building without metals. Nails, screws, fittings and metal reinforcement are needed. In certain constructions, steel structural elements are required as well.

Masonry, such as brick and sandlime brick used in load-bearing construction, needs to be supplemented with insulation. Other materials such as expanded clay and lightweight concrete, both support and insulate if the buildings aren't too high.

Sandlime brick may be grey and consists of quicklime and natural sand. White sandlime brick consists of white lime, burned flint and solid granite. Grey sandlime brick is manufactured from quicklime (5–8 per cent) and quartz sand (92–95 per cent), which is mixed with water and pressed into brick-shaped blocks. This is hardened under steam pressure for 4–8 hours in a kiln at a temperature of 200–300°C (autoclave process). This process releases silicone dioxide that together with the slaked lime builds crystalline compounds to bind the grains of sand together. Small amounts (<1 per cent weight) of colour powder may be included.

Sandlime brick requires less energy to produce than brick. The material's moisture-absorbing ability is a third that of brick, but together with plaster it nonetheless has good heat- and moisture-buffering properties. Sandlime brick is marketed as building brick, building block and cladding. It has a long life and the bricks are, in principle, maintenance free.

Lightweight sandlime brick is manufactured in Nordbayern, Germany. The material is produced in the same way as sandlime brick, but can be composed of up to 90 per cent expanded clay. Light sandlime brick has approximately the same properties as expanded clay blocks. The difference is that the binding agent in lightweight sandlime brick is not cement but lime, which is more environmentally friendly. The thermal conductivity λ for lightweight sandlime brick is 0.13 W/mK.

Lightweight concrete is made of finely ground sandstone/quartz sand (70 per cent weight), lime (7–20 per cent weight), cement (7–18 per cent weight), natural gypsum (3–7 per cent weight), water, and a little sand. Aluminium powder (0.1 per cent weight) is added as a raising agent. When the aluminium powder is added the mixed raw materials produce hydrogen gas which makes the mixture expand and produce air bubbles. When the compound hardens, it is cut to the desired dimensions and hardened in an autoclave. Silicone and cellulose derivatives may be used as additives. Lightweight concrete is made into blocks or slabs and used in walls, ceilings and floor structures. It has average noise insulation properties, poor moisture resistance, good moisture absorption, average heat accumulation properties and good compressive strength. Compared to other materials, walls need to be made thicker in order to provide good insulation. Their life is very dependent on the quality of the binding agent (the cement). Lightweight concrete constructions often contain steel reinforcement.

Expanded clay is made by firing clay materials at a temperature of 1250°C in a rotary kiln. The clay expands and forms small pellets with a hard exterior and an interior of air-filled cells. Expanded clay is used loose as insulation in foundations.

Expanded clay blocks and panels consist of expanded clay pellets moulded together with cement. They contain expanded clay (75 per cent by weight), Portland cement (23 per cent by weight), and water (2 per cent by weight). Reducing the fraction of expanded clay makes the material stronger. Sometimes small amounts of porosity builders are added. The expanded clay blocks have a relatively good compression strength, good noise insulation and heat storage properties, tolerate moisture but are moisture-repellent. They are used in interior, exterior and cellar walls. Expanded clay blocks can be used in load-bearing walls in structures of up to three storeys. Reinforcement is placed in the joints. The blocks are also available as sandwich elements with a core of insulating material for use in exterior walls. The energy used in production is relatively high.

Brick (see Brick in 1.1.2 Knowledge of Materials)

Wood (see Wood in 1.1.2 Knowledge of Materials) has been used for load-bearing construction in 'log-cabin' construction, timber framing and timber-clad buildings. Currently, wood is used for load-bearing in framing, both as solid joists and as light joists. Solid wood may also be used for ceilings, walls and floors.

Wooden beams and joists are limited in terms of both length and thickness, and large beams have a tendency to fail. Methods have been developed where thin pieces of wood are glued together into beams that have a high load-bearing capacity. There are glulam beams, lightweight beams and trusses. On the one hand designs can be freer and on the other the load-bearing capacity is greater for glulam beams than for solid wood beams.

Renovation of an old castle ruin in Koldinghus, Denmark. The new load-bearing construction is built of glulam.

Source: Johannes Exner, Architect

Wooden roof trusses and frameworks are commonly used in building roofs for single-family houses. They are usually manufactured using nail plates. It is possible to build wooden frames that span long distances, up to just over 20m. Framed wooden roof trusses are also used in building machinery halls, barns and animal stalls. Many frames can be built entirely of wood, while others require steel rods or cables.

Lightweight studs and lightweight beams are another way to use wood effectively. They can be made as I-beams or box beams consisting of wooden studs and hard fibreboard. There are also laminated veneer timber beams and veneer strip beams which are designed in a similar manner. Such beams have a high load-bearing capacity in relation to their weight. They are used where thin beams are required. In roof trusses they may have an I-beam or T-beam section.

Glulam beams are produced by glueing pieces of board (most often spruce) together under pressure into beams. Glulam beams of different lengths and thicknesses are produced. They can be made up to 30m long and up to 2m high. A glulam beam has a significantly higher load-bearing capacity than a solid wood beam of the same dimensions. Their only disadvantage is the binding agents used in their production. The binding agents are usually phenol formaldehyde resin or polyurethane (contains isocyanates). Phenol formaldehyde resin is a mixture of phenolic resin adhesive and formaldehyde adhesive (see also Adhesive Compounds). The amount of binding agent used is less than 3 per cent and the area where the binder comes into air contact is small. In Switzerland the material is evaluated so that a product containing less than 3 per cent adhesive may be recommended by the evaluating agency. If products contain between 3–5 per cent adhesive, they are acceptable. Sometimes a lattice beam with reinforced screwed joints can be used instead of glulam beams.

OSB-beams (oriented strand board) is a product where wood chips are glued together to form beams with different profiles. Whole building systems with beams and

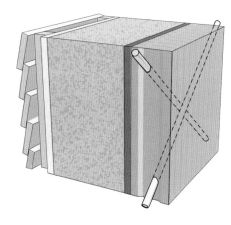

Construction where load-bearing solid wood elements are joined together diagonally with wooden dowels. No glue is used. On the outside is façade material, followed by studs with a 1.5 inch air gap, a water resistant hard board, seven inches of insulation, plywood (5/8 inch) and three inches of diagonally pegged solid wood.

Source: Sohm, Austria.

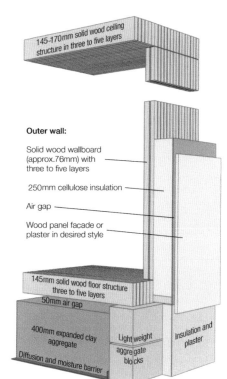

Building system using solid wood technology according to Ekologibyggarna, Vadstena, Sweden.

Source: Det Hållbara Huset, finalist i Miljöinnovation 2002. Maria Block, Varis Bokalders, architects

Construction Materials

Recommended	Acceptable	To be avoided
Sandlime brick	Concrete*	Aluminium
Clay material	Glass	Steel**
Lightweight sandlime brick	Lightweight concrete	Wood, pressure impregnated
Stone, local	Expanded clay blocks	Stone, imported
Wood	Glulam	
Wooden framework	Foamed concrete	
Wood-wool cement	Brick	
	Cellular glass blocks	
	OSB-beams	

Notes: * Reinforced concrete is energy-intensive and its use should be minimized. Reinforced concrete can sometimes be replaced by other construction materials such as solid wood or brick.
** Steel production is both energy-intensive and destructive for the environment, but it is not possible to build a modern building without steel. Nails, screws, fittings and metal reinforcement are needed. In certain constructions, beams and columns of steel are required as well. It is important to minimize iron and steel use or use other materials instead of steel. Steel bolts can be replaced by wooden dowels and iron reinforcement can be replaced by bundled fibreglass reinforcement, etc. Continuous steel structures as well as steel fabric reinforcement and metal studwork should be earthed in order to minimize their effect on the building's electromagnetic fields.

Detached house in Staffanstorp, Sweden, made of wood-wool cement.
Source: Architect: Mattias Rückert

columns are made in this way. One type of board uses chips from aspen, a cheap material. With this technology strong beams may be constructed with a slim profile.

Solid wood elements are modern methods of joining wood materials together to form beams or boards. The boards can be nailed, screwed, dowelled or glued together. From an environmental point of view elements without glue are preferred. Solid wood elements are manufactured both as beams and as solid boards that can be used for ceilings, walls and floor structures. Solid wood boards are so strong that a 12cm-thick plank can be used as a joist spanning 5m. Solid wood boards are relatively fire-resistant, sound-insulating and moisture-buffering. For walls separating dwelling units and in floor and ceiling joists, boards need to be complemented with other materials to increase sound insulation. There are also companies producing solid wood blocks that can be used as large bricks to build houses.

Wood-wool cement (see Thermal Insulation)

Thermal Insulation

Most insulating materials are based on the principle that stationary air insulates, so they contain many small air pockets. They can be divided into insulation materials that tolerate humid environments and those that must be kept dry. It is essential to know whether the material stores heat and moisture, is compact or permeable, or whether it tolerates pressure. Insulating materials are available as loose filling, bats, boards and blocks. Special products are also available, such as pipe insulation.

In some countries new insulating materials are available, including snail-shell insulation which is used in foundations, hemp fibre insulation, as well as insulation of thin aluminium foil with stationary air sandwiched between the layers. The lower the λ (lambda) value, the better the insulation capacity.

Cotton (λ = 0.04W/mK) has good insulating properties and is used in the form of mats and as loose insulation. It does not require moth protection, but boric acid is added as a flame retardant. Transport from source to use is long and energy-intensive. Cotton producers use large amounts of poison in the cultivation and processing of cotton, but traces of poison have not been found in the finished material.

Calcium silicate sheets (λ = 0.045–0.065W/mK) are made of lime, silicon oxide and 3 to 6 per cent cellulose.

The sheets can withstand extremely high temperatures, buffer moisture and counteract mould. The sheets can be cut, sawn and drilled. Production requires relatively high energy consumption.

Cellular glass boards (λ = 0.042W/mK) are made by blowing air and pulverized coal into melted glass, creating a closed structure with small air bubbles. The material is waterproof and non-porous, does not burn or attract mould, and has a long life. Energy consumption during production is high. It is suitable for foundations. The high compressive strength of cellular glass means that if it is used on the ground, a concrete slab is not needed. If it is used underneath a concrete slab, the concrete slab can be thinner than if cellular glass were not used. The material enables the complete elimination of thermal bridges; and it is strong enough for a car to be driven on to a construction insulated with it without incurring damage. Wall and roof elements are also available.

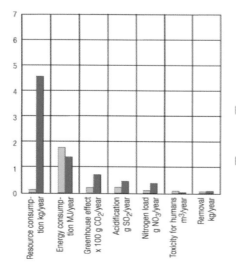

Environmental profiles for 1m² insulation (thick enough to provide a U-value of 0.2W/m²K) of cellulose fibre and polystyrene. The comparison shows that plastic insulation has a significantly greater environmental load than cellulose fibre insulation.

Source: *Arkitektur og miljø – form konstruktion materialer – og miljøpåvirkning*, Rob Marsh, Michael Lauring, Ebbe Holleris Petersen, 2000

A cellular glass foundation where the insulation material is the ground slab. Concrete casting is not needed.

Source: Foamglas

Cellular glass blocks (λ = **0.15W/mK**) are made from loose cellular glass (85 per cent) and cement (15 per cent). They are used as load-bearing and insulating building blocks. It holds its shape, withstands frost, mould and fire, and contains no hazardous additives.

Cellular glass loose fill (λ = **0.1–0.11W/mK**) is made from recycled waste glass by melting it and blowing air into it. Normal grain size is in the range 10–60mm. Typical use of cellular glass loose fill is as a thermal insulation layer in the foundation, i. e. under floor slabs on the ground. This lightweight fill also works as a water capillary barrier, a drainage layer and a frost-insulating layer. During installation of cellular glass loose fill in load-bearing structures, the loose fill material is compacted to obtain a stable and optimal load-bearing capacity. Loose fill is considerably cheaper than boards.

Cellular plastic (λ = **0.036–0.04W/mK**) is made of either EPS (cut blocks of expanded polystyrene) or XPS (extruded polystyrene blocks with 'skin'). EPS is made of 91–94 per cent by weight polystyrene, 4–7 per cent by weight pentane and 1 per cent flame retardant. Cellular plastic is manufactured as

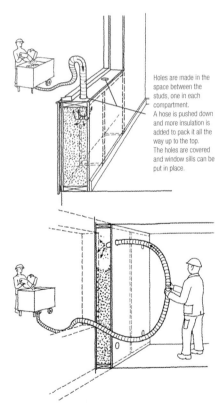

Holes are made in the space between the studs, one in each compartment.
A hose is pushed down and more insulation is added to pack it all the way up to the top.
The holes are covered and window sills can be put in place.

Cellulose fibre is an insulation material that is used in a lot of eco-buildings. It is sprayed into box constructions. Cellulose fibre is made from recycled newspaper or new pulp, and it is impregnated with boric acid or ammonium polyphosphate.

Spraying loose cellulose fibre.

insulation boards. They are broken down by ultraviolet light and so should not be used on exteriors. Cellular plastic absorbs only small amounts of heat and moisture, and cannot contribute to evening out the interior climate. Production is complicated, energy-intensive and environmentally hazardous but takes place in closed processes. Benzol is produced from petroleum, is carcinogenic and affects the immune system, and is used to produce styrene (which is hormone-disturbing). Xylene and styrene are released during production as well as from the finished material for the first two to three months. There is also cellular plastic that contains brominated flame retardants. Density >20kg/m^3. The λ-value can be reduced to 0.032 by the addition of graphite. If cellular plastic is on fire, it is burned down very quickly.

Cellulose fibre (λ = 0.04 W/mK) insulation is made from old newspapers or newly produced cellulose, and is delivered loose, in boards or in packing strips. The paper is ground up and boric acid, borax, sodium silicate or ammonium polyphosphate is added to make the material less flammable and attractive to fungi and insects. Additives may account for up to 14–25 per cent of the weight of the material. In the EU, additives containing boric acid and borax are questionable from a health point of view, and cellulose fibre with, for example, ammonium polyphosphate is preferred. The material has a good thermal insulation value, and can absorb and buffer moisture.

When applied correctly, cellulose fibre is highly airtight, which counteracts air movement (convection) in the insulation. Cellulose fibre is relatively heavy, which results in good sound insulation. The material is sprayed in with compressors under pressure, which fills spaces efficiently without settling occurring. The cellulose makes dust during spraying and breathing masks should be used. Production of cellulose fibre from recycled paper is energy efficient, while newly produced cellulose requires much more energy.

Sheep's wool (λ = 0.04W/mK) insulation often has boric acid added to it at about 3 per cent by weight, and can be mixed with up to 18 per cent polyester fibre to stiffen the boards. It is fire-resistant and has a good capacity to absorb, store and release moisture (30–40 per cent of its own weight). Hard compressed bats of sheep's wool are also used as impact sound insulation. As sheep's wool is vulnerable to attack by moths, it is impregnated with halogen-derived organic compounds such as eulan and mitin FF. Sheep's wool without anti-moth protection is subject to insect attack.

Straw (λ = 0.07–0.085W/mK) can be used both as insulation and as a load-bearing material in walls. It is most important to keep the straw construction dry by having a proper foundation and a substantial roof projection. Insulating with straw can be done using loose fill, boards or bales. Straw bales measure about 35 × 35 × 60cm and weigh about 20kg each. Bales must be properly compressed, dry (10–16 per cent moisture content) and free of all signs of mildew.

Hemp fibre (λ = 0.038–0.04W/mK) bats are made of compressed hemp fibre with a thickness of 30–180mm. Since hemp is naturally fungus- and bacteria-resistant, this insulation material does not need to be impregnated. Hemp bats are moisture-buffering and provide good sound insulation. Hemp is a hardy plant that grows well without either fertilizers or pesticides. The type of hemp (industrial hemp) grown for fibre contains extremely small amounts of narcotic substances. Hemp insulation is often mixed with polyester fibre (15 per cent) for stability and soda is added as fire protection.

Coconut fibre (λ = 0.045–0.05W/mK) bats or boards consist of pure coconut fibre that is felted together. The material is flammable and easily ignited and is therefore impregnated with ammonium phosphate, boric acid or sodium silicate. Coconut fibre is moisture-repellent and resistant to rotting and bacterial attack. It is elastic and durable. Coconut fibre is used as packing material and thermal insulation. However, transport routes from source to use are long.

Cork (λ = 0.045W/mK) is made from the bark of cork oak trees grown in Spain, Portugal and North Africa, which are mature for harvesting about 25 years after planting. Cork trees are peeled every 8–15 years and at the most one-third of the bark is taken. It is marketed in boards or in granular form, is resistant to moisture and rot and is not attacked by pests. If it is exposed to moisture for a long period it can begin to go mouldy. Cork has minimal temperature fluctuations. To increase its insulating capacity, it is common to expand cork with steam at 380°C in pressure tanks. It is then compressed under high pressure into boards or preformed pipe insulation. Cork's own adhesive substances prevent disintegration. Boards are relatively strong, have good recovery capacity and are therefore used for insulation exposed to pressure, e.g. exterior insulation of terrace roofs (warm roofs). Granulated cork is used as insulation in floor structures.

Wood shavings (λ = 0.06–0.08W/mK) were commonly used in the past as loose fill in walls and ceilings. Wood shavings are dried so that the moisture content is below 20 per cent, put into walls and compressed to pack it well. The construction must allow for adding more shavings, as wood shavings settle over time. The material absorbs and releases moisture in the same manner as wood. About 5 per cent slaked lime can be added to reduce the risk of attack by insects and pests. In addition, flammability can be reduced by adding 5–8 per cent borax or magnesium chloride.

Flax fibre (λ = 0.04W/mK) bats are made of flax fibre that is too short for fabric production. The material is naturally resistant to insect pests, tolerates moisture and doesn't burn

Insulating materials for sale at an ecoconstruction products exhibition in Germany. Shown here are insulation materials made of wool, flax and cotton.

Hemp fibre insulation has a very good environmental reputation. In Germany, it is for sale at the many shops selling ecoconstruction materials. It is also available in Denmark.

very well. It has good moisture-buffering (up to 25 per cent of its own weight) and sound-insulation properties. To stiffen bats, polyester fibre is added (2–18 per cent by weight). Flax cultivation does not require intensive use of fertilizers or pesticides. The fibres are bonded together during a short warming process and formed into bats. Flax fibre insulation is also available with added fire retardant.

Lightweight concrete (λ = 0.1–0.20W/mK) (see table Construction Materials) is used in walls, ceilings and floor structures as both load-bearing and pressure-tolerant insulation material. The Xellas Multipor slab brand (max. 200mm) is made of sand, lime, cement and water. It is specially developed to have a better than normal insulation value (λ = 0.045W/mK).

Expanded clay (λ = 0.13W/mK, loose fill; λ = 0.15–0.20W/mK, blocks) (see table Construction Materials) blocks and slabs are used in walls, ceilings and floor structures as both load-bearing and pressure-tolerant insulation material.

Clay and straw mixtures (λ = 0.2–0.4W/mK). Mixtures of clay and straw, lightweight clay constructions, have been used as a construction material for hundreds of years. The clay preserves the organic material and the straw insulates. Buildings with walls of lightweight clay require a supportive framework of wood. The straw is soaked

Flax fibre insulation in the form of a mat.

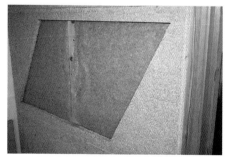

Flax insulation in the form of boards mounted in a demonstration wall.

carefully with slip. Walls of lightweight clay are compressed into shuttering or laid using prefabricated bricks. The harder the mixture is compressed, the lower the insulation value. It is important that the walls dry out properly. Thick walls can be built to make well-insulated buildings. Expanded clay can be used instead of straw in this mixture.

Mineral wool (λ = 0.033–0.045W/mK) can be either glass wool or rock wool. It is sold loose, in boards, in bats and in packing strips. Mineral wool has good sound insulation properties and is fire-resistant. Buildings with mineral wool insulation should be airtight so the fibres cannot escape into the interior air, and diffusion resistant so that moisture cannot damage the construction. Since the material cannot buffer moisture, moisture can run down and damage wooden elements. When moist, emissions of formaldehyde can occur, and in a fire, dangerous gases are released, primarily phenol. Damp mineral wool can smell bad due to the urea content. The material may contain a considerable number of fibres that can cause lung damage, i.e. fibres that are thinner than 5µm and longer than 3mm. Controversies have forced mineral wool producers to develop new products. Some producers have switched production to fibres that are less damaging to the lungs.

Rock wool is made of diabase and dolomite. Small amounts of urea as well as phenol formaldehyde resins are used as binding agents. In addition, 1 per cent by weight silicone or mineral oil is added to reduce dust and increase moisture resistance. The raw material is mixed and melted at 1350–1500°C. The molten material is blown through a nozzle against a rotating surface, whereby mineral fibre is formed. The fibres are bonded together with phenol formaldehyde resins, which harden at 220°C. Production is energy-intensive.

In glass wool, stone is replaced with quartz sand (30 per cent by weight), feldspar and dolomite (30 per cent by weight), soda (10 per cent by weight), as well as recycled glass (30 per cent by weight). Lately, the proportion of recycled glass has increased substantially.

ISO-soft is an American installation material with long, elastic glass fibres, that are bound together with acrylic fabric and contains no formaldehyde or urea. This installation feels and looks like cotton but has the same insulating properties as mineral wool. ISO-soft does not itch or dust as much as rock or glass wool and is an alternative to these for people with allergies.

Perlite (λ = 0.044–0.053W/mK) is made by heating natural glass of volcanic origin to about 1000–1100°C. The material then expands to a volume 15–20 times larger than the original. The raw material is mined in Iceland, the Greek island of Milos, Hungary and Turkey. Perlite does not burn and doesn't react chemically with other construction materials. It allows diffusion and has such a high compressive strength that it can be used as underlay for floating floors.

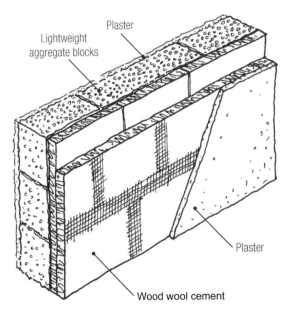

If expanded clay is used in construction, walls must be made extra thick or supplementary insulation used, because expanded clay itself is not a good insulator. Shown here is an expanded clay wall that has been insulated with wood-wool cement board.

In its granulated form, it is used as insulation in floors and cavity walls. In Europe, perlite is also sold in boards. Hyperlite is perlite that has been sprayed with silicate (0.2 per cent by weight) at 400°C, which makes it moisture repellent. Perlite treated with bitumen can release emissions and should not be used indoors.

Foamed concrete (λ = 0.1W/mK) is used in solid constructions as both load-bearing and insulation material. Foamed concrete requires a lot of binding agents and is made up of cement grout containing 50 per cent cement, 50 per cent merit (a slag product), water and a frothing agent (tenside-based and degradable), as well as expanded clay balls as ballast. It is a relatively light and solid material. In factories it can be cast into blocks or sections. The sections can be load-bearing but are reinforced with galvanized steel net. The reinforcement is needed primarily to enable transport of the sections. The material cannot be used as a moisture barrier. Foamed concrete is a relatively new material, which means that experience with it is limited.

Peat (λ = 0.08W/mK, loose fill; λ = 0.038 W/mK, boards) was used in the past as an insulation material in much the same way as wood shavings. Today, it is not commonly used. Peat can be used as loose fill, in boards or as blocks. Peat is extracted from the uppermost layer in bogs where the decomposition process is not well advanced and the fibres remain intact. Loose-fill peat is dried and ground and about 5 per cent lime is often added. Peat has a low pH which inhibits bacterial and mould growth. Even in countries where there are large peat reserves, growth takes place slowly. Peat settles to some degree in walls and therefore has to be replenished. Dust is created during installation.

Textile fibre mats are made in various thicknesses of textile waste bound together with polyester. Production requires relatively low energy consumption and the lambda value can be compared to mineral wool and cellulose fibre.

Translucent insulation is a futuristic plastic insulation material that lets light through. Both the material's structure in the form of a honeycomb, fibre structures or small tubes, as well as the material's characteristics (it contains a lot of air) are used to advantage. It is often made of polycarbonate or aerogels. The material is currently expensive and used mostly to improve energy transfer in passive solar heating systems and solar collectors.

Vacuum-fused fumed silica insulation Vacupor® (λ < = 0.005W/mK) has extremely good insulation properties. It is a

A wall with translucent plastic insulation lets the light through, at the Ecotopia training centre, Aneby, Småland, Sweden.

Textile fibre mats are made of textile waste and bound with plastic.

Source: www.ultimat.no

microporous inorganic material composed 85 per cent of silicon dioxide (SiO_2) and 15 per cent silicon carbide (SiC), which is vacuum sealed in thin sheets with a metal barrier film of aluminium. Removal of air to form a vacuum prevents any thermal transmission by convection. The core material is produced without any emissions, is not flammable and usually does not contain any organic components that may release harmful decomposition products when exposed to heat. The product can be recycled and, according to industrial hygiene experts, does not pose any health hazard. The material is expensive but extremely space-saving.

Cellulose fibre is in the compartment on the left, perlite is in the middle and on the right is mineral wool.

Wood fibreboards (insulation boards, λ = 0.037W/mK, porous boards, λ = 0.045W/mK, hardboards, λ = 0.17W/mK) are made of wood that is chipped and ground (defibred), diluted with water, and then compressed using heat and pressure. The lignin in the wood acts as a binding agent. Softwood fibre insulation boards (with a thickness of 40–200mm) contain polyolefin as a binding agent and ammonium polyphosphate as a flame retardant. German hardwood fibre insulation boards often contain some polyurethane resin (about 4 per cent), a substance that is not recommended. Fibreboard is available as hard, medium-density and porous boards. Porous fibreboards (12–40mm) are used primarily for thermal insulation, as impact sound insulation and as a sound absorber. In some cases, aluminium sulphate (1–3 per cent) is added as mould protection and ammonium sulphate is added to increase fire resistance. Some boards have a surface layer of wax. Oil-cured wood fibreboard is more moisture repellent and usually contains pine resin. Wood fibreboard with adhesives should be avoided for environmental reasons.

Wood-wool cement (λ = 0.09–0.15W/mK) boards are made of wood-wool (35 per cent by weight) with cement or magnesite as a binding agent (65 per cent by weight). Calcium chloride (0.2 per cent by weight) can be added to control hardening. The material is moisture-repellent, moisture-buffering, sound absorbent, pressure-proof, non-flammable, and has a high pH value, which inhibits mould growth. It is often used for ceilings in indoor swimming pools, gymnasiums and bathrooms. The boards are heavy and so insulate against sound. Due to their coarse surface, the boards are excellent for

Honeycomb brick with a mineral granular filling, made of basalt, from the German company Unipor. Unipor makes a product called Coriso that is specially developed for low-energy and passive houses, with a lambda value of 0.08 W/mK.

Insulation: Environmental Evaluation

Recommended	Acceptable	To be avoided
Cellulose fibre*	Cellular glass	Cellular plastic EPS, XPS
Hemp fibre	Lightweight concrete	Glass wool
Coconut fibre	Expanded clay aggregate	Rock wool**
Cork	Perlite	Polyurethane (PUR)
Wood chips	Polyester	
Flax fibre*	Foamed concrete	
Straw	Sheep's wool*	
Wood fibre insulation board	Vacuum-fused fumed silica insulation	
Wood-wool cement boards	Vermiculite	
Peat	Calcium silicate sheets	
Corrugated cardboard	Honeycomb brick***	
Foam glass gravel	Textile fibre mats	
Shells		

Notes: * The environmental impact of cellulose fibre, flax fibre and sheep's wool insulation depends on the additives they contain. These materials usually contain fire retardants and mould-impeding substances such as boron salt, boric acid, ammonium polyphosphate or sodium silicate. Boric acid is notified by the European Commision as hormone disturbing, category 1. Flax fibre insulation is sometimes mixed with polyester to increase the stiffness of the material.
** Rock wool is valuable for fire protection.
*** Also available with mineral granular filling.

plastering. There are low-rise buildings built of wood-wool cement boards, where the boards are both load-bearing and insulating. Also available are wood-wool cement sections reinforced with softwood supports, less than 3m wide. Prepainted wood-wool boards of varied wood-wool fineness are available. Production of cement is energy-intensive.

Vermiculite (λ = 0.053–0.065W/mK) is made of mica. The material is heated to 800–1000°C, when it divides into thin boards and curls up into a light and porous mass that is used for thermal insulation, either as loose fill or boards. It can also be used as an ingredient in concrete in a 6:1 ratio to make the concrete lighter and improve its thermal insulation properties. Expanded vermiculite is chemically inert, incombustible and withstands high temperatures. Vermiculite absorbs more moisture and there is a greater risk of settling than with perlite. Production is energy-intensive.

Cladding

Wood is the cladding that causes the least environmental stress. If a building is constructed using bricks, the construction material can also serve as cladding material for both homogeneous and compound wall constructions. Some board materials can be used as cladding. Many materials can be plastered or painted to give the cladding a certain appearance.

Concrete (see 1.1.2 Knowledge of Materials and Construction Materials)

Cement based boards (see Boards)

Glass is made from quartz sand, soda or potash, and limestone. It melts at about 1400–1550°C. There are many types of glass. Old glass can be recycled into new glass. Production of glass is energy-intensive. Mechanical mounting should be used in order to avoid environmentally hazardous sealants.

Flat glass is used for windows, as sheets for patio roofs and walls, for cladding and interior partitions.

Energy-efficient glass is coated with a thin layer of stannic oxide, which is hard and can be used outdoors, or thin coatings of silver or gold, which are so easily scratched that they can only be used within multiple panes. The layers impede long-wave radiation (warmth) from escaping through the window. Phosphorus oxide (P_2O_3) can be added to allow UV radiation to pass through.

Insulating windows are made of two or three glass sheets separated by hermetically sealed spaces filled with clean, dry air. Aluminium and flexible packing are the most commonly used spacers. To reduce energy loss, the air in the sealed space is replaced with a heavy gas, such as argon, which slows convection. There is also new technology that uses a vacuum.

Sun protection glass serves to reduce the discomfort indoors from glare and high temperatures. Such glass is normally absorbing or reflecting, but combinations exist. There is also a type of sun protection glass where the level of protection from the sun can be controlled using an electric signal.

Glass block ('glass concrete') is an example of a shaped glass product. It is used for exterior and interior walls. Glass block can also be used in constructions that receive quite heavy traffic, such as courtyard decks. Since a glass concrete construction consists of many glass sections, heat loss is relatively high.

Wired glass is flat glass with wire mesh inserted into it, which provides fire protection. Wired glass is less durable than an equivalent sheet of homogeneous glass.

There are two types of safety glass, tempered glass and laminated glass. Tempered glass is made by warming ordinary glass up to 600°C and then quickly cooling it down. If broken, tempered glass fractures into many small cube-shaped pieces that people cannot easily cut themselves on. Laminated glass is made by bonding two or more layers of glass with a thin plastic sheet between the layers. This type of glass is used for burglary protection or bullet-proof glass. More energy is required to produce tempered glass than non-tempered glass.

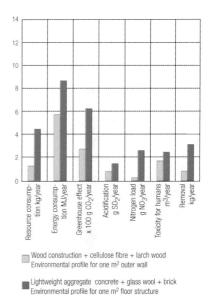

Environmental profiles for two similar exterior wall constructions with different cladding. One is clad with wood, the other with brick.

Source: *Arkitektur og miljø – form konstruktion materialer – og miljøpåvirkning,*, Rob Marsh, Michael Lauring, Ebbe Holleris Petersen, 2000

Calcium silicate boards (see Sheets table)

Sand-lime brick (see Construction Materials)

Plywood (see Sheets)

Lightweight concrete (see Construction Materials table)

Expanded clay aggregate (see Construction Materials table)

Plaster (see Plaster and Mortar)

Stone (see Flooring). Stone cladding is installed with special stainless steel metal anchors and is ventilated on the inner side.

Residential area in Hestra, Borås, Sweden, with both cladding and roofs of black corrugated cement-based boards (fibre cement boards).

Source: Architect: Jens Arnfred, Vandkunsten, Denmark

The town centre and continuing education academy in Mont-Cenis Sodingen, Herne, in the Ruhr region of Germany. The centre is made up of various buildings built inside a large glass building. The glass building is supported on a wooden construction. It has solar cells on the roof and is ventilated with natural ventilation via underground pipes. So a microclimate has been created in the enclosed centre.

Source: Architect: Büro Jorda et Perraudin

Untreated larch cladding (when the building was newly constructed) at the National Parks' building in Tyresta, near Stockholm.

Source: Architect: Per Liedner, Formverkstan arkitekter AB

Wood cladding treated with inexpensive green vitriol at the Understenshöjden eco-village, Stockholm. Some parts of the cladding are brick, behind which are bathrooms built entirely of mineral materials.

I HEALTHY BUILDINGS | I.I MATERIALS AND CONSTRUCTION METHODS | I.I.3 ASSESSMENT OF MATERIALS

Cladding: Environmental Evaluation

Recommended	Acceptable	To be avoided
Calcium silicate board	Concrete block	Concrete, reinforced***
Stucco	Cement fibreboard	Metal (sheets)***
Sand-lime brick	Glass	
Clay plaster	Ceramic tiles	
Stone	Expanded clay**	
Wood*	Lightweight concrete	
Wood-wool cement	Plywood	
	Brick	

Notes: * Pressure-impregnated wood should be avoided.
** Sandwich panels made of expanded clay blocks with embedded cellular plastic, should be avoided.
*** Reinforced concrete and sheet metal claddings are energy-intensive and have a large environmental impact, but cannot always be avoided.

Brick (see 1.1.2 Knowledge of Materials and Construction Materials). Brick is used as cladding, either as a load-bearing structure or as a simple cladding.

Wood (see 1.1.2 Knowledge of Materials and Construction Materials table). Wood cladding lasts as a rule for 30–50 years. A significant amount of room should be left between the cladding and the ground.

Wood-wool cement board (see Thermal Insulation). The boards have a very good surface for plastering and can be used if a plaster cladding is desired on a wood building.

Roofing

Choice of roofing material depends on the slope of the roof. In damp climates roofs require a good slope to protect buildings from rain and snow. If a relatively flat roof is chosen, sod or sedum roofing are possibilities.

Cement tiles (roof slope >14°) are made from sand (75 per cent by weight), cement (25 per cent by weight) and water. Sometimes lime slag, silicate and iron oxide (less than 1 per cent by weight) are added. They have good load resistance and frost resistance. Cement tiles require less energy to make than ceramic tiles, but cement production is both energy-intensive and environmentally destructive. There are also surface-treated cement tiles with an outer layer of acrylic. Locally made roof tiles should be prioritized in order to reduce transportation.

Fibre cement sheets (roof slope >14°) used as roofing can be corrugated or flat. Since the boards are relatively light, the dimensions of roof trusses can be smaller than for heavy roofs. Cement-bound fibre sheets consist of cement (65–80 per cent by weight), filler (limestone or dolomite) and cellulose fibre (6 per cent by weight). The cellulose fibres are impregnated with silicic acid (8 per cent by weight) so that they do not break

Sedum roof.

Source: VegTech AB

Grass roof with 15cm of soil.

Source: Tyresta Museum of the Swedish National Parks. Architect: Per Liedner, Formverkstan Arkitekter AB

down in the alkaline environment. The silicates used are amorphous. The sheets are also available with 2 per cent PVA (polyvinyl alcohol) fibre and with fibreglass reinforcement. The sheets can be painted with acrylic paint or dyed with pigment. They are mounted with screws. Cement production is energy-intensive. Smaller sheets are produced without fibre reinforcement.

Glass (see Cladding) is sometimes used for roofing in glass-walled rooms and patios.

Green roofing (roof slope >3–45°) (see also Vegetation and Cultivation) is a generic term for grass and sedum roofing. The thickness of the soil layer can vary from 3–15cm, or may even be thicker in the case of roof gardens.

There are many advantages to a green roof: good noise reduction, improvement of the local climate due to evaporation from plants, cleaner air since the vegetation layer absorbs and binds dust and fluff and releases oxygen. The vegetation layer binds up to 50–80 per cent of rainwater, so relieving pressure on the runoff drainage system. It offers ecological habitats for birds and insects. Sedum roofs are almost maintenance-free while grass roofs require more maintenance and risk drying out during

Close-up photo of a green roof of mixed sedum species.

Source: VegTech AB

warm summers. A grass roof with 15cm soil has a high thermal capacity, and therefore is not as warm in summer or as cold in winter as an ordinary roof.

On completely flat roofs, water runs off poorly and vegetation can be damaged. Gentle slopes (3–7°) require a drainage layer. Steep slopes require a water storage layer. A roof slope of 7–10° is ideal. When the slope of a sedum roof is greater than 14°, the vegetation mats must be fastened to the underlay. At these slopes and greater, overlapped polyethylene sheets could be used as a waterproof layer. Grass roofs can be laid on slopes of up to 45° but if the slope is greater than 20°, the roof must be equipped with a mat that anchors the substrate (e.g. braided jute that keeps sand and soil from running off). Steep grass roofs require rainwater gutters along the eaves.

Thin sedum roofing (moss-sedum mat) is 3cm thick, but with slight slopes an additional 3cm drainage layer must be added. Thicker herb roofing (sedum-herb mat) has about 3cm mineral soil, about 3cm drainage and a 3cm water retention layer (total 9cm).

A true grass roof (sedum-herb-grass mat) is 12–15cm thick. It has at least 3cm soil, and the division between drainage and water retention layers depends on the slope.

Grass roofs have a long life and are almost maintenance free. The layer protecting against root penetration is especially important for grass roofs. This protective layer should prevent plant roots from damaging the vapour barrier. Since the protective layer provides a second waterproof layer, it acts also as an extra vapour barrier, which is good since flat roofs often have problems with leaks. A cold roof means that the insulation is underneath the vapour barrier and that there is an air gap between the insulation and the panel on which the vapour barrier lies. A warm roof is one where the insulation is above the vapour barrier and the insulation material should therefore tolerate a moist environment.

Sedum roofs have the same properties as grass roofs. As they are thinner, they cannot absorb as much water and have a smaller heat storage capacity. A sedum roof is 2–6cm thick above the waterproof layer. Succulents (sedum) are especially suitable plants for these roofs, e.g. English stonecrop. With sedum on the roof, no watering is needed as sedum is a succulent. About 40 different sedum species in a variety of colours are used.

There are waterproof layers suitable for use under green roofs made of polythene, polypropylene, polyolefin, synthetic rubber (EPDM and butyl rubber) and plastomer-modified asphalt (APP and ECB). Polythene (bubble sheets that form a vapour barrier) cannot be joined or glued together. The sheets must be overlapped on a sloping roof (minimum 15°). Polyolefin mats are custom-made with notches and flanges for chimneys and pipes. EPDM rubber is more expensive than plastic materials and contains some environmentally hazardous solvents. The best material from an ecological perspective, according to the German Ökotest, is plastomer-modified asphalt, a mixture of asphalt and elastic plastics. This material is easy to work with, but is most suitable for small roofs.

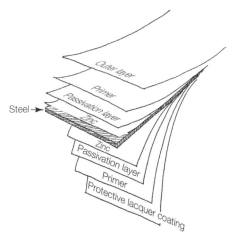

Different layers on a typical metal sheet for roofing.

Source: VegTech AB

A grass roof is heavy (120–150kg/m² in a saturated state) and may require reinforcement of the roof construction. A herb roof weighs 110–120kg/m². Thin sedum roofs weigh 40–50kg/m² in a saturated state, which doesn't normally require reinforcement of the roof.

Metal roofs (copper roofing >3°, double-seam lock metal >6°, single-seam lock metal and profiled sheet metal >14°) are produced using very energy-intensive extraction and production processes which release many dangerous substances. Nonetheless, because thin metal is used, the energy consumption per m² of roof material is moderate. However, environmentally conscious building contractors should use metal sparingly.

Steel sheets are made of iron (98 per cent by weight), which is the least energy-intensive of metals, but it rusts and must be protected with a surface layer. There are, however, few environmentally friendly anti-rust paints.

Steel sheets coated with zinc and/or aluminium are used for roofing. The metal is coated with a thin layer of zinc and/or aluminium for corrosion protection. There is a certain amount of zinc runoff from hot-dipped galvanized surfaces exposed to precipitation. Aluminium-zinc coatings result in less zinc runoff than hot-dipped galvanized steel sheets. In large amounts, zinc is an environmental hazard, especially for aquatic organisms. The metal galvanization process can produce emissions of nickel, chromium, cyanides and fluorides.

Steel sheets colour-coated with a layer of PVC, polyester or polyvinyl idene fluoride (PVF2) are used for roofing. The colour coating has a life expectancy of at least ten years. The sheets used are thin, hot-dipped galvanized sheets. Before coating the sheets with a PVC layer, they are painted with primer, such as acrylate, epoxy and polyester. The back is also painted with a thin layer of epoxy. Plastisol paint is made up of PVC, softening agents, organic solvents, metal salts and pigment. Problems occur because of the paint flakes. Currently there are plastisol coatings that are totally phthalate-free and that use a polymer or plant-oil-based softener. Dioxins can form during uncontrolled burning of PVC (if oxygen is limited and the temperature is not high enough).

Stainless steel is defined as an iron-based alloy that contains at least 10.5–20 per cent chromium and a maximum of 1.2 per cent carbon. Depending on the environment where the metal is used, it also contains varying levels of nickel (0.5–14 per cent) and molybdenum (<3 per cent) (the latter improves corrosion resistance properties). The forms of chromium and nickel used in stainless steel are not dangerous and are not allergenic. For example stainless steel used in coastal areas and polluted town or industrial environments is composed of about 17 per cent chromium, 11 per cent nickel, 3 per cent molybdenum and the remaining part iron. The life of the product is equivalent to the life of the building, which is, however, highly dependent on how aggressive the environment is. Stainless steel should be used so as to facilitate it being recycled.

Aluminium production requires large amounts of energy. During production, emissions of poisonous fluorine compounds and large amounts of sludge are produced that are harmful to the environment. Producing recycled aluminium requires less than 5 per cent of the energy required to produce new aluminium from bauxite. As a rule, the life of aluminium is equal to the life of the building. Normally, aluminium can be left untreated, but in cities with bad air and industries it corrodes faster and requires surface protection, i.e. lacquering or anodizing.

Zinc alloys are relatively stable, but production of zinc is energy-intensive and results in the release of many hazardous substances into the air. Production of 1 tonne of zinc typically releases about 10kg zinc, 1.8 tonnes lead and 0.1kg arsenic. Large amounts of zinc are an environmental hazard, especially for aquatic organisms.

Copper sheeting is soft and tough, which makes it suitable as a covering for complicated roof shapes. Other uses for copper sheeting are gutters, drainpipes and doors. Copper releases copper ions which are poisonous to aquatic organisms. Copper concentrations in urban waste water are derived typically as 1 tonne from industry, 1 tonne from copper roofs and 6 tonnes from the corrosion of copper pipes in buildings. (In addition 4–5 tonnes derive from brake linings.) The use of copper may be restricted to buildings where it has a cultural-historical value. Copper sheeting contains primarily pure copper (>99 per cent). Copper sheeting for roofs and façades normally has a long life. In urban environments, copper surfaces corrode at a rate of about 0.9μm per year. The City of Stockholm has, as a landowner, a policy of not using copper in roof and façade material in new construction in areas where water runoff is not treated.

Seaweed (*Zostera marina*) may be used as a roof and insulation material on historical buildings in districts where roofs were steep and very thick (600–800mm). They lasted for several hundred years because the seaweed contains a lot of silica. Some seaweed can be cleaned, dried, chopped and used as an insulation material in buildings.

Slate (roof slope, double slate >25°) roofing comes in thin tiles (though not thinner than 6mm) of split slate. Slate is waterproof, frost and heat resistant, as well as resistant to air pollutants. Slate roofs are regarded as one of the most durable roofs (lasting at least 100 years). They are easy to drill holes in and nail onto the roofing structure. The tiles must be laid with a lot of overlap (in three layers). Old slates can be reused on new roofs. Production of slate tiles is energy-intensive. However, the greatest environmental impact takes place when the product is transported from the mine to the consumer; so local stone should be prioritized.

Skifferit (the term indicates high slate content) is an artificial roof material made of about 75 per cent slate granules, calcium carbonate, and the binding agent, polyester

Both the roof and cladding of Gunnarsnäs manor in Dalsland, Sweden are made of slate.

Straw roof on the Open Air Museum Rocca al Mare in Tallinn, Estonia.

resin. The surface of *skifferit* is similar to that of natural slate. It is not, however, as durable as natural slate.

Thatched roofs made of reeds, great fen-sedge or straw (roof slope >45°) were common in the past. Reeds are the best material for thatched roofs. Rye or wheat straw is usually used, though oats, barley and flax have also been used. The straw should be ripe, well-grown and threshed. A big disadvantage of these roofs is that they are flammable, which leads to high insurance costs. Fire retardants are available for straw, but they are not especially environmentally friendly. Reeds and great fen-sedge are not as flammable as straw. Straw roofs laid by professionals are fairly durable (last about 30 years) and provide some heat and sound insulation. The straw is bound together or held down using wooden spars.

Roofing-felt (roof slope, three layers of felt >0.5°, two layers of felt >3°) (asphalt roofing felt) is made from synthetic felt impregnated with polymer-modified asphalt. The felt, either polyester felt or mineral-based felt (e.g. fibreglass), is impregnated with polymer-modified asphalt (polymer bitumen) (60–70 per cent by weight) and styrene-butadine-elastomer (10 per cent). The surface is protected from the sun's UV light by applying crushed minerals in a variety of colours, e.g. ground limestone, sand or slate flakes. Roofing felt is available in sheets or rolls that are laid in an overlapping pattern. Sheets are attached to the underlay with nails and then stuck to each other.

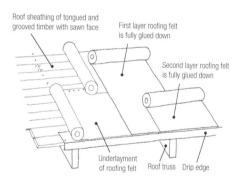

Felt roofs may seem environmentally friendly, but these days felt roofing is often made of synthetic materials, such as bitumen-impregnated woven fibreglass. Bitumen-impregnated woven wool is also available, but it can be hard to get. Bitumen is an asphalt product.

Source: adapted from Träinformations *Träbyggnadshandbok* (Timber Information Timber Building Handbook)

Roofing felt lasts about 30 years. There are various types of polymer-modified asphalt with different properties, such as APP (atactic polyolefin) and SBS (styrene-butadiene-styrene). The best asphalt from an environmental perspective is plastomer asphalt (APP and ECB). Asphalt products should not be exposed to interior air since they can secrete polycyclic aromatic hydrocarbons (PAHs). A roofing sheet certified by Sunda Hus AB that doesn't contain any substances hazardous to the environment or health is Sarnafils TG66-15. It is made of flexible polyolefins and is used for single layer sheeting with ballast or as a membrane.

Roof tiles (roof slope >22° for unrebated, or >14° for rebated) are made from clay that is fired at 900–1200°C. The quality of the clay determines the durability, frost resistance and colour of the tiles. Roof tile is laid in an overlapping manner on battens, and can be fixed to the battens with clips if there is a risk from strong winds. Roof tiles last for a long time and replacing individual tiles is relatively easy. Firing roof tiles is very energy-intensive. Roof tiles can be glazed or treated with pigment (iron oxide) to increase their life and to add colour. Producing surface-treated tiles requires more energy than producing tiles without a surface treatment. Locally produced tiles should be prioritized to reduce transportation.

Wooden roofs [roof slope for board roofs >27° (in some cases e. g. larch >15°), shingled roofs >45°] made from split logs, boards or shingles have been common in well-forested areas. Timber from spruce, pine, oak, aspen, cedar and larch is used, and should be as straight-grained and knot-free ('clear') as possible. Heartwood and resin-rich types of trees such as larch are the most durable. Wood roofs must always be ventilated. Pressure-impregnated wood should be avoided. Use of imported timber means that energy consumption to produce and deliver the product greatly increases due to energy used for transport.

Hand-split roof shingles can last a long time (30–100 years). Shingles are split from logs lengthwise along the grain. Machine

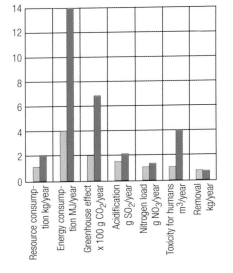

Comparison of environmental profiles for two types of roofs, a tile roof and a zinc roof construction.

Source: *Arkitektur og miljø – form konstruktion materialer – og miljøpåvirkning*, Rob Marsh, Michael Lauring, Ebbe Holleris Petersen, 2000

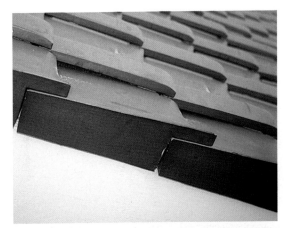

Clay tile roof in China. In China there are three kinds of roof tiles: ridge tile, roof tile and verge tile. Verge tile is made in order to provide a harmonious finish to the roof. Wooden bargeboards at the outer edge of a tile roof are not used in all countries.

splitting involves cutting through the grain of a tree, and therefore such shingles are less durable than properly hand-split shingles. Shingles should be laid with an overlap so that there is triple coverage. Thick shingles (shakes), which are often found on old churches, are very durable. The steeper the roof, the longer its life. Rainwater must be able to run off quickly to avoid damage to wooden shingles. Shingles impregnated with oil or regularly tarred last the longest.

Roof Material: Environmental Evaluation

Recommended	Acceptable	To be avoided
Grass, sedum	Concrete tiles	Sheet metal*
Shale, local	Cement fibre sheets	Felt roofs**
Thatching	Glass	
Clay tiles		
Wood		

Notes: * Sheet metal roofs can be a useful alternative in some situations.
** Felt roofs can be better or worse depending on their composition, but they usually contain environmentally hazardous chemicals.

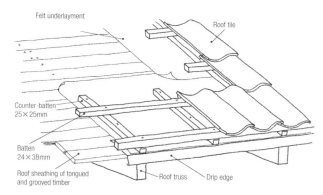

Clay tile roofs are considered to be the best choice from an environmental point of view according to the environmental priority strategies in product design (EPS) method in the life-cycle analysis done by Martin Erlandsson at the Department of Building Materials, the Royal Institute of Technology in Stockholm. A comparison between clay tile, concrete and sheet metal roofs was carried out.

Source: adapted from Träinformations *Träbyggnadshandbok*

Shingled roof on a Norwegian stave church.

Wood roof with notched boards, that is boards with grooves so water can run off easily, on a test building in the eco-village of Engeshöjden near Gävle, Sweden.

Source: Architect Anders Nyquist

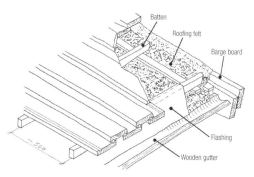

Construction of a boarded roof.

Board roofs last as a rule for 30–50 years; boards are overlapped and grooved boards (lengthwise) are used to improve rainwater runoff.

Air- and Water-Proof Barriers

There are different types of air- and water-proof barriers:

- airtight layers for walls and roofs with varying degrees of water permeability;
- weatherproofing in wet areas is used in constructions that don't tolerate moisture, e.g. wood constructions;
- waterproof layers on the ground and on roofs that are used for green roofs and ponds (see also Geomembranes).

Waterproof layers for wetrooms, the ground, green roofs and ponds are compared in the same table (Weatherproofing – Watertight).

Airtight Weatherproofing

Buildings should be constructed so that they are air- and wind-proof to avoid draughts, energy loss and moisture damage. In buildings that are not airtight, moisture damage can occur where warm air leaks out. Therefore, it is important that layers of weatherproofing overlap each other and are glued, taped or clamped together. It is important to achieve a tight seal everywhere, and special attention should be given to sealing around gables, chimneys and ventilation ducts. Wind protection is even used inside claddings and in roof constructions to prevent air movement in insulation. To get a durable construction it's important that the tape has a long life and is tested.

When using organic insulation materials airtight layers should not have too high a capacity for stopping diffusion since diffusion through the construction is then prevented, thereby shielding moisture-buffering properties. Although it is true that most interior air moisture escapes through airing and ventilation, about 2 per cent goes out through the outer walls if they do not have too high a water-vapour resistance.

The ability of a material to let through moisture is measured using water-vapour resistance. The lower the level, the more moisture a material lets through. Building engineers believe that the inside of a construction should have a five to ten times higher water-vapour resistance than the outside so that moisture can escape. No matter how a building is constructed, it is important that it is easier for moisture to get out than in. The interior water-vapour resistance equals the total water-vapour resistance of the inner surface layers, i.e. building paper, interior wall-covering and surface treatment (paint or wallpaper).

Many constructions are themselves air and wind tight, such as concrete or solid wood constructions. For such buildings, it is especially important to make sure the joints are sealed by using packing and sealing compounds. In lighter constructions, an air- and wind-tight layer is needed to prevent draughts, reduce energy losses and prevent moisture damage. Plastic sheeting is commonly used, but it is often vapour-tight. In ecological building, the goal is to have a layer that is air- and wind-proof but vapour permeable, so that the material's moisture-buffering capacity can contribute to a better interior climate.

Plasterboard (see Sheet Materials) is windproof, and is used for external wind protection. In order to get a good windproof layer with plasterboard, proper sealing of the joints and edges is important. Water-repellent plasterboard is coated with a wax emulsion made of paraffin wax or silicone sealant to prevent ingress of moisture. In some cases it may contain fungicides.

Hard fibreboard (see Sheet Materials) is windproof. Hard fibreboard (3–6mm) can be used for weatherproofing on the inside of roofs and as an interior airtight board in wall constructions. For moist environments, wood fibreboard oil-tempered with tall oil or asphalt impregnated boards is used. Some boards have an exterior coating of wax.

Building paper/felt used to be a commonly used weather barrier, but the market has been taken over by plastic sheeting. In any case, building paper/felt remains of interest in ecological building. There are different kinds available, such as cellulose and textile. Uncoated building paper/felt is available, but impregnating the material can make it more or less vapour-tight. Oils, resins or asphalt are used to impregnate the paper/felt. Building paper/felt with either one or both sides coated with polythene, polypropylene and/or polyester, or aluminium foil is also available. The last is vapour-tight. Wool paperboard and jute felt coated with a thin layer of polythene are relatively permeable and act as moisture barriers, not as vapour-tight layers. For certain types of windproofing, information about additives in the form of titanium dioxide and UV stabilizers is available. Kraft liner board is tear-resistant as it contains woven fabric. In order to achieve a good windproof layer with building paper, the joints and edge seals are important. The paper/felt is laid in an overlapping manner, folded and jammed between mouldings. Vapour resistance for unimpregnated building paper is about 1×10^3 s/m (seconds per metre). Life is estimated to be 40–50 years. No maintenance is required.

Wool-based felt is made of recycled paper and at least 15 per cent wool. It is soft and porous, and is often used in floors to reduce noise.

Plastic sheeting is usually made of polyethylene (PE) with additives, e.g. pigment, titanium-oxide, UV stabilizers, etc. PE windbreaks are used inside façade coverings to block air movements within insulation. Other areas of use for vapour-tight plastic sheeting are moisture barriers in foundations, walls and roofs; vapour barriers

Water- and wind-resistant building paper/felt at a ecological building outlet in The Netherlands.

between concrete and wood-based flooring; as vapour barriers on the ground in crawl spaces; as well as for building materials during the construction phase. PE vapour barriers are normally 0.2mm thick. In some cases plastic sheeting is reinforced with fibreglass or polyamide (nylon). Common plastic sheeting is vapour-tight, but there is also windproof and vapour-permeable plastic sheeting, e.g. windproof fabric made of polyolefin. The latter has pores that are too small to let through water droplets, but large enough to let through water vapour. All plastic sheeting is a petroleum product. All joints should be sealed, e.g. with butyl tape. Life is estimated to be 50 years.

Vapour resistance for vapour-permeable plastic sheeting is 2.7×10^3 s/m. Vapour resistance for vapour-tight plastic sheeting is 2000×10^3 s/m.

Waterproof Layers

Wetrooms must be built so that moisture problems do not occur. It is best to build wetrooms using mineral materials that can withstand moisture, and so avoid having to use waterproof layers.

When wetrooms are built in wooden buildings, waterproofing in the floor and walls is required, but there are no really good alternatives from an environmental perspective. The alternatives available are polymer-based primer, rubber or plastic sheets that are protected by glazed tile and expanded clay, or bitumen coating or sheets. (See 1.1.4 Wetrooms.)

Waterproof layers are used in several places in and around buildings to prevent water ingress, for example, waterproofing for green roofs, on the ground and in foundations. Waterproofing on the ground is used to line dams or to keep water from undermining embankments and foundations. Whether or not a waterproof layer is used when building a dam depends on the type of soil at the site; sometimes it is not needed to keep the water in.

Bentonite (sodium bentonite) is a clay that expands when it comes into contact with water and becomes a hard and waterproof clay-like mass. The material is used in landscaping, but must be covered with at least 40cm soil to provide enough counterpressure. Panels and mats made of bentonite clay encapsulated between polypropylene or polythene geotextiles are available. There are also joint-sealing tapes containing bentonite used in concrete casting. Bentonite clay is only found in a few places in Europe, primarily Southern Germany and Greece (Milos). The major bentonite producer in the world is the US.

Bitumen is a by-product of oil refining. Asphalt is made by mixing bitumen with crushed rock. Some bitumen is suspected of being carcinogenic as certain types of coal tar with tar components and cracking products contain large amounts of polycyclic aromatic hydrocarbons (PAHs). On the other hand, carefully distilled and 'pure' bitumen does not contain any unhealthy substances. If bitumen is protected from sunlight, temperature variations, and large amounts of moisture or acids from the soil, it has a long life. Production of bitumen is energy-intensive and environmentally hazardous, but is considered to be less environmentally hazardous than production of plastic. Bitumen exposed to sunlight and/or heat can release unhealthy emissions (PAHs), which is why it should not be exposed to interior air.

Bitumen coating (asphalt coating) is used in foundations and basement walls to prevent intrusion of moisture, and as a vapour barrier between a foundation wall

Weatherproofing – Airtight: Environmental Evaluation

Recommended	Acceptable	To be avoided
Building paper/felt	Building paper/felt, bitumen	Aluminium foil
Hard woodfibre board	Polythene sheeting (PE)	Building paper/felt, aluminium
Plasterboard	Plasticized polyester fabric	Polybutene
Plasterboard treated with silicone sealant	Polypropylene (PP)	PVC
Hardwood fibreboard, latex	Hard woodfibre board, bitumen	
Corrugated cardboard		

and a building. Floors, walls and ceilings in wetrooms can be coated with bitumen to prevent intrusion of moisture. Bitumen can also be used as an adhesive when installing foam glass sheets. Oxidized bitumen is made by blowing warm air (about 250°C) through distilled bitumen. This process results in a bitumen that is harder and more rubber-like than distilled asphalt. Oxidized bitumen is heated (to a maximum of 180°C) before being applied as a coating (warm asphalt). Bitumen products that contain solvents (white spirit, benzene, petrol and carbon disulphide) to improve ease of use (e.g. cold asphalt) are also available. Asphalt primer used as a primary coat when using mastic asphalt contains bitumen, some type of polymer, some type of amine, as well as organic solvents (which can account for up to 60–70 per cent). Asphalt primer should be avoided due to the amines and solvents.

Asphalt-coated paper, textile and felt can have a basic structure of ragboard, mineral fibre, polyester or polypropylene felt that is coated with bitumen and ground limestone or sand. Asphalt-coated paper is also used as a water drainage penetration protection material under roof tiles and must be protected from sunlight. It both absorbs and emits moisture and has a moisture-balancing effect. It can last about 50 years. There are several types of asphalt-coated paper, each with an identifying letter or letters, as shown in brackets below. To waterproof foundation walls, a bitumen (asphalt)-impregnated (A) or bitumen-coated (Y) backing of building paper, polyester (P) or mineral fibre (M) may be used. In some cases the product is coated with polypropylene film or slate granules (S). The product can also have a gap-building granule coating (Ko) on the underside. Such products may consist of a fibreglass textile coated with 3–4mm asphalt. The additives used should be checked. Bitumen barriers are used for moisture-proofing in wetrooms. Such barriers consist of 65 per cent bitumen and 24 per cent ataxic polypropylene (APP). A load-bearing material such as a fibreglass textile and an underside of polyester felt (nonwoven) hold the moisture barrier together. Wood fibreboard may also be impregnated with bitumen (asphalt-coated pasteboard) to increase its moisture resistance.

Rubber mats can be suitable for use in wetrooms, though joints must be sealed as they are not waterproof. As the material is waterproof, it is used in halls and entrances that receive a lot of pedestrian traffic. It has a high wear and slip resistance. Both mats and tiles are available. Rubber can also act as a barrier layer on level (or slightly sloping) roofs. Synthetic rubber (SBR and EPDM) is used to make sealing rings for

During restoration of the Katarina Church in Stockholm in the 1990s, birch bark was used wherever wood came into contact with masonry.

plumbing and weatherstripping for doors and windows.

Natural rubber floors, in addition to rubber, consist of 30 per cent by weight flowers of sulphur, pigment, and kaolin and chalk filler. They also contain vulcanizing agents, stabilizers, fire retardants (usually zinc oxide) and lubricants in the form of stearin (2.5 per cent by weight). Natural rubber flooring may also be mixed with 3–50 per cent by weight of synthetic rubber to improve its characteristics.

Synthetic rubber flooring consists most often of styrene-butadiene rubber (SBR). SBR has additives such as stabilizers, fire retardants, vulcanizing agents and softeners. It tends to shrink and during its production poisonous nitrosamines are released. Butadiene, released in the production of SBR, is a carcinogen and can cause damage in the gene pool.

Ethylene-propylene rubber (EPDM) is completely ozone-resistant, has good heat and chemical resistance and is thus used for outdoor rubber parts, e.g. weather stripping and glazing beads. It is also available as waterproof sheets for green roofs. Mats of EPDM-based rubber granules are made from recycled tyres. These are mostly used for impact sound reduction and weatherproofing.

Butyl rubber (IIR) is especially noted for its high diffusion resistance. It is used as a waterproof barrier in roofs and water reservoirs.

Birch bark is a traditional material in Norway and Sweden and has been used as weatherproofing for green roofs (3–20 layers), as moisture protection between ground plates and foundations, and where wood comes into contact with masonry. Birch bark is durable and rot-resistant. It is moisture balancing and is more suitable than asphalt roofing felt for protecting the ends of wooden beams set into masonry. Today, birch bark is mostly used as a protective layer in the restoration of old buildings. Birch bark was used wherever wood came into contact with masonry.

Plastic (see the introduction to the Synthetic Materials section as well). Much plastic weatherproofing is made of PVC, but there are more environmentally friendly alternatives such as polyethylene (PE), polypropylene (PP) and mixtures of these called polyolefins. Air-gap-building moisture barriers of PE and PP are available in the form of plastic sheeting. They cannot be melted or glued together, but must be installed overlapping each other. The sheets are used on sloping roofs (at least 15°), on the exterior of basement walls and sometimes as a moisture barrier in floors built on concrete foundations. The intention is to create both a waterproof layer and an air gap that can be ventilated and prevents capillary action. Such sheets consist of at least 0.35mm-thick high-density polyethylene (HDPE) which is dimpled in order to create an air gap. Polyolefin mats are made to measure,

Weatherproofing – Watertight: Environmental Evaluation

Recommended	Acceptable	To be avoided
Bentonite	Polyester	Bitumen coating
Birch bark	Polythene (PE)	Polyisobutene
	Polyolefin	Polymer-based primer
	Polypropylene (PP)	PVC
	Rubber*	Roofing felt**

Notes: * It is difficult to give general criteria for synthetic rubber and mixtures of natural and synthetic rubber. Rather, the ingredients in the rubber mixture need to be examined.
** Roofing felt can be better or worse depending on the ingredients used, but usually contains environmentally hazardous chemicals

with notches and flanges to fit round chimneys and pipes. Plastic film made of low-density polyethylene (LDPE) and PP can be used as a vapour barrier for wetroom floors and foundations. Sheets are available that are reinforced with fibreglass or polyamide (nylon). Plastic film is taped at the seams. Plastic barriers may contain UV stabilizers and other additives such as preservatives, antioxidants and stabilizers, as well as flame retardants.

Plastomer-asphalt board, textile and felt are materials made by mixing bitumen with certain synthetic materials, and thereby improving technical characteristics significantly. Such mixtures are called polymer-modified bitumen (plastomer asphalt). They can be divided into three categories: styrene-butadiene-styrene (SBS bitumen), APP (atactic polypropylene) and ethylene copolymerized bitumen (ECB). Plastomer asphalt should be protected from sunlight, heat and oxygen. The production process is environmentally destructive, but there are no health dangers from the finished products. SBS improves the characteristics of a barrier material, particularly at low temperatures. The mats may consist of 60–70 per cent SBS bitumen and 10 per cent styrene-butadiene rubber (SBR). They are reinforced with fibreglass, polyester or jute and also contain filler. Polyester reinforcement is most common. The mats may have an outer coating of sand or stone powder. According to the German Ökotest, the best barrier material for green roofs from an ecological perspective is plastomer asphalt (APP and ECB). The material is easy to work with, but is most suitable for small roofs. Mats made of APP bitumen are reinforced with polyester felt.

Polymer-based primer is used to prevent moisture from penetrating down through the joints between ceramic tiles or the underlying sheet material or concrete foundation. Polymer or rubber dispersion is brushed or rolled onto the wall. When it dries, a waterproof elastic film is created that also covers cracks. The primer often consists of polymer dispersion (e.g. acrylate dispersion). For wetrooms, liquid waterproof barrier material based on rubber solutions and containing up to 50 per cent mineral filler is also available. Common fillers are limestone and dolomite. Additives include preservatives and thickeners. Most of the common preservatives are allergenic. Sealants for tiles and expanded clay may contain chlorinated paraffins. The polymer primers considered to be the least environmentally harmful are those based on acrylate polymers. Their life is about 40 years. The raw material for polymer and synthetic rubber dispersions is petroleum.

Sheet Materials

Sheet materials are used both for interiors and exteriors, in dry areas and wetrooms, for external surfaces and interior constructions, for weatherproofing or as load-bearing structures.

Asphalt sheets are wood fibreboards impregnated with bitumen. Raw asphalt (bitumen) makes up about 12 per cent of the weight of a sheet. Bitumen-impregnated wood fibre sheet also contains paper, and alum and bentonite (clay) additives. The glue contains a certain amount of free formaldehyde (about 0.1 per cent), which is equivalent to about 7mg/kg sheet. In some cases, aluminium sulphate (which protects against mould) is added. There are hard, medium-hard and porous impregnated wood fibre sheets. Asphalt sheets are used as a water repellent and windproof layer in environments exposed to moisture. Asphalt (bitumen) is a product from oil refining and should not be used indoors due to the risk of polycyclic aromatic hydrocarbon (PAHs) emissions.

There are several types of **cement-based boards** available. A feature they all have in common is that they contain Portland cement as a binding agent. There is cement-based chipboard and cement-based fibreboard with cellulose, glass or plastic fibre reinforcement. They tolerate moisture and frost and are fire-resistant, but cement production is energy-intensive.

Cement-based chipboard is flat. Chipboard with magnesite as a binder is also available. Although more expensive, it is an environmentally friendly alternative to glue-bound chipboard. It is heavy and used as building sheets for walls and floors such as cement chipboard tongue-and-groove flooring.

Cement-based fibre sheeting may be corrugated or flat. Corrugated sheeting is used as roofing or cladding. Flat sheets are used in wetrooms, on façades and as wind protection. Cement-based fibre sheets consist of cement (65–80 per cent), filler (limestone or dolomite) and cellulose fibre (6 per cent by weight). To prevent the cellulose fibres from breaking down in the alkaline environment, they are impregnated with silicic acid (8 per cent by weight). Some sheets contain a mixture of about 2 per cent synthetic fibre (polyvinyl acetate, polypropylene or calcium silicate) and there are sheets that are reinforced with fibreglass. The sheets may have a surface coating of acrylic and they may also be coloured with pigment. Aluminium stearate is sometimes used as an additive.

Gypsum fibreboard consists of gypsum paste and cellulose fibre from recycled paper that has been sprinkled with water and compressed under high pressure. Gypsum fibreboard may contain potato starch binding agents and/or have a silicone sealant surface coating (0.3 per cent by weight). Gypsum fibreboard is stronger than gypsum board, is windproof, fire-resistant, heat-retentive, moisture-buffering and sound-absorbent. It can be used in damp areas without extra additives. Gypsum fibreboard is available as tongue-and-groove elements for subfloors.

Gypsum board consists of a core (about 95 per cent) of gypsum (calcium sulphate) with pasteboard glued to both sides to give the board adequate tensile and flexural strength. There are two forms of gypsum: natural gypsum and industrial gypsum. The additives used in production are mostly consumed in the process, i.e. tensides, dispersing agents, retarders and accelerators. Either starch or polyvinyl acetate (PVAc) glue (about 1 per cent by weight) is used between the gypsum and the pasteboard. Gypsum sheets are windproof, fire-resistant, heat-retentive, moisture-buffering and sound-absorbent. They are used in interior walls and ceilings, but also as exterior wind protection. Fire-resistant gypsum board contains kaolin (clay) and vermiculite, and may be reinforced with fibreglass. Water-repellent gypsum board has a wax emulsion coating consisting of paraffin wax or, alternatively, silicone sealant. The boards sometimes contain fungicides; they have low emission values.

Straw board can be made from wheat, barley or flax. The straw is glued together with its own lignin. When moist, fungal growth may occur.

Calcium silicate board consists of calcium silicate (produced from quicklime and highly quartzose sand or crushed sandstone), cement and filler (mica, perlite or vermiculite). The board is reinforced with a little cellulose fibre. They are autoclaved (hardened) at a high moisture level and high temperature to obtain good shape retention. The board is durable, moisture- and mould-proof, and fire-resistant. They are used both inside and outside, in wetrooms, in ventilation ducts and where a high standard of fire resistance is required. They are manufactured in England, Belgium and Scotland, among other places.

Plywood is a strong sheet material used both indoors and outdoors. It is used as wall panelling, for concrete shuttering, for carpentry, as well as for construction purposes. Plywood sheets consist primarily of veneer, thinly peeled sheets of wood from both softwood and hardwood that are cut to a suitable size, dried and glued together into sheets. Plywood is made of at least three layers of veneer glued together with the fibres in each layer lying at right angles to fibres in the next layer. The most commonly used adhesive is phenolic resin glue (PF), which makes up 5–10 per cent by weight. A PF-glued plywood sheet releases <0.01mg/m^3 formaldehyde.

Laminated core board consists of a wood strip core with an outer layer of veneer. The core is made is such a way that moisture movement is reduced to a minimum. The veneer is thin peeled sheets of wood glued together. Blockboard consists of wood strips and glue. A distinguishing characteristic of the sheets is that they are highly durable and flexible. Laminated core board is used for carpentry and interior fittings. The main environmental problem with laminated core board is the formaldehyde in the phenolic resin glue. However, as these sheets only contain glue in thin joints, the gas release is much lower than from chipboard.

Laminated sheet (plastic laminated chipboard) is used for floors, furniture, ceilings and worktops where a high standard of hygiene is required. The chipboard is made of sawdust (about 87 per cent) and glue (UF resin). UF resin may contain small amounts of free formaldehyde. The outer surface consists of pressed paper with a surface coating of melamine plastic. Other additives (<1 per cent by weight) include wax, urea, ammonia and ammonium sulphate. The measured level of formaldehyde from laminated chipboard is about 0.07mg/m^3. Formaldehyde may also be present in the melamine coating, but in a hardened state the risk of emissions is small.

Clay sheets, reinforced with reeds, are made primarily of aluminium silicates. Clay mixed with fine quartz sand and small mica flakes makes a good construction material with a good moisture-buffering capacity. In Germany, construction sheets of strong reed mats plastered with clay are available. They are used primarily indoors to increase the moisture-buffering property of inner walls.

MDF, HDF, LDF, etc., fibreboard (dry process wood fibreboard) is closely comparable to chipboard with regard to chemical composition and production method. The glue content in dry process fibreboard is 5–10 per cent by weight. Dry process wood fibreboard consists of wood and urea-formaldehyde glue. It may also contain small amounts of wax, urea, ammonium sulphate and iron sulphate additives. Phenolic resin may contain small amounts of free formaldehyde. Wood fibreboard is used for wallboard, subflooring, roof underlay, ceilings, laminated flooring and in the furniture industry. Oriented strand board (OSB) is made from long, thin wood strands (0.6–65mm) that are dried and aligned parallel to one another and then coated with glue. In every panel, three such wood strand layers are glued turned through 90° to each other. Because of the orientation of the long wood strands, OSB board is considerably stronger than normal chipboard. It also swells and shrinks little and holds its shape. Because of its strength, it is used in load-bearing constructions such as beams, but also for interior partitions. Formaldehyde in the phenolic resin glue is the main environmental problem with OSB board. The proportion of glue is, however, only 2–3 per cent by weight (normal chipboard has about 7 per cent by weight) and it therefore emits less formaldehyde and is more suitable than regular chipboard for interior use.

Hemp fibreboards are made from hurd, a waste product from hemp textile manufacturing. This medium-density fibreboard is free from formaldehyde and has zero volatile organic compound (VOC) emissions but does contain a binding agent. It looks like OSB board and is marketed in Germany.

The interior walls in the Toresdotters' home in Skärgårdsstad, Sweden, were built using hard fibreboard on a layer of gypsum board. The walls are covered with fabric. All the walls can thus be used as bulletin boards.

Critical toxicologists believe that this level is too high. If a small room is filled with furniture made of chipboard and the chipboard is unpainted or there is air contact via drill holes and sawn edges, the formaldehyde level in the room can exceed the recommended value. Additional additives (<1 per cent by weight) are wax, urea, ammonia and ammonium sulphate. Chipboard is used for flooring, walls, ceilings and furniture.

Chipboard with isocyanates, such as MDI board, contains diphenyl-methane-4,4-diisocyanate. Isocyanates are made using a dangerous and energy-intensive chlorine production process in which poisonous phosgene is made to react with carcinogenic aromatic amines.

Wood fibreboard (wet process) is produced by chipping and grinding (defibring) wood, diluting it with water, and compressing it using heat and pressure. The wood lignin acts as a binding agent, which means that very little (<2 per cent by weight) or no glue is needed. The glue used is phenolic resin glue or phenolic-formaldehyde glue. Wood fibreboard produced using the wet process releases very low levels of formaldehyde. Hard, medium-density and porous boards are available. Additives used (about 1–3 per cent) include aluminium sulphate (for mould protection), ammonium

Chipboard with formaldehyde consists of wood chips and glue (about 7–10 per cent by weight). The binding agent used is primarily formaldehyde resins or UF glue (urea-formaldehyde). The glue contains low levels of formaldehyde and phenol. The formaldehyde does not harden completely and the excess releases gas over a long period of time. If the chipboard is exposed to moisture, some formaldehyde may also be released. E1-labelled chipboard releases up to 0.1ppm formaldehyde, which is equivalent to about $0.015 mg/m^3$ interior air.

Sheet Materials: Environmental Evaluation

Recommended	Acceptable	To be avoided
Gypsum plasterboard	Asphalt board	MDF board*
Gypsum fibreboard	Cement-based boards	Chipboard, formaldehyde
Straw board	Laminated wood board*	Chipboard, isocyanates
Calcium silicate board	Laminated sheet	
Clay board, reed reinforced	Plywood	
Chipboard, magnesite	OSB board*	
Wood fibreboard		
Wood-wool cement board		
Reed mats		
Corrugated fibreboard		
Hemp fibreboard		

Note: * In Switzerland there are levels for formaldehyde glue content. Sheets with less than 3 per cent glue are recommended (laminating sheet), 3–5 per cent glue is acceptable (OSB board), and greater amounts of glue content should be avoided (chipboard and MDF board).

sulphate (to increase fire resistance) and iron sulphate. These salts do not pose a health danger. In environments exposed to moisture, wood fibreboard oil tempered with tall oil or impregnated with asphalt is used.

Wood-wool cement board (see Thermal Insulation)

Reed mats are a traditional construction material. They have been used for a long time as a plaster base on wood walls to hold plaster in place. Reeds have relatively good thermal insulating properties. They are moisture repellent and due to their high content of silicic acid are sufficiently fire-resistant so that impregnation with fire retardant is not necessary. The reed stalks are laid tightly and parallel to each other and held together with galvanized iron wire or hemp string to create flexible mats with a thickness of 2–10cm. They are easy to cut and work with.

Pasteboard is used along the base of a roof to keep insulation in place so that the air gap isn't obstructed. Pasteboard can also be used on the inner side of roofs when insulating attics and lofts.

Plaster and Mortar

Plaster and mortar consist of a mixture of binding agents, aggregate, water and additives. Plaster is used to give a wall an aesthetically pleasing weather protection. Plaster usually consists of three layers: a thin scratch coat, a roughcast plastering layer and finally the finishing coat. Washing with a thin layer of plaster is a method that brings out the texture of the substrate. A plaster should never be stronger than the substrate, otherwise it can easily fall apart. Plastering is labour-intensive, but a good lime plaster application has a life of 40–60 years.

Cement plaster is mostly used on exteriors. It can be used on concrete walls, concrete blocks or expanded clay and lightweight concrete blocks. First, all cracks and unevenness must be repaired or smoothed with cement mortar (with a ratio of 1 part cement to 3 parts sand by volume). The surface is brushed with a cement grout made in the same ratio. Finally, the surface is plastered with cement plaster with a ratio of 1:1 cement/sand for solid concrete and 1:3 cement/sand for blocks, lightweight aggregate (LWA) or lightweight concrete. With a few coats of the top layer of plaster applied, the surface becomes almost waterproof. Production of cement is energy-intensive.

Cement mortar has a ratio of 1 part cement to 3/4 parts sand by volume, plus water. The mortar is strong but not elastic. It has a minimal moisture-absorbing capacity, is frost-resistant and goes off slowly. Cement mortar is used primarily in the laying of ceramic tiles. Since cement mortar is so strong, ceramic tiles or bricks laid using cement mortar cannot be reused.

Lime cement plaster is used a lot outdoors. It is somewhat stronger than lime plaster and more elastic than pure cement plaster. Exterior wall plaster consists mostly of a cement or lime mortar containing sand and various additives. Usually, the binding agent in cement constitutes 30–50 per cent by weight. If much cement is used in exterior wall plaster, the wall is denser but also acts as a vapour barrier. Too much cement in the plaster makes it impossible for moisture in the wall to escape, so that parts of the mortar may loosen. Lime cement plaster has a lower moisture-buffering capacity than lime, gypsum and clay plaster. Adding 25 per cent perlite results in an insulating plaster.

Lime cement mortar consists of lime, cement and sand in a volume ratio of 1:2:10, 1:1:7 or 2:1:11. Lime cement mortar must contain at least 35 per cent by volume cement. Lime cement mortar is strong, elastic and frost-resistant. It has a relatively good moisture absorption properties and hardens relatively slowly. It can be used for all types of interior and exterior masonry. Since lime cement mortar is weaker than brick, it is possible to reuse brick laid using it.

Lime gypsum plaster and lime gypsum mortar consist of 10 parts by volume of lime, 1–5 parts gypsum and 30–40 parts sand. Less sand is used when mixing is done by hand and more sand is used when machine mixing. For better viscosity, starch may be added, which is especially important for machine-mixed plaster. Lime gypsum plaster is not waterproof, but can be used under tiles in wetrooms. The plaster can be damaged if water is sprayed directly on it. The plaster regulates air moisture well. This takes place via mineralization, whereby lime is reformed from calcium hydroxide through the absorption of CO_2 from the air. When painting the plaster with lime paint, a pretreatment is necessary.

Lime plaster contains slaked lime, sand (in the ratio 1:3 by volume) and water. This plaster both absorbs and releases moisture. The lime is fired at 1000°C to make quicklime, and is then slaked with water to make slaked lime. Less energy is required to make lime than to make cement. Lime has a disinfectant effect and is therefore especially good as a binding agent in joints and mortar. Lime plaster is applied in several layers until it is about 1.5cm thick. Pigment additives must be alkali-resistant and must not exceed 10 per cent by weight.

Lime mortar is made up of 1 part lime to 2/3 parts sand, plus water. The mortar is elastic and relatively hard, but it isn't very resistant to water or frost. The mortar hardens quite quickly, can absorb moisture and contribute to balancing humidity. Since lime mortar is weaker than brick, it is possible to reuse bricks laid with lime mortar or lime cement mortar.

Hydraulic lime mortar consists of hydraulic lime and sand mixed in a ratio 1:2–4, plus water. A distinguishing characteristic of hydraulic lime is that it hardens through a reaction with water into a product that is water repellent. The mortar is elastic, is relatively strong, balances humidity and resists frost, as well as hardening quickly. It can be used for all types of interior and exterior masonry.

Hydraulic lime plaster is made from a particular type of limestone that, besides calcium carbonate, contains silicic acid, aluminium oxide and iron oxide. Hydraulic lime plaster (or lime-pozzolana cement mortar) is stronger than ordinary lime plaster.

Clay plaster consists of 75 per cent sand and 20 per cent clay by weight. Straw, flax, hemp and animal hair may be used as reinforcement. It has a significant moisture-buffering capacity. Clay plaster can be dyed with earth pigment or painted with emulsion paint or lime paint. Cracking can occur if drying takes place too quickly. For clay walls, it is important that the moisture content of the base layer is lower than 5 per cent.

Clay plaster in a variety of natural colours.

For outdoor plastering, the result is greatly influenced by the weather. The layer should be 2–3cm thick. Many old recipes for clay plaster and wattle-and-daub walls include ant hill, moss and horse or cow manure. Animal glue is also sometimes used in clay plaster. In Germany, ready-to-use clay plaster is sold in powder form to be mixed with water on site. Clay plaster requires an uneven substrate in order to bond to a wall. Wood walls can either be roughened or have reed mats fastened to them for the clay to bond to. Clay plaster and mortar can be reused by dissolving them in water.

Clay mortar consists of 5 parts clay to 1 part sand, plus water. Clay mortar is elastic but weak and has poor water- and frost-resistance. Clay mortar is used to lay blocks of clay or clay straw and can be used to lay lightly fired brick. Clay mortar is also used to make tiled stoves, which makes it easy to dismantle them.

Silica plaster is a 1–3mm thick plaster for outdoor use on mineral substrates. The binding agent in the plaster is a potassium silicate solution. It is strongly alkaline and therefore requires no preservatives. Sodium silicate reacts with the mineral substrate to create a water-repellent but vapour-permeable exterior surface. It has good durability. The plaster can only be dyed with alkali resistant pigment, e.g. iron oxide and titanium oxide. There are also silica plasters on the market that contain synthetic resin dispersions.

Woven Wallcoverings and Wallpaper

Most cellulose-based wallpaper has a plastic coating so that it can be wiped clean. If a paintable wall lining with a surface structure is required, fibreglass-free cellulose fabric can be used. Jute fibre wallpaper has a coarser and more uneven structure. Woven wall coverings can also be made from other natural fibres such as cotton, flax, hemp and wool, but these are less common. They are sometimes reinforced with fibreglass.

Cellulose fabric is used as a paintable wall lining and contains cellulose, binding agents and in some cases polyester. Information on binding agents is inadequate. Cellulose fabric may be treated with flame retardants. The material has a felt structure and is relatively soft and easy to work with.

Glass textile wallcovering is made from fibreglass impregnated with polyvinyl acetate (PVAc) and starch. During construction, fibreglass may cause discomfort from itching and skin irritation.

Jute fibre wallcovering usually has a relatively coarse structure. Cultivation, harvesting and care of the plant is labour-intensive, but jute harvests are twice as big as hemp and five times as big as flax harvests. About 96 per cent of the world's production is grown in Asia. Jute wallcovering is extremely absorbent and can therefore be treated with an oil-based paint as a primer and then painted with another type of paint.

Plaster and Mortar: Environmental Evaluation

Recommended	Acceptable	To be avoided
Gypsum plaster	Cement plaster	Synthetic resin plaster*
Gypsum lime plaster	Lime cement plaster	
Lime plaster		
Clay plaster		
Silica plaster		

Note: * e.g. styrene acrylic

Woven Wallcovering and Wallpaper: Environmental Evaluation

Recommended	Acceptable	To be avoided
Cellulose fabric	Synthetic fibre wallcovering*	Glass textile wallcovering
Natural fibre wallcovering		Plastic wallpaper
Paper wallpaper		

Note: * Depending on the kind of fibres used.

Wallpaper may be made from cellulose, mineral fibre or plastic with an application of printing ink. Most wallpaper, including cellulose wallpaper, is coated with plastic (acrylate/vinyl acetate) for protection and so it can be cleaned.

Flooring

The floor is the part of a building most exposed to wear and dirt. When choosing flooring, it is important to assess ease of cleaning, maintenance and life. A distinction is made between flooring for dry and wet areas, hard and soft floors, floors for public areas and for homes. Floors often consist of three layers: the structural system, the screed and the surface material. Screeding can be accomplished using levelling compounds, mats or sheets. A floor can be floating (on sand), laid on a subfloor or on joists. How a floor is fixed affects its environmental impact. Glues are often problematic from an environmental point of view. Therefore, flooring and floor constructions without glue or with small amounts of glue should be chosen. Carpets in small rooms can lie loose, kept in place with skirting, but are usually glued to the subfloor. It is also possible to fasten them using staples hidden behind skirting.

Bamboo is actually a grass that, among other things, is cultivated in China in fields until reaching its full height of 35 metres. Bamboo is extremely quick growing and grows at least seven metres per year. It is often grown without chemical means of control. It is harder than oak and tolerates light well. Bamboo floors can be sanded. Colour can vary from quite light to dark caramel tones depending on how much it is heated in an oven during the manufacturing process.

Concrete is used as an outer surface for industrial floors or in secondary areas such as garages. Normally, hardened cement and concrete are durable with regards to stationary water, alkaline substances, oil and weak acids. Concrete tiles are produced to look like imitation stone and coloured with iron oxide. To avoid problems with dust, concrete floors are treated with paint or dust binding agents such as silicate or concrete oil.

Cement mosaic (terrazzo) flooring is made of cement paste, crushed rock and pigment. Marble or limestone are usually used, but granite, feldspar or quartz can be used to make a harder floor. Cement mosaic is durable and skid-resistant but doesn't tolerate heavy loads. The surface is machine-polished, emits less dust than a concrete floor and is easy to clean. Cement mosaic can be poured in place

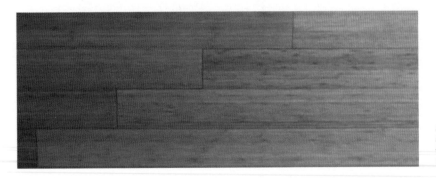

An example of an oiled bamboo floor, sold by Nordic Bamboo.

Three types of flooring: linoleum, wood and ceramic tiles.

(15–20mm) or obtained as tiles (40–60mm). It is used in public places such as shops, stations and schools, as well as in wetrooms. As a surface treatment, silicate or soap is suitable.

Rubber (see 1.1.2 Knowledge of Materials and Weatherproofing)

Clinker slabs are hardwearing and tolerates liquids, are in principle maintenance-free and easy to clean, but their production is energy-intensive. Clinker slabs are made from special clays based on aluminium silicate, quartz and chalk. The tiles are fired at 1200–1300°C. They are available glazed or unglazed. Glazing is carried out by firing ground mineral pigment onto the surface. Unglazed tiles can be press-moulded using high pressure prior to firing. This method provides an unglazed surface with the same density as a glazed tile, and with ten times better scratch resistance. Clinker slabs can be set in cement mortar or glued with cement-based glues and joined using mortar. A clinker slab surface is not totally watertight because of the joins, therefore the substrate must be either mineral and tolerate moisture, or be equipped with a moisture-proof layer. For floors that require good slip resistance, special tiles with hard mineral particles on the surface can be used. Tiles with a lead glaze should be avoided. Domestic tiles should be

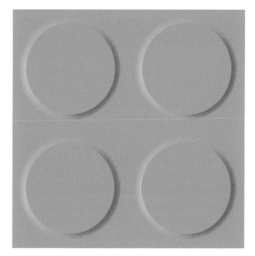

Nora® flooring material is made up of rubber and does not contain either phthalates or halogens (e.g. chlorine). A step in the right direction.

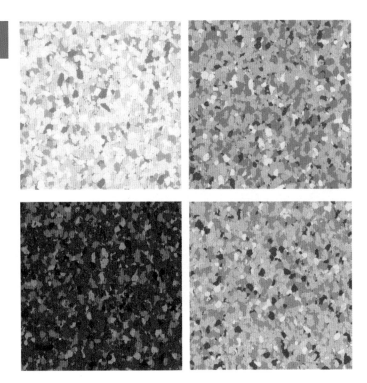

Calcium carbonate flooring LifeLine™ from Upofloor Oy.

A cork floor that has been treated with oil and wax.

prioritized, as ceramic is heavy and transportation is very energy-intensive.

Cork (see Thermal Insulation) provides a hard-wearing, soft and easily cleaned floor. Most cork tiles have a surface coating of plastic resin, such as PVC plastic. They may even have a PVC coating on the undersurface. Cork tiles with PVC should be avoided. Cork tiles without a surface coating are treated with oil and wax. The tiles are glued to the subfloor.

Calcium carbonate flooring contains 50–80 per cent by weight calcium carbonate (depending on the type), which is a natural mineral, and 20–45 per cent by weight thermoplastic polymers, as well as pigment and acrylic polymers for surface reinforcement. A calcium carbonate floor for public areas, called LifeLine™, was developed by and is produced by Upofoor Oy, the leading floor manufacturer in Finland. The flooring does not contain any hazardous substances and is installed with environmentally friendly glue. The very low level of total volatile organic compounds (TVOCs) of <5µg/m^2 are present after four weeks. LifeLine™ is intended for use in all types of public environments. The product is recommended in the *Sunda Hus* (Healthy Buildings) database.

Laminate (see Sheet Materials)

Linoleum consists of a wear surface with a woven jute backing. The wear surface is made of binding agents (oxidized linseed oil and natural resin), filler (wood flour, cork flour, ground limestone) and pigment. The linoleum mixture must be thoroughly hardened to avoid too many emissions. Linoleum is hard-wearing and muffles impact noise but is sensitive to moisture. Zinc carbonate is sometimes used as an additive. The floors usually have an exterior surface of polyurethane varnish, polyacrylate, or PVC plastic and are glued to the subfloor with EVA adhesive (ethylene vinyl acetate). Linoleum floors should be regularly waxed

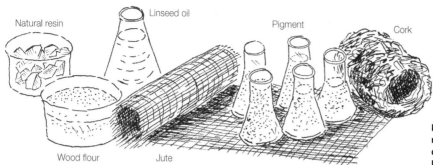

Linoleum is made from natural materials, but it is often coated with a thin plastic layer to protect it.

and cleaned using dry-cleaning systems. Avoid linoleum mats with foam backing. Colophony, which is an allergen, is a natural resin that is often present in small amounts in linoleum.

Plastic floors are made from PVC, polyurethane or polyolefins. PVC and polyurethane should be avoided for environmental reasons. Polyolefins (polyethene, polypropylene and mixtures of these) are relatively eco-friendly plastics, depending on the additives used. Polyolefin floors are more difficult to maintain and not as hard-wearing as other plastic floors; they also require greater preparation of the subfloor. Better plastic floors are being researched. The polyolefin floors on the market may contain sand, limestone and aluminium compounds, flame retardants, an anti-static finish, oxidation protection and pigment. Both mats and tiles of polyolefin are available, often with a wear surface of polyurethane (PUR) or acrylate. The floors are glued and the seams sealed with polyolefin weld rod. They do not tolerate heavy, wet traffic and aren't used in wetrooms as they are difficult to seal tightly. Plastic floors often generate static electricity, but can be treated with an anti-static finish.

Stone floors are hard-wearing. Slate, limestone, marble, soapstone and granite are suitable stone materials for flooring. Marble is naturally antiseptic, which makes it appropriate for use in dining rooms, kitchens and bathrooms. Stone floors can be ground and polished, which facilitates cleaning but may result in a slippery floor. Natural stone can be laid in mortar on concrete. In outbuildings, winter gardens, etc., stone can be laid in sand or earth without caulking. Use of domestic stone minimizes energy-intensive transport. A few stone materials emit radon. Stone floors can be oiled, soaped or treated with wax to avoid grease spots. Stone can be damaged by strong alkaline or acidic substances.

Brick (see Construction Materials table) as flooring is laid in mortar or sand. Bricks can be reused if they are laid in sand or laid using lime mortar. The brick surface can be treated with linseed oil.

Terracotta tiles are made from alumina (aluminium oxide) which is fired at the relatively low temperature of 900°C. They may be surface-treated with wax and/or oils. They are used for home flooring, but are not watertight. The material is porous and has a humidity-balancing effect on the indoor atmosphere.

Textiles used for most wall-to-wall carpeting are made of synthetic materials such as polyamides, polyester or polyacrylonitrile. The carpeting has a backing of woven polypropylene or woven polyester as well as a foam plastic underlay of synthetic rubber (SBR) or polyurethane (PUR). Synthetic carpet contains additives such as anti-static agents, anti-dirt agents, disinfectants and flame retardants. Natural materials used to make carpets include sheep's wool, goat hair, coconut fibre, sisal and hemp. Woven jute is used as a backing and textile, latex or polyurethane (PUR) as underlay. Avoid natural materials treated with flame retardants or mothproofing, and carpeting with a polyurethane (PUR) underlay. There are also carpets made from mixtures of synthetic and natural fibres. Sometimes textile carpets are glued to the floor; to avoid glue you can fix the carpet with fabric hook-and-loop fasteners (Velcro) instead. Wall-to-wall carpets absorb dust, are difficult to clean and have a relatively short life.

Solid wood floors are hard-wearing, can be renovated and may consist of boards, parquet segments or end-grain wood blocks. Wood flooring can be nailed, screwed, snapped together, glued or fastened down with screwed baseboards or metal strips. Glued or nailed flooring is difficult to reuse. Suitable timbers for flooring are pine, spruce and aspen (soft); birch, beech, oak and ash (hard-wearing), as well as elm and maple. Aspen does not splinter.

Laminated wood flooring consists of several layers of wood material glued together. The core material may be solid wood, chipboard or wood fibreboard. To be called laminated parquet, the outer layer must be at least 2.5mm thick. Glue is the main environmental problem. Formaldehyde resin water-based glue (urea-formaldehyde resin glue) or polyvinyl acetate (PVAc) are commonly used. UV acrylic lacquer is usually used for factory lacquered flooring. An environmental evaluation of laminated parquet has to consider the core material, the amount of glue and the surface treatment.

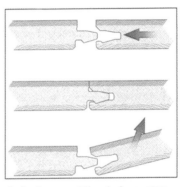

A simple parquet floor laying system.
Source: Tarkett

Oiled wood block floor.

Flooring: Environmental Evaluation

Recommended	Acceptable	To be avoided
Cork without PVC or PUR	Cement, dust-binding agents	Plastic mat, pvc
Coconut fibre, sisal, jute carpets	Cement mosaic	Synthetic rubber
Linoleum	Ceramic tiles	Textile, synthetic
Stone, local	Natural rubber	Textile, foam-backed
Textile, wool*	Plastic mat, polyolefin	Wood, laminated, chipboard
Solid wood	Wood, laminated, not chipboard	
	Bamboo	
	Calcium carbonate	
	Brick	

Note: * Watch out for fire-retardant and insect-proofing additives.

Screed Material

There is a difference between screed that has to be spread and liquid screed (self-levelling). Concrete floors often need a surface screed before laying the floor. Liquid screed based on either Portland cement or aluminous cement with some or no Portland cement added is available. There is also liquid screed made from a mixture of aluminous cement, Portland cement and granulated blast-furnace slag. Production of cement is energy-intensive. Products used on an inorganic underlay, e.g. concrete or stone, should be based on inorganic substances.

Portland cement-based liquid screed consists of Portland cement (30–40 per cent), sand (45–50 per cent) and filler, i.e. fine mineral material (5–15 per cent). In the past, the use of liquid putty with casein additive in combination with Portland cement could cause odour problems. The high pH of Portland cement in combination with a high level of moisture in concrete may cause emissions from glue and carpets.

Aluminous cement-based liquid screed consists of aluminous cement (30–40 per cent), sand (45–50 per cent), and filler, i.e. fine mineral material (5–15 per cent). There is low-alkaline aluminous cement-based liquid screed that contains casein. It poses considerably less risk of odour problems compared to earlier Portland cement-based liquid screed with casein additive.

Aluminous cement contains lime, marl and sandstone, as well as gypsum. Ground limestone, dolomite, fly ash and slag are used as fillers. Polymers used in screed material are mostly vinyl acetate based. Additives include anti-foaming agents (<1 per cent) in the form of mineral or silicone oil, inorganic salts, cellulose and melamine-based liquefiers (<1 per cent). Liquid screed both with and without liquefiers is available. At present, primarily liquid screed with a melamine resin-based liquefier is used, which is stable in an alkaline environment. Melamine-based liquefiers may release very low levels of formaldehyde.

Screed Material: Environmental Evaluation

Recommended	Acceptable	To be avoided
Aluminous cement*		Cement mixtures
		Portland cement

Note: * Avoid screed material with superplasticizer.

Filler (Wall, Roof and Wood Filler)

Filler usually comes in the form of a paste, but is also available as a powder to be mixed with water. It can be applied manually or sprayed. There is two-part filler and one-part filler. The health risks of working with one-part filler have to do mostly with allergenic substances in the preservatives and binding agents. Oil filler should be used under oil paint. Its use could be significantly reduced by concentrating on careful groundwork instead. Water-based filler often contains biocides. If filler contains lime or cement, the use of biocides can be avoided.

Gypsum filler consists of slaked lime, gypsum and sand.

Light filler and sand filler consist mainly of dolomite. Other raw materials that may be included are cellulose, gypsum, aluminium compounds and plastic-based additives. There are often

Filler: Environmental Evaluation

Recommended	Acceptable	To be avoided
Gypsum filler		Two-component filler
Light filler*		
Sand filler*		

Note: * If possible, avoid fillers with biocides.

biocides in this type of filler. Other admixtures that may be present include anti-foaming agents, pigment, additives, etc.

Two-part filler consists of a reactive substance (e.g. epoxy), solvents, filler and hardeners. Two-part filler is used for filling wood, concrete, plastic and metal. The health risks of working with two-part filler have to do primarily with allergenic substances in the filler, e.g. epoxy hardeners and benzoyl peroxide. Epoxy hardeners are corrosive. Other problematic substances possibly present include isocyanates (allergens), of which toluene isocyanate (TDI) is also carcinogenic. Solvents can affect the nervous system. Styrene is, in the EU, questionable from a health point of view (hormone disturbing). Preservatives are often poisonous to aquatic organisms. Of the softening agents, dibutylphthalate is classed as environmentally hazardous.

Adhesive Compounds

Adhesive compounds used to fasten tiles are available for both outdoor and indoor use. Water-based adhesives usually contain biocides. However, if the adhesive contains lime or cement, biocides are not necessary.

Cement-based adhesive compounds contain Portland cement, sand, limestone flour, small amounts of cellulose as well as dry dispersion. The products may also contain a plastic additive of polyvinyl acetate, acrylates and calcium formate. Adhesives with plastic additives contain biocides. Avoid cement-based adhesive compounds with plastic additives.

Plastic-based adhesive compounds consist of two-component epoxy. The base is primarily composed of resin. The hardener may contain substances from groups such as amines and phenols, and the filler can be made of glass particles or wood flour. Plastic-based adhesives contain biocides. Avoid these adhesive compounds with a plastic base.

Paint, Surface Finishes

Paint consists of binding agents, pigment, solvents and additives. Since each paint has its own particular recipe, it is not possible to say that a certain paint is good or bad for the environment. In order to find out, it is necessary to know the paint's ingredients. The EU flower is an environmental label for paint and surface treatments. EU flower criteria for paint with regard to health and the environment are, among other things:

1. There are special restrictions on the use of white pigment.

2. The level of volatile organic compounds must not exceed certain levels, e.g. 15g/L (including water) in matt wall- and ceiling paints, and 60g/L (including water) in glossy paints for walls and ceilings.

3. The level of volatile aromatic hydrocarbons in wall paint must not exceed 0.1 per cent by weight in the end product.

4. Ingredients or preparations must not contain cadmium, lead, chromium VI, mercury, arsenic, barium (except barium sulfphate), selenium, antinomy or compounds containing them. Cobolt, to a certain amount, is accepted in the siccative.

Adhesive Compounds: Environmental Evaluation

Recommended	Acceptable	To be avoided
Adhesive compound,* Portland cement		Adhesive compound, plastic-based

Note: * If possible avoid adhesive compounds with biocides.

5 The product must not be classified as very toxic, toxic, hazardous to the environment, carcinogenic, reprotoxic, mutagenic or irritating (exception risk phrase R43) in accordance with Directive 1999/45/EG. There are specific requirements for some ingredients, such as formaldehyde, isothiazolinone compounds, glycol ethers, etc.

Alkyd oil paint contains alkyd oil, organic solvents, pigment and additives. Alkyd oil is made from boiling together an oil, an acid and an alcohol. The oil can be a linseed oil, soya oil, tall oil, Chinese wood oil or castor oil. The oil molecules are enlarged in the process, thereby shortening drying time and making the binding agent more resistant to chemicals and weathering. Alkyd paint has a function similar to linseed-oil paint. The white pigment, however, consists of titanium dioxide instead of zinc oxide. Inert carbon compounds are used as filler. The same type of siccative is used as in linseed-oil paint and fungicides are almost always included. Also included are anti-skinning agents (about 0.3 per cent), and usually methyl ethyl ketoxime (which is an allergen). A final coat of alkyd oil paint usually requires about 35 per cent organic solvents. Today's alkyd oil paints are not a good environmental choice due to their high levels of organic solvents and additives.

Fireproof paint has in some cases become more eco-friendly than earlier. There are a lot of different contents, and watch out for additives. Methods of construction using fireproof materials, for example using large dimensions or a gypsum covering to provide protection for wood structures, afford fire protection with not too much environmental impact. Steel structures can also be covered with gypsum. The use of sprinklers is a modern method to protect, for example, multistorey buildings of wooden construction, without the use of fireproof paint.

Emulsion paint (tempera paint) is usually made from oil emulsified in water with pigment/filler additives (10–50 per cent). Examples of emulsifiers used are egg (egg oil tempera) or casein (casein tempera), which also act as binding agents (10–50 per cent). Prepared paint may contain small amounts of preservative, small amounts of a drying agent (0.1 per cent) and a thickening agent (cellulose-derived, a few per cent). Linseed oil, tall oil or alkyl oil are the commonly used oils. Castor oil is sometimes added to prevent yellowing. Emulsion paint is considered eco-friendly but may contain environmentally hazardous additives. Paint containing alkyl phenol etoxylates, the drying agents cobalt/zirconium salts (which are allergens and poisonous to aquatic organisms), and isothiazolone preservatives such as Kathon, Bit or Bronopol should be avoided. Without preservatives, the shelf-life of paint in a can is limited, but paints without preservatives are available. Drying requires bright light and good air circulation, since several substances are released during the drying process. Painters sometimes complain about headaches and eye and bronchial irritation while working with egg oil tempera, but the substances released do not pose a long-term health or environmental hazard. The paint may be diluted with water and may be used on most surfaces. It can be applied as a cover or a glaze. The finish varies from flat to high gloss. Egg oil tempera takes time to dry, about 24 hours to surface-dry and three weeks to dry completely. The paint is difficult to touch up without it showing. The most eco-friendly alternative is to mix egg oil tempera (egg or egg powder, oil, water and pigment) on site.

Fibre paint is paint that is reinforced with fibre resulting in an appearance that resembles a plaster surface. Fibre paint may consist of fibre, water, binding agents, filler (such as calcium carbonate and talc) and preservatives. An example of a fibre paint is paint with cellulose as a binding agent which is strengthened with cellulose fibre.

Priming oil and primer may contain high levels of organic solvents (usually white spirit) but there are water-based alternatives. When painting exterior wood, the first step is to impregnate end-grain surfaces, joints and nail holes with a penetrating priming oil. After that, the wood is primed using primer, and then one or two top coats of paint are applied. Water-based oil mixtures or oil mixtures that are almost exclusively linseed oil of varying viscosity can be used. The primer should, besides penetrating and being water repellent, provide a good surface for the top coat to adhere to. Primer consists of penetrating oils, solvents and pigment or filler. Common primer usually contains fungicides. Primer without organic solvents is preferable.

Impregnating wood in order to protect it from fungi and infestation by insects can in principle be done in two ways, as a preventive measure or an active treatment.

Glaze paint in the church at Saltå mill, just outside Järna, Sweden.

- Use high-quality wood in vulnerable locations, e.g. use pine or larch with a high proportion of heartwood for cladding, and oak heartwood underground.

- Use wood in structures where the wood is kept as dry and well-ventilated as possible. Fungi only grow on moist wood. For example avoid moisture absorption at the ends of boards by leaving a significant margin of at least 30cm between vertical wood panelling and the ground. Combining building elements made of different materials can cause problems where the materials meet, e.g. wood and plaster, or wood and metal.

- Work with passive impregnation, e.g. use preventative methods before a tree is felled and in the timber yard. Passive impregnation is also possible by using protective surface treatments (e.g. linseed oil), or by burning the wood's surface as a charred surface layer protects the rest of the wood. An alternative to using wood preservatives is heat treatment (Heatwood, LunaWood). These alternatives make the lumber resistant enough to be used above ground. Heat treatment, however, reduces the strength of the wood, which means that the lumber cannot be used for constructions exposed to large weights.

- Treatment using active impregnation, i.e. some form of poison, should be used as little as possible. According to architect Björn Berge, the least poisonous but also the least effective active impregnation substance (where repeated treatment is required) is tar, soda+potash and green vitriol.

Conventional preservatives are made up of copper compounds, often in combination with amine and organic fungicide, or chromium and possibly arsenic. These water-soluble preparations, which are used for most impregnated wood, bind to the wood (most often pine) and result in the characteristic green colour. Inspected

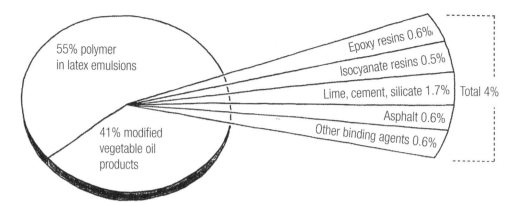

Ingredients of a modern paint. The pie chart shows that the binding agents (4 per cent) may pose environmental or health risks. Additives may also constitute environmental and health risks, e.g. the biocides. Some solvents and pigments are also problematic.

Source: Alcro-Beckers brochure, Målarfärg och miljö

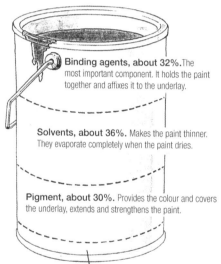

Paint consists of:

Binding agents, about 32%. The most important component. It holds the paint together and affixes it to the underlay.

Solvents, about 36%. Makes the paint thinner. They evaporate completely when the paint dries.

Pigment, about 30%. Provides the colour and covers the underlay, extends and strengthens the paint.

Additives, about 2%. Helper substances that in small amounts affect the paint's qualities, e.g., faster drying time, prevention of skin formation or keeping the surface mould free.

pressure impregnated lumber is labelled by the Nordic Wood Preservation Council (NTR). Waste must be managed according to municipal regulations.

Lumber is now also available that is pressure impregnated with linseed oil (Linotech). There is also non-toxic acetylated wood (Titan Wood), and furfurylated wood that is treated with a furfuryl alcohol solution, which is a renewable raw material (Kebony). These alternatives are currently more expensive than conventionally impregnated lumber.

Green vitriol is a hydrate of iron sulphate ($FeSO_4 \cdot 7H_2O$) that has anti-fungal and mould properties. It is a completely colourless liquid when applied (if a small amount of pigment is added, it is possible to see the area already covered). After application, the surface turns a silver-grey-green colour and then a chemical reaction takes place with the wood and there is a colour change to darker or silvery shades. Green vitriol can also be used on mineral materials resulting in an ochre yellow or rust red colour tone. Green vitriol can cause skin irritation and is hazardous to aquatic organisms but it doesn't bioaccumulate and is considered to have an insignificant impact on the environment. A well-proven method is to leave the façade untreated for a year to be exposed to weather and wind so that the surface tension in the wood disappears. The wood is then treated with green vitriol, though at half-strength compared to normal (which is usually 5 per cent).

Examples of paint and surface treatments recommended for gypsum-, paper- or wood-based materials on ceilings or interior walls, as well as treatments for more vulnerable wood panelling and carpentry work.

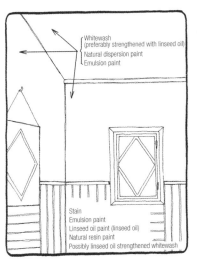

Whitewash (preferably strengthened with linseed oil)
Natural dispersion paint
Emulsion paint

Stain
Emulsion paint
Linseed oil paint (linseed oil)
Natural resin paint
Possibly linseed oil strengthened whitewash

Examples of paint and surface treatments for a room with wood panelling. Distemper may flake off.

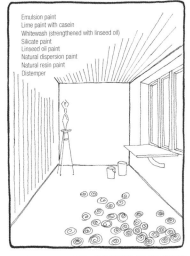

Emulsion paint
Lime paint with casein
Whitewash (strengthened with linseed oil)
Silicate paint
Linseed oil paint
Natural dispersion paint
Natural resin paint
Distemper

Examples of types of paint and surface treatments suitable for timber cladding.

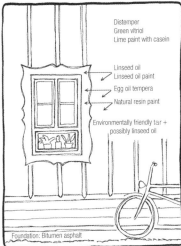

Distemper
Green vitriol
Lime paint with casein

Linseed oil
Linseed oil paint
Egg oil tempera
Natural resin paint

Environmentally friendly tar + possibly linseed oil

Foundation: Bitumen asphalt

Suitable surface treatments for floors of mineral-based materials.

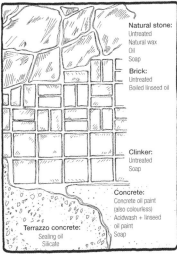

Natural stone:
Untreated
Natural wax
Oil
Soap

Brick:
Untreated
Boiled linseed oil

Clinker:
Untreated
Soap

Concrete:
Concrete oil paint (also colourless)
Acidwash + linseed oil paint
Soap

Terrazzo concrete:
Sealing oil
Silicate

Examples of paint that can be used for eco-friendly surface treatment of exterior brick walls.

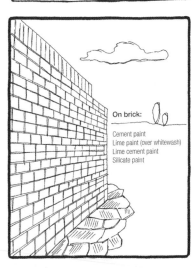

On brick:
Cement paint
Lime paint (over whitewash)
Lime cement paint
Silicate paint

Examples of environmentally friendly surface treatments for exterior cladding of lightweight concrete, plaster and/or natural stone.

On lightweight concrete or plaster:
Cement paint
Lime cement paint
Lime paint (on plaster)
Green vitriol (on plaster)
Silicate paint
Oil paint following neutralization

On natural stone:
Lime cement paint
Lime paint
Silicate paint
Egg oil tempera

Lime paint and lime cement paint consist of slaked lime (binding agent and pigment) and water, and in lime cement paint there is also cement. Dolomite may be used as a filler. If a colour other than white is required, limeproof pigment must be used (e.g. earth colours). Factory-produced lime paint may contain styrene, cellulose and latex. In Germany lime paint containing casein is even suitable for use on rough-sawn wood. Additives containing styrene should be avoided. This type of paint is strongly alkaline, which means that eye and skin protection is required when working with it. Lime paint has a long history of use and is environmentally friendly. The first coat should be pure lime water. Factory-produced lime paint and lime cement paint require only two coats. The paint should not be applied in strong sun or if there is risk of frost. The paint weathers with time, and so repainting must take place relatively often. Lime paint breaks down more quickly in an acid environment. Lime cement paint is less sensitive to acid environments and results in a stronger, more robust surface.

The binding agent in **casein paint** is based on natural lactic acid casein. Casein paint also contains calcite and chalk filler, as well as pigment (e.g. white titanium dioxide). It may also contain essential oils (e.g. thyme, lavender and eucalyptus), linseed oil, beeswax, shellac, borax, potash, zeolite or white slaked lime. The paint may be diluted with

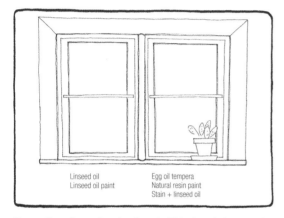

Suggestions for surface treatment of interior window woodwork. The easiest maintenance is oiling about every five years. To prevent yellowing, the wood surface must first be luted or white pigment may be added to the oil.

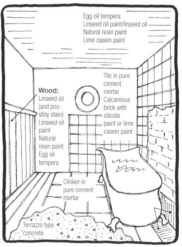

Some examples of materials, paints and surface treatments to be used in wetrooms.

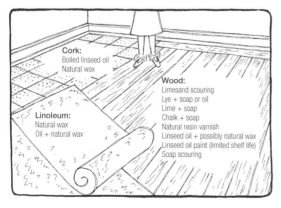

Suitable surface treatments for organic flooring materials.

A home near the sea in Gothenburg, Sweden, cladding treated with green vitriol.

Source: Architect Gert Wingård

water. Casein paint is an interior paint suitable for a variety of surfaces and can be used as a primer and as glaze paint with e.g. beeswax glazing. Casein paint should be applied thinly, otherwise there is a risk of cracking. The paint produces a matt finish that is waterproof and doesn't attract dust. Casein paint in the can is considered perishable and has a limited shelf-life: it lasts about six months and cannot be used once it starts to turn rancid, easily detected by the smell. Casein paint is available in powder form for mixing on site. Citrus peel oil (citrus turpentine) is used as a solvent in some of these paints. There is a risk of an allergic reaction from citrus turpentine. When using paint containing citrus peel oil, skin should be protected, there should be good ventilation, and a charcoal filter mask should be used.

Clay paint is, just as it sounds, made of clay. This type of paint is common in Germany and was introduced in Sweden by Ekologiska Byggvaruhuset AB. The paint is sold in powder form to be mixed with water on location. It binds to all absorbing surfaces indoors. It is made up of clay and natural earth pigment. No plastics or preservatives are used.

Varnish, e.g. UV-hardened acrylate varnish, is very important for the furniture industry. The binding agent may have a polyester, polyurethane or epoxy resin base. As a rule, it does not contain organic solvents. Acrylate monomers can, however, cause allergies, which is why this type of varnish is not used for manual spraying in factories. Polyester varnishes, urethane varnishes and cellulose varnishes are other commonly used varnishes. It is difficult to find varnishes without evaporating solvents or other dangerous substances. Water-based varnishes have been developed to improve the work environment, but their technical qualities are often inferior to solvent-based varnishes for treatment of table tops, worktops, kitchens and bathrooms. There are varnishes that prevent penetration of UV radiation and so counteract yellowing

A wall of unburnt clay bricks, coated with casein paint.

of wood. The raw materials are based on products from the petrochemical industry. Varnish provides a durable surface but is difficult to touch up. Surfaces often have to be sanded down prior to revarnishing.

Le Tonkinois, an old French company, sells a high-quality lacquer with the following contents: 70 per cent linseed oil (raw cold pressed and refined) and 30 per cent heavy oil (Chinese wood oil). Le Tonkinois contains no aromatic, chlorinated or essential oil solvents, nor anything that is dangerous in contact with the skin or when breathed in.

Soft distemper (glue paint) may consist of about 1 per cent glue (binding agent), 33 per cent water (solvent) and 66 per cent chalk (pigment). Cellulose or animal-based glue may be used. The white pigment in soft distemper is chalk ($CaCO_3$), though coloured pigment may also be used. A small amount (less than 1 per cent) of casein (as an emulsifier) may be added to soft distemper to reduce chalking. Soft distemper without preservatives is available (it has a limited shelf-life) so the production date should be checked at time of purchase. Soft distemper is also available in powder form, which saves on transport costs of water in the finished paint. Linseed oil-reinforced soft distemper is available. The addition of linseed oil makes it possible to clean the painted surface. Avoid soft distemper with environmentally hazardous additives. Some on the market contain additives such as plastic and preservatives, e.g. isothiazolinones such as Kathon or BIT or Bronopol (can cause allergies and is highly toxic to aquatic organisms). Soft distemper is used on interior walls and ceilings on most substrates. It leaves a lustrous, even, matt finish. Soft distemper should be applied in thin coats. It is difficult to touch up invisibly. The paint is sensitive to mechanical damage and can rub off (if it isn't fortified with linseed oil).

Linseed oil paint consists of 40–50 per cent linseed oil (binding agent), 30–40 per cent pigment (in the form of zinc oxide or titanium oxide) which is sometimes partly substituted with filler (calcium carbonate or barium sulphate), as well as a tint colour (often earth colours). Drying agents (siccatives in the form of metal compounds) are used (about 0.1 per cent) and biocides (about 1 per cent). Up to 5 per cent organic solvents can be included in the final coat to achieve good brushability.

The first two undercoats are usually diluted with greater amounts of solvents, but linseed oil-based primer containing practically no organic solvents is available. The linseed oil may contain a certain amount of linseed stand oil. Linseed oil paint containing more than 5 per cent solvents is to be avoided, as is linseed oil paint containing environmentally hazardous substances. Cobalt or zirconium compounds are sometimes used as drying agents. The biocide may be carcinogenic. Small amounts of zinc oxide are acceptable. During drying, various substances are released that may smell and cause irritation but they do not pose a health risk. Linseed oil paint has been used since the 18th century on both interior and exterior woodwork. The paint is durable, withstands wear and has a low moisture transfer capacity. Good weather conditions are required for paint application. Timber cladding is normally painted with three layers: two undercoats and a final coat. Linseed oil takes a long time to dry. Heat treatment (boiled linseed oil), light treatment or drying agents shorten the drying time. With linseed oil there is a risk that oil-saturated rags and cotton waste may self-ignite, and they should thus be stored in closed glass jars.

Lye and whitener change the appearance of various kinds of wood. Lye solutions can be used to lighten or darken the colour of wood. Caustic soda (sodium hydroxide solution or caustic soda), sodium carbonate (washing soda or soda ash), sodium hypochlorite (chlorine) and borax are examples of substances contained in lye used for floors. Solutions usually contain 1–10 per cent of these substances. After treating a floor with lye, it is either given a soap treatment, oiled and/or waxed. As lye is corrosive, protective equipment should be used. Lye treatment of pine flooring is most common. A sodium hydroxide solution results in a light tone being retained by the sapwood, while the heartwood becomes reddish brown. Spruce flooring treated with a sodium hydroxide solution turns a light greyish colour. Oak and beech treated with sodium hydroxide solution become brownish. Other types of wood take on different hues, which is why a small test area should be tried before treatment. Different shades of lye are available for treating floors. A tone that gives a hint of the required shade on the finished floor should be chosen. The blondness of pine and spruce can be further enhanced by adding a whitener to the lye, e.g. titanium dioxide, lime, chalk or talc. Whitener is generally included in the oil, soap or wax used to protect the tree after lye treatments.

Solvents and thinners are differentiated. A thinner (or diluent) is a liquid that thins paint to the desired consistency without dissolving any of its ingredients. A solvent, however, dissolves a solid substance in the paint. Nonetheless, the same liquid is often used as both a solvent and a thinner.

The most common thinner and solvent currently used for surface treatments and paints, other than water, is white spirit. Other mineral solvents made from petroleum have product names such as dilutin, varnolen, thinner, xylene, toluene, etc. In low aromatic white spirit or isoaliphate, the most dangerous hydrocarbons (aromatic hydrocarbons) have been removed.

Turpentine is a name for different types of solvents. It was the most commonly used organic solvent until white spirit came on to the market. Gum turpentine, which is distilled from the resin of coniferous trees, and citrus turpentine (made from citrus peels), are used by companies that want to use renewable raw materials. In gum turpentine, it is mainly the substance delta-3-carene that is an allergen. Citrus turpentine contains limonene, which is poisonous to aquatic organisms, bioaccumulates, is an allergen and affects the liver.

All organic solvents cause ground-level ozone formation, which damages vegetation, irritates respiratory passages and increases the greenhouse effect. Long-term use can damage the central nervous system. Paint and surface treatments without organic thinners and solvents or with as few as possible should be used.

Natural paints originate in the German environmental movement. An important sales argument is that paints made from natural products are favourable from an environmental and health perspective. However, natural products can pose a health risk, so it is important to be careful even when choosing natural paint. Natural paints consist of binding agents (such as beeswax, natural resins, essential oils and casein substances), solvents, filler, pigment and sometimes siccatives (drying agents). The natural resin colophony, used in some paints, is an allergen. Common to all these paints is that they do not contain softeners or strong solvents such as white spirit. In some products, however, a synthetic isoaliphate product or lemon peel product is used as a solvent. These solvents are said to be less dangerous to inhale than white spirit, but there is a risk of allergic reactions from citrus products. The long-term risks for exposure to isoaliphate and citrus turpentine are similar to the long-term risks of exposure to white spirit. Some paints contain cobalt-zirconium, which is carcinogenic if breathed in the form of grinding dust.

Carpentry oils of many different varieties are available. Usually wood oil contains a mixture of alkyd, resin (natural resin = gum from different kinds of trees), balsam

Wood floor treated with lye in architect Mattias Rückert's own house in Skaffanstorp, Skåne, where tha walls are built with wood-wool cement blocks.

Pigment is the most important ingredient in a paint as it provides the colour. In the construction sector, pigment may be divided into three main groups: inorganic natural pigment (such as earth colours), inorganic synthetic pigment and organic synthetic pigment. It is important to know how they react in different mixtures. In lime mixtures, only lime-resistant pigment may be used. Certain pigments are more suitable for oil mixtures. Earth pigments are usually suitable for use in most types of paint. The light stability of different pigments varies. Earth pigments are the most reliable in this respect. Earth pigments are usually iron oxides or ferric hydroxides, whose environmental impact is minimal.

Inorganic synthetic pigment is produced from chemical reactions between various metal oxides and minerals. Pigments made in this way produce stronger colours than inorganic natural pigments, especially in the blue and violet range. Metal oxide pigments are often named after the metal which is the main ingredient, e.g. chrome yellow, chromium oxide green or zinc white. Copper, chromium, lead, cadmium, zinc and cobalt can cause environmental damage. Lead, cadmium and some forms of chromium are especially dangerous. Titanium oxide is by far the most commonly used pigment. In the case of titanium oxide, it isn't the pigment itself that is problematic, but rather the two processes used to produce the pigment. The sulphate process involves considerable emissions to the air as well as sulphurous waste, and the chloride process creates waste containing chlorine. The EU's environmental labelling directive for interior paint restricts emissions from production of titanium dioxide.

Organic pigments can have a very strong chromaticity and brilliance. Many have a very complicated composition, such as azo pigments, especially with bright yellow, orange or red colours. Light stability and heat resistance vary. Of the organic pigments used in paint production, a few, such as

(= mixtures of resins and volatile oils) and solvents. To make a product dry quicker, siccatives (drying agents) that may contain cobalt and/or manganese may also be added. Wax is added to some wood oils to provide extra protection from water penetration. Some oils can penetrate the wood slightly and provide deeper protection; other oils stay on the surface and protect against wear. Oils may be pigmented and used as a glaze. Wood oils commonly contain too much organic solvent. Oils can be emulsified in water, a process which makes it possible to get rid of the dangerous solvents. There are wood oils available that do not contain any organic solvents or water. Where the oil is emulsified in water, attention should be paid to the biocides and preservatives. Oils for surface treatment products mostly come from plants, though there are animal and mineral oils. Synthetic resins and alkyds are, for example, produced from the latter.

azo pigments, give rise to aromatic amines, which are a health and environmental risk. Countries may have various bans on pigments containing lead, chromium/chromates and cadmium.

Plastic paint (latex and acrylic paint) consists of polymers (binding agents), water, pigment, filler and additives. The polymers (plastic particles) are dispersed in water. The advantage of these paints is that they do not contain organic solvents; however, they may contain additives that are unhealthy and environmentally dangerous, though in small amounts. Polymers are of petrochemical origin. Plastic paints are easy to apply, dry quickly, do not yellow and can be wiped clean. They are available in a range of finishes and as both top coats and glazes. Locally produced paints with low emission levels, and without coalescing agents, drying agents and environmentally hazardous tensides should be chosen. Research is under way to make available plastic paint in powder form.

Anti-rust paint is used on metal. In order to achieve a long-lasting result it is important to remove all rust first. Generally, both primer and top-coat anti-rust paint fall into the category of non-aqueous alkyd oil paint, but epoxy-based and water-based anti-rust paint is also available. The least environmental impact is achieved by using closed processes, factory lacquering with epoxy lacquer, and paint with zinc phosphate, linseed oil and a minimum of solvents; or alternatively some form of natural graphite paint or natural anti-rust paint. Linseed oil curing is a treatment with pure, heated linseed oil.

Silicate paint formerly used pigment suspended in water, which was then fixed with potassium silicate (also known as water-glass). This type of paint can still be purchased. Currently ready-mixed silicate dispersion paints are most commonly used. They consist of 10–30 per cent potassium silicate (binding agent), 25–45 per cent chalk, talc or titanium dioxide (pigment and filler), as well as 5 per cent acrylate copolymer (stabilizer), and 20–60 per cent water. Silicate paint may also contain a coalescing agent (usually 0.5–1 per cent glycol

Paint pigments.

ether). The acrylate copolymer is usually a styrene/butyl acrylate copolymer, but due to the small amount used in the paint, silicate dispersion paints are regarded as environmentally friendly. The paint is strongly alkaline, which means that precautions and protective measures must be taken by those using it. Silicate paint is applied to mineral-based materials. Chalk as well as strongly coloured pigments are available for use in this kind of paint. The binding agent bonds chemically to the substrate, resulting in a strong paint coating. The paint is water-vapour permeable, durable and waterproof. From an environmental perspective, it is best to mix the paint on site using the old method, so avoiding the use of acrylate co-polymers and coalescing agents.

Distemper consists of about 65 per cent water, 5 per cent flour (binding agent), 20 per cent pigment and 1 per cent green vitriol (fungicide). Nowadays 8 per cent linseed oil is often added to increase the binding capacity and prevent it from flaking off. In its basic form it is a relatively environmentally friendly paint that has been used for several hundred years. There is some lead in Falun red paint (<0.15 per cent after being mixed), which is not recommended according to the Swedish Chemicals Agency guidelines for substances to be phased out.

Environmentally certified distempers in the *Sunda Hus* (Healthy Buildings) database include products from Mäster Olof Naturfärger, Kulturhantverkarna and Auro.

Ready-to-use canned paint also contains a preservative (a biocide) so the paint won't rot in the can. Biocide use can be avoided by using paint powder and mixing on site. Biocides used include allergenic isothiazolinones, hydroxymethyl urea, etc. If linseed oil is added, a tenside is required for the linseed oil to mix with the water. Distemper containing polyvinyl acetate (PVAc) is also available. Distemper is applied outdoors on rough-sawn timber, and one coat may be enough. The paint is microporous and completely matt. It is available in various colours: light red, dark red, brown, black, white, blue, yellow and green. Light-coloured distemper paint is more sensitive to mildew and bleeding. The pigment used in genuine Falun red paint is exclusively Falun red pigment. Breathing protection should be used when rubbing down Falun red paint due to its quartz content (long-term inhalation of quartz can cause silicosis). On-site mixing eliminates the need for preservatives.

Soap is used as a surface protection on untreated, lye-treated and oiled wood. Soap for use on stone and concrete is also available. Soap used for treating floors should, however, not be mistaken for washing soap used for cleaning. Floor treatment soap may contain saponified vegetable and drying oils. This soap is available both as a liquid and a paste. Avoid soap containing environmentally hazardous ingredients. Soap may contain tensides which are dirt removers, and some of the many tensides available are poisonous. Green soap should not be used as washing soap on wood floors as it eventually turns them a light greenish colour. When ordering new floors that will

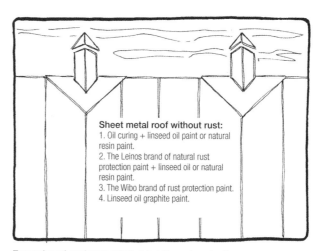

Examples of environmentally friendly rust protection.

be scrub-treated, the heartwood should face upwards, otherwise the boards may move and 'cup'. Floors should be scrubbed with soap a few times per year using cold water since warm water turns floors grey and dull. Even a newly laid floor soon lightens and becomes lustrous, but it takes some years before it becomes a real soaped floor.

Wood oil is used to protect exterior untreated or impregnated wood from cracking, mould and rot. It is common for wood oils to have an excessive solvent content, most often white spirit (up to 80–85 per cent). Choose wood oils that do not contain organic solvents. Completely oil-based wood oils contain fungicides, e.g. tolylfluanid (less than 1 per cent). Organic metallic salts are sometimes added, such as zinc decanoate, to protect against rot (1+ per cent). Water-based oils usually contain 20–50 per cent oil. If the oil is emulsified in water, it is important to watch out for tensides, co-solvent substances such as glycol ethers, and biocides. Water-based wood oils often contain different biocides than those in other wood oils, e.g. iodine propynyl butylcarbamate. Isothiazolinone-type preservatives are usually used. Some wood oils contain wax to provide extra water penetration protection. The oils come from renewable raw materials. White spirit is a petrochemical product. Wax can be obtained from plants as well as produced from petrochemicals. Surfaces, which should be treated annually, should be dry when oiled. Some wood oils can be warmed to improve surface penetration.

Wood tar is distilled from wood. Wood tar was produced in the past from tar piles in charcoal stacks. It is resin-rich and water-repellent. Tar is now produced through pyrolysis in retorts where charcoal is also produced. Coal tar presents a greater health risk than wood tar. Wood tar has been used since the Middle Ages as a wood preservative for boats, docks, jetties and buildings (e.g. on shingle roofs). It protects wood well from rot and is an alternative to pressure impregnation. Tar brings out the structure of the substrate, is glossy to start out with but dulls quickly and darkens with time. When warmed-up wood tar is brushed onto a warm wood surface, it has good penetration ability. Wood tar mixed with equal amounts of

A variety of distemper paint tests on timber cladding at The Gysinge Centre for the Preservation of Historic Buildings (between Gävle and Uppsala, Sweden).

linseed oil and gum turpentine easily penetrates the wood and results in a dry, water-repellent surface. This mixture has a long tradition in the treatment of wooden boats. Examples of tars in Scandinavia that do not contain substances hazardous to health or the environment are Becker's wood tar and Claesson's wood tar.

Waxes are fatty substances that can be polished. They are produced from both plant and animal substances, but may also be of mineral or synthetic origin. All waxes have a relatively low melting point (40–90°C) and can be dissolved in various types of solvents. When the solvent evaporates, almost all waxes harden. Wax is dispersible, i.e. can be suspended in water, which is often used as an alternative to organic solvents. When pure beeswax is used, it is usually mixed with up to 10 per cent solvent, e.g. turpentine.

Avoid synthetic wax and wax containing a lot of organic solvents. Wax should not smell too strong and it is not a suitable treatment for wetrooms.

Wax can be used to protect wood surfaces that have already been treated with floor oil, and on wood surfaces that are varnished or painted. The surface is polished with a polishing machine or rag and takes on a silk matt lustre. The surface is dirt-repellent and easy to maintain. Beeswax is known for its ability to reduce electrostatic charges. Washing wax is used to maintain waxed floors so that they remain shiny.

Paint and Surface Treatments

Area of use	Recommended surface treatment	Examples and comments
Exterior woodwork	Distemper paint	On rough-sawn timber, without lead and PVAc
	Priming paint	Priming oil, oiled substrate
	Linseed oil paint	<5% solvent
	Emulsion paint	Egg oil tempera, oil emulsion paint
	Wood oil	Linseed oil
	Green vitriol	Results in greyish brown shades
Plaster and concrete	Silicate paint	KEIM, Färgbygge AB (paint company in Sweden)
	Lime and lime cement paint	Lime from Gotland
	Linseed oil paint	On a neutralized substrate
	Green vitriol	Results in yellowish red shades
Interior	Emulsion paint	Egg-oil tempera, oil emulsion paint
	Soft distemper (glue paint)	Reinforced with linseed oil
	Casein paint	Primer
Woodwork	Linseed oil paint	<5% solvent
	Emulsion paint	Egg-oil tempera, oil emulsion paint
	Natural paint	Natural resin paint, German environmentally friendly paints
Flooring	Wood oil	Linseed oil
	Soap	Floor soap
	Lye and whitener	Caustic soda, lime, chalk
	Wax	Natural wax
Concrete flooring	Sodium silicate	Dust binding
Interior sheet metal	Rust protection paint	Warm linseed oil, graphite paint

PAINT INGREDIENTS TO BE AVOIDED

Binding agents:

- urea-formaldehyde resin, two-part, cold-setting (releases formaldehyde)
- epoxy, two-part (very allergenic if in contact with the skin)
- isocyanates, polyurethane paints (PUR) based on isocyanates (allergenic when inhaled or in contact with the skin)

Pigment, etc.:

- lead compounds, lead naphthenate/lead octoate (drying agents), red lead (anti-rust paint); lead is a heavy metal, bioaccumulative and a fertility hazard
- chrome compounds, heavy metal (carcinogenic and allergenic)
- zinc compounds, heavy metal (poisonous to aquatic organisms)
- titanium, titanium dioxide (environmentally destructive production process)
- cadmium, heavy metal
- copper naphthenate, anti-rot protection (classified as a biocide, requires special permission)

Solvents: It is preferable to use paint without organic solvents. Avoid paint containing more than 5 per cent organic solvents. Volatile organic solvents cause nerve damage and create ground-level ozone.

- white spirit, aromatic white spirit (hydrotreated naphtha is considered preferable as it contains fewer aromatics)
- turpentine, balsam turpentine which contains less delta-3-carene is considered preferable (less allergenic) and does not contribute to the creation of ozone
- sulphate turpentine (contains high levels of delta-3-carene)
- citrus turpentine (contains limonene which is allergenic, poisonous for aquatic organisms, affects the liver)
- xylene (affects human health and has a negative effect on aquatic organisms)
- toluene (affects human health and has a negative effect on aquatic organisms)
- glycol ethers, solvents and coalescing agents (certain glycol ethers are a serious health hazard, e.g. 2-ethoxyethanol, 2-methoxyethanol, 2-butoxyethanol, butyl glycol, diethylene glycol monomethyl ether)

Additives to be avoided:

(Biocide is an umbrella term for preservatives and fungicides.)

Preservatives (to prevent paint from rotting in the can):

- Kathon, an isothiazolinone (allergenic and poisonous for aquatic organisms)
- 1,2-Benzisothiazolin-3-one (BIT), an isothiazolinone (allergenic)
- 1,3-Propanediol (Bronopol) (very poisonous for aquatic organisms)

Fungicides (to prevent algae and fungi from growing on the finished painted surface):

- folpet (carcinogenic and allergenic, very poisonous for aquatic organisms)
- fluorfolpet (carcinogenic and allergenic, very poisonous for aquatic organisms)
- dichlofluanid (allergenic and bioaccumulative, very poisonous for aquatic organisms)
- tolylfluanid (allergenic, very poisonous for aquatic organisms)
- chlortalonil (carcinogenic)
- carbendazim (carcinogenic)
- diuron (carcinogenic)

- arsenic-based wood preservative (carcinogenic, bioaccumulative, poisonous for aquatic organisms)

Dispersing agents (tensides) (to finely disperse the pigments in the paint):

- alkylphenol ethoxylates (environmentally hazardous), e.g. subgroup nonylphenol ethoxylates
- nonylphenol ethoxylate (nonylphenol), oestrogen-like substances (a fertility hazard, mutagenic, bioaccumulative and poisonous)

Softeners (keep the paint from becoming brittle):

- phthalates (many are bioaccumulative and poisonous for aquatic organisms)
- chlorinated paraffins (bioaccumulative and poisonous for aquatic organisms, emit dioxins when burned)

Anti-skinning agents (to prevent skin formation on paint in the can):

- methyl ethyl ketoxime, also known as 2-butanonoxim (allergenic)

Pipes, etc

Various types of pipes are found in and around a building. There are water pipes for tap water, sewage pipes for waste water, downpipes for rain/snow, drainage pipes for ground drainage, and heating pipes. Gutters and downpipes are used for roof runoff. Insulation is required for heating pipes.

Sewage pipes in the ground are mostly made of concrete. If the ground conditions are corrosive, plastic pipes may be preferable, e.g. polypropylene pipes (PP pipes). Ceramic pipes are common: glazed pipes joined with polypropylene sleeves are used. Inside buildings the most eco-friendly choices are polypropylene pipes, cross-linked polyethylene (PEX) and cast iron pipes. Cast iron has a minimal environmental impact, though cast iron pipes nowadays often have an interior plastic epoxy coating. PVC pipes are most common, but should be avoided for environmental reasons.

Runoff drainage pipes in the ground may be made of ceramic (glazed stoneware) or concrete. Cross-linked polyethylene (PEX), polypropylene (PP), unglazed earthenware and polyester are alternatives.

Pipe insulation made from mineral wool or cellular rubber is common. Mineral wool contains troublesome fibres and cellular rubber may contain flame retardants. Pipe insulation made from cork or sheep's wool, used in Germany and Switzerland, is more environmentally friendly.

Sanitary fittings such as mixers and taps are most often brass. Brass usually contains 62 per cent copper, 36 per cent zinc and 2 per cent lead. Taps are usually chrome-plated. The amount of chrome and nickel, however, is small (0.1 per cent chrome and 0.5 per cent nickel). Use of chrome-plated sanitary fixtures has been questioned since the surface may give rise to nickel allergy. Powder lacquered sanitary fixtures do not last as long as chrome-plated ones. The lead in the brass is the greatest environmental problem. Sanitary fittings of stainless steel are also available.

Roof runoff may be carried using traditional methods such as wooden gutters; these are sometimes lined with waterproofing (EPDM or bitumen). Instead of a downpipe, a chain or wooden pole can be used. A more modern approach is to use gutters and downpipes made of polyester. Aluminium and galvanized steel plate are also used. Production of aluminium is energy-intensive and environmentally hazardous. Zinc leaks from galvanized steel plate if the surface is not treated.

Water pipes should be chosen according to local conditions since the chemical composition of water varies from place to place. Consideration must also be given to how the pipe material affects drinking water quality. Polythene (PE) pipes can be used for underground water pipes. Cast iron is another possibility, but iron has a more limited life. Stainless steel is probably a good environmentally friendly choice if the water is corrosive and ground conditions won't corrode the steel. For indoors water pipes, environmentally friendly alternatives to copper are polypropylene (PP) pipes and polythene pipes (cross-linked polythene PEX pipes and high-density polythene HDPE pipes). In The Netherlands, hot and cold water pipes made of polybutene (PB) are common, where the plastic is regarded as environmentally friendly. Stainless steel is an alternative but has a greater environmental impact than plastic. Copper pipes should be avoided for environmental reasons, as their use results in high levels of copper in drinking water and sewage waste. If drinking water pipes consist of 'pipe-in-pipe' systems, the outer pipe (the empty pipe) should be made of polyethylene.

Photo of a suspended foundation, insulated with foam glass, showing sewage and clean water pipes made of polythene, a more eco-friendly choice than PVC or copper.

Roof drainage using a wooden gutter and downpipe. Heartwood timber must be used, installed in an overlapping manner and with a spout to carry the water away from the building. Gutters may be lined on the inside with a material that makes them more durable.

Pipes: Environmental Evaluation

Recommended	Acceptable	To be avoided
Concrete	Cast iron**	Asbestos
Ceramic, for sewage	Polythene (PEX)	Lead
Steel, for heating*	Polypropylene (PP)	Copper
Earthenware, for drainage	Polybutene, for tap water	PVC****
Wood, for gutters	Polyester, for drainage	Steel, galvanized
	Stainless steel, for tap water***	
	Aluminium, for gutters	

Notes: * Steel pipe (galvanized) must not be joined to copper pipes, because then a galvanizing process occurs whereby amongst other substances zinc is released.
** Cast iron with epoxy coating on the inside reduces the evaluation category to 'acceptable'. In the past, cast iron pipe with an interior coating of asphalt was available (should now be avoided).
*** Stainless steel is preferred in corrosive environments, but not if other materials last just as long.
**** PVC contains 57 per cent chlorine, which is problematic from an environmental perspective. PVC piping made in the Nordic countries now uses calcium/zinc instead of lead stabilizers.

Interior heating pipes made of steel and installed so that they are easily accessible for inspection and repair are in general a good solution in all respects. Polythene pipes (PEX) are used to distribute heat, e.g. in floor and wall heating systems. Cross-linked polythene is suitable for both cold and hot water pipes. Polybutene is used in The Netherlands where it is considered an environmentally friendly plastic.

Interior Fittings and Furnishings

Interior fittings and furnishings can be the source of many environmentally hazardous substances, such as flame retardants in textiles and electronic equipment, furniture stuffing that releases poisonous gases if it catches fire, and cotton that has been sprayed and treated with dangerous chemicals during cultivation and production. The first choice should be solid wood and wood fibreboard that is bound together with the wood's own lignin (see Sheet Materials).

The Healthy Bedroom

In Germany, those concerned with health issues believe that it is of primary importance to have a healthy bedroom where one can recover at night. Such a bedroom has furniture made of solid wood, a wall-to-wall carpet of natural materials, and a bed frame of wooden slats supporting a mattress of natural rubber containing no metal. Mattresses have a core of natural latex, cotton stuffing on one side for winter use and horsehair on the other side for summer use. The mattress cover and bed linen are made of organic cotton or linen. Pillows may be stuffed with wool, cotton from kapok trees, buckwheat husks or combinations of these. Furthermore, the electricity is turned off at the fuse box to minimize electrical and magnetic fields in the bedroom.

Permanent Fixtures

For permanent fixtures such as fitted cupboards, wardrobes and kitchen fittings, it is especially important to avoid chipboard. The total amount of formaldehyde in a room should be kept as low as possible. Therefore, in general, fixtures made of solid wood, plywood, or with a wooden framework covered with hard fibreboard should be chosen.

Furniture Upholstery

Furniture upholstery accounts for a large amount of the material in our surroundings. Foamed plastics such as polyurethane are often used for stuffing, which may be the greatest source of emissions to interior air from fittings and fixtures. During a fire, burning stuffing may emit deadly gases. Natural materials such as horsehair, cotton or buckwheat husks are alternatives to plastic materials.

Textiles

About half of all textiles are made from cotton and about half from synthetic materials. A small amount are made from wool or viscose (made from wood pulp). Organically grown cotton should be chosen. Locally produced wool is preferable. Linen is a good material. Preferred synthetic materials are polyester and polyamide. Almost all textile fibres are spun, then woven or knitted, bleached, dyed and given a finishing treatment. How this is done has an environmental impact, and whether the process used is better or worse does not relate to the fibre used (natural or synthetic).

Carded wool and warm soapy water is needed to make felt.

Examples of **natural fibre** are cotton, linen, wool, silk and hemp. Cotton is one of the most commonly used natural materials, but if it is conventionally produced, 1kg of cotton cloth requires a total of 1kg chemicals for its cultivation, manufacture, final treatment and storage. Many of these chemicals adversely affect the environment or health. More than half of all cotton is grown using artificial irrigation. There are textile factories that work with organic cotton using environmentally friendly and healthy methods. An international organization called the International Federation of Organic Agriculture Movements (IFOAM) is trying to establish and maintain basic standards and labelling for organic cotton.

Wool itself is non-flammable. However, no environmentally friendly method exists to mothproof wool. As with cotton, chemicals are used during the manufacturing process.

Modern flax cultivation uses both pesticides and artificial fertilizers, but not in especially large quantities. In the past, flax was retted by laying it out in lakes and it was bleached by laying it out on snow and exposing it to intense sunlight. Today, chemicals are used for retting and bleaching.

Pesticides cannot be used on silk worms, which is why there are no big environmental problems associated with the production of silk fibre.

Viscose is made from bleached wood pulp and is also called rayon. To transform the wood pulp into textile fibres, large amounts of poisonous, sulphur-containing carbon disulphide are used. Almost all finished viscose fibres are bleached, even though they are produced from wood pulp that has already been bleached. So most viscose is unnecessarily twice chlorine-bleached. In some countries, the consequences of poorly cleaned discharges from viscose production may be catastrophic. Nonetheless, the idea of transforming cellulose to fibre is reasonable since wood is a renewable resource. Alternative, chlorine-free methods exist for bleaching wood pulp. In the US there is a factory that uses a completely sulphur-free method of producing viscose.

Synthetic textiles are made from crude oil or recycled plastic. The most common fibres are polyester, polyamide and acrylic. Polyester and polyamide are a little better than acrylic from an environmental perspective since acrylic fibres, when stretched into threads, require both organic solvents and undesirable chemicals.

The processing of textile fibres starts in the spinning mill where the fibres are spun into threads and yarn. Environmentally hazardous spinning oils are often used, even though there are alternatives. Cotton, viscose and wool can be spun completely without or with only small amounts of spinning oil. Synthetic material requires more oil. Avoid chlorine-bleached cotton, flax or viscose. Do not choose textiles treated with optical whiteners. Textile dying often causes environmental problems. The fibres are first washed, then dyed, and then washed again. All this takes place in various industries and hundreds of chemicals are used, of which only small amounts remain in the cloth, so most ends up in sewage or air. Detergents may contain nonylphenol ethoxylate (which is hormone disturbing), even though substitutes are readily available. The Swedish, Danish and German textile industries have to a large extent stopped using nonylphenol ethoxylates, though they are still common in the rest of Europe and elsewhere in the world. There are thousands of different

Healthy buildings should contain furniture made from healthy materials. In Germany and The Netherlands, bedsteads and mattresses that do not contain synthetic materials, glues or metals are commonly available for purchase.

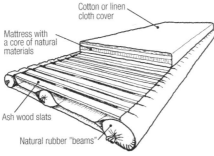

Cotton or linen cloth cover
Mattress with a core of natural materials
Ash wood slats
Natural rubber "beams"

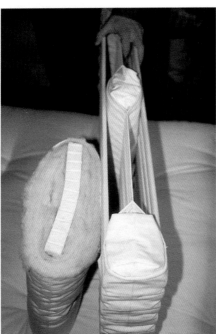

Bedding without metal or synthetic materials.

dyes and almost all of them are synthetic. Since a good dye should be able to tolerate sunlight and many washes, dyes should be difficult to degrade. Therefore, there is a great risk that they are also environmentally hazardous. Avoid textiles that are dyed with benzidine and dyes that contain heavy metals. The amount of energy and water consumption, as well as the release of dangerous substances into sewage systems, is highly dependent on how technically advanced the dying equipment is and how the sewage water is cleaned.

Treatments to make cloth crease-resistant leave formaldehyde residues in the cloth. The most common impregnating agent contains fluorocarbons. Probably the actual production of fluorocarbons has the greatest environmental impact. In order to make silk cloth heavier and drape well, metal salts are sometimes added. Weighted silk may contain 25–60 per cent tin phosphate. All flame retardants used to treat cloth are a source of emissions. Textiles treated with flame retardants may continue to release dangerous substances over a long period of time. Brominated flame retardants are the worst from an environmental perspective.

Anti-mould substances can be released and make people sick.

Joints

Caulking compounds often contain dangerous chemicals and their use should be minimized. Caulking compounds are used to seal joints between various structural elements and to seal around penetration points. By building carefully and covering joints with mouldings, caulking compounds and caulking foam do not need to be used. Oakum can also be used instead of caulking.

A function of some caulking compounds is to absorb movement from surrounding structural parts without impairing the seal. Caulking compounds can be divided into two categories: elastic, which regain their shape after being deformed; and plastic, which are less elastic and do not regain their shape after being deformed. Expansion joints are needed in most buildings. They make it possible for building parts to move in relation to each other. The greatest area of use for caulking compounds is in prefabricated elements.

Packing

Packing for sealing joints between frames and walls is often made from mineral wool, polyurethane foam (PUR foam) or similar materials. These can be replaced with oakum made from natural materials.

Cellulose fibre (see Thermal Insulation) in strips is used as packing material.

Sheep's wool (see Thermal Insulation) can be used unwashed as packing material without further treatment.

Hemp fibre (see Thermal Insulation) is used for oakum, and consists of clean, compressed hemp fibres. Since hemp is naturally resistant to fungus and bacteria, hemp oakum does not need to be impregnated unless fire protection is required.

Jute belongs to a genus of the linden family comprising about 40 species. It contains strong fibres that can be used as oakum. Most jute is cultivated in Asia, where 96 per cent of production takes place.

Sisal is fibre from the *Agave sisalana* plant. The fibres, extracted from the leaves, are flexible, strong and yellowish-white. Sisal is used to make rope, hammocks, and as oakum. The plant is Mexican in origin and is cultivated in tropical areas.

Flax fibre (see Thermal Insulation) is one of the strongest natural fibres and an excellent packing material around window and door frames, in boats and pipe joints, because it does not rot in a moist environment.

Coconut fibre (see Thermal Insulation) is a product that requires long-distance transport. There is a surplus of coconut fibre on the world market.

Mineral wool (see Thermal Insulation) may be either glass wool or rock wool. It is used in strips for packing.

Sealing Strips

Sealing strips are used for concrete casting. They usually consist of softened PVC (polyvinyl chloride). It is better to use bentonite sealant strips instead.

Bentonite sealant strips (sodium bentonite) consist of clay that expands when it comes into contact with water and becomes a hard and watertight clay-like mass. There are sealant strips for concrete casting that contain bentonite. These are placed in the formwork, and the concrete is poured up to the movement-absorbing part of the sealing strip. When the first pour hardens, casting continues. In this way, despite the shrinkage that later takes place, a tight construction joint is achieved.

Sealing Compounds

Sealing compounds should be avoided as much as possible.

Acrylate sealing compound is also called latex or painter's sealant. It consists of about 30 per cent acrylate polymer (latex) finely dispersed in water, surface-active dispersing agents, about 55 per cent filler (e.g. calcium carbonate), and about 8 per cent softener and pigment if required. Additives include thickeners, anti-foam agents and bonding improvement agents. Common monomers are butyl methacrylate, vinyl chloride and hydroxyethyl acrylate. These do not cause problems as long as they are bound within the polymer but residual monomers may remain and affect health. Solvents such as white spirit or butyldiglycolacetate may be present in small amounts (about 2 per cent). Latex sealants always contain biocides. Acrylate-based sealing compounds are not as elastic as many other sealing compounds. They have good UV radiation resistance and stick to most surfaces without initial priming. Their tolerance of moisture and temperature differences is poor, however, so they are used almost exclusively indoors. They can usually be painted.

Butyl-based sealing compounds are one-part sealants with solvents (alcohols). The polymer is often made by copolymerizing isobutene with isoprene. These sealing compounds may also contain softeners, fillers and tacking agents. Polybutenes are often used as softeners. Butyl sealing compounds harden and cannot absorb much movement. An area of use for these compounds is sealing insulating glass.

Cork granule sealant is an elastic sealant made of small pieces of cork and natural resins. It is available in Germany and is recommended by German building biologists.

Modified silicone (MS) polymer-based sealant is based on polyether and consists of alkoxy silicones, filler and thixotropes in the form of hydrogenated vegetable oils or alternatively polyamide wax. Chalk (calcium carbonate) is commonly used as filler. Additives consist of siloxanes (drying agents, about 1 per cent) and organic tin compounds (catalysts, in small fractions of 1 per cent). During hardening, methanol and siloxanes are released. Phthalates (20–30 per cent) are often used as soften-

Coconut fibre packing in a lightweight concrete wall.

ers. There are MS-polymer sealants without phthalates, although these are hard and unsuitable for construction joints. MS polymer sealants are an isocyanate-free alternative to polyurethane (PUR)-based sealants. Areas of use are similar to those for polyurethane-based sealants, i.e. exterior joints between prefabricated concrete elements instead of packing for windows and doors. However, they adhere rather poorly to concrete outdoors when applied without primer and their function in a moist environment is inferior to that of PUR sealants.

Oil-based sealants are included in the plastic sealants category. Oil-based sealants consist of either vegetable or synthetic oils, resins and synthetic rubber. They may also contain organic solvents such as white spirit and xylene in small amounts. Dolomite and chalk may be used as fillers. The sealants contain drying agents based on cobalt compounds. They have a relatively short life and a limited movement absorption capacity, which limits their range of application. Window putty made of linseed oil and chalk can be painted and does not present environmental problems. It eventually cracks and must be maintained.

Polyurethane-based sealants (PUR sealants) consist of 30–35 per cent prepolymer-diisocyanates, 30–45 per cent filler and 15–35 per cent softeners. Commonly used fillers are calcium carbonate and dolomite. They usually contain 3–5 per cent organic solvents such as toluene, xylene and white spirit. Organic tin compounds are added as a catalyser. Polyurethane is produced using isocyanates which can cause allergic reactions. PUR sealants contain lower levels of isocyanates than polyurethane foam sealant. Phthalates are often used as a softener in polyurethane sealants. If polyurethane sealants are heated or burned, isocyanates and other substances hazardous to health, such as phenols, hydrocyanic acid, nitrous gases, etc., are released.

PUR sealants are perhaps the most commonly used sealants in the construction sector. They remain elastic after hardening and can absorb large movements. They can be painted and adhere to most surfaces, often without primer. They are commonly used in façade joints between prefabricated concrete elements and as window and door sealants.

Rope made of natural materials such as hemp and sisal has traditionally been used in joints in boat building to seal and absorb movement. Rope can also be used in building joints to absorb movement between materials with different expansion coefficients, e.g. between wood and stone. There are no specific products; common rope of an appropriate thickness is used.

Silicone sealants consist of silicon polymers, filler, some type of reactive cross-binder and often a catalyser. Silicone sealants are made of siliceous sand, carbon, methanol and chlorine gas. Organic tin compounds may be added as catalysers. Organic tin compounds are environmentally hazardous. Softeners used in silicone sealants include various types of silicone oils consisting of silicate polymers,

Rope is used as the joint material where wood and stone floors meet. The National Park building in Tyresta, near Stockholm.

Source: Architect: Per Liedner

oxygen and organic molecules in the ethyl, methyl or phenyl groups. Silicone oils are thought to be stable and bioaccumulative. Some silicone sealants may contain up to 20 per cent organic solvents. Biocides are added to almost all silicone sealants, e.g. organic arsenic compounds, carbendazim, BIT and Kathon.

Various substances are released during hardening, in the range of 1.5–4 per cent of the respective substance. Methanol or ethanol is released from alkoxy-cured silicone sealants. Methyl ketoxime is released from neutral cured sealants. Methyl ethyl ketoxime released from oxime-cured (neutral cured) sealants can cause allergies. Amine-cured silicone sealants release benzamide or amines or combinations of these, as well as some substances already mentioned. Acetic acid is released from acetic-acid-cured silicone sealants. Acetic acid has a strong smell but is not a health hazard in the concentrations released. Acetic-acid-cured products do not work well with cement-bound materials such as glazed tile sealant. Silicone sealant is used to seal wetrooms as well as wet areas in kitchens. Good ventilation is required when using silicone sealant.

Foam Sealant

Foam sealant should be avoided. Use packing instead.

Polyurethane sealant foam (PUR sealant foam) is purchased in spray cans. When the product reaches the air it hardens with the help of the moisture in the air, expands 20–30 times and forms a foam-like mass. PUR sealant foam consists of polyurethane. Polyurethane is produced with isocyanates posing a risk of allergic reactions during their production as well as during the post-hardening joint finishing process. Organic tin compounds, which are very toxic to aquatic organisms, are added as a catalyser. Phthalates are often used as softeners. Either butane or propane is used as the propellant gas. The risks with foam sealant are that isocyanates may be released to the air as an aerosol and inhaled, especially when the sealant is improperly used, for instance at too cold a temperature or when they are poorly mixed.

Urea-formaldehyde (UF) foam is based on uric acid and formaldehyde. It releases formaldehyde gas. Relatively small doses of formaldehyde may cause acute problems such as eye irritation, itchy nose, dry throat and sleeping difficulties. The substance is

Joints: Environmental Evaluation

Recommended	Acceptable	To be avoided
Bentonite	Acrylate sealant*	Mineral wool packing
Cellulose fibre packing	EPDM rubber moulding	PUR sealant
Felt strips	Glazed tile sealant	PUR foam
Sheep's wool	MS polymer sealant*	PVC sealant strips
Hemp fibre packing	Silicone sealant*	UF foam
Jute and sisal packing		
Lime mortar, lime cement mortar		
Coconut fibre packing		
Granulated cork paste		
Flax fibre packing		
Natural latex sealant		
Oil-based sealant		
Rope of natural material		

Note: * Watch out for biocides and phthalates.

carcinogenic and allergenic. It may expand if affected by moisture.

Weather Stripping

Weather stripping is used to seal between frames, windows and doors. Textile weather stripping used to be common. It is currently more commonly made of EPDM rubber or silicone rubber. The environmental impact depends a lot on the additives used.

EPDM rubber (ethylene-propylene rubber) is entirely ozone-resistant, has good heat and chemical resistance, and so is used for rubber details outdoors, such as weather stripping and glazing beads.

Silicone rubber consists of silicon, vinyl and filler. The raw material for silicone is silicate sand, among other things. The filler is usually chalk.

Glues and Pastes

Glue is usually avoided in environmentally friendly building, not least because it makes recycling more difficult. Mechanical joints provide an alternative to glueing. There is no clear line of distinction between the terms glue and paste. However, pastes provide weaker adhesion and are primarily used for putting up wallpaper, woven fabric, foils and similar materials. Water-based glues/pastes often contain biocides.

Animal Based

Animal-based glues come from protein-rich substances such as milk, blood and tissue. They are water soluble. Good quality animal-based glue often contains wood flour. They are excellent glues for wood, but they are not moisture-repellent and should only be used indoors.

Blood albumin glue is made of blood protein, ammonia, slaked lime and water. It sometimes contains fungicides. It is used to glue veneer. The object to be glued must be warmed.

Animal glue is made of tissue (e.g. slaughterhouse waste), and sometimes calcium chloride and water. It is used to glue furniture and veneer. Bone glue and skin glue are used warm and require pressure to fasten. Fish glue can be used cold.

Casein glue is made of milk protein, lime and water. It sometimes contains fungicides. It is used to glue plywood and laminated wood, including large glulam beams. It is a strong glue that can be used in load-bearing structures indoors. A fresh batch of casein glue must be mixed every day.

Mineral Based

Mineral-based glues are made from readily available raw materials. Their production is relatively energy efficient and environmentally friendly.

Cement glue consists of Portland cement, stone powder and water. This glue is used to glue ceramic tiles and lightweight concrete. It is very strong when mixed with acrylate. Acrylate may cause problems indoors during hardening.

Sulphite lye glue consists of lye and water. It is used to glue wood fibreboards, bitumen felt and linoleum.

Sodium silicate glue is made of sodium silicate, lime, stone powder and water. It is used to glue ceramic tiles, paper, chipboard and in putty. Sodium silicate is a skin and eye irritant.

Synthetic Based

Synthetic-based glues have fossil origin. Production is energy-intensive and environmentally hazardous. The synthetic glues that pose the least environmental problems are polyvinyl acetate glue (PVAc) and ethylene vinyl acetate glue (EVA).

Most synthetic glues, such as urea-formaldehyde glue (UF glue), phenol-formaldehyde glue (PF glue), resorcinol-formaldehyde glue (RF glue) and phenol-resorcinol-formaldehyde glue (PRF glue), contain formaldehyde, which is a health hazard even in small amounts. Formaldehyde resin is used as a binding agent in chipboard. It consists primarily of carbamide formaldehyde resin, which emits formaldehyde gas, and water-resistant phenol formaldehyde resins, which emit some gases containing phenol. Building materials made with these glues release formaldehyde over a long period of time. Relatively small doses of formaldehyde may result in acute eye irritation, itchy nose, dry throat and sleeping problems. The substance is carcinogenic and an allergen. The level of free formaldehyde in ready-to-use glue is low and the level in the glued, finished product is very low. Nonetheless, sensitive people may have problems, especially if the room contains many exposed glued joints. According to Swiss environmental recommendations, material containing more than 5 per cent formaldehyde binding agents should be avoided.

Most of the other components in these glues are also problematic. Solvents pose a health risk during application, especially for the nervous system. They can also create surface ozone and photo-oxidants, and so damage forests and crops. It is important to have good ventilation when working with these products. Some products contain colophony and nonylphenol.

Almost all water-based products contain biocides. Isothiazolinones are commonly used biocides. Several isothiazolinones are classified as allergens. The biocide Bronopol, which is not a isothiazolinone, is also used. It is classified as very toxic to aquatic organisms.

Acrylate glue is petroleum-based and often contains chalk as a filler. It contains copolymers of vinyl acetate, ethylene and acrylates. Common acrylates are acrylonitrile, ethylhexyl acrylate and butyl acrylate. The glue is used as floor glue for linoleum and plastic mats, to glue wood, ceramics and fibreglass. It can be either water-based or solvent-based. Emissions may occur during application. If in contact with moisture, the glue may saponify and release chemical substances that irritate mucous membranes. The glue may release residual monomers that can cause allergies, and levels of these should be checked to ensure that they are as low as possible.

Asphalt glue is made from crude oil and contains bitumen, large amounts of organic solvents, as well as chalk. It is used in foundations, in wetrooms, for glueing roofing felt and in insulation material such as expanded polystyrene. Its composition varies. Either an emulsion of bitumen with water or with a solvent such as naphtha is used. Some asphalt glues need to be heated prior to application. Amines irritate the throat and mucous membranes for those working with the heated mass. In its hardened state it has low-level emissions.

Epoxy glue contains epichlorohydrin, phenol and alcohol. Water-based epoxy glue is also available. It is used to glue concrete, stone, glass, metal, plastic, ceramic tiles and may be used in spackle (filler). Epoxy is one of the strongest allergens. Epichlorohydrin is a carcinogen and an allergen. Epoxy glues also contain alkyl phenols and bisphenol A, which are oestrogen-mimicking substances. Precautionary measures are required in the work environment (according to directives for thermosetting plastic). Hardened epoxy is, however, inert.

Ethylene vinyl acetate glue (EVA glue) consists of ethylene, vinyl acetate and water. It sometimes contains fungicides. It is used to glue plastic mats and linoleum on walls or floors. The glue may also contain softeners such as DBP (dibutylphthalate), which should be avoided as it affects the nervous system, immune system and fertility.

Phenol-formaldehyde glue (PF glue) consists of phenol, formaldehyde and organic solvents. The glue is used in mineral wool bats, waterproof plywood and laminate. Phenol is poisonous. Small quantities of phenol and formaldehyde may be found in the resin even after it has hardened. In that case emissions are usually small.

Isocyanate glue (EPI glue) contains isocyanates and styrene-butadiene rubber or polyvinyl acetate. It is used to glue plywood, doors, windows, furniture and metals. Isocyanates may cause skin allergies and asthma. Emissions may be present in buildings where this glue was used and did not harden completely.

Chloroprene and neoprene glue are contact glues consisting of acetylene, chlorine, and large quantities of organic solvents such as toluene, xylene and n-hexane. They are used to glue plastic. Chloroprene is believed to affect fertility. The danger is greatest when the glueing takes place, though there may also be emissions in the finished building if the reaction was not complete.

MS polymer glue has similar areas of use to PUR glue; however, it doesn't contain isocyanates, which PUR glue does. MS polymer glue is a somewhat better option than PUR glue from an environmental viewpoint. MS glue requires softeners, which may be phthalates. There are, however, alternatives. MS polymer in general requires the addition of organic tin compounds as catalysers, e.g. dibutyltin compounds.

Polyurethane glue (PUR glue) is based on isocyanates, a group of substances with allergenic characteristics. The glue also contains polyhydric alcohols. It is used for glueing wood used outdoors, metal, plastic, concrete and is used in some strawboards. Isocyanates pose a risk of allergic reactions, even during processing after hardening. The solvents pose health risks, particularly for the nervous system, when the glue is applied. They can also create surface ozone and photo-oxidants, and so damage forests and crops. Dangerous substances (including isocyanates) may be emitted when the material is heated. Organic solvents may also be present.

Polyvinyl acetate glue (PVAc glue) is often referred to as wood glue or white glue, and is suitable for use on furniture, windows and for glueing wood indoors. It is also used as floor glue for linoleum and plastic mats. It contains acetylene, acetate acid, polyvinyl, alcohol and water or organic solvents, and chalk is often included as a filler. PVAc glue is the only synthetic glue that does not contain formaldehyde. Dangers with the use of PVAc glue depend primarily on the residual levels of free monomers and additives, biocides such as Kathon and BIT, softeners such as DBP (dibutylphthalate), colophony, nonylphenol and sulphonamides. Chromium-III-nitrate, a commonly used hardener, is corrosive and may cause eczema, and so skin contact should be avoided. If in contact with moisture, the glue may saponify and release chemical substances that irritate mucous membranes. PVAc glue may also contain sulphonamides that may damage the immune system.

Resorcinol and phenol-resorcinol-formaldehyde (PRF) glue contains resorcinol, formaldehyde and water. The glues sometimes contain phenol. They are used in load-bearing timber structures such as glulam beams, window frames and outside doors, as well as in glue-laminated core board. Resorcinol may cause skin allergies, can damage the genetic system and is an environmental hazard. Phenol is poisonous. Residual levels of resorcinol and formaldehyde are very low.

Styrene butadiene glue (SBR glue) is a contact glue. It contains butadiene, styrene and organic solvents. It is used to glue chipboard, gypsum board, wood, concrete and wall-to-wall carpeting. Butadiene is carcinogenic. Styrene is, in the EU, questionable from a health point of view (hormone disturbing).

Urea-formaldehyde glue (UF glue) consists of urea, formaldehyde and water. The glue is also referred to as carbamide glue. It is used to glue veneer, in some plywood production, in carpentry, for parquet, doors, mats on subfloors and in chipboard. Formaldehyde is released from construction materials glued with it. UF glues usually emit more formaldehyde than PF, PRF and similar glues.

Plant Based

Wallpaper paste and textile adhesives are usually based on starch or cellulose and are water-based. The paste is usually distributed as a powder that is dissolved in water. Nevertheless, many of these pastes contain preservatives (biocides). Each time the glue is needed, only the amount required should be mixed. Biocides are only used when the mixed paste is going to be stored for an extended time.

Cellulose paste consists of methylcellulose and water. It is used to fix wallpaper, linen fabric, linoleum and in spackle.

Cellulose glue consists of a derivative of cellulose and organic solvents. It is used to glue linoleum. Cellulose glue may be used in moist environments. The solvent used in cellulose glue is usually either turpentine or alcohol, sometimes up to a level of 70 per cent. The cellulose glue production process is environmentally hazardous.

Natural rubber glue consists of natural rubber and organic solvents. It is used to glue tiles and linoleum.

Natural resin glue is a dispersion glue based on natural latex. It consists of lignin, shellac or copal and usually organic solvents (or water). The glue is used to glue linoleum and timber. The solvent used is either turpentine or alcohol at a level of up to 70 per cent.

Potato flour paste consists of potato starch and water. It may also contain fungicides. The paste is used to hang wallpaper. Potato flour paste is used only in dry areas.

Rye flour paste consists of rye flour starch and water. It may also contain fungicides. The paste is used to fix wallpaper, jute-fibre wallcovering and linoleum. Rye flour paste is the strongest of the plant-based pastes.

Soya paste consists of soya protein and water. It may also contain sodium silicate or fungicides. The paste is used to glue plywood. Soya paste is used only in dry areas.

Glues and Pastes: Environmental Evaluation

Recommended	Acceptable	To be avoided
Animal-based*	EVA	Acrylate
Cement	PVAc	Asphalt
Sodium silicate		EPI
Plant-based*		Epoxy
		PF
		Chloroprene/neoprene
		MS polymer
		PUR
		RF
		SBR
		UF

Note: * Watch out for fungicides and biocides.

1.1.4 Choosing a Construction Method

Even if environmentally friendly materials are chosen, the consequences vary depending on how the materials are combined in various constructions and where the materials are placed in the building. The interior climate is affected by whether heavy or light material is chosen. Moisture-buffering material may result in a more even interior climate and is advantageous for use in bathrooms. Some materials do not work well together. Some glues, for example, cannot tolerate moist concrete. Moisture damage may result if moisture movement is not thought through with regard to where materials are placed and how permeable they are.

Roof structure: Roofing sheet/single roof tile W-truss 37%

Cellulose fibre insulation, Secondary spaced boarding Gypsum wallboard

Walls: Glazing paint, beeswax

Inner doors

Intermediate floor: Cellular floor unit concrete 185mm + concrete 60–70mm

Floor: Linoleum Impact sound insulation: Board Oiled wood floor

Windows: Triple-pane glass

Foundation: Expanded clay aggregate 400mm Foamed plastic 50+80mm Concrete 220mm Foamglas 50mm

Kitchen woodwork: Laminated core board with synthetic varnish Oiled laminated beech board

Radiators: Slow flow water system

Outer walls: *Outside:* Lime plaster Wood panel with air gap Light concrete 450mm *Inside:* Smooth plaster

Outer door: Plywood Wood panel

Bathroom: Walls: tile Floor: clinker

Walls between flats: Lightweight aggregate blocks, 150mm+150mm + insulation between

Inner walls: Gypsum with wooden studs 96mm

The Sjö district of Vadstena, Sweden (Erik Asmussen, architect) is built using 45cm-thick solid lightweight concrete. A homogeneous, relatively heavy wall was wanted, and lightweight concrete was chosen despite its low insulation value. The roof structure is insulated with cellulose fibre and there is foam glass under the ground slab.

Source: adapted from Leif Qvist

Heat Storage

Lightweight buildings can be heated quickly, but the heat is also rapidly lost when the heating stops. Heavyweight buildings store heat or cold and even out temperature fluctuations, especially over a 24-hour period. Lightweight buildings are often well insulated, and heavy materials have a poorer insulation value. A building with light, well-insulated outer walls and heavy, heat-regulating inner walls is a good combination.

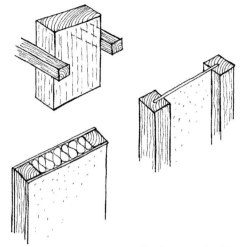

One way to build well-insulated walls and roofs with a minimum of building materials is to use lightweight studs.

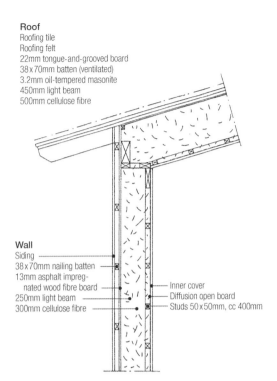

Roof
Roofing tile
Roofing felt
22mm tongue-and-grooved board
38 x 70mm batten (ventilated)
3.2mm oil-tempered masonite
450mm light beam
500mm cellulose fibre

Wall
Siding
38 x 70mm nailing batten
13mm asphalt impregnated wood fibre board
250mm light beam
300mm cellulose fibre
Inner cover
Diffusion open board
Studs 50 x 50mm, cc 400mm

Lightweight beams are made from studs and wood fibreboard, which means that a small amount of building material supports the building. There is a plenty of space for insulation. This kind of structure often leads to reduced costs.

Source: Bertil Jonsson, BOJARK Arkitekt & Konstruktion AB, Täby, Sweden

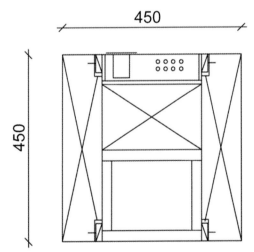

Architect Beat Kämpfen's office building, made using special solid-wood technology for Marché, Zurich, has a pillar system that is part of the load bearing system, and contains all the installations for ventilation, water, electricity and data.

Heavy Construction

Positive characteristics of heavy structures such as brick, lightweight concrete and clay are durability and heat- and moisture-storage capacity. There are also heavy wood structures where sawn wooden boards are joined together to make solid boards. This system was developed in Switzerland by Professor Julius Natterer, and the method is used increasingly elsewhere.

Light Construction

As highly insulated buildings require thick walls and roof structures, a lightweight stud system is often used. In this way, lightweight and well-insulated structures are achieved using a minimum of materials. Lightweight studs are made from wood and wood fibreboard.

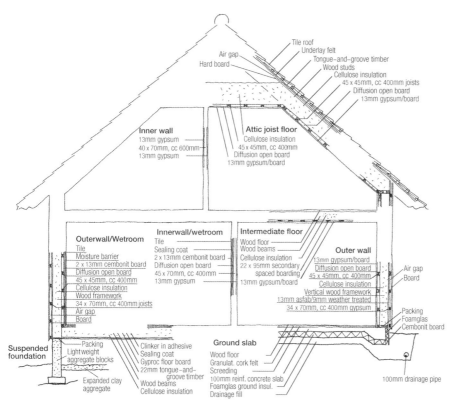

Wood is an eco-friendly and good building material. Timber frame construction is often used in order to economize on wood. This construction is by Bertil Johnsson, BOJARK Arkitekt & Konstruktion AB, Täby, Sweden.

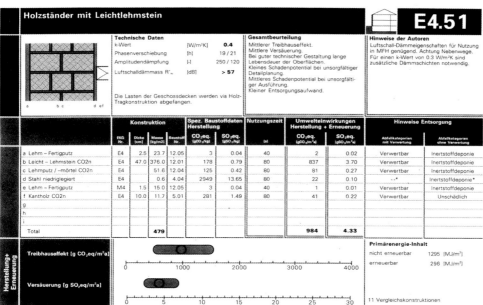

The Swiss compare equivalent structures instead of materials, and account for environmental impacts in the various phases of the life cycle.

Source: Hochbaukonstruktionen nach Ökologischen Gesichtspunkten, Schweizerischer Ingenieur- und Architekten-Verein, Dokumentation D0123, Septemper 1995

Environmental Profiles

In Switzerland, life-cycle analysis has been carried out on different parts of a construction, such as the inner and outer walls, floor structure, etc., so that the environmental impact of various types of construction may be compared. A list is made of the materials in 1 m² of a building component and the sum of their environmental impact is calculated, including greenhouse effect in grams CO_2/m^2, acidification in grams SO_2/m^2, and energy consumption in MJ/m^2. Furthermore, building components are subjected to the following eight criteria: emissions, building process, toxicity, maintenance, repairability, recyclability of the waste, waste value and other comments. Similar environmental profiles have been carried out by the Danish Building Research Institute (By og Byg/Statens Byggeforskningsinstitut).

Combined Materials Tests

How do various materials work together? Even though individual materials may be environmentally approved, under certain circumstances they may affect a construction negatively. This is particularly the case if moisture is involved. If, for example, flooring is going to be glued onto concrete, the concrete has to be dry enough, the glue must be of a type that is not damaged by the alkalinity of the concrete, etc.

Moisture Properties of Materials

Another consideration when choosing materials is whether they will be exposed to moisture. Materials placed outdoors and in wetrooms must be able to tolerate moisture. Moisture migration and relative humidity also need to be considered. Warm air can carry more moisture than cold air, which means that structures dividing warm from cold must be thought through (see also 1.3.1 Damp).

Moisture Migration

In a heated building, moisture migrates from the inside out, from a warmer space to a colder one. The moisture level must not exceed the wall's capacity to handle moisture. Well-insulated walls should therefore be made with an inner layer that is more vapour-tight than the outer layer so that it is easier for moisture to get out than in. This is especially important in buildings with a large amount of evaporation from people, washing, drying, bathing, plants, etc. Some of this moisture can escape the building through ventilation. Another method of reducing the moisture load is to install a dehumidifier.

Many construction handbooks assume that sheet metal will be used for both top and bottom window edge flashing. However, this is a modern assumption as sheet metal was not used in older buildings. Heartwood with a protective coating such as good quality linseed-oil paint works well to protect windows from both above and below.

Source: Photo: Hus Torkel, Kovikshamn, Bohuslän, built 1997–2001. Architects: Maria Block and Varis Bokalders

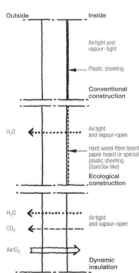

Different types of construction for walls and roofs with regard to moisture movement.

Vapour and Airtightness

Distinction is made between different types of constructions according to moisture movement. Examples are: vapour-tight with a plastic or metal vapour barrier that prevents both air and water from passing through, vapour permeable but airtight where the airtightness is achieved by using paperboard, hard fibreboard or special plastic, as well as air permeable, which is used with dynamic insulation where throughput of air is regulated with perforations in pasteboard.

Moisture-Buffering

There is often a surplus of moisture indoors when people are showering, preparing food or doing laundry, and a deficit during a large part of the day. Some materials can buffer moisture, i.e. absorb moisture when it is humid and release moisture when it is dry, and so even out indoors humidity. Clay is regarded as the best material in this regard and is sometimes used indoors for walling or plaster, usually with the purpose of improving the interior climate. Wood also has a relatively good moisture-buffering capacity. Wood-wool cement board in bathroom ceilings buffers moisture, and mirrors in such bathrooms do not mist up. Silica gel is used in museum display cases since it has an extremely high moisture-buffering capacity. It is also used in medicine containers and for transport of electronics.

When considering 24-hour buffering, only a few centimetres of material are involved. In annual buffering from summer to winter, the entire solid wall is involved, e.g. solid timber or brick walls.

Wetrooms

Wetroom design requires extra care in order to avoid moisture damage. Good practice is to have local ventilation, i.e. moisture is vented out directly from a shower cubicle with a ceiling. The most environmentally friendly method of building a wetroom is to make it using only mineral materials, which don't go mouldy or rot, so no waterproof layer is required. In light constructions, however, a waterproof layer cannot be avoided.

Bathroom with marble splash areas and clay walls as moisture-buffering material. Marble has a high pH value and is antiseptic.

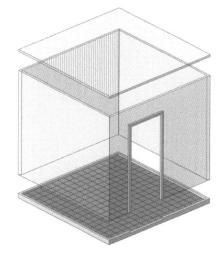

This wetroom is a moisture-repellent unit within the building, a watertight cement element with an embedded floor drain and concave mouldings. The walls can be covered with ceramic tiles or glass sheets, clinker slabs or cement mosaic (terrazzo) can be put on the floor and wood-wool cement board, which buffers moisture, can be used on the ceiling.

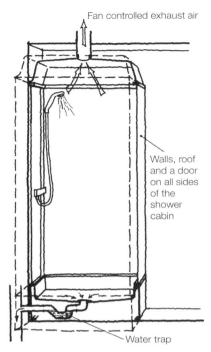

A shower cubicle can be made as a separate room within the bathroom. It can have its own ceiling with local ventilation since it has such a high water vapour load.

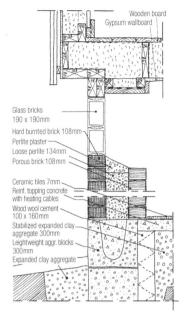

Outer wall of the wetroom at Grorud building, Oslo, Norway, designed by architect Björn Berge. The wall is a brick cavity wall insulated with loose perlite. The façade brick is highly fired and the interior brick is porous (lightly fired). So the wall has no waterproof layer. Instead the porous brick absorbs and releases moisture. Only the bathroom is made of mineral materials. The rest of the building is made of wood.

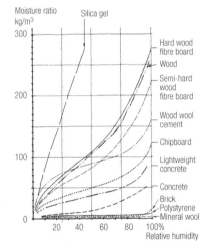

The moisture-buffering capacity of a material affects the interior climate. Therefore, an attempt is made to choose materials that can store moisture. Waterproof finishes or surface coats should be avoided so that the moisture storage capacity is not limited.

Source: This figure was put together by the authors using information from *Fukthandboken*, Nevander and Elmarsson

A shower stall designed as a room within a room may be good approach in a bathroom. It is optimal if the exhaust air is ventilated directly from the shower area.

Due to the many moisture problems in wetrooms in light constructions the regulations have been changed. The waterproof layer must have a water-vapour resistance greater than 1 million s/m (measured under realistic conditions). Gypsum boards with a paper layer should not be used in wetrooms. Floors must slope towards the floor drain (at least 1:100, preferably 1:50). The waterproof layer should not be put directly on floor boards (though if it is, screed must be placed between the boards and the waterproof layer).

Foundations

There are in principle four different kinds of foundations: basements, suspended foundations, slab foundations on the ground and plinth foundations. From the moisture protection point of view, plinth foundations are safest as they result in the least ground contact. From an energy conservation point of view, slab foundations on the ground are preferred as the floor framework is not exposed to outside air. All these methods have improved in recent years. Conditions have changed as buildings are better insulated and there is a greater awareness of moisture problems.

Basements often have moisture problems due to moisture penetrating through basement walls. A correctly constructed basement is properly drained and protected from moisture by a vapour barrier of plastic sheeting that creates an air gap on the outside of the basement wall. Insulation should be placed on the outside of the load-bearing construction.

Suspended foundations have been regarded as a good type of foundation, though in recent years they have had moisture problems. This is due to better insulated floor structures, which lower the temperature in the suspended foundation. In modern suspended foundations, the ground is insulated and a vapour barrier is laid to prevent ground

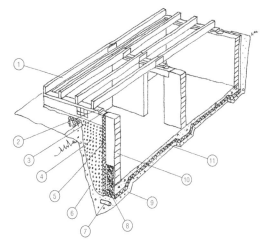

Basements must be well drained. If insulated, the construction must be insulated from the outside. Basements are considered expensive, especially if blasting is necessary, and if it isn't otherwise necessary to excavate material from under the building.

1   Floor joists of structural timber
2   Ground plate of structural timber
3   Moisture barrier
4   Air-gap-creating plastic sheet
5   External thermal insulation, moisture tolerant
6   Backfill
7   Drainage pipes
8   Exterior moisture protection (asphalt coating)
9   Draining and capillary-breaking layer under the bottom slab
10 Basement wall of concrete or lightweight concrete block, alternatively reinforced concrete cast in situ
11 Bottom slab of reinforced concrete

Source: adapted from Träinformations *Träbyggnadshandbok*

moisture from coming up. Organic material should be avoided in a suspended foundation. The underside of floor structures should be made of moisture-tolerant material, such as foam glass sheets or wood-wool cement.

A tried and tested method of making suspended foundations is to ventilate them with interior air instead of outdoor air. This is referred to as a warm suspended foundation and requires insulating the suspended foundation from the ground and at the sides. There are warm foundations that are actively ventilated with exhaust air, and there are shallow, warm foundations where the air in the gap under the floor is in contact with interior air via small cracks and leaks. In the latter case, the floor is not insulated and the air gap under the floor in the foundation is at the same temperature as the interior air.

Slab foundations on the ground used to be insulated above the cement slab, which often resulted in moisture damage. These days, slabs are insulated underneath to keep out moisture. Insulation around the edges is extra-important and sometimes the ground around the slab is insulated so that frost cannot penetrate under the slab and damage it. Well-designed drainage with a capillary breaking material is extra-important for slab foundations on the ground.

If the bottom slab is made of stiff insulation material, such as foam glass, concrete is not needed in the construction. The building, that is the floor and walls, may then be built directly on the foundation.

Plinth foundations usually consist of concrete plinths with wooden joist, on top, on which the building is constructed. An advantage of plinth foundations is that air can move freely under the building and so there is no risk of moisture damage. But the floor is exposed to outside air which creates an additional surface through which heat is lost. In recent years, for constructions in hilly terrain, high plinths or pillars have been used in order to avoid intrusive impact on the terrain.

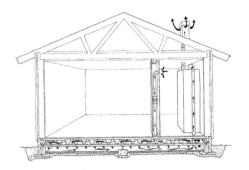

A foundation system with a warm suspended foundation. Heated exhaust air is directed down into the suspended foundation to repel moisture, and so the foundation is protected from moisture.

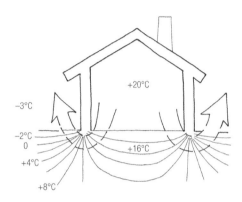

A heat cushion builds up under a building with a slab foundation on the ground. With such a foundation, heat can only escape from under the outside walls.

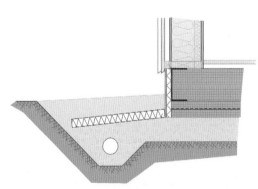

The Koljern foundation method, with a ground slab made of cellular glass panels. As a thin aluminium sheet is part of the construction, it is totally impervious to radon and ground odours.

Source: www.koljern.se

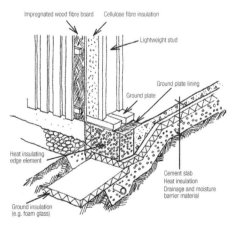

Slab foundations on the ground are considered to be reliable if they are well drained and there is insulation under the slab. This is the most energy-efficient foundation system, especially with successful avoidance of thermal bridges.

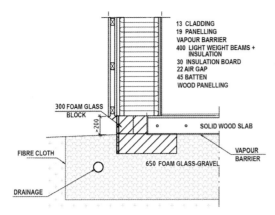

A basic construction using foam glass-gravel and blocks where the thermal bridge in the side beams has been eliminated.

Source: Status preliminary drawing, Simone Kreutzer, Tyréns.

Structures for Dismantling

Catarina Thormark (PhD in Technology, Division of Construction Management, Lund University) has in articles and a book described the advantages of designing buildings for dismantling and recycling. Development of this concept has taken place since the late 1980s and mostly in Europe (Denmark, Germany, The Netherlands), the US and Australia. The advantages in brief follow.

Economics:

- the cost of waste management and raw materials is increasing;
- the recycling potential of buildings will be included in the environmental evaluation of buildings;
- buildings that can be dismantled will have a higher residual value.

Flexibility:

- large changes have taken place and are taking place in population development and household structures;
- there is a tendency for buildings to be torn down before the end of their technical lifetime.

Environment:

- shortages of raw materials;
- about one-third of all waste produced in society is made up of construction and demolition waste;
- pure material fractions are easier to recycle;
- materials production makes up a large part of a low-energy building's resource consumption;
- the influence of climate change on location of settlements.

Examples of some systems made for dismantling are screw piles that can be screwed down into the ground instead of being pile-driven, building blocks (similar to large pieces of Lego), timber connectors that use click systems, beam and joist systems that can be dismantled for installations and flooring, and tape for gypsum wall board and panelling that makes dismantling and reuse possible.

Following is an example of designing for dismantling. The Norwegian architect Björn Berge has developed a flexible construction system that can be dismantled, rebuilt and extended. It is based on three principles: (i) the various construction elements are separate; (ii) every component is easy to dismantle; and (iii) every component has a standard size and is made of homogeneous material.

Nothing is glued or nailed. All joining is carried out mechanically using bolts, nuts, screws and clip-on brackets. The components consist of solid material and not glued composites.

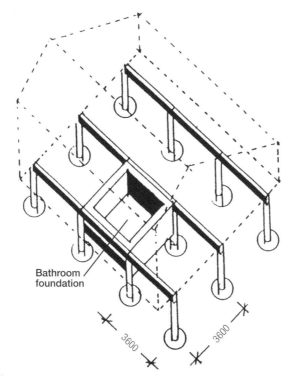

Bathroom foundation

The various parts of a building in Norwegian architect Bjørn Berge's system are divided into the foundation, supporting structure, shell and services.

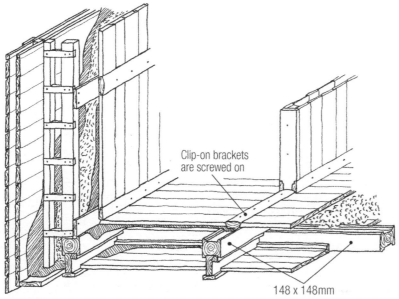

Clip-on brackets are screwed on

148 x 148mm

Example of architect Bjørn Berge's building system.
The outside walls consist of:

pine heartwood battens 21/98mm
pine heartwood exterior boarding 21mm
studs 48/48mm
boarding 21mm
unimpregnated cellulose board
cellulose insulation or wood shavings 250–450mm
cellulose board, linseed oil treated
interior boarding 21mm

Art installation at the Building Exhibition, BO01, in Malmö, Sweden, 2001

1.2 Services

Development of services has meant they are now more numerous, more expensive and more complex. They are installed in the following order: heating, electricity, water and sewage. Ventilation is added rather late in the process, particularly in smaller buildings. The latest addition is information technology (IT) services for both outside communication and for controlling other services. The aim must be to choose services that provide a good interior climate while simultaneously conserving resources. The system should function without too much management or special expertise.

1.2.0 Interior Climate

There are many factors affecting interior climate, and constructing a healthy, comfortable building must be well thought through. Consideration needs to be given to the location of the building, especially with regard to air pollution from traffic and industry. Building materials may release unhealthy emissions. The building's services affect interior climate, with the ventilation system exerting the greatest influence. Electrical systems affect the electro-climate. Heating and cooling systems affect comfort. Water and sewage systems cause no problems until they leak. Even activities that take place inside a building, such as smoking, significantly influence the air quality.

Comfort

Comfort can be divided into the following categories: thermal comfort; humidity; air, sound and light quality; electro-climate and ease of adjustment. With regard to thermal comfort, the term operative temperature is used, which takes into consideration room temperature, surface temperature draughts and temperature variations within the room. Relative humidity has to do with air moisture levels that are neither too dry nor too moist. Optimal comfort is achieved between 40–60 per cent relative humidity. Air quality addresses fresh air and quantities of air, as well as smell, dust, particles, emissions and fibres. Sound quality has to do with sound level, speech comprehension, reverberation time, echo, as well as noise, impact sound, infrasound and vibrations. Light quality has to do with lighting, light strength, colour reproduction, availability of daylight and sunlight, as well as glare and reflections. The electro-climate depends on electrical and magnetic fields as well as static electricity. Ease of adjustment has to do with being able to control the interior climate, i.e. temperature, ventilation and light, an important comfort factor for many people.

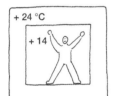 

The temperature of the walls influences how the room temperature is perceived more than the air temperature. If the walls are warm but the room air cold, the room is experienced as warmer than if the air temperature is warm and the walls are cold.

Source: Architects Eva and Bruno Erat, Finland

The most important comfort parameters are temperature and humidity. It's impossible to please everyone; 10 per cent always complain and therefore it is good to be able to control our own interior climate.

Control and Regulation Systems

Interior climate may be adjusted either manually or automatically. Manual regulation includes opening and closing windows and ventilation outlets, as well as turning the heating up and down. This can be done using simple methods but requires participation by the occupants. Automatic regulation has developed considerably in recent years, concurrent with computer development. Current regulation systems are usually connected to a local computer control system that controls everything affecting interior climate such as ventilation, heating, cooling, lighting, etc. In principle, a building is equipped with various sensors that monitor temperature, air humidity, pressure, light conditions, etc. All the sensors are connected to the computer control system programmed to control dampers, vents, lamps, etc. The goal is to achieve a good interior climate while conserving energy and other resources. New, efficient regulation equipment can save up to 10 per cent of energy consumption in multi-family dwellings, and 20 to 30 per cent of energy consumption (heat and electricity) in office buildings in

Interior comfort is influenced by many different factors, including air cleanliness, temperature and humidity as well as light conditions, sound levels and electro-climate.

- Radon
- Mould
- Dust
- Tobacco smoke
- Fibres
- Formaldehyde
- Bacteria
- Viruses

Interior air may be polluted by a variety of sources, including unhealthy and unpleasant contents.

INTERIOR CLIMATE REQUIREMENTS

Ventilation experts Torkel Andersson and Håkan Gillbro, specialists in reinforced natural ventilation, have made the following list of requirements for interior climate in schools and offices:

- equivalent room temperature 20–21°C
- relative humidity 30–40 per cent during the cold season and 40–60 per cent during the warm season
- low static electricity

- no audible or inaudible sound
- low concentration of particulate substances
- low concentration of gaseous substances
- good and pleasant lighting and lighting level
- good natural lighting

Sweden. For every degree that the average interior temperature can be lowered, there is a 4–5 per cent savings in total energy consumption.

When functions in a building are computer-controlled, these functions can be extended. Buildings controlled by such systems are referred to as 'smart' or 'intelligent'. Functions that can be added include burglar alarms, fire alarms, overflow alarms, humidity alarms, motion detectors to adjust ventilation or lighting, magnetic switches that indicate whether or not windows or outer doors are open, timers, etc. A display screen can be connected where it is possible to regulate the system, and to observe electricity and water consumption, interior and

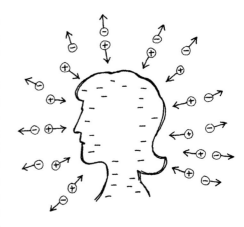

A human being can be electrostatically charged. A negatively charged person attracts positively charged dirt particles. A shock is received when an electrostatically charged person touches an earthed object.

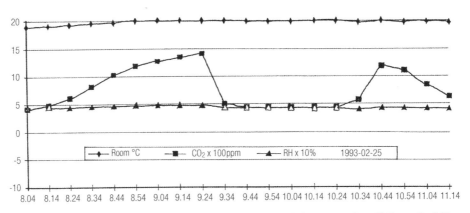

Fredskulla school in Kungälv, Sweden (designed by architect Christer Nordström) has natural ventilation, adjustable as required. Climate monitoring shows that they succeeded in keeping the room temperature at about 20°C and the relative humidity at about 50 per cent. During the winter, the level of CO_2 sometimes exceeded 1500 parts per million (ppm). That level does not affect health but exceeds earlier recommendations. When staff and pupils take regular breaks and open the windows, the CO_2 level can be quickly lowered.

Source: Ventilation consultants Torkel Andersson and Håkan Gillbro, DELTAte, Gothenburg

outdoor temperature, who's standing at the front door, and to reserve a slot in a shared laundry room, etc. When spring sun shines in through a window, a good control system would immediately detect the free heat and lower the output from the radiators. Opening windows for a short time may cause cold air to sink and trip a thermostat that regulates heating. A smart system can be programmed so that heating is not turned up during this short period. In a waterborne system, old thermostats can be replaced with new ones that are part of the control system.

Such systems open up completely new possibilities for property managers. Systems can be connected together, functions checked, problems detected, and comparisons made between buildings to determine where various energy-efficiency measures should be used.

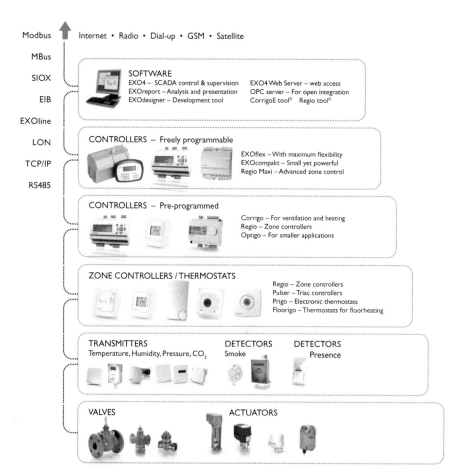

A computerized control and regulation system.

Source: Copyright AB Regin.
Illustration by Klas Eriksson,
www.wendelbo.com

Regulation System Requirements

A control and regulation system should meet strict expectations for comfort and energy efficiency. Here is a checklist of requirements for a good system.

- The output and level of heat from radiators should adjust quickly and precisely to changes in the surroundings.

- The system should have window switches to turn off radiators when windows are opened, i.e. a function that allows the room to be aired without the radiators needlessly turning on to compensate for a sudden drop in temperature.

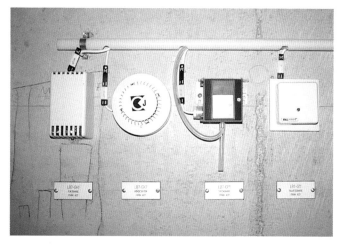

Various sensors that monitor indoor climate, including sensors for air humidity, smoke, pressure and temperature.

- It should be possible to set a weekly schedule for desired room temperature that includes daytime and night-time decreases and weekend requirements. It should also be possible to set monthly schedules for occupants' long absences, with a low temperature that is automatically raised before their return home.

- It should be possible to divide the building into at least four zones which can be regulated or to regulate each room with different temperatures. For example, bedrooms could be kept at 18°C, the living room at 19°C and the kitchen at 21°C during the time each respective zone is occupied.

- The control system should be easy to install and programme. It should not require daily inspection.

- An agreement should be reached with the electricity supplier to set a ceiling on electricity consumption during temporary grid overloading and consumption peaks. This is called 'load control' and is managed by the supplier. The advantage of load control is that the electricity grid and the whole production apparatus does not need to be adjusted for temporary consumption peaks. Home owners who agree to load control receive a lower, set electricity rate.

Computer control system for ventilation and heating at the office of pharmaceutical company Novartis, Täby, Sweden.

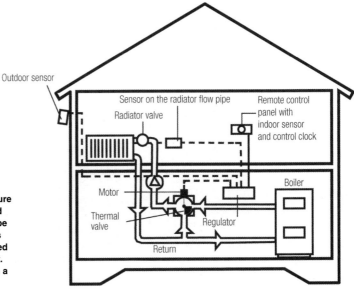

There are regulation systems with temperature reduction, weekly programming, interior and outdoor sensors, or with a sensor on the pipe supplying the radiators. Interior thermostats should be placed so that they are not affected by airing. Simple systems usually work best. For example, indoor and the flow sensors in a single-family house.

1.2.1 Ventilation

Ventilation is the service system that has the greatest effect on interior climate. Ventilation also creates most irritation and gives rise to most complaints. Solutions to this problem, i.e. chosen ventilation systems, have varied over time with in principle a new strategy every decade. This is an indication that ventilation is difficult and that in order to achieve a good interior climate in an energy-efficient and sustainable way the ventilation system must be carefully planned.

Why Ventilation?

Buildings are not ventilated in order to meet human oxygen requirements. It is difficult to construct a building that is so airtight that an oxygen shortage would arise. This only occurs in such spaces as submarines and bank vaults. Ventilation is not necessary because of CO_2, though the CO_2 level is often monitored. It has been recommended by worker safety organizations that levels not exceed 1000ppm. Monitoring of CO_2 only shows air turnover per person in relation to the volume in a particular room. However, research has shown that people and health are not significantly affected at levels up to 5000ppm CO_2, and up to 10,000ppm CO_2 is allowed in submarines. Today, natural ventilation systems are not only assessed according to CO_2 levels.

So ventilation is primarily required to:

- adjust the relative air humidity (preferably 40–60 per cent);
- remove excess heat (especially in schools and offices);
- remove body odours and emissions (especially from materials in the room).

Pollution of Interior Air

The Danish researcher Ole Fanger has studied sources of interior air pollution. It can be concluded from his studies that the ventilation system itself and the air intake ducts pollute the most, followed by inappropriate activities such as smoking. Emissions from materials are also a large source of pollution.

Conclusion: if healthy construction materials are used, smoking indoors is avoided, and the air intake valves in ventilation systems are not polluted, good interior air can be achieved despite a decrease in the ventilation requirement.

Reducing the Need for Ventilation

Heat surplus can be reduced by using sun blinds, efficient lighting and heat-storing materials. Moisture surplus can be reduced through water-use management, local ventilation (e.g. in a shower cubicle or kitchen cooker hood) and use of moisture-buffering materials. Chemical emissions can be reduced by carefully selecting materials, surface coatings and cleaning chemicals. Body odours can be reduced through good hygiene and by having flowers and green plants in a room.

If these steps are taken, ventilation is required primarily to remove surplus heat and regulate relative humidity. Both these factors are affected by the climate, weather and seasons as well as the number of people in a room. So ventilation should be adjusted as necessary in order to avoid over-ventilation (which is a waste of energy).

Air Movement in Rooms

Warm air rises, and so there is an ascending air current at a radiator. Cold air sinks, so there is usually a cool downdraught at windows. It is common to place radiators under windows to counteract the downdraught. In such a room air moves like a wheel, up the outer wall and down at opposite inner wall, which mixes the air well. If windows have a low U-value, the radiators can be placed on the opposite inner wall and air will circulate in the opposite direction. It takes a long time for air flows of different temperatures to mix.

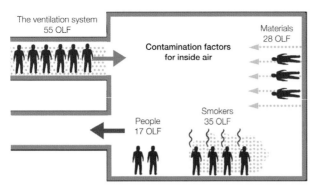

Ole Fanger has developed a unit, called OLF, which shows a person's emissions to the interior air. The figure shows average sources of pollution in 15 offices in Copenhagen. An average of 17 people worked in each office.

Source: Professor Ole Fanger's research from the 1980s. (*ASHRAE Journal* November 1988)

How this takes place is affected by where the intake air is added and the exhaust is extracted. Distinction is drawn between displacement ventilation where air intake is at a low level in a room and mixing ventilation where intake air is introduced at a high level.

There are many factors affecting air movement. An open door, for example, provides much more air exchange than a fresh air duct. It is important to prevent 'short circuits', where incoming air goes straight out the exhaust air vent. A ventilation system that functions well distributes fresh air evenly throughout the room.

Ventilation Methods

Natural ventilation was the most common type of ventilation system during the 1950s and 1960s. A problem with natural ventilation is that it is uneven. Sometimes there is too little and sometimes too much. Mechanical exhaust air ventilation systems became common during the 1960s and 1970s. Ducts were connected together and a fan was installed in the exhaust vent. This guaranteed a minimum amount of ventilation, but there was too much ventilation when natural ventilation and the fan worked together. In addition, fan noise could be problematic. After the oil price increase in the 1970s these systems were considered too energy-intensive due to increased heating costs.

An open door contributes more to air exchange than ventilation ducts.

Mechanical ventilation systems with an air-to-air heat exchanger were introduced in the 1970s and 1980s in order to reduce energy consumption. The systems were expensive, complicated and had many problems, such as fan noise, intake ducts becoming dirty and affecting air quality, soiled filters affecting flow adjustment, air leakage, condensation and freezing in the heat exchanger, etc. A report at the end of the 1980s showed that it rarely paid to put such systems in single-family homes.

Exhaust air heat pumps became common during the 1980s and 1990s. Exhaust air ventilation was reintroduced but the system was supplemented with a heat pump that took heat from exhaust air and used it to heat water in the hot water heater. These systems were also expensive and complicated. Furthermore, many of the heat pumps contained freons, which are environmentally hazardous.

Fan-reinforced natural ventilation was introduced in the 1990s. Natural ventilation returned but was designed so that flow adjustments could be made. Thermostatically controlled air terminal devices were added to reduce over-ventilation during cold periods and fans were installed that forced the flow during warm periods or cooled the building at night in order to achieve a good interior climate during the day. In large buildings, such as schools and offices, sensors and computer systems are used to control ventilation flow. Ventilation can be adjusted according to climate and building population requirements. Natural ventilation must cooperate with a building, the architect and the ventilation consultant must cooperate, and architectural design (such as ceiling height, room position, roof design, monitors, chimneys and towers) becomes part of the ventilation system.

Ventilation Systems

Previously, ventilation systems were purely natural, but in recent decades fully mechanical ventilation systems have become common. The latest approach is to combine natural and mechanical ventilation. These systems are called reinforced natural ventilation, fan-reinforced natural ventilation or hybrid ventilation systems.

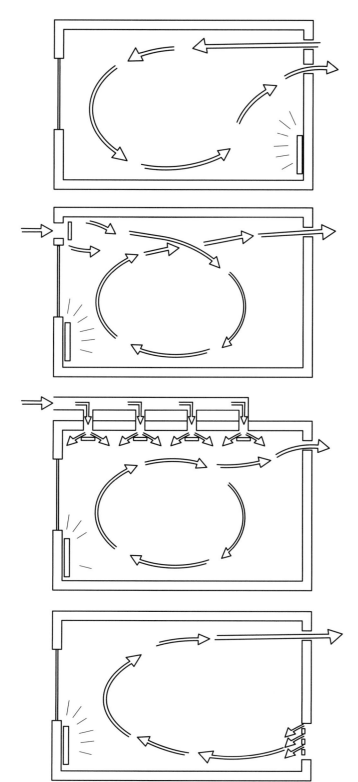

There are different points of view about what is best, displacement or mixing ventilation. The top figure illustrates air movement if intake air is introduced at an inner wall, if the windows are energy-efficient and the radiator is on an inner wall. The second figure down shows air movement when intake air is introduced through vents above the windows. The third figure down illustrates a variation of mixing ventilation where low-temperature fresh air is 'sprinkled' into the room from several fixtures in the ceiling. This results in good mixing and very cold intake air can be added to a room without there being a draught. The system is called the Mölndal model. The bottom figure illustrates displacement ventilation where air enters at a low level on an inner wall, radiators are located under windows and the exhaust is removed at high level.

1.2

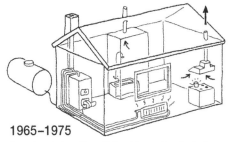

1965–1975
Natural ventilation with ventilation chimneys from the bathroom and kitchen plus a kitchen fan.
Waterborne heating system with an oil heater.

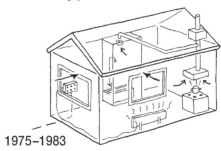

1975–1983
Mechanical, continual exhaust air ventilation with direct electric heating.

1981–
Mechanical intake and exhaust air system with an air-to-air heat exchanger. Direct electric heating.

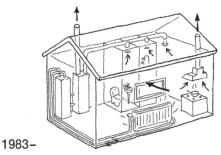

1983–
Mechanical ventilation with an exhaust air heat pump. The heat pump extracts energy from the exhaust air and provides heat to the hot water heater and the heating system.
Waterborne heating with an electric emersion heater for additional heat when required.

Typical ventilation systems for houses in Sweden.

Source: Lee Schipper

PRINCIPLES OF ENVIRONMENTALLY FRIENDLY VENTILATION SYSTEMS

- It should be possible to adjust the air flow in every room according to season, weather and temporary needs.
- Air in the intake air system should be treated as little as possible.
- It should be possible to add intake air without creating a draught, even during the cold season.
- The heating system should only heat if the temperature in a room is below the desired level.
- The heating and ventilation systems should be soundless.
- Interaction between the building and the ventilation system should work well.
- The system should include a minimum of technical components that can cause problems.
- The system should be easy to inspect, clean and wash.
- The system should work together with natural ventilation.

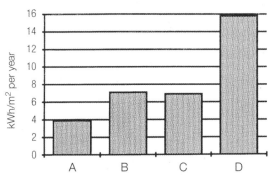

A Mechanical exhaust-air ventilation
B Mechanical ventilation with an exhaust-air heat pump
C Mechanical intake and exhaust-air ventilation
D Mechanical intake and exhaust-air ventilation with an air-to-air heat exchanger

Electricity consumption for different ventilation systems in blocks of flats.

Source: Effektiv ventilation, BFR rapport T11:1994)

1.2

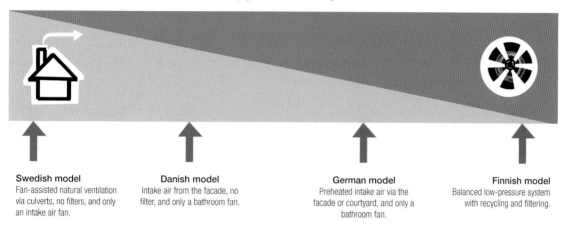

| Natural ventilation | Natural and mechanical ventilation (hybrid ventilation) | | Mechanical ventilation |

Swedish model
Fan-assisted natural ventilation via culverts, no filters, and only an intake air fan.

Danish model
Intake air from the facade, no filter, and only a bathroom fan.

German model
Preheated intake air via the facade or courtyard, and only a bathroom fan.

Finnish model
Balanced low-pressure system with recycling and filtering.

Previously either natural or mechanical ventilation was used. Currently, more and more systems combine natural ventilation with mechanical ventilation. These are sometimes called hybrid systems. The more to the right a system falls in the above figure, the more technology is involved, with associated energy consumption, noise, problems and risk of contamination of intake air.

Source: Bygningsintegrert ventilasjon – en vejleder, ØkoByggprogrammet – program för en miljøeffektiv byggebransje, Norge, 2003

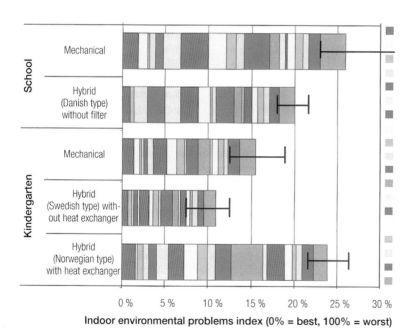

Indoor environmental problems index (0% = best, 100% = worst)

A Norwegian study of ventilation systems showed that users were most satisfied with fan-reinforced natural ventilation systems and least satisfied with purely mechanical ventilation systems.

Source: Bygningsintegrert ventilasjon – en vejleder, ØkoByggprogrammet – program för en miljøeffektiv byggebransje, Norge, 2003

I HEALTHY BUILDINGS | I.2 SERVICES | I.2.I VENTILATION

Natural Ventilation Forces

Natural ventilation is driven by thermal and wind forces. Thermal forces arise partly due to the difference in pressure between colder, heavier outside air and warmer, lighter interior air (warm air rises), and partly due to pressure differences at different heights (stack effect). Wind forces arise partly due to the presence of positive pressure and negative pressure on different surfaces of the building (wind pressure), and partly due to the suction power created by the wind flowing over the chimney opening (the Venturi effect).

The amount of thermal force depends on the difference between the inside and outside temperatures, and the difference in height between the top of the chimney and the air intake. The amount of wind force depends on wind speed and direction, the shape and size of the building, as well as the design of the chimney opening.

Natural ventilation works when there are small pressure differences. For purposes of comparison it can be pointed out that a circulation pump works at about 2000mm wc (water column), a fan at about 200mm wc, while a natural ventilation system works at about 0.2–1mm wc. About half the thermal force is generated from difference in temperature and about half from the height of the chimney. Wind force is usually the dominating force in a natural ventilation system. However, it is not difficult to create enough flow in a natural ventilation system if the ducts are big enough. The difficulty is adjusting the ventilation. It is a question of limiting the excess ventilation produced by wind and thermal forces during the winter, and having a good interior climate on warm summer days with the help of cross-draughts, open windows and monitors, as well as underground ducts, night ventilation and other methods of 'passive cooling'.

Older Natural Ventilation Systems

High ceiling heights together with warm chimney stacks from tile stoves in each room and leaks around windows resulted in good ventilation. Advanced natural ventilation systems were developed during the late 1800s and early 1900s. These used both intake and exhaust air. Intake air was preheated in special chambers in the winter and humidifying was sometimes used in the summer to cool air. Heating and ventilation systems were often combined. The ventilation systems had many ducts, often embedded in walls, that were adjusted with dampers. The intake air fixtures, which consisted of cast-iron grates in the floor, walls and ceiling, were often beautifully designed and exhaust air exited through elegant ventilation chimneys and towers on roofs.

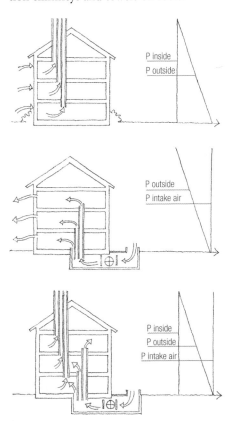

A problem with natural ventilation systems is that there is better draught on the ground floor than on the floor above. This can be compensated for by taking in all air to an intake chamber in the basement, from which intake air to all the floors is distributed. 'P' refers to pressure.

Source: P. O. Nylund, *Kulturmiljövård* 1/91

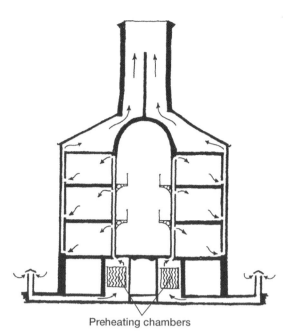

At the end of the 1800s and in the early 1900s it was common for large premises to have natural ventilation systems with preheated intake air and preheating chambers. The drawing above shows Pentonville Prison, London, England in 1844. During the summer, thermal forces were intensified by lighting coal-burning stoves in the attic.

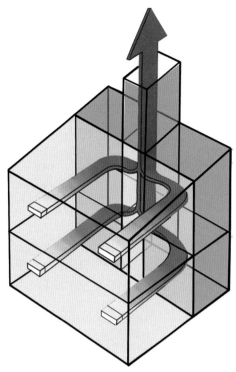

When designing natural ventilation, an architect must think through how air moves through a building.

Source: *Bolig og naturlig ventilation – indeklima energi driftssikkerhet*, edited by Rob Marsh and Michael Lauring, Denmark, 2003

Beautiful ventilation grates for exhaust air at Operakällaren in Stockholm.

Source: Photo: Maria Block

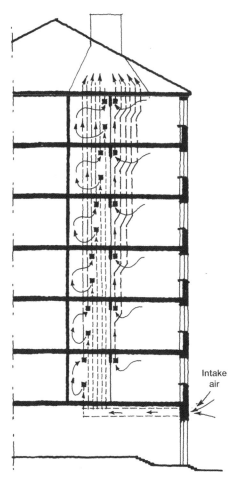

The 'Stockholm ventilation' is a natural ventilation system used in blocks of flats. A duct takes fresh air from the ground floor up to an intake duct at low level in the bathroom. The exhaust air is taken from an exhaust duct at high level in the room via the same duct up to the chimney. The duct is divided up into intake and exhaust sections without contact between the two. Flats may also have additional exhaust ducts.

Designing Natural Ventilation

It is possible to construct natural ventilation systems that work well, but there is a lot to consider.

Airtight buildings with high ceilings facilitate ventilation. Heat-storing and moisture-buffering materials even out the interior climate. Sunscreening and efficient electricity use reduce the amount of internal thermal stress requiring ventilation.

A well-thought-out architectural design can minimize the requirement for ventilation ducts. The floor plan affects air flow in a building, where air is taken in, how interior air flow occurs, and where air is extracted. Air is often taken in in bedrooms and work rooms. The air then goes as transferred air to communication spaces, and is extracted from kitchens and bathrooms. Slots in doors is one way of directing air from one room to another. Relatively large ducts are required for natural ventilation, about 150cm^2 (half bat × half bat brick) in order to ventilate a bedroom. Long ducts and horizontal ducts should be avoided. The ducts should be easy to clean.

Fan-reinforced Natural Ventilation

Fan-reinforced or otherwise controlled natural ventilation means that a natural ventilation system is equipped with automatic control technology such as thermostats, dampers or fans in order to provide better control over ventilation flow. For example it is possible to have thermostatically controlled valves in the air intake ducts in bedrooms, with the air intake placed high up or behind radiators to avoid draughts. Exhaust air chimneys fitted with exhaust air fans that can be adjusted as necessary direct air out of kitchens and bathrooms. At the top of the ventilation chimneys there are hoods to prevent reverse air flow, and/or a spinner that increases and adjusts the ventilation flow. Adjustment can take place manually, for example with a damper, with an 'away' button (when no one is home the ventilation is reduced), or by turning on a fan to force ventilation (when cooking, during parties, etc.). During the summer, ventilation can be increased even more by opening windows.

During the 1990s, a large number of **Swedish schools** were built with natural ventilation systems. In principle, they are built in the following way: air is taken in through an underground duct where it is brought to room temperature – it is

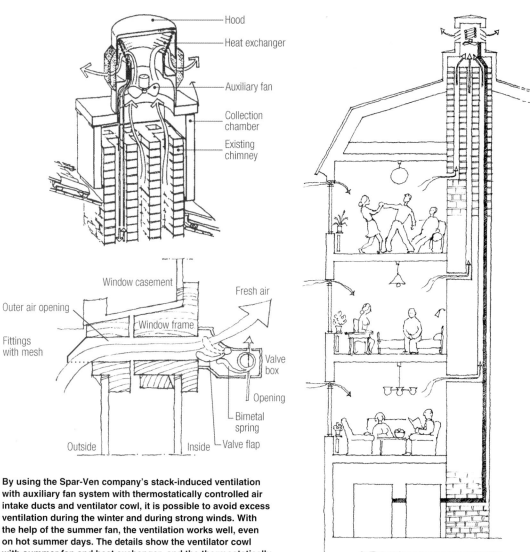

By using the Spar-Ven company's stack-induced ventilation with auxiliary fan system with thermostatically controlled air intake ducts and ventilator cowl, it is possible to avoid excess ventilation during the winter and during strong winds. With the help of the summer fan, the ventilation works well, even on hot summer days. The details show the ventilator cowl with summer fan and heat exchanger, and the thermostatically controlled air intake duct.

preheated in winter and cooled in summer. The underground ducts which are often designed as culverts, and where cables and pipes for other services are placed, are so big that air movement is slow. The slow air movement results in sedimentation of particles, that is they fall to the floor where they can be cleaned away. So the system does not require filters. The culverts are equipped with mesh to keep out small animals. Furthermore, the ducts are installed at an angle and drain so that condensation can be removed, and they are designed so that they can be easily inspected and cleaned. Sometimes there is a fan for ventilation at night, a fire damper, and radiators for preheating intake air in the intake air system. Air is directed up to the classrooms, sometimes through a muffler, and in such a way that draughts are avoided. The ceilings in the classrooms are high, exhaust air is removed through ventilation chimneys and monitors in which windows or hatches can be opened to varying degrees depending on

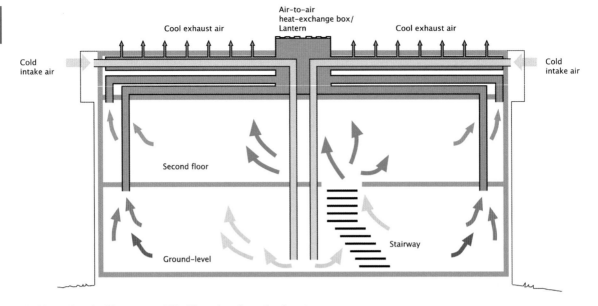

Architects Mauritz Glaumann and Ulla Westerberg have developed a method to reuse heat in a natural ventilation system, a solution they have in their own home in Vidja outside of Stockholm. There is a pipe-in-pipe system in the ridge where the cold fresh air in the inner pipe is heated by the hot used air in the outer pipe.

Detail from the home of Glaumann and Westerberg

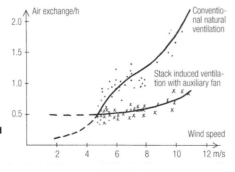

Air exchange rate as a function of the wind speed over the top of the chimney (summertime example) with and without fan reinforced natural ventilation.

the season. In the summer, cross-draughts can be achieved by opening windows. The ventilation is adjusted with dampers and by opening windows. Adjustments can be made manually or automatically.

Interview studies have shown that most people in the schools with natural ventilation are satisfied with the interior climate.

Danish office buildings with reinforced natural ventilation have received a lot of attention. Intake air comes through the outer wall, either through ventilation windows placed at high level or via air intake fixtures. The ceilings are high and since the air intake is high up in the room, the incoming air mixes with the inside air and creates a draught. In some cases, air intake radiators placed by the air intake fixtures preheat the air. From the offices, the air moves through open doors or air transmission fixtures out into the communal areas. Stairwells or courtyards function as exhaust air ducts, and air is released through openable windows or roofs located at the top of the building. The buildings are often constructed using heavy materials in order to even out the interior climate. Adjustments take place automatically via dampers and windows that open

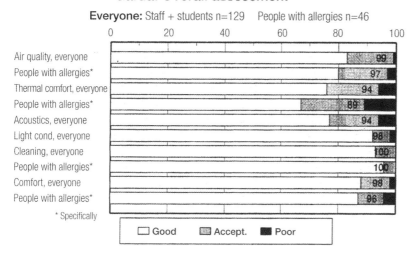

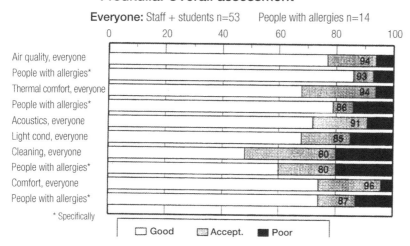

The diagrams show people's experience of the interior environment in Swedish schools Garda in Visby and Fredkulla in Kungälv.

Source: *Skolor med ventilation där självdrag används – Exempel på lösningar och resultat*, Anslagsrapport A11:1997 Byggforskningsrådet, Marie Hult, White architechts AB

and close as required and according to wind direction. Among other things, there are pressure-adjusted air intake ducts. There are also several exhaust air fans located on the roof in case it is necessary to force ventilation, e.g. for night cooling.

Several **German office buildings** have been built so that intake air enters through a double-glazed façade. Sometimes the sheets of glass are placed so far apart that they form a glassed-in space used as an entrance and stairwell. The glassed-in spaces are often designed as greenhouses and filled with green

Schematic section of a building with fan-reinforced ventilation in a culvert (the Swedish model)

Source: Bygningsintegrert ventilasjon – en vejleder, ØkoByggprogrammet- program för en miljøeffektiv byggebransje, Norge, 2003

I HEALTHY BUILDINGS | I.2 SERVICES | I.2.I VENTILATION

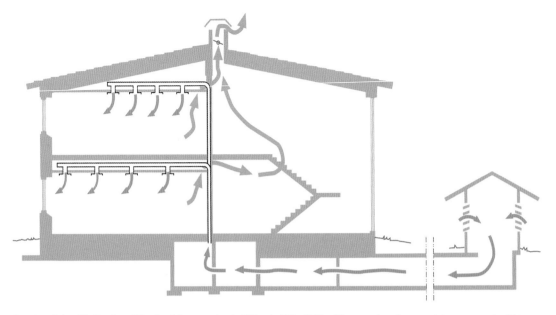

Vargbro School in Storfors, Värmland, is a passive building built in 2008, with several environmental components. The hybrid ventilation system, with air is brought in via a culvert that starts 30m from the façade, is energy efficient and soundless. Exhaust air goes out through ventilation openings in the roof. Fans in the underground culvert can help with air movement if needed. The school is super-insulated and is equipped with a solar cell system that is a little more than 130m² in size and provides 18,000kWh per year. Further, the facility has an advanced computerized operation system.

Architects: K-Konsult Arkitekter I Värmland AB. Ventilation: Torkel Andersson, DELTAte.

plants that clean and moisturize the intake air. From there, the preheated air is taken into the offices. The architecture is dominated by a number of high chimneys that carry the exhaust air away from the building. Chimneys and towers have once again become common architectural features.

Swedish office buildings are usually built with completely mechanical ventilation systems. To save energy, office buildings with large, built-in, heavy, heat-storing frames that balance heating and cooling requirements have been developed. To increase the heat-transmitting surface, cavity walls are sometimes used for intake air and cavity floors for exhaust air. There are systems for cooling and heating in every room that function cooperatively. In such office buildings it is possible to accurately control the interior climate, but the system is expensive and complicated. Even in this type of office building there is nothing to prevent the use of underground pipes for intake air and of chimneys, stairwells or glassed-in courtyards as ducts for exhaust air.

Mechanical Ventilation

It isn't always possible to use natural ventilation systems, e.g. in large kitchens and laboratories, and so mechanical ventilation must be used. Ultra-energy-efficient buildings with heat exchangers for ventilation also require mechanical ventilation. There is an ongoing discussion about what is best: prioritizing energy efficiency and choosing mechanical ventilation, or prioritizing the interior climate quality and choosing controlled natural ventilation, which consumes somewhat more energy. To achieve a good interior climate with mechanical ventilation, extra effort is needed when designing the system.

Air-to-Air Heat Exchangers

An air-to-air system with a heat exchanger is an option in a building designed to be so energy efficient that no heating system is required. In this case, the ventilation system requires a heat exchanger. In such a system, it is especially important to have quiet fans, an easily accessible heat exchanger (so that it is easy to change filters and keep clean), and air intake ducts that are accessible for cleaning. The heat exchanger itself must be highly efficient.

The size of the heat exchanger in passive homes is usually the size of a common tall cupboard.

Intake Air

Intake air is taken from outdoors where it is cleanest, but there are many different ways to introduce air to a room. It is best to locate and design air intake fixtures so that they are affected by the wind as little as possible. Air can be taken directly into a room via slot air vents or air intake fixtures which can be thermostatically controlled to avoid over-ventilation in winter. To avoid draughts, the air intake can be situated high up in the room so that air mixing takes place, or behind a radiator so that the air is preheated. There are also systems that sprinkle in air from several fixtures in the ceiling, and so very cold air can be added without creating a draught. Sometimes intake air is introduced through an air intake duct, either to increase the stack effect (distance between intake and exhaust) then the intake is placed low down, or to introduce clean air and then the air intake can be placed high up. To preheat the air, intake air can be brought in through a glassed-in space or glassed-in veranda, a double-glazed façade, or a solar panel for heating air. Another choice is to introduce air through an underground pipe where it is preheated during winter and cooled during summer. Intake air can also be cooled with the help of evaporative cooling if the air is brought

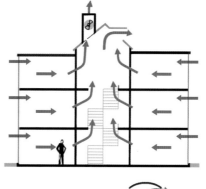

Schematic section of a building with the air intake entering directly via the façade (the Danish model).

Source: Bygningsintegrert ventilasjon – en vejleder, ØkoByggprogrammet-program för en miljøeffektiv byggebransje, Norge, 2003

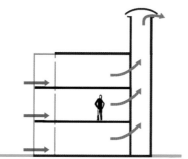

Schematic section of a building with the air intake entering through a double-glazed façade (the German model).

Source: Bygningsintegrert ventilasjon – en vejleder, Øko-Byggprogrammet- program för en miljøeffektiv byggebransje, Norge, 2003

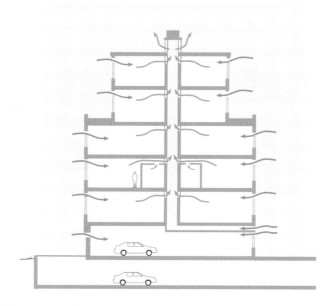

The company Window Master constructs natural ventilation based systems. Advanced window openers (that can be set according to desire) supply fresh air via gaps near the inner ceiling. Exhaust air goes out via ventilation chimneys. Cooling during the night ensures a pleasant indoor climate during the summer. The heavy mass of the building contributes to evening out the indoor temperature.

Source: Office building for Akzo-Nobel in Stenungsund by architect Arne Algeröd, Mats & Arne Arkitektkontor. Ventilation: Torkel Andersson, DELTAte.

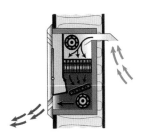
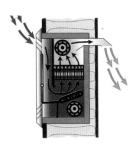
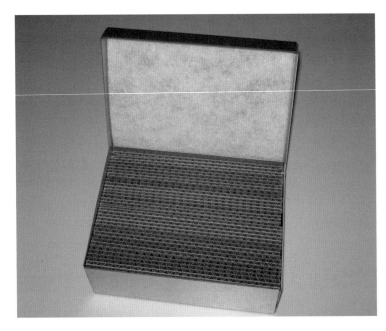

Air Star is a decentralized ventilation system with a unit in each room. It alternately conveys in fresh air, which is heated by the heat exchanger, and conveys out air that heats up the heat exchanger. The heat exchanger is made up of corrugated cardboard and stores that heat through sorption. Water is gathered from the moisture in the exhaust air and absorption and condensation heat is created.

Source: Clean Air System.

in across a sprinkled expanded clay bed, an artificial waterfall or a fountain.

Different methods for air intake can be combined or used differently in summer and winter. A natural ventilation system that is adjusted as necessary must be regulated in some way. Regulation can be carried out manually by opening and closing windows and dampers or by using entirely computer-controlled systems where sensors monitor temperatures, humidity and wind direction, and a computer control system automatically adjusts fans, dampers and preheating. There are mechanically controlled dampers. The most common are thermostatically controlled air intake fixtures. There are also dampers controlled by moisture and pressure.

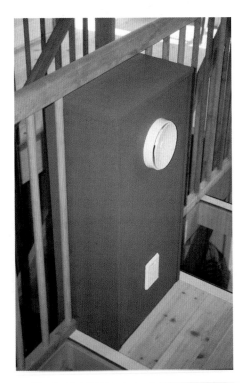

An air intake duct centrally placed in the building. The socket for a built-in central vacuum system is also visible in the photo.

Source: The Solid Wood House at the Building Exhibition, BO 01, in Malmö, Sweden, White Arkitekter AB and Ekologibyggarna AB

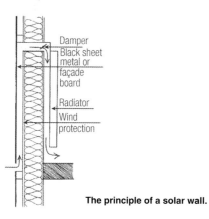

The principle of a solar wall.

The air intake fixture called 'Flipper' from Acticon Ltd is self-adjusting, depending on pressure and amount of air. Counterweighted flaps open under pressure and make sure the air flow is rapid enough to distribute the cool intake air.

Bengt Hidemark is an architect who has applied the preheated intake air method. The spaces between the bottom of the windows and the floors in this home for the elderly in Märsta, Sweden are designed as solar walls.

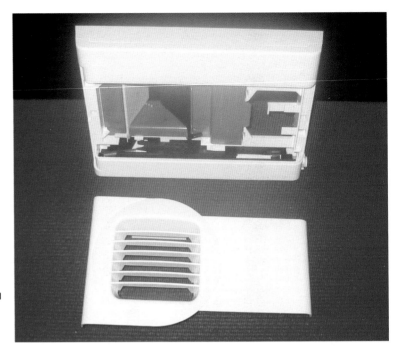

The Exhausto brand moisture-controlled air intake fixture is mechanical. A nylon strip responds to air humidity.

Source: Leif Kindgren

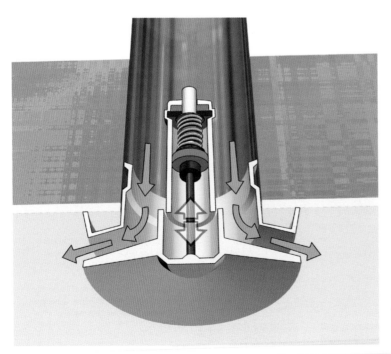

The Velco brand valve has a built-in thermostat that regulates air intake according to outdoor temperature.

Source: Leif Kindgren

The main office of the Pihl & Søner construction company in Copenhagen, Denmark, is naturally ventilated. Intake air is drawn through narrow, high slit windows that adjust automatically. The regular windows are opened and closed manually as required.

Natural ventilation via an underground pipe involves bringing intake air in through an underground pipe to an air intake chamber under the building. From there, separate air intake ducts (that are easy to clean) lead to all the bedrooms and the living room. Air is exhausted from the bathrooms, laundry room and kitchen. Air exhaust ducts go straight up to ventilation chimneys on the roof. Pinwheels at the top of the chimneys force and regulate the draught. Manual regulation with a damper is also used. As the intake air is brought in at a central point in the building, it is not affected by wind pressure, and a more even ventilation is achieved. Furthermore, because downdraughts and draughts from air intake fixtures on the outer wall are not

I HEALTHY BUILDINGS | I.2 SERVICES | I.2.I VENTILATION | 133 a concern, it is possible to install a simpler and cheaper heating system with radiators on inner walls.

Natural ventilation with air intake from above is an option if underground ducts cannot be used or are not desirable. Intake air is brought in via an air intake duct with an opening above the ceiling. The duct works as a 'snorkel' and leads fresh air down to an air intake chamber. Otherwise, the system works in the same way as natural ventilation with an underground duct.

Preheating of intake air can take place in a space designed as a solar wall between the bottom of a window and the floor. Fresh air enters via a gap between the wall's insulation and exterior black sheet metal covered with glass. During the summer, it must be possible to block the air flow so that it does not get too warm indoors. In industrial buildings built of sheet metal, intake air can be brought in behind the metal on the south side of the building (as well as on the roof). The metal is warmed by the sun, especially if it is a dark colour, and warms the intake air in the gap between the metal and the insulation. Radiators can be designed so that they preheat incoming air. In some radiators, warm room air is circulated with a fan and mixed with outside air. This method avoids draughts even when there is a large amount of ventilation.

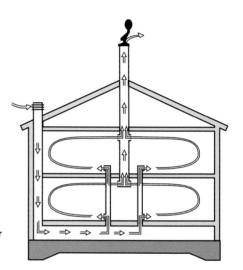

Snorkel ventilation with an air intake above the roof that pulls air down to an air intake chamber under the rooms/premises to be ventilated.

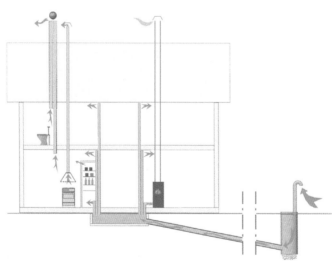

Natural ventilation via an underground duct in a house. The fresh air is taken to an air chamber in the foundation. Easy to clean straight ducts lead from that chamber to every room as well as to the fireplace and larder. Exhaust air is removed through chimneys from toilets, bathrooms and kitchen.

Source: Det Hållbara Huset [the sustainable house] finalist in MiljöInnovation 2002, architects Maria Block and Varis Bokalders

Intake air can be cooled using evaporative cooling. The photo shows the air intake to an underground duct filled with expanded clay pellets which are sprinkled with water in the summer. The intake air is thus cooled first by evaporation and then by the cool underground temperature.

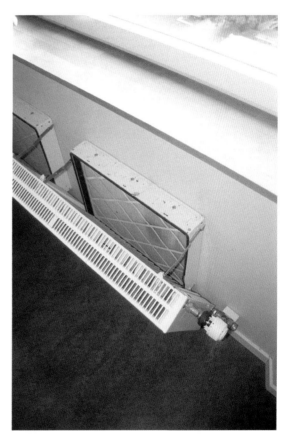

There are various methods for preheating intake air. Intake air can be introduced via a box behind a radiator or through a special air intake radiator. The photo shows such a radiator made by Clean Air System, which is designed to withstand damage from freezing.

Modern ventilation slots are often combined with a metal box behind the radiator where air is preheated. If necessary, a filter can be placed in the box. The photo shows this type of radiator, which can be opened to facilitate changing filters and cleaning the air intake duct.

Exhaust Air

It is most common for air to be exhausted from the kitchen, laundry room and bathroom. A cooker hood in the kitchen removes cooking odours via natural ventilation. Cooker hoods often have a separate duct since it accumulates grease and should be especially easy to clean. In the bathroom, a shower cubical can act as a room within a room and the bathroom can be ventilated through it. Both these methods are a type of local exhaust where pollutants, food odours and moisture are removed as close to the source as possible. This natural ventilation can be regulated with a damper. It is of course possible to install a fan to periodically force ventilate, but the design should not prevent natural ventilation when the fan is turned off.

Ventilation Chimneys

Chimney top design affects ventilation. There are caps to prevent backdraughts caused by the wind. There are wind-induced downdraught-preventing caps that adjust according to the wind, prevent backdraught and increase the Venturi effect. There are turbines that increase ventilation with the help of the wind's force. There are turbines (Savonius rotors) that increase ventilation up to a certain point where they are self-regulating so ventilation is stabilized. There are also turbines that are

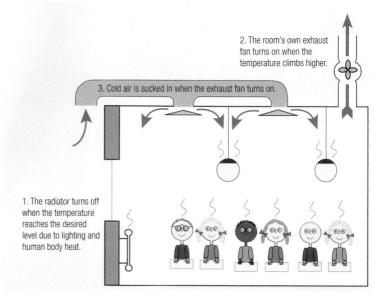

In the Mölndal ventilation model, air flow is controlled by room temperature. Body heat eventually starts the exhaust fan. There is then negative pressure in the room and outside air (cold) is sucked in through valves in the roof.

Source: *Bra luft i skolan – Om ventilationens betydelse för inneklimatet i Mölndals skolor*, Svenska kommunförbundet, 2002

A wind vane on the ventilation chimney adjusts to the wind direction. It counteracts backdraughts and increases natural ventilation via the Venturi effect.

self-regulating and restrained by a thermo-element when it is cold out and the forces created by temperature difference increase ('the clean turbine').

A solar chimney is a chimney designed as a solar collector. The sheet metal chimney is painted black and a glass tube encloses the sheet metal pipe. The exhaust air is thus heated and the temperature difference increases, thereby increasing the natural ventilation force as well. Of course, a 'summer fan' can be installed in the chimney and turned on as needed.

Super-turbine.

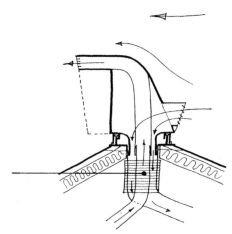

There are advanced types of combined-function rotary terminals that act both as air intake and exhaust air vents.

Source: *Architecture in a Climate of Change – A Guide to Sustainable Design*, Peter F. Smith, 2001

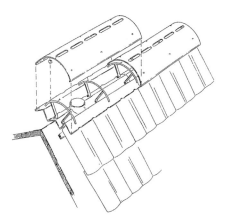

Instead of a ventilation chimney, exhaust air can be removed through a natural ventilation fixture in a roof ridge ventilator. Wind blowing over the roof ridge creates a Venturi effect that intensifies the natural ventilation.

Source: Architect Mauritz Glaumann, The Royal Institute of Technology (KTH), Gävle, Sweden

The 'Rena Snurren' rotating cap on the Torkel building in Kovikshamn, Bohuslän, is made by Håkan Gillbro, DELTAte, Gothenburg, Sweden

Tofors Ltd markets humidity- and demand-controlled exhaust air fixtures with a small, electronically controlled motor. The electronics respond to humidity, temperature and motion.

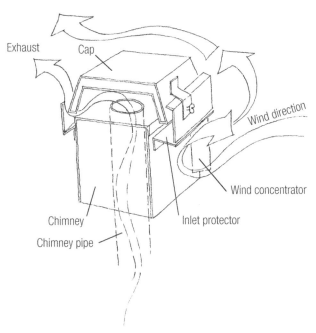

Ventilation consultant Kalle Magnusson, Saltsjö-Boo, Sweden has developed a ventilation cap with inlet protection and wind concentrator, called the Kalle hood, to improve natural ventilation.

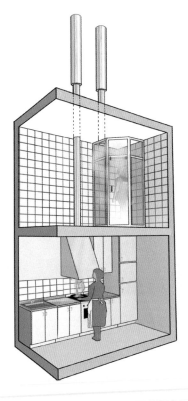

Local exhaust ventilation from a shower cubicle and a cooker hood.

Source: *Bolig og naturlig ventilation – indeklima energi driftssikkerhet*, edited by Rob Marsh and Michael Lauring, Denmark, 2003. Illustration: Leif Kindgren

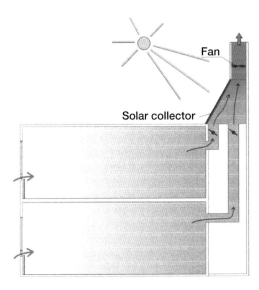

The principle for ventilation in Tånga school in Falkenberg, Sweden. Intake air is introduced through radiators in the outer wall and exhaust takes place through solar chimneys.

The solar chimney, Tånga school, Falkenberg, Sweden.

Source: Architect Christer Nordström

Solar chimney.

Ventilation chimney at the camping site, Själland, Denmark. The wing shape above the ventilation chimney increases natural ventilation by creating negative pressure on the underside of the wing.

Savonius rotor.

EXAMPLES OF SWEDISH BUILDINGS WITH NATURAL VENTILATION

Most new buildings in Sweden use mechanical ventilation, but there is a revival of natural ventilation in sustainable buildings. Modern natural ventilation systems are regulated according to need and season. This can be done manually or by computer. More complex buildings combine natural ventilation with fans. These kinds of ventilation systems are called hybrid systems. Experience shows that the interior air climate in most naturally ventilated buildings is appreciated by the occupants.

The Hus Torkel building in Kovikshamn, Bohuslän, is ventilated with fresh air through three earth-pipes, and the exhaust air is removed through a ventilation chimney with a rotating cap. One of the earth-pipes cools the larder.

Source: Architects: Maria Block and Varis Bokalders. Ventilation: Torkel Andersson, DELTAte

Glaumann/Westerbergs house in Vidja, outside Stockholm, has a natural ventilation system with heat exchanger. In the roof ridge there is a large pipe with a slot in the top. There is a spoiler over the slot that creates negative wind pressure. This pipe is for the exhaust air. Inside this large pipe there is a smaller pipe for fresh air with intakes at the top of the gables. The exhaust air in the bigger pipe heats the fresh air through the metal in the inner pipe. The exhaust air is extracted through the ceilings. The fresh air is delivered to the ground floor.

Architects: Mauritz Glaumann and Ulla Westerberg

Multi-family houses, 'Birgittas Trädgårdar' in Vadstena. Fresh air is introduced through an earth-pipe to an air chamber in the foundation. Individual ducts take air to every room. Every kitchen and bathroom has its own ventilation chimney with a rotating cap.

Architect: Peter Kark. Ventilation: Torkel Andersson, DELTAte

The Mikaeli school in Nyköping is one of the many naturally ventilated schools. The air is introduced through an earth-pipe to a corridor under the school. From here there are ducts to every classroom. The classrooms have high ceilings and at the top of the gables there are flaps for exhaust air. The system is manually regulated.

Architect: Asmussen Arkitektkontor.
Adviser: Varis Bokalders.
Ventilation: Torkel Andersson, DELTAte

An educational building of the Royal Institute of Technology in Haninge, Stockholm. The air is introduced through an earth-pipe into a common space. This open space goes through all the floors and is ventilated through windows at the top. The individual rooms are ventilated through this open space. There are also ducts from the classrooms to fan-equipped ventilation chimneys on the roof. The system is computer-controlled.

Architect: Temagruppen Sverige AB. Ventilation: Göran Ståhlbom, Allmänna VVS-byrån

Office building for Akzo Nobel in Stenungsund. This office building is built using heavy materials. The heavy mass is used to balance heating and cooling needs. The building is ventilated through small windows. These windows, made by Window Master, are closed and opened as necessary by self-regulating sensors, which regulate both temperature and relative humidity. This concept significantly lowers energy needs.

Architect: Arne Algeröd, Mats & Arne Arkitektkontor. Ventilation: Torkel Andersson, DELTAte. Photo: Mats & Arne Arkitektkontor

1.2.2 Electric Services

We surround ourselves with electricity cables, electric gadgets, lighting, etc. The alternating current that passes through electric wires and devices creates electric and magnetic fields, sometimes at strengths that may pose health hazards in the long term. It is important to know how to install electricity so as to minimize these fields. Electric services in buildings are not the only cause of electromagnetic disturbances. There are many other causes including: overhead lines, underground cables, mobile telephone communication base station towers, radar stations, electric fences and TV transmitters, to name a few.

Electric and Magnetic Fields

Electric services produce both electric and magnetic fields. Since electricity is today used almost everywhere, people are continually surrounded by various fields. If steps are taken to improve the electric environment or make it easier to make improvements at a later date, future requirements can be met with less difficulty. For a new construction, such measures do not cost much. However, electrical renovations made at a later date are much more expensive. Electric fields from electric services and devices can be limited by using earthed thin metal foil or metal net. Common causes of increased magnetic fields are stray currents and overtones (disturbed or disharmonic currents)

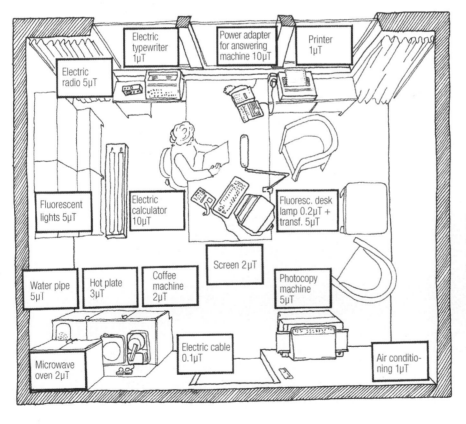

In offices, people are often surrounded by many devices that emit electromagnetic fields. Be aware of where electric cables run and that fluorescent lights in the room below may be only 30cm under your feet. The figures indicate values for up to a distance of 10cm and may vary greatly.

Source: adapted from drawing by Inger Blomgren in the periodical *Arbetsmiljö* 3/89

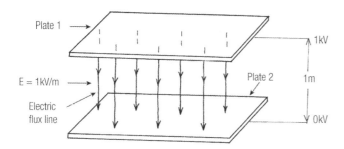

Plate 1
E = 1kV/m
Electric flux line
Plate 2
1kV
1m
0kV

Electric fields occur between objects that have different voltages. E (electric field strength) = V (voltage difference)/m (distance between objects).

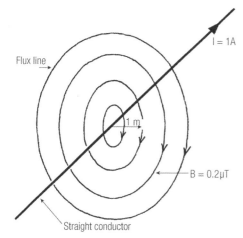

I = 1A
Flux line
1 m
B = 0.2µT
Straight conductor

Magnetic fields are created around charged conductors and are measured in tesla (T). B (magnetic flux density) amounts to 0.2µT (microtesla) 1m from a conductor with a current of 1 ampere (1A).

Magnetic fields can be limited, for example, by increasing the distance from the source, by placing magnetic fields in opposition so that they neutralize each other and by using five-conductor cable systems instead of four-conductor cable systems.

Hypersensitivity to Electricity

It is not possible to determine who is hypersensitive to electricity by doing a blood test or some other medical examination. Those who are classified as hypersensitive to electricity are those who themselves feel that they are. By measuring, among other things, pulse, blood pressure and brain reaction, researchers have shown that the nervous systems of people hypersensitive to electricity are more sensitive than those of other people, i.e. they respond more to various stimuli, for example light and sound.

The largest group of people hypersensitive to electricity are those who experience problems working in front of monitors. Many are helped by altering the electric environment in their homes and workplaces. Other measures include reducing the time spent in front of monitors, using a monitor with lower field intensity, replacing fluorescent tubes, improving air quality and reducing psychosocial stress.

In recent years, mobile telephone networks have developed. So turning off the main circuit doesn't help to free us from electric and magnetic fields – radiation comes in from outside our homes.

Electro-conductive paint that shields from electromagnetic radiation is marketed by Caparol (ElectroShield). The paint contains ingredients that block electromagnetic radiation and low-frequency fields that surround electric devices and electric cables. The painted surface should be grounded and the radiation is thus directed down into the ground.

Power Lines

According to the EU regulations the recommended value that should not be exceeded for the exposure to the magnetic field from powerlines of a population over a considerable time is 100µT [micro (= 10^{-6}) tesla] and 5kV/m.

People hypersensitive to electricity may react to fields as weak as 20–30nT (nanotesla = 10^{-9}). Levels lower than this are found several hundred metres from a 400kV line. Research has shown that childhood leukaemia can occur to a somewhat higher degee than normal if the magnetic field flux density in the home exceeds 0.4μT.

Overhead power lines can be replaced with underground cables. This radically reduces the strength of the magnetic fields, but is quite costly. A less expensive approach is to install the three phases of the line in a triangular shape. This reduces the magnetic field but not as effectively as underground cable.

Transformers

One source of magnetic fields are enclosed substations. In the spaces directly over and beside such stations there may be increases of up to tens of T. Substations should not be located in or near buildings where people live or work.

Supply Lines

Underground supply lines that supply a building with electricity should be placed 5–10m from the building and laid to the electrical distribution board in such a way that the route within the building is as short as possible.

Electrical Distribution Boards

Electrical distribution boards should not be located near spaces regularly frequented by people. Distribution centres should be connected directly to primary distribution boards to minimize the intensity of current in the respective main cables. All centres should be metal-enclosed and earthed. All main cables should be connected with separate wires for neutral and earth to minimize currents arising in the earth protection when there is asymmetrical loading.

Distance (m)	0.1	0.5	1.0
TV	1.5–4μT	0.2–1 μT	0.1–0.2μT
Stove	1–3μT	0.1–0.6 μT	0.05–0.2μT
Hairdryer	2–12μT	0.1–0.3 μT	0.05–0.1μT
Vacuum	15–35μT	0.4–1.5 μT	0.1–0.5μT

Strong magnetic fields can exist close to electric devices, but the strength of the field diminishes very quickly away from the source. From a health perspective, one should be cautious about spending much time near devices that are used over long periods of time.

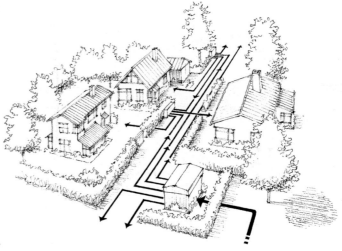

While planning electrical lines the magnetic fields created by transformer stations and underground electrical cables should be taken into consideration.

Magnetic Fields from Various Power Lines

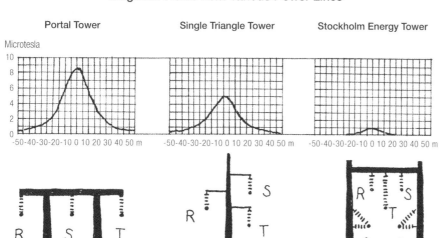

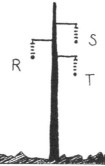
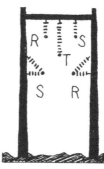

Magnetic fields around power lines can be reduced by 90 per cent, depending on how the lines are arranged. Lars Jonsson at Stockholm Energy has developed a tower with five instead of three phase lines, which reduces magnetic fields considerably.

Artistically designed powerline supports, Esbo, Finland.
Source: Fingrid Oyj. Photo: Esa Kurkikangas.

Artistically designed powerline supports, Jyväskylä, Finland.
Source: Fingrid Oyj. Photo: Risto Jutila.

1.2 PROPOSED STEPS TOWARDS A GOOD ELECTRO-CLIMATE

- Shield installations: cable nets, appliances, connector devices.
- Use twined FK-cable in piping systems in order to reduce magnetic fields.
- Place meter panels, electrical boxes and such where they are least intrusive.
- Use five-wire electric cables as much as possible.
- Potential equalization.
- Prevent stray current and overtones.
- Avoid 230V equipment the creates overtones.
- Install a device that breaks the current in an electric line when the last load is disconnected.
- A low-voltage bus system combined with a 230V system can be a good alternative.

Source: Eljo El-San

Isolation Transformers

An isolation transformer receives and delivers the same voltage. Its only purpose is to separate a service from the rest of the electricity supply mains so that they are not directly connected, only via the magnetic field in the core of the transformer, so the name 'isolation transformer'. The idea of this arrangement is that disturbances from the electric mains cannot enter the supply. Supply lines from energy companies' transformers are often too long and this results in high inductance and high capacitance. The inductance causes earthing to function poorly, and the capacitance results in disturbances jumping from one conductor to another. An isolation transformer makes it possible to make a fresh start, with an new earth point, short cables and a new five-wire system. It is also possible to direct incoming electricity through a network filter that prevents disturbances being introduced from outside.

Earthing

Possibilities for establishing a good electric environment can be greatly improved if a building is equipped with its own earth system. An earth system usually consists of two parts: an earthing rod, usually buried below the frost line; and a earthing conductor to connect the earthing rod with whatever needs to be earthed. In Germany, all new buildings must be equipped with a earthing system consisting of an iron band embedded in the concrete foundation below the vapour barrier. The iron band should go around the entire foundation in a closed circle. In Sweden, this method cannot be used since winter ground frost requires that the foundation be insulated from ground moisture. Instead, an earth line can be buried around the building. All earthing should connect to the same earthing point to avoid displacement current.

Earth Leakage Circuit Breakers

Earth leakage circuit breakers release and shut off electricity if all the current does not return from a device. This is a good precaution against injury from short circuits.

Main Circuit Breaker

A main circuit breaker is a high-voltage relay with accompanying electronics located in the electric distribution box ('fuse box'). Main circuit breakers can be used with advantage in bedrooms where we spend eight hours every night. The idea is to sleep without being exposed to electric current. Many people like the idea and feel that they sleep better when free from exposure to electric current. When the last light is turned off, the circuit is broken, and all cords and electric wires stop emitting electrical alternating fields. With the help of direct voltage the circuit breaker receives a signal if a light or other electric device is turned on, and then lets current through. Direct current does not give rise to biologically hazardous fields.

The material in wooden buildings does not moderate electric fields. If the current is broken to all the electrical lines in a bedroom, electric fields can still enter from surrounding rooms, from the floor below, and from neighbouring flats. Buildings with stone walls and concrete beams are, from this point of view, much more favourable. In wooden buildings, main circuit breakers in addition to shielded electric cables are most effective for reducing electric fields.

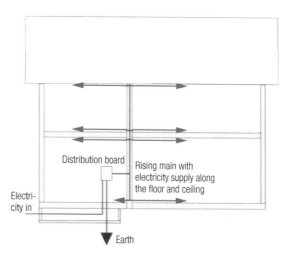

Electrical services should be installed so that electric and magnetic fields are minimized. Protected, earthed electric wiring is put through a centrally situated main cable and is distributed in a branch-like pattern behind skirting boards. Installation of electrical services is flexible, and computer, speaker and telephone lines may be installed behind skirtings and other mouldings.

Source: Det Hållbara Huset, finalist i Miljöinnovation 2002, architects Maria Block and Varis Bokalders

To reduce the effects of electromagnetic fields produced by metal in building components (caused by capacitance), these may be earthed.

Halogen-free electric wires

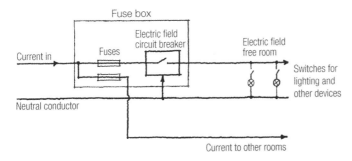

The electric field circuit breaker receives a signal when the last light is turned off and then breaks the current to all the devices in the bedroom, so that it is possible to sleep in a room free from electric fields.

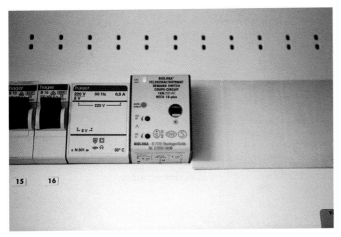

Main circuit breaker located in a central distribution board (fuse box).

Regulating Equipment for Heating, Water and Sanitation

Electrical equipment is gathered together in a single cabinet and connected with separate neutral and protective earth wires in order to minimize the ground potential differences that may arise when there is asymmetrical loading. The electrical cabinet is placed in an area not regularly frequented by people. If speed controls for fans and motors are used, the manufacturer and quality of the speed controller should be carefully chosen. The speed controller should be equipped with a harmonic filter.

Potential Equalization

Potential equalization and grounding are important for both personal safety and the electro-climate. Electric potential is the voltage between two objects where one of the objects is the ground. When measuring the voltage between a grounding rod and the protective earth there can quite likely be an alternating current of one volt or more. Voltage can also be found when measuring between other metal objects: the telephone line's grounding wire, iron reinforcing bars, a central heating system and ventilation shafts. The aim of potential equalization is to eliminate all these voltages. Potential equalization prevents a difference in potential between electrical conducting wires at different locations if a problem or disturbance occurs. It should then be possible without any risk to hold one hand on the water tap and the other on the freezer at the same time as standing barefoot on the floor drain. Thanks to potential equalization all metal objects have the same electrical potential, so that there is zero voltage between them.

It is best if electricity services, water lines and the connection to the earth electrode enter the house at the same place, so that potential equalization can take place at the entry point using short, thick copper conductors. The potential equalization currents then travel the shortest possible route and their stray current in the building's installations is limited. A properly functioning potential equalization system means that the reinforcing iron and other large metal constructions in the building are grounded, and that incoming electrical and telephone lines as well as district heating services are also grounded. If different objects have different electric potential an electric alternating field is created between them. Electronic alternating fields disturb both people and electronic devices, not least because exposure to stray pulses is always a possibility. The grounding and potential equalization of electronic installations must be properly carried out in order for shielding to function.

Five-wire Systems

The basic principle is that three-phase systems should be designed as five-wire systems (TN-S system) to eliminate magnetic fields, and that all wiring should be protected in order to limit electric fields. Older three-phase current in a four-wire systems running up to a central distribution board, with a three-phase line to the electric cooker, washing machine and other devices requiring strong current, should be replaced with five-wire systems.

Shielded Electrical Systems

In modern systems, electric wiring is carried in plastic ducting that does not shield fields. So electric fields from electric wiring may be found in walls. Choosing a protected electrical system is one way to reduce the antenna effect of the wiring. The strength of electric fields is often higher in residential environments than in offices, as electric supply in offices is usually earthed.

Electric Wiring, Junction Boxes and Sockets

Electric wiring in close proximity to people is a problem. There should be enough electric sockets so that long 'field-radiating' electric cords are not required. A shielded electric cord is covered with a woven metal sleeve that makes the cord more flexible. The end of the sleeve is designed so that it can be earthed. At present, unshielded cords are common. It is also desirable to have twisted-pair wires in electric cords in order to reduce magnetic fields.

Supra LoRas (global patent by Tommy Jenving) is the only fully certified shielded mains cable in the EU. The short-pitch twisting shields from both magnetic radiation and magnetic fields. Electrical and magnetic alternating fields are efficiently blocked.

Sockets, junction boxes, switches and plugs should be earthed. These days, fittings are made of stainless steel among other

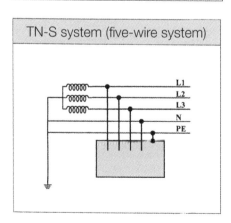

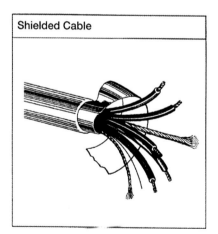

To achieve a better electric environment, shielded wiring can be used for five-wire systems. The figure shows Wasan Flex cable where the protective foil is in contact with a bare copper wire to guarantee contact, even if the foil is broken. An alternative is shielded twisted cable.

In existing buildings in Sweden the four-wire system is most usual. In these, the protective earth (PE) and neutral (N) wires are in the same PEN conductor. In a five-wire system, the neutral and PE conductors are separate.

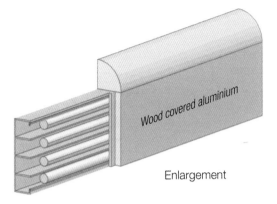

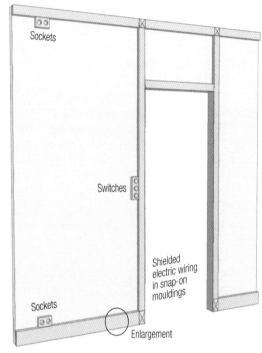

Earthed, halogen-free wiring in shielded cable ducting can be replaced. The snap-on moulding is easy to put on and remove.

Wooden wall socket (ash).

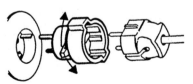

An intermediate plug with a two-pin current-breaker. Disconnecting both wires ensures that no current emitting an electromagnetic field remains when the device is turned off.

Source: *Bostad och hälsa – En praktisk handbok för ett sundare hem*, Ragnar Forshuvud, 1998

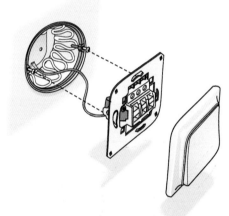

This solution shields from electromagnetic fields forwards, backwards and to the sides. A shielded metal sheet is connected to the box with self-adhesive tape and then the side-protecting spiral lines the inside walls.

Source: Brochure from Eljo El-San.

things. Older, built-in junction boxes with metal covers are ideal in these contexts. A screw is put in the bottom of the box and the earthing wires are attached, and the electric field disappears. Plastic junction boxes are more difficult to use. A shield is connected on top of the box with a separate top fastener. It is also possible to enclose the cables in the junction box with aluminium foil, and earth the foil. Single-pin plugs should be replaced with two-pin types, so that all phases as well as neutral can be disconnected.

Lighting

Low-energy light bulbs (compact fluorescent) create stronger magnetic fields than filament light bulbs. Fluorescent tubes with high-frequency fittings usually create weaker magnetic fields than ordinary fluorescent tube fittings, but the high-frequency electric fields are stronger. If high-frequency lighting is going to be installed, the possibility of shielding the high-frequency electric field should be examined.

Shielded light fittings can be used in rooms where people spend a lot of time. They should consist of a metal body for discharging electric fields and an anti-glare shield that is made of metal or filter glass, earthed in order to discharge electric fields.

Electromagnetic disturbances are generated both by lighting and devices and can be both conductive (carried by cables) and radiating. Electrical installations act as an antenna and transform conductive disturbances into radiating disturbances. Conductive disturbances also come from outside the electricity supply through the cable supplying the current.

Computers

Electromagnetic disturbances affect computers adversely but not instantly, since computers are programmed to ignore incorrect input. The process takes time and becomes noticeable when computers slow down. Furthermore, these disturbances cause hypersensitive people to react with symptoms of disease, and there is reason to believe that many people are affected negatively without knowing the cause. In a healthy building, where an attempt is made to eliminate all negative influences on the environment, electromagnetic disturbances are also reduced as much as possible.

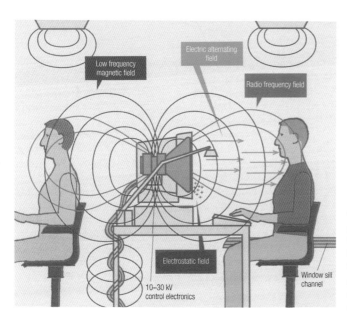

The different fields that surround a computer terminal work station. The importance of sufficient distance between work stations is also illustrated.

Source: *Elöverkänslighet – Hur hanterar vi den?*, Maggie Miller, Arbetarskyddsnämnden, 1995

Metal Building Elements

Capacitive voltage means that metal construction parts without direct contact with any electric conductor are live from an electric field. Well-known examples are metal fences and sheet-metal roofs near power lines, which have to be earthed to prevent shocks when they are touched. If the metal construction is earthed, the field disappears almost completely. A metal desk frame can have considerable capacitive voltage and should be earthed. Other elements that should be earthed are metal beams and reinforcement in concrete. Instead of steel studs in walls around junction boxes, switchboards and large computer rooms, wood should be used.

Geomagnetic Fields

In many cultures and at various times, the presence of geomagnetic fields has been investigated when selecting a building location. Geomagnetic fields are believed to have an effect even high up in multistorey buildings. There are different views on whether or not this is significant. Those who believe that geomagnetic fields affect our well-being believe that bedrooms and workplaces should not be sited close to the following:

- underground watercourses;
- fault zones in bedrock;
- Curry lines that form a NW–NE grid pattern with lines about 4m apart; and
- Hartmann lines that form a N–S grid pattern with lines about 3m apart. It is most important to avoid points where lines intersect.

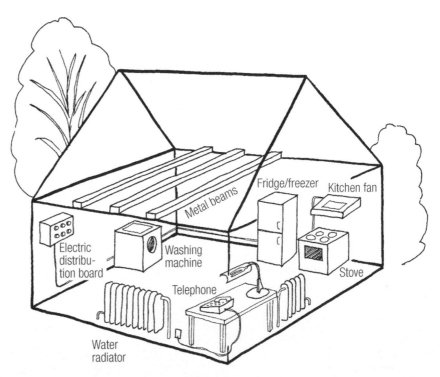

Stray ('vagabond') current creates magnetic fields. Stray current is when the current from an electric device does not return to the electric distribution board but takes a short cut through metal objects in a building.

1.2.3 Plumbing

Insurance companies pay out large sums of money every year to cover damage from water leakage. How can water damage be prevented? Installation of fresh water, heating and sewage pipes both underground and in buildings should seek to minimize impacts in the ground and facilitate easy maintenance and replacement.

Water-damage-proof Construction

Water damage in Sweden is estimated to cost five billion SEK a year. A large part of the damage is caused by leakage from piping systems due to corrosion, freezing or improper installation. Faulty installation also risks legionnaire, bacteria and burn damage. The trade association for heating, ventilation and sanitation workers has a set of standards that addresses important health and safety issues (see www.sakervatten.se).

Wetrooms and services should be designed so that water leaks are quickly discovered. If water does leak, it should not immediately cause damage. Services should be designed so that a repair affects as few rooms as possible. The most expensive room to repair is the bathroom. If all the services can be replaced for a reasonable price, they can be replaced before they are worn out and damage occurs.

Pipe Laying

When laying pipes, plans should be made for dealing with both leakage and repairs.

- As many pipes as possible should be replaceable. Factor in the replacement of all piping without greatly interfering with the building structure.
- All plumbing should be installed so that it can be inspected. It should be possible to remove any pipe casings.
- It should be possible to quickly detect water from a leaking pipe.
- The siting of the plumbing should suit the building structure and internal layout.
- Partitions and internal walls should be designed so that water leaks become visible without immediately causing damage.
- Drainage ducts for water, sewage and heating systems should be designed so that any water leaks can be seen.
- Joints in tap water pipes should be placed in rooms with waterproof flooring.
- Joints in tap water and heating pipes should be replaceable and not put in hidden locations. Placing joints in slots and ducts should be totally avoided.
- There should be labelled stopcocks for hot and cold water in every flat.
- Plumbing for dishwashers should be equipped with sink taps that have a dishwasher shut-off valve and a trap solely for the dishwasher drain.
- Outside water taps should be frostproof and protected from freezing if a garden hose is left connected.

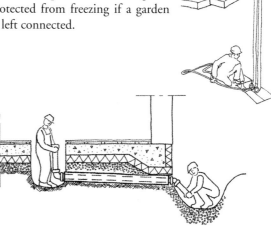

Laying pipes inside empty pipework makes it possible to replace pipes and at the same time minimize the risk of leakage. Laying pipes in drained culverts is a flexible and safe water and sewage installation method.

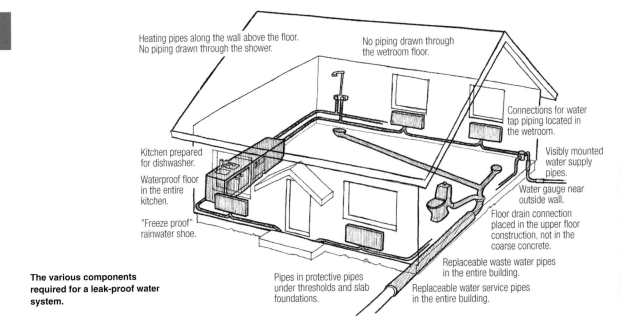

The various components required for a leak-proof water system.

Floor Drains

- Type-approved floor drains should be prioritized.

- Do not use floor drains with raised rings on a floor with a plastic floor covering.

- Do not use floor drains with extra side intakes below the floor's waterproof layer.

- Follow assembly instructions for floor drains. Important details in the instructions, e.g. attachment to the floor and joins to the floor waterproofing, should be shown in the assembly instructions.

- Measure the floor drain for width and height. The floor drain should be installed level and placed so that it is accessible for cleaning.

- Make sure that all tradespeople involved are aware of the floor drain system instructions.

Pipe Ducts and Screw Fixings

Plan bathroom layout and piping so that screw fixings and pipe penetration points are not in places exposed to splashing water.

- Wetrooms should be constructed so that tap water and heating pipes do not penetrate the floor.

- Pipes penetrating waterproof layers should be sealed according to a prescribed method.

- Do not place screw mountings in the floor or near the shower area.

- Choose taps, shower fixtures and other products (that will be placed in the shower area) that require drilling as few holes as possible in the waterproof layer. Shelves, soap dishes, hooks, etc. that must be screwed into the wall should not be placed in the shower area.

- Surface-mounted pipes in wetrooms should not be clamped in the shower area or behind a bathtub.

All screw fixings and other holes in the waterproof layer in the wetroom should be sealed according to a prescribed method. Any fixings that are required should be carried out during the construction phase.

Floors and Walls in Wetrooms

- Floors in wetrooms should be made so that they slope towards the floor drain. If possible, choose a slope of at least 1:100 for the whole floor.

- Waterproof flooring and wall coverings should suit the substrate and local conditions.

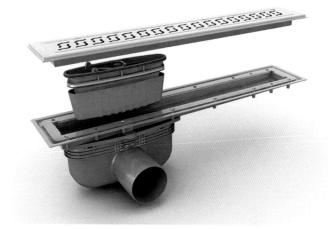

A differently designed floor drain, Purus line, which includes a smell trap. With this product the shower floor need only drain in one direction.
Source: Purus AB

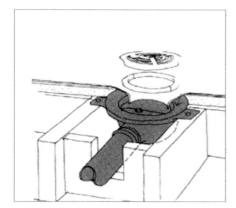

Floor drains account for 25 per cent of total water damage in wetrooms.
Source: VASKA-projektet, Umeå

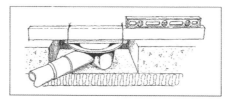

Hollow out a space, don't plumb the floor drain directly into the concrete. Level the floor drain in the right position before fixing it.

Kitchen Floors and Walls

- Kitchen flooring should cover the whole floor and be laid before installing fixtures. Flooring should be laid up the wall to a height of at least 50mm behind fixtures and the dishwasher, and at least 5mm elsewhere.

- It should be possible to inspect the floor under the sink.

Junction Boxes

The possibility of water damage can be reduced by using pipes and hoses without joints since leaks often occur at joints (these pipes and hoses are recommended by insurance companies). Several pipe systems with bends, flanges and collars made directly

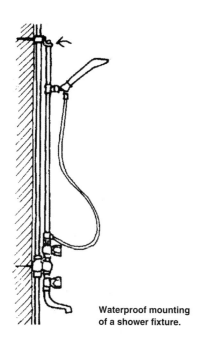

Waterproof mounting of a shower fixture.

of the pipe material have been developed. There are also tap-water systems made of polythene placed inside special protective pipes.

Instruction Signs

Equip flats or houses with 'user instructions' that specify:

- where the main water stopcock is located;
- where any system draincock is located and what it means if water runs out of it;
- where it is permissible to make holes in bathrooms;
- instructions for installing dishwashers and washing machines;
- cleaning instructions for exhaust air fixtures in wetrooms; and
- emergency telephone numbers.

These instructions should be permanently attached to the wall, for example at the central electrical distribution board.

Foundations and Shafts for Services

By using a foundation designed to contain building services that has a centrally placed, drained inspection pit, all pipes and wires can be gathered together in an easily accessible, inspectable and waterproof location, so that it isn't necessary to have built-in pipes or cables, and repairs and maintenance are easily undertaken. A problem is that such structures for services require more site preparation, which is expensive, as accessibility requires headroom in some places. On the other hand, installation is simplified and so costs somewhat less.

For example, in schools with natural ventilation with an underground intake air system, pipes and cables can be put in the large air intake shafts. It is also possible to put all pipes and wiring in shafts, which facilitates replacement. The first major renovation that usually needs to be carried out in a building is the replacement of mains water and sewage pipes. Designs with pipes and cables in shafts last longer and attract reduced costs during renovation.

Buried Pipes

Most of the environmental impacts of laying pipes in the ground occur during the laying phase. About 80 per cent of resource consumption takes place during laying.

In the Birgitta's Garden flats in Vadstena, all services are confined to a few vertical shafts. Access to the service shaft is via a small door in the bathroom, where the meters for cold and hot tap water, heating and electricity are situated.

Furthermore, excavation places a great strain on the surrounding environment with open trenches, closed roads and problems for pedestrians and shopkeepers. It is almost always possible to lay pipes without digging a trench, going under or around a sensitive area. Today, trenchless technology is a realistic and competitive alternative to open trenching. The methods are commonly referred to as 'no-dig'.

Auger boring and microtunnelling involve pushing a pipe through the ground with jacks, and removing excess earth through the pipe using a drill. Pipes can also be ploughed in, but this is actually a type of 'digging'. Controlled drilling is done using steel pipe and a nozzle and can be used in practically all ground conditions other than rock. An entrance pit and receiving pit are dug. The drill head rotates, spraying drilling fluid at high pressure and moving straight ahead. When the rotating stops, a directional change takes place as the drill head is eccentric. While the drill head drills the hole, the hole is widened with a rotating reamer (a type of conical drill) that also pulls the pipe along. Diamond or hard metal drill bits are used for drilling in rock.

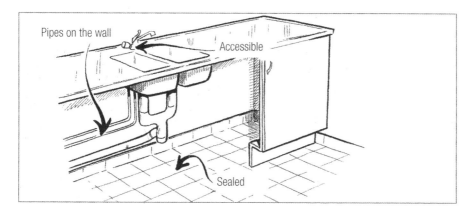

It is important for services to be waterproof so that moisture damage and mould are avoided. Kitchen flooring should run under cupboards and up walls so that water leakage is easily noticed when water runs onto the floor. Piping should be accessible and holes in the floor and draining boards should be avoided.
Source: VASKA-projektet, Umeå

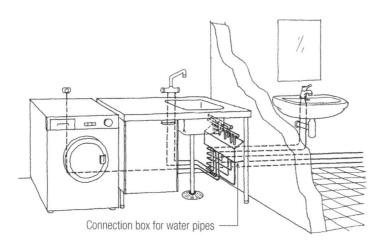

To avoid concealed joints, nowadays, water pipe connection boxes are often used. All connections are installed joint-free to the box.

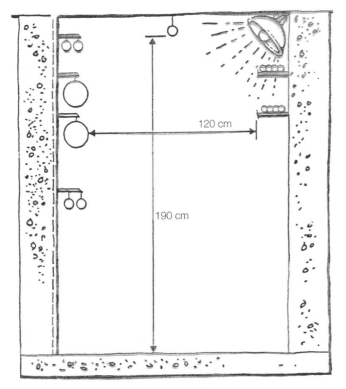

It is often possible to repair old pipes without digging and there are many no-dig installation methods. Flexible casings can be inserted inside damaged pipe; they can be inserted using air pressure and hardened with steam. Slip-lining involves inserting a new pipe inside an existing pipe. An expanded lining involves folding a plastic pipe and inserting it into an old pipe. Then, with the help of heat and pressure, it is expanded against the walls of the old pipe. Panels or pipe segments can be placed in large pipes and cast in place. Pipe bursting involves using a pipe-bursting head to hammer through a host pipe, resulting in fragmentation of that pipe, while simultaneously pulling a new pipe into place.

Sometimes a pipe only needs to be cleaned, and there are many methods available for this purpose. High-pressure rinsing is the most commonly used method, which can be carried out using a variety of rinsing nozzles. Tree roots are cut with a root cutter that, like different scraping tools, spins around inside the pipe. A cleaning pig is a device that is pushed through a pipe using water pressure.

A well-designed culvert permits quick assembly and simple future additions, as well as easy service and replacement. What is more, overall costs may be lower. The culvert can also be used as an air intake (an earthpipe) that preheats air in winter and cools air in summer.

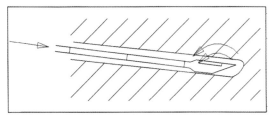
Rotation and pressure. There is no change in the pitch and direction of the drill.

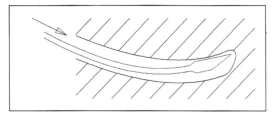
Pressure without rotation. A change in direction takes place when there is pressure without rotation.

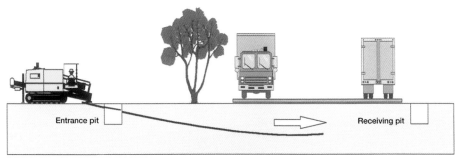
Drilling a pilot hole.

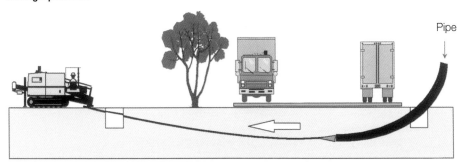
Pulling a pipe through.

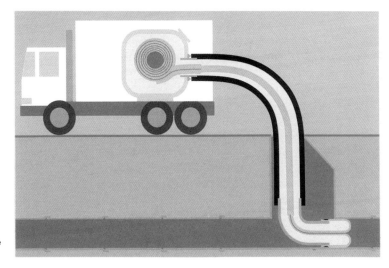
Threading a pipe in using air pressure and hardening with steam.

1.2.4 Heating Systems

Generally, there are three ways to heat buildings: waterborne systems, warm air central heating systems, and radiant heat from electric radiators or separate wood-burning appliances. In Scandinavia in the past, separate wood-burning appliances were most commonly used for heating, so there might be a tiled stove in every room. Direct electric heating is not addressed in this book as it is not compatible with sustainable development. Warm air systems have often resulted in a poor interior climate: large volumes of air circulate in ducts and rooms, carrying particles with them, sometimes resulting in large temperature differences between rooms. With these systems, ventilation requirements rarely correspond with heating requirements, so they are not described in this book. Ecologically constructed buildings usually use waterborne heating systems and/or radiant heat from separate wood-burning appliances.

Heating Methods

Radiant heat (like heat from a tiled stove) provides the most pleasant heat. Most comfort is attained by getting as much heat as possible from radiant sources. Preferably, radiation should come from a large area at a low temperature. Warm-air heating systems are not nearly as pleasant as radiant heat. Some radiators, such as convectors, emit a lot of convective heat, i.e. warm air currents, which are also not as comfortable as radiant heat. For a heating system with a large proportion of radiant heat, low-temperature waterborne systems should be chosen, such as underfloor heating, wall heating, skirting heating or radiators. Waterborne ceiling heat is not as pleasant as it results in greater temperature differences in a room and it can easily become too warm at head height. There are also closed-air heating systems that provide radiant heat, such as warm-air floor heating or hypocaust systems. Hypocaust stoves are wood-burning stoves equipped with a closed vent system for warm air that heats large areas in a room. They are common in Germany.

Waterborne Heat

It is important to clarify what is meant here by a heating system. Every heating system has an energy source which is connected to a distribution system, and the heat is distributed to the various rooms via the heating system. The distribution system includes storage tanks, hot water heaters, and an operating, control and adjustment system. The waterborne heating systems examined here are radiators, underfloor heating, wall heating and skirting board heating. All parts of a heating system should be connected by insulated pipes.

Storage Tanks

Heating systems using renewable heat such as solar and biomass often need a storage tank to store the heat. Solar heat must be stored from day to night, and efficient heating with biomass is done by heating intensely for short periods and replenishing the heat in the storage tanks. Storage tanks provide a flexible heating system. Heat can be added to tanks from solar, biomass or electric sources, and heat can be drawn on for heating space and/or tap water. Storage tanks that fulfil all these functions are often equipped with three heat exchanger coils: one in the bottom of the tank where it is coldest and solar heat is added, one in the middle to preheat hot tap water, and one at the top where it is warmest for final heating of tap water. Heat from a water-jacketed wood-burning appliance is added in the middle of a storage

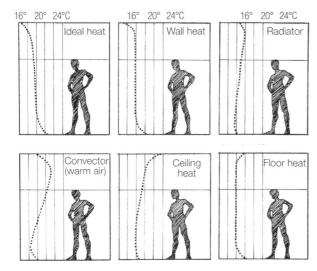

Different methods of heating provide different temperature distributions in a room. Wall heating is probably the method closest to the ideal, i.e. a large surface area with a low temperature that radiates from the side.

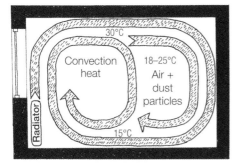

Convection heat is carried out into a room by air heated by a radiator (convector). A convector is a radiator designed to produce more convection heat than radiant heat.

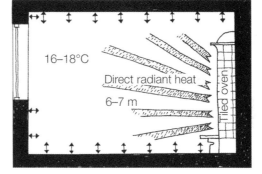

Radiant heat is the heat that radiates out into a room from a warm mass, such as a tiled stove.

tank and heat to radiators is removed from the top of the tank. Storage capacity can be increased by adding another tank to the system. The second tank is not equipped with heat exchanger coils, is referred to as a slave tank, and can be disconnected in summer. Storage tanks are made of steel and should be pressure-certified. Tanks insulated with cellulose fibre are available. For safety, in case the water starts to boil, the tanks must be equipped with an expansion and safety release valve.

Charging System

A charging system is needed between the storage tank and the boiler. The charging system functions as follows:

1 Start phase: circulation to the storage tank is halted so that water only circulates in the boiler in order to reach working temperature as quickly as possible.

2 Charging phase: when water in the boiler heat exchanger reaches the thermal valve's opening temperature, the thermostat gradually adjusts so that cold water is allowed to mix with the now warm water in the boiler heat exchanger. If the water temperature in the boiler heat exchanger drops, the amount of cold water that is mixed in is reduced, and vice versa. The thermostat thus acts as a mixing valve during the charging period and ensures that water in the boiler heat exchanger maintains a high and constant temperature and that the flow through the

The principle of placing the storage tank at the heart of a system. The boiler heats water that flows into the top part of the tank. From there, the water provides heat to the central heating system, thereby becoming colder, then returns to the boiler to be reheated. So that the water isn't too cold (at least 70°C) when it returns to the boiler, some of the heated water leaving the boiler is mixed with cool water entering the boiler. In the upper part of the tank, also, the hot water heater is situated.

Source: adapted from *Bioenergi Villa*, 1/2003

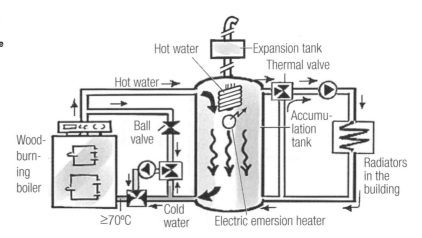

storage tank is as small as possible. This is important for preventing water temperature stratification. With this system, the top of the storage tank is charged with warm water and the border between hot and cold water is very marked. Charging can be stopped even if the storage tank is not fully charged.

3 Closing phase: once burning in the boiler stops, the boiler cools down and the thermal valve closes off the flow between the boiler and the storage tank. So, there is no heat loss during the boiler's idle period.

Operating Equipment Packages

For best cost and operating efficiency the various components of a heating system should be compatible. So many heating system manufacturers, for example of solar heating systems, market operating packages that include pumps, pressure tanks, safety-release valves and all the additional equipment needed for a system to operate adequately. Complete packages are available as it can be difficult for an inexperienced plumber to collect all the parts and find all the right components.

Thermostatic mixing valves (or shunt valves)

There are three faults that are commonly found in heating systems:

1 rapid flow systems that are difficult to adjust and noisy;

2 systems that are often oversized; and

3 shunt connections that are not properly constructed.

Shunt connections are used to mix hot water with cooler water to achieve a suitable temperature for water entering the heating system. Improperly constructed shunt

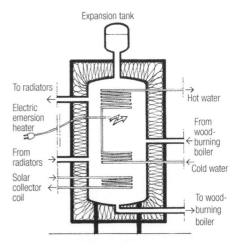

This is how a storage tank connected to a wood-burning boiler (at the bottom) and a heating system (at the top) can look. The tank has three coils for hot water production. The bottom coil is connected to solar collectors, the middle one preheats, and the one at the top provides top heat for the hot water. The electric heater is situated so that it can help with the top heating of hot water.

Accumulation tank.

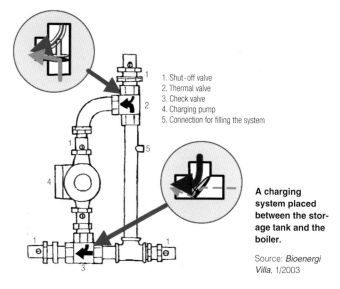

1. Shut-off valve
2. Thermal valve
3. Check valve
4. Charging pump
5. Connection for filling the system

A charging system placed between the storage tank and the boiler.

Source: *Bioenergi Villa*, 1/2003

connections result in a primary-side pressure that is too high so that the pumps work against each other. In homes, a shunt is not required if there are thermostats on the radiators.

Heating systems should be designed so that they require a minimal amount of pump energy. The Kiruna method is a system designed without a primary pump. It has a pump in the hot water tank cycle and a three-way valve connected as a mixing valve installed ahead of the pump that pumps water to the radiators. So the two pumps in the system work together and less pump energy is required.

Rapid Flow – Slow Flow

Many modern heating systems are rapid flow systems, capable of taking advantage of low water temperatures. However, such systems are difficult to control as they are not easy to adjust. Rooms easily become too hot and there is a gurgling and flowing sound in the pipes. Preferable are slow-flow systems where the flow and pump pressures are much lower. The temperature difference between the feed and the return water is much higher, so increasing the possibility of adjusting the heat so that a uniform room temperature is maintained in the entire system. The low pump pressure and slow water flow result in a quiet heating system with good thermostat function.

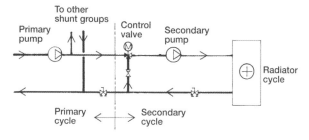

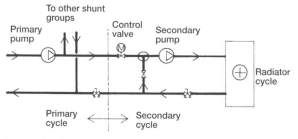

In many heating systems the pumps counteract each other, resulting in high energy consumption. One way to avoid this is to install a three-way shunt connection, as shown in the top diagram.

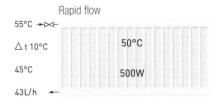

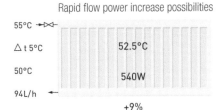

Waterborne radiators are a suitable heating method, but in order to achieve the best possible heat adjustment, they should be designed as slow-flow systems. In addition, slow-flow systems are quiet.

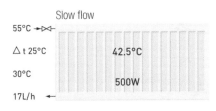

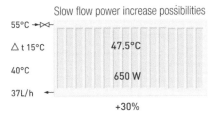

Gravity Systems

In the past, waterborne heat was distributed by gravity. In certain eco-buildings, natural circulation is preferred in order to avoid disruption due to power cuts. However, for the heating system to work without any electricity, there must be natural circulation to the radiators as well as the storage tank. Natural circulation systems require large pipes and carefully designed plumbing. Heating systems that work using natural circulation if the pump stops are also available, although they work more efficiently with the pump.

Most wood-burning boilers are dependent on electricity since they include pumps and fans. In areas with frequent power cuts, naturally ventilated boilers that work without fans are available. The storage tank can be connected to the boiler so that heat is transferred by natural circulation (without a pump). The water in a heat exchanger in a boiler dependent on a pump to transfer heat to the storage tank may start to boil if there is a power cut after the boiler has been lit. Another alternative is to have backup batteries in case of power cuts so that pumps and fans can be run with current from the batteries.

Radiator Heat

There are various types of radiators: flat radiators, convectors that consist of heating pipes covered with several layers of convector plates, and old-style as well as newly designed radiators. Only 20–30 per cent of the heat emitted by convectors is radiant heat and the rest is warm air. Furthermore, dust collects between

The heat radiating surface of large, waterfilled wall radiators is large, which improves comfort.

the plates. At higher temperatures, convection is stronger and dust is carried into the air. Radiators emit about 40 per cent of their heat as radiant heat. Flat radiators without convector layers are the best type of radiator since the proportion of radiant heat is greater than 60 per cent. They are also the most energy efficient since they are light and contain a small amount of water, and therefore can quickly react to changes in the system. Large flat radiators work with fairly low temperatures and are thus suitable for use with low temperature systems using storage tanks.

Underfloor Heating

Waterborne floor heat can consume up to 30 per cent more energy than radiators since it adjusts slowly and there are large heat losses through the floor. Floors require better insulation with underfloor heating systems than with other heating systems. The floor temperature should not be higher than 24°C. At 27°C the heat feels uncomfortable, convection occurs, and the air fills with dust. German building biologists also believe that too much foot warmth may cause varicose veins, and they therefore recommend that floor heat not be used in rooms where people spent a lot of time. Underfloor heating in bathrooms and halls is usually appreciated, but it should be possible to shut it off to avoid unnecessarily high energy consumption. Some builders are hesitant about installing underfloor heating because of the life of the plastic piping and the concomitant risk of leaks. Airborne floor heat adjusts even more slowly than waterborne floor heat but is not as problematic should a leak occur.

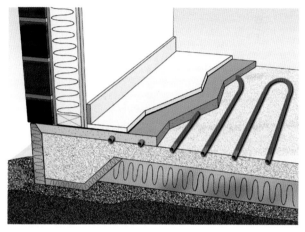

Waterborne underfloor heating is a popular method of heating. Its disadvantage is that it adjusts relatively slowly, and so it is difficult to take advantage of free heat. Furthermore, heating pipes are built in, which in the long term can mean leakage problems.

Source (illustration): Leif Kindgren

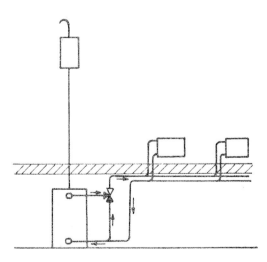

Natural circulation requires careful boiler and radiator siting as well as larger diameter heating pipes. It should, however, be pointed out that such systems are more difficult to adjust than pumped systems and are therefore more energy-intensive. Their advantage is that they work even during power cuts.

Source: VVS-teknik, Tord Markstedt, Sten Åström, 1972

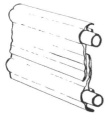

Lacquered metal heating baseboard, 5.5m long, 15mm diameter pipe x 1mm

Waterborne skirting board heating acts like a radiator. The radiator is designed as a metal skirting. This provides a relatively large radiator surface area that does not interfere with furniture placement.

Skirting Board Heating

There are two variations of skirting board heating: flat, metal mouldings that look like skirtings, and long, low radiators with fins that run along the skirting. In buildings with minimal energy requirements, skirting board heating along the inside of outer walls can be sufficient. Flat skirting boards function as flat radiators and emit 60 per cent of their heat as radiant heat. On the other hand, the long radiators with fins produce hot air that rises and warms the walls above them.

Wall Heating

Waterborne wall heating consists of pipes containing low temperature hot water that are recessed into the walls. Wall heating systems are popular among eco-builders in Germany. They require a large surface area and make some walls unsuitable for furniture, cupboards or paintings. The better insulated a building is, the smaller the wall heat surface area required. Energy losses for wall heating of inner walls are smaller than for underfloor heating. Wall heating systems placed in light inner walls made of gypsum board are the most energy-efficient and the easiest to regulate. However wall heating of heavy outer walls is considered to provide the most pleasant warmth, e.g. externally insulated solid wooden walls where heating pipes are plastered into clay plaster on the inside of the wall. The clay distributes the heat quite a distance, so pipes do not need to be placed close together.

Airborne wall heating systems consist of long, laminated pipe radiators built into the lower edge of a wall that heat up the air in a void in the wall, e.g. behind plasterboard. Also available are special hollow concrete blocks with air channels that are used to heat the air. Inner walls are thicker than usual in such systems.

Rappgo has developed a multi-function panel, the FFP panel, which is a wall panel with functions for heating systems and cabling.

Source: Rappgo.

 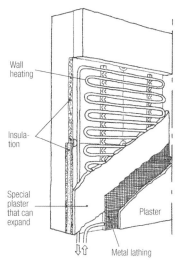

Waterborne wall heating is popular in Germany where people feel it produces an excellent interior climate. There are several reliable designs. However, the system is relatively slow and in the long term there can be problems with leaks.

 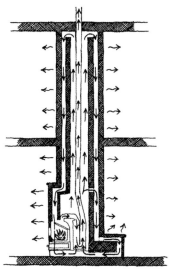

In Germany, the hypocaust system is common in eco-buildings. It is a closed hot air system built around a wood-burning appliance. Such a system is relatively expensive and limits floor plan options since the system must reach every room. However, it is considered to provide a very good interior climate.

Radiant Heat from Wood-burning Appliances

In highly insulated buildings with minimal heating requirements it is possible to get by with a single wood-burning appliance, and in many other buildings having a wood-burning appliance is a way to make use of renewable energy. They do, however, require an open floor plan so that the heat can spread throughout the building. In order to more easily distribute heat from a wood-burning appliance, ducts with fans can be installed, but then there are the problems of airborne heat. Heat distribution via long flues is also possible. For example, in Finland, partition walls where the chimney channels can be drawn to and fro are common and are designed in such a way that the flues are swept. In Germany closed airborne systems, hypocaust systems, are used, where heat from flue gases is transferred in a heat exchanger to heat air. The hot air then circulates via natural circulation in brick channels in the building. The wood-burning appliance, chimney and brick channels make up a sculptural unit that is placed centrally in the building. Wood-burning appliances can of course be complemented by small radiators in more distant rooms.

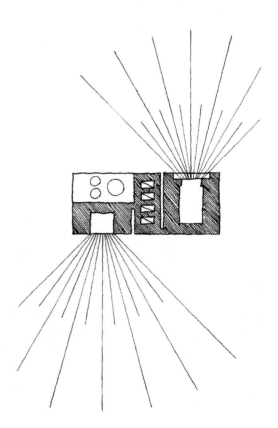

Buildings with an open floor plan can be heated using radiant heat from wood-burning appliances placed around a chimney. The figure shows a brick chimney unit with a wood-burning stove, a baking oven and a tiled stove placed round it.

The Grorud house in Oslo is a space-efficient building with a wetroom built entirely of mineral materials, which are visible in the façade. Mineral materials are used to avoid moisture and mould damage in the bathroom, which is subject to a lot of moisture.

Source: Architect Björn Berge, Norge

1.3 Construction

To build a healthy building, it is not sufficient to choose healthy materials. The materials must be combined appropriately in structures suited to their intended function. If mistakes are made, problems with moisture, radon, noise and ease of cleaning, for example, can easily arise. Moisture is a common cause of sick buildings. Noise creates stress and discomfort. Ease of cleaning and reasonable maintenance are important factors in healthy buildings.

1.3.0 Build Correctly from the Start

It is important to keep up to date with research and development. Environmental development moves quickly. New materials, construction methods and environmental techniques are coming onto the market all the time. It is equally important to receive feedback from practical experience, since many aspects of building ecologically are relatively new and untried.

Care and Maintenance

Care and maintenance are affected both practically and economically by how various methods are carried out and how details are designed. Ecologically constructed buildings should consider care and maintenance during the planning phase. Much of the information in this section (1.3.0) is taken from *Fastigheten som arbetsmiljö – bygg rätt från början*, Gudrun Linn, 1997.

Outdoor Care and Maintenance

Today, vehicles handle most care and maintenance tasks. Accessibility is required, yet it must not be an invitation to break traffic regulations. Access points are needed for disabled people, taxis, ambulances, fire engines, etc. Care and maintenance vehicles may require parking places in order to carry out their tasks. Differences in levels and curbs should be well thought through, especially with regard to snow removal. Ground cover should be planned with maintenance in mind, e.g. weeding.

Green areas should be designed to facilitate maintenance. Grass surfaces can be intensive (require cutting with a lawn mower) or extensive (such as meadows). Grass surfaces that require machine mowing should be connected and not too steep for a mower. Vegetation should not block views and outer walls can be protected from earth spattered by rain by placing a strip of gravel alongside them. Bushes that do not need to be pruned often may be preferred from a maintenance point of view.

Cutting hedges with hedge shears is a strain on arms and shoulders and prickly bushes can be a problem. Close-planting covers the soil surface and counteracts weeds. As the root systems of some trees, such as fast-growing poplar and willows, may invade water, sewage and electrical lines and lift hard ground surfaces, tree choice should take this into consideration. Some trees also drop more 'litter' than others.

Care and Maintenance of Buildings

Roofs should be accessible for snow removal, and roof hatches and roof ladders should be designed with this in mind. Windows and hatches that are motor operated, e.g. ventilation and smoke hatches, should be accessible for maintenance without involving risky climbing. Fans should be accessible from inside the building. Safety must be a consideration if they have to be accessed from outside – long roof walkways and ladders should be avoided. Glass façades should be equipped with permanent fixtures for both interior and exterior cleaning. Alternatively, accessibility should be arranged with moveable scaffolding or a cradle both inside and outside.

Care and Maintenance Indoors

Building maintenance is facilitated by controls that regulate services and point up problems. The equipment should be readily visible and accessible, and should not be more extensive than necessary. Fans and filters, including fans outside on roofs, should be easily accessible for servicing.

There should be readily accessible water stopcocks for every mains pipe and every tap. Water traps, including those in floor drains, should be accessible for cleaning. Culverts should be designed so that it is possible to stand up in them. Piping and other services should be placed on a wall where they are easily accessible. Crawl spaces should be avoided. Floor-level changes linked only by stairs hinder accessibility for cleaning machines, vacuum cleaners and wheelchairs. Ceilings below ducts and cables should be easy to take down and replace. Conventional thresholds present an obstruction for cleaning machines, wheelchairs, deliveries, etc. Lofts containing services, ducts, inspection panels and roof hatches should be easy to access for building maintenance staff, along with any equipment and spare parts that may be required. Stairs to fan rooms should be designed so that assembly parts and equipment can be easily taken in and out. Service access to windows and fittings high up in a room must be carefully thought through. In offices where there are many living plants, the requirements and maintenance of the plants need to be considered.

Refuse Rooms

The space and facilities for refuse management should be suited to the amount and type of the refuse as well as to the management system and how often the rubbish is collected. Accessibility for refuse collection vehicles and bins requires careful consideration.

1.3.1 Damp

The most common causes of damage due to moisture are:
- soil dampness (due to poor drainage)
- leaks (from services or flat roofs)
- poor construction
- building moisture (due to carelessness or insufficient drying times)

Moisture problems can also occur because of poor ventilation, capillary suction in materials, and heavy rain. They can also occur due to energy conservation measures such as weatherproofing which reduces ventilation, and increased showering which increases moisture load.

Moisture Damage

Moisture can cause many problems, such as mould, rot, increased emissions from construction materials and furnishings, and mites. Many homes, especially single-family dwellings, are subject to moisture damage. Signs of moisture problems include mould odour, condensation inside windows, moisture stains, rot in wood or decomposing stone and plaster. The sooner moisture damage is discovered, the easier it is to deal with. A leak is not always located close to the visible damage since water can run a long way before it becomes noticeable.

Faulty construction is the cause of 90 per cent of moisture problems.

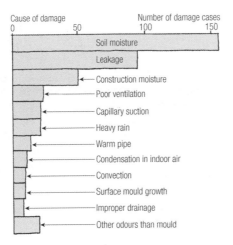

Damage caused by damp according to the Swedish National Testing and Research Institute.

Pay attention to the following to avoid moisture problems: Use construction materials made for the local conditions. Don't build in a way that, according to the laws of physics, will result in hazardously high levels of moisture. Don't close in construction materials that still have a high moisture content from when manufactured or constructed. Build using forgiving construction systems, so that any eventual moisture can dissipate or be taken care of without the occurrence of mould damage.

Damp, Mould and Rot

Mould is a common form of damp damage, especially in homes built in the 1970s and 1980s. Ever more hasty and careless building in the 1990s and beginning of the 2000s has its consequences. Mould develops at lower temperatures than rot, and therefore has become a common problem. As mould frequently occurs in hidden areas, it is often first noticed because of a bad smell. Mould and rot fungus spread via spores that are normally found in all air, both outdoors and indoors. For fungi to grow, access to moisture, heat, oxygen and a food source are required. Mould spores require at least 75 per cent relative humidity or more to survive, and the temperature must be higher than 5°C. Sugar or nitrogen sources are required as food, which the spores get from organic materials, such as wood, paper or textiles.

Soil Moisture

Soil moisture combined with inappropriately designed foundations and floors above suspended foundations are the most common causes of damp problems.

Rain and melted snow should be directed away from foundations. Good drainage requires a gradient of 15cm within 3m of foundations. In steeply sloping terrain, cut-off ditches above the building are often necessary.

Drainage under buildings should use washed gravel, crushed aggregate, or foamed glass (of the Hasopor or Glasopor type). The gravel or aggregate layer should be at least 150mm thick. Foamed glass has both drainage and insulating properties, and in such cases a 600mm thick layer is used.

Foundations should be well drained using a damp-proof course that prevents water from being drawn into the structure, and they should be connected to a drain that carries away water. It should be possible to inspect the drain and to rinse it clean when needed. Gulley beds should slope towards the drainage line.

Foundation systems vary in their susceptibility to damp problems. Plinth foundations are the safest alternative. Moisture problems can arise with suspended foundations if they are poorly ventilated. In order to avoid moisture problems, three measures are required: (i) avoid exposing organic material in the suspended foundation; (ii) prevent moisture from seeping up from the ground by laying a waterproof plastic sheet or a layer of expanded clay on the ground; and (iii) increase the temperature in the suspended foundation by insulating its walls.

For slab foundations: (i) place insulation under the slab; (ii) make sure there is a capillary break under the insulation; and (iii) provide good drainage.

Considerations for basement walls are as follows: (i) insulate the exterior of basement walls; (ii) make sure there is a ventilated weatherproof layer on the exterior of the basement wall; (iii) provide good drainage and make sure there is a capillary-breaking layer outside the weatherproof layer on the basement wall.

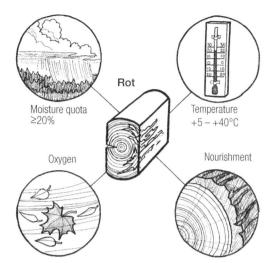

The occurrence of rot in organic material requires the simultaneous presence of: a food source, oxygen, a certain temperature and moisture.

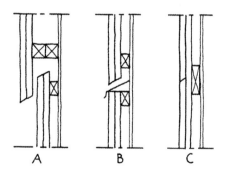

Construction techniques and design details are important for avoiding moisture problems. Three different ways to join boards are shown above:

A results in a board without cracks that is easy to paint;

B results in few cracks and good protection for the end grain; and

C results in slight cracks where nailed, the endgrain can't be treated when the panel is mounted; this method is, however, currently the one commonly used.

Source: *Beständiga träfasader*, Trätek

In rainy climates, buildings should have sloping roofs and proper drainage. Surfaces nearest a building should slope away from the building.

Leakage

There have been many moisture problems with flat roofs. It is only a quesion of time before a flat roof starts to leak, and it can be a major problem to remove snow. For the Nordic climate, roofs should slope properly. Good roof overhangs and proper window and door components reduce the risk of moisture damage. Most leakage problems are due to improper installations. Information on waterproof installations is in Section 1.2.3 Installations.

New Moisture Problems

It is important to build 'tight', that is airtight and waterproof, to keep warm and moist air from penetrating the structure. Making sure that joints are tight is of the utmost importance. This can be achieved by overlapping weatherproofing material and by covering joints with studs. Vapour-tight plastic is usually used for weatherproofing, but when building ecologically, it is common to use a diffusion-open weatherproofing layer, e.g. Gore-tex material, pasteboard or hard fibreboard. The inside seal should be tighter than the outside seal so that moisture can migrate outwards in the construction.

In some buildings insulated with cellulose fibre without an airtight plastic vapour barrier, mould has been found inside the loft roof. The roof insulation in these buildings has not been 'tight' enough. This, in combination with only slight excess loading on the upper floor, results in warm moist air leaking through the insulation and condensing on the cold inner side of the roof, where mould begins to form. This is not caused by use of cellulose fibre insulation, but by insufficient airtightness. Breathable construction requires airtight construction with pasteboard, plasterboard, hard fibreboard or special plastic. Particular attention must be paid to sealing joints, corners and connections.

Condensation can occur during cold winters even in airtight loft ceilings. A contributing factor is that during cold nights, especially when there is a clear sky, the inside of the outer roof can have a lower temperature than the outside air resulting in formation of moisture. Research by the Development Fund of the Swedish Construction Industry (SBUF) found that a solution is using diffusion-open underlay in cold attics in combination with reduced ventilation in the winter.

Well-insulated suspended foundations are subject to condensation and mould beneath the subfloor, especially during the summer. In the past, poor insulation and warm chimney stacks warmed up the suspended foundation and so moisture damage was uncommon.

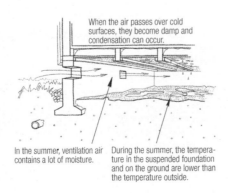

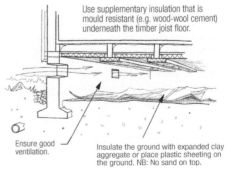

In recent years many wooden houses have been built with plaster directly on cellular plastic insulation without an air gap. In the wooden construction there are gypsum boards with paper. Moisture has penetrated the plaster at fixtures, gone through the insulation and caused mould on the paper of the gypsum boards. Today, it is recommended that plaster be put on mineral-based boards and that there be an air gap between the façade and the insulation. Due to mould problems gypsum boards with paper now are forbidden in bathroom walls.

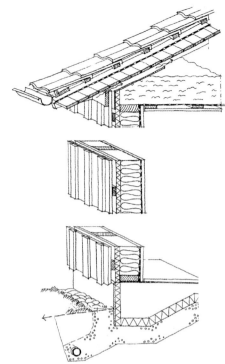

In order to avoid moisture problems, buildings must be airtight. In conventional constructions, plastic sheeting is used that is vapour-proof as well as airtight. In ecological constructions, airtightness is achieved using pasteboard sheets (require extra careful workmanship at joints and corners) or vapour-permeable plastic. The inner structural layer should be more moisture-proof than the outer layer, which should guide the choice of materials.

Source: From 'Bygg fuktsäkert', BFR, 1995

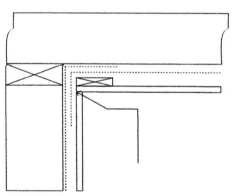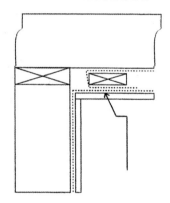

Where the wall meets the roof, windproof pasteboard is used as the inner airtight layer. It is important to overlap the layers and fasten them securely.

Source: *Lufttäthet i hus med träregelstomme och utan plastfolie*, Eva Sikander, Agneta Olsson-Jonsson, SP Rapport 1997:34

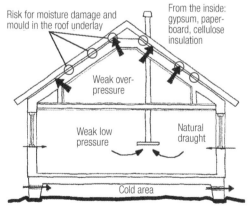

Risk for moisture damage and mould in the roof underlay

From the inside: gypsum, paperboard, cellulose insulation

Weak overpressure

Weak low pressure

Natural draught

Cold area

In well-insulated lofts with poor ventilation, condensation and mould have formed on the underside of the roof. This has occurred most in structures that were not airtight and where there was excess loading. Care must be taken to ensure that the inner roof's inner layer is airtight and that loft ventilation is good.

Source: *Ny Teknik* 4/1997

1.3.2 Radon

Radon may cause many lung cancer deaths per year. Radium, a decay product of uranium, decays into radon gas. When radium in the ground decays, radon is released into any air or water found in the ground. As there is usually lower pressure in buildings than in the ground, air from the ground with radon in it can be sucked in if the foundations are not airtight. Some building materials contain more radium than others, e.g. blue lightweight concrete made from uranium-rich alum shale. Such concrete emits radon. Radon can also be released into groundwater and come via the tap water into buildings using deep wells.

Guidelines and Monitoring

Radon is an inert, odourless gas that cannot be seen or felt. Radon gas decays into highly dangerous radon decay (radon daughter) product, particles that may get stuck in the respiratory system if either radon or radon daughters are inhaled. The four radon daughters in the uranium decay series all decay within about 50 minutes, and each releases dangerous radiation. It is especially dangerous to smoke and live in a building with a raised level of radon.

The occurrence of radon gas varies. Municipal environmental offices may have maps showing high-risk areas. Ground radon is the most common source of radon gas in homes. In blocks of flats, blue lightweight concrete used in construction is the main source of radon.

Radon levels are measurable, and are therefore not considered to be a hidden defect when selling a building. Radon levels should be measured prior to purchasing a building if there is the least suspicion of a radon problem. Accurate radon measurements are easily carried out, but take at least two months during the time of the year when heating is used. At least two measuring kits should be ordered from a laboratory or municipal environmental office and placed according to instructions. Kits may then be sent back and results are received after a few weeks.

The Swedish Radiation Safety Authority has estimated that there are about 500,000 homes in Sweden with levels of radon over 200Bq/m³.

Legend

High-risk area
- Granite and pegmatite with elevated (high) levels of uranium (and radium 226) and elevated gamma radiation.
- Moraine, sand or gravel with elevated levels of radium 226 or with radon levels over 50,000Bq/m³.

Local high-risk area
- Gravel or course sand where local radon levels can be over 50,000Bq/m³.

Normal-risk area
- Land areas consisting primarily of ground and bedrock with normal levels of radioactivity.

Low-risk area
- Areas consisting of fine sand, silt or clay.

Many municipalities in Sweden have published maps that show high-, normal-and low-risk areas for ground radon.

Source: Radon map of Ekerö Municipality, Sweden

Precaution with New Buildings

Radon problems can be avoided in new buildings. The sources of radon problems in new buildings are ground radon and tap water in a building with its own deep well. Land is divided into three categories with regard to radon risk: low-radon land requiring no special precautions, normal-radon land requiring protective design, and high-radon land requiring radon-proof design.

Protective design against radon should be applied in areas with normal levels of radon in the ground. Lowest levels of interior radon are achieved in a building that does not admit air from the ground and that has properly adjusted mechanical intake and exhaust air ventilation.

- There should not be obvious leaks between the ground and the floor and walls.
- Avoid slab edge insulation that allows ground air in along the edges of the concrete slab.
- Build so that settlement is avoided.
- Seal areas where pipes penetrate a concrete slab foundation.
- Ensure that joints and places where pipes penetrate floor structures over suspended foundations are airtight.
- The building should also be airtight above the ground.

Radon-proof design is required in high-risk areas. It is important to ensure that ground air will not leak into the building. To meet these standards, some of the following combinations can be used:

- edge-reinforced, airtight concrete slab or sealed cellular glass foundation. Use of a radon-proof sheet is an alternative;
- airtight pipe penetration points;
- concrete exterior basement walls;
- drainage hoses in the damp-proof course under the building are connected to one pipe

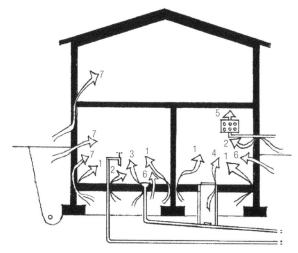

1 Crack between concrete floor and wall.
2 Leaks at pipe penetration points as well as in protective pipes.
3 Leaks at floor drains.
4 Leaks around drain inspection hatch.
5 Leaks around ducting for electric and telephone lines.
6 Cracks in the floor or wall due to settlement.
7 Leakage through air-permeable construction material.

Common leaks in a building's basic construction.
Source: Radonboken, BFR, 1992

that goes up through the building or to the outside edge of the concrete slab. If the air pressure under the building needs to be lowered, a fan should be installed in the pipe.

One alternative is to build in ventilated air gaps in the floor and on basement walls (if there is a basement). Air coming in from the ground is sucked away in the ventilated air gap and does not get into the interior air. In addition, pipe penetration points should be airtight. Other alternatives are to build an unencased plinth foundation, since air from the ground will be blown away, or a well-ventilated suspended foundation. However, note that extra insulation is then required for pipes, plinths and suspended floor structures in order to prevent damage from freezing and cold floors.

It is important to examine the design and prepare for simple additional measures that can be carried out if the level of radon daughters in the interior air turns out to be too high.

Reconstructing Buildings with Elevated Radon Levels

For existing buildings, there are a number of ways to deal with the problem of radon from the ground. Leakage points can be sealed and/or the direction of air flow between the ground and the building can be changed. If nothing else helps, interior ventilation can be increased, but intake air should be radon-free and ventilation should occur with higher pressure in the building.

- Seal cracks and gaps against ground radon.
- Change the air pressure between the ground and inside the building.
- Ventilate the ground under the building or the suspended foundation.
- Increase interior ventilation.

It can be difficult to seal cracks and gaps with caulking material so that the seal holds for extended time periods. Ventilation slots between flooring and walls should not be sealed in order to avoid a risk of moisture damage to floors.

Radon suction, radon drainage and the air-cushion method are examples of preventative measures that affect air pressure.

Radon in Construction Materials

Blue lightweight concrete, used in both walls and floor structures since the end of the 1920s but primarily between 1945 and 1975, can cause radon levels more than five times greater than is acceptable. It is not always easy to know what materials are hidden in walls, floors and roofs. Gamma radiation measurements can determine if lightweight concrete is present. A properly applied sealant layer may at best lower the level of radon in a room by 50 per cent. Ventilation can also be increased to get rid of radon. Ventilation can be increased indoors by installing, for example, a balanced air intake and exhaust ventilation system with a heat exchanger. The heat exchanger keeps energy consumption low. Installation of a fan-driven ventilation system often lowers the pressure indoors, and thus also the leakage of ground radon into a building, and so should not be used as a preventative measure against ground radon.

Radon in Tap Water

It is primarily water from deep wells in areas with uranium-rich granite and pegmatite that may have high levels of radon. When a water tap is turned on, as during a shower, radon enters the interior air. According to the Swedish Radiation Safety Authority the greatest health risk of radon in water is that the radon is transferred to indoor air.

International radiation protection research has shown that the health risks of drinking water with high levels of radon in it are much greater than earlier believed, particularly for children. There is technology available to remove radon from water (see Section 2.3: Clean Water, in Conservation

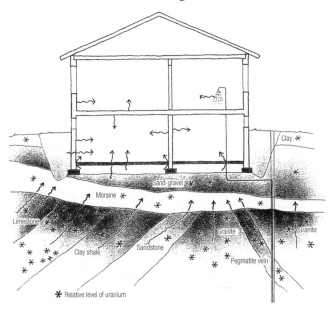

Radon is emitted by mineral particles in the ground to the air or groundwater between the particles. Radon can be sucked into a building together with air in the ground as air pressure indoors is lower than in the ground. The greater the pressure difference, the more leaks there are and the more air-permeable the ground is, the greater the amount of radon sucked into a building.

Source: *Radon i byggnader*, Text häfte till video, Bertil Clavensjö et al, BFR, 1999

of Resources). If radon levels are high in water, it is likely that interior air also contains radon. It may be possible for municipal environmental offices to answer questions about carrying out an analysis.

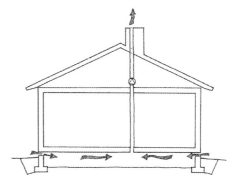

A method of building in a radon-safe manner on ground emitting radon is to have a separate exhaust air ventilated crawl space. The channel is insulated within a heated space so that moisture does not form. It may be necessary to equip the construction with a fan.

Source: Radonboken, BFR, 1992

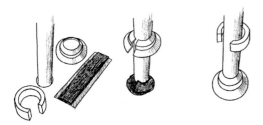

Pipe penetration points are sensitive areas with regards to radon penetration. There are several good systems to seal penetration points, which often are the cheapest preventative measures.

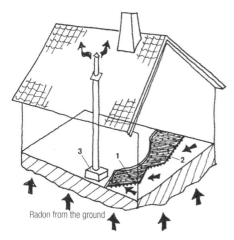

Radon from the ground

1. A dense plastic mat (e.g Platon) is placed over the cement slab (the mat also stops moisture and mould).
2. The radon remains in the air space between the mat and the concrete.
3. A fan sucks out air with radon in it.

In an existing building on radon-emitting ground, the radon can be sucked away if an extra airtight floor is added so that there is an air gap between the two floors.

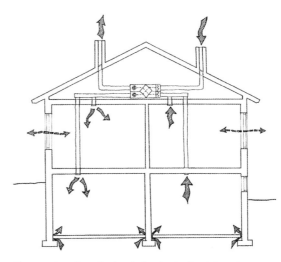

One way to reduce the level of radon in the air is to increase ventilation. In this case a building can be equipped with an intake and exhaust air ventilation system with a heat exchanger so that too much energy is not lost.

Source: Radonboken, BFR, 1992

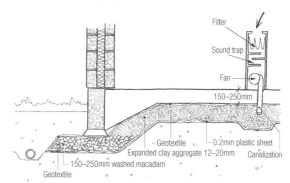

An alternative is to equip existing buildings with a radon exhaust system that expels air through the drainage layer and so prevents radon from entering the building.

Source: Radonboken, BFR, 1992

1.3.3 Sound and Noise

Noise problems are common in everyday life. People are disturbed by traffic and unwanted sounds from ventilation and heating systems. People are also affected by low-frequency sound, infrasound, that is not consciously perceived, such as noise from fans. People may be bothered by impact sound and air sounds in buildings. These problems can be prevented by using heavy construction materials, discontinuous construction, special window designs, etc. In a noisy locality, sound insulation measures should be included in any new construction.

Noise

The unit decibel, dB(A), is often used to describe sound volume. The 'A' refers to the different frequencies being weighted to correspond to how the human ear interprets the sound volume. An increase of 3dB(A) is equivalent to experiencing a doubling of the sound volume.

In Sweden, slightly more than two million people are exposed to noise that influences health and almost one million report being bothered within the vicinity of their home, often by traffic noise that is over 55 decibels. Noise from all sources together should not exceed 55 decibels. The sound from a normal conversation is 60 decibels. EU countries are required to map noise problems and establish remedial measures in cities with a population of more than 100,000 by 2013.

The National Board of Housing, Building and Planning (*Boverket*) has set the following long-term goals according to the noise guidelines supported by the Swedish Parliament:

By the year 2020 noise levels in home environments will not exceed:

- 30dB(A) equivalent indoors;
- 45dBA maximum level indoors at night;
- 55dBA equivalent level outdoors (at the façade);
- 70dBA maximum level in outdoor areas associated with homes.

These are guidelines for new construction or substantial roadway reconstruction.

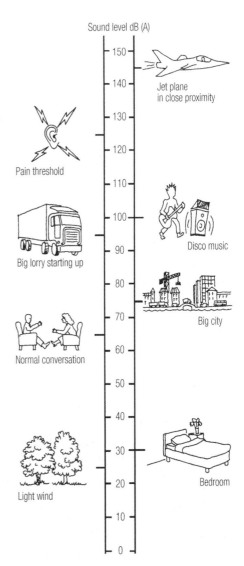

Sound level is measured in decibels on a logarithmic scale.

Sound Reduction

In our stressful society more and more people strive after more tranquil and quieter environments. Several studies unequivocally show that a good sound environment is high up on residents' wish lists. In Sweden in 1999 a new sound standard came into force that takes into consideration frequencies down to 50Hz, in other words the base level that sound systems emit. It can be difficult to comply with the standard using common light construction techniques. It is easier to meet the requirements using heavy building systems.

Following are the Swedish sound classes according to standard SS 252 67. The difference between classes is normally 4dB.

Sound class A: corresponds to very good sound conditions.

Sound class B: corresponds to significantly better sound conditions than sound class C.

Sound class C: provides satisfactory conditions for the majority of residents and can be applied as a minimum requirement according to the National Board of Housing, Building and Planning regulations.

Sound class D: corresponds to conditions meant to be applied when sound class C cannot be achieved. An example is older homes that for particular reasons cannot be renovated in a way that meets the requirements of sound class C, e.g. when careful restoration work is being carried out.

Sound class B is usually applied for new construction.

The Quiet Building

It is possible to build quiet flats. With the help of common sense, holistic thinking and collaboration, housing free from noise disturbances has been built, for example in a project called 'The Quiet Building' in Lund, Sweden. Neither traffic, neighbours, plumbing nor fans disturb those who live there. The quiet buildings were constructed using thicker concrete, floor plans and services designed to minimize sound disturbances. The cost of constructing buildings that are experienced as quiet can be 2–3 per cent higher, which may be considered to be within reasonable limits.

Separating Walls and Floors in Flats

There are three ways to reduce sound transfer: one is to use heavy materials; a second is to use double rather than solid structures; and a third is to fill empty spaces in structures with insulating material. Increasing the thickness of walls also improves the sound climate. Impact sound can be muffled through proper choice of floor covering or by putting insulation between the floor covering and the load-bearing construction.

1. Direct sound transmission
2. Flanking transmission
3. Overhearing
4. Noise leakage

Today many people live in a poor sound environment. The need for sound insulation has grown due to an increase in new sources of disturbing sound, such as heavy traffic, powerful stereo systems and fans. Sound transmission through the air between two spaces is usually given as a total of direct transmission, flanking transmission, overhearing and leakage.

Timber frame buildings with more than three storeys make it especially important to consider sound insulation between flats. Several new construction methods for this have been developed. In 'The Quiet Building' in Lund, walls separating the flats were made using 240mm concrete instead of the usual 160mm. The thickness of the floor structure was increased from the usual 200mm to 290mm, and the outside walls' inner concrete layer from 150mm to 240mm. It is important to know that one consequence of increasing the amount of concrete is an extended drying time.

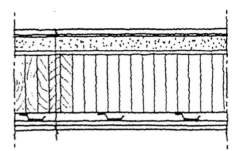

A solid timber floor between flats. Under the floor joists, metal acoustic sections are attached to plasterboard. The outer floor surface consists of parquet flooring laid on top of impact sound-absorbing material.

Sound-Insulating Windows

Special sound-insulating windows are available for buildings exposed to a lot of outside noise. Better windows are considered the most appropriate preventative measure in at least half of all buildings exposed to noise. An unfortunate consequence is that many desirable sounds from outside are also excluded, such as birdsong, children playing and the patter of rain. Renovation to improve noise reduction can include adding inner windows (with a glass thickness of about 1cm) onto existing windows, which can reduce noise by about 35dB.

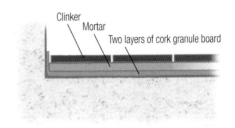

An example of impact sound insulation with cork granule board under a expanded clay floor set in mortar.

Source: Cementa

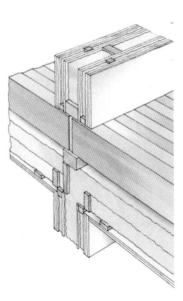

For walls separating flats, using discontinuous construction does not usually provide sufficient noise reduction. Double walls are often built and the floor structure should also have separated sections. Solid timber floors between flats should be insulated on the underside if the solid wood acts as a floor surface without any other floor covering.

Source: *Massivträ*, Teknisk beskrivning, Trätek, 2000

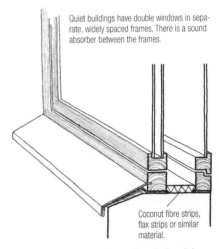

Window with reinforced sound insulation. It is now most common to use laminated glass with sound-absorbing foil in the insulating glass.

Source: Projektet 'Det tysta huset', JM Byggnads och Fastighets AB, Lund, 1989

Quiet Services

Since dB(A) values do not give a true picture of disturbance by low-frequency sound, requirements for the highest sound level in dB(A) should be supplemented by requirements in dB(C). If the C-level is more than 15dB higher than the A-value, the sound can be said to be dominated by low frequencies. If the C-level is above the A-level by 25dB units or more, it is a question of more serious sound disturbances.

There is a lot to consider on this subject. Circular channels are better than rectangular ones. Pipes can be laid in well-insulated ducts and in a vibration-absorbing manner. Circulation pumps and fans can also be mounted so as to prevent vibration. In addition, pressure-reduction valves can be placed in heating pipes.

The following is a checklist of things to consider in order to achieve a quieter ventilation system:

- location of fans and ducts;
- sound absorption;
- sound insulation;
- choice and location of fans;
- fan connection;
- ducting and sound insulation;
- duct mounting;
- ventilation fixtures;
- vibration insulation.

R°_w/ D_{rTw}	Murmur	Normal speech, Office machines in calm environment	Normal speech, Office-machines	Loud conversation	Shouting	Sound from a loudspeaker, moderate level	Discomusic
35							
40							
44							
48	YELLOW – audible						
52	WHITE – inaudible						
60	GREY – audible but not diturbing under normal conditions						

Subjective interpretation of sound with various types of sound insulation.
Source: 'Bullerskydd i bostäder och lokaler', Boverket, 2008.

	Sound insulation measure	Sound reduction
	a) No insulation	0dB(A)
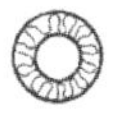	b) 50mm mineral wool preformed pipe section (glass wool 50kg/m^2 or rock wool 150kg/m^2)	12–14dB(A)
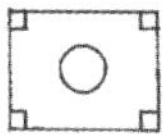	c) Box of 13mm plasterboard (dimension: 300 X 400mm)	14–18dB(A)
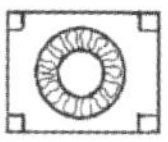	d) Box as in c) and mineral wool as in b)	25–30dB(A)
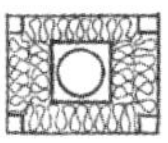	e) Box as in c) and 50mm mineral wool sheets (glass wool 36 kg/m^2 or rock wool 75kg/m^2)	24–28dB(A)

To reduce the noise from services, ducts can be encased and insulated. The illustration shows an example with a vertical plastic (high density polythene) pipe.

Source: 'Regnvatteninstallation', Svensk Byggtjänst, 1978

Choose quiet fans that are installed so that vibrations aren't transmitted. For a quiet system, make sure that motors (not requiring high-speed operation), fans, ducts and fixtures are overspecified in relation to the norm. Disturbing sounds from fans can be muffled using sound traps with baffles.

Joining Structural Elements

Sound can be transferred from one structural element to another through continuous construction. This can be avoided by breaking the continuity in the construction with muffling material. Steel supports on walkways and balconies can be insulated with rubber spacers to reduce impact sound. Garages located beneath residences can be made with a floor slab separated from the structural frame with special insulating material. The ground-floor joists can be constructed so that vibrations are directed towards the garage underneath. Stairwells with expanded clay floors can include a cushioned layer on top of the concrete.

Lifts

Lifts may be muffled with rubber. To avoid rattles in lift doors, side-hinged doors should be chosen instead of sliding doors with rattling door panels. Side-hinged doors are equipped with dampers, which prevent them from being shut with such force that direct sound disturbance is created. The door frames are fastened to the studwork with rubber plugs in the frame so that there is no direct contact with the frame.

Sound Muffling

Sound in a room can be muffled with sound absorbers made of wood-wool cement, perforated plaster, acoustic plaster or thick textiles.

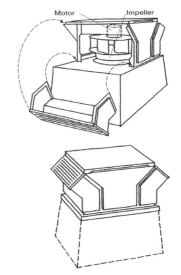

Some fan companies have developed quiet fans designed for easy maintenance.

Source: Gebhardts genovent

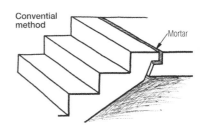

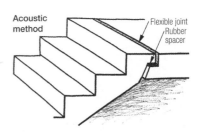

For acoustic reasons, prefabricated concrete steps should be installed with flexible joints and rubber spacers.

Source: Projektet 'Det tysta huset', 1989, cooperation between BFR, JM and SABO.

To avoid transfer of sound in ducts, the ducts can be equipped with sound traps. The sound trap in the illustration is insulated with cellulose fibre.

Source: Acticon AB

1.3.4 Ease of Cleaning

Cleaning costs constitute a significant part of maintenance costs. It is important to choose easily cleaned materials and to clean regularly. Environmentally friendly cleaning methods should be used (preferably dry methods). The most efficient methods are preventative, because rough surfaces, especially vertical ones, collect more dirt than smooth surfaces. Electrostatic charges also influence the build-up of dirt. The chief source of information for this section was Gudrun Linn's book, *Bygg rätt för städning och fönsterputs*, and the folder *Renhold og indeklima begynder på tegnebordet*, published by the Danish-based company ISS.

Three Simple Principles

There are three simple, interrelated principles for achieving an easily cleaned building:

- Dirt should be prevented from coming into the building.
- Polluting activities should be efficiently separated from other activities.
- It should be possible to remove dirt easily.

Dirt Traps

One of the most important measures to facilitate cleaning in a building is to prevent dirt from coming in. The entrance should have a proper protective roof, and the ground should be paved and slope gently away from the entrance. There should be an enclosed porch, sufficiently long, and the doors should open in the same direction to make it easy to go in and out. Every doormat should be big enough to take two steps with each foot on it. A rule of thumb is that doormats should be at least 2m long. The risk of slipping on floor grates can be avoided if they are made of grid-patterned flat iron.

Accessibility

Accessibility for cleaning is crucial for the easy removal of dirt and dust. Some examples of easy accessibility are as follows:

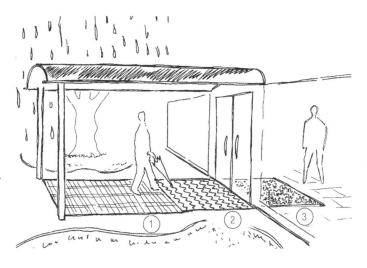

Dirt trap entrances to offices can be designed as shown:

1) floor grate, 2) rubber mat, 3) natural fibre mat made of coconut fibre.

In addition, the ground adjacent to the entrance can be heated, and the lobby can have a stone floor.

Buildings should be easy to clean.

- Glass roofs require special arrangements for maintenance. Permanent fixtures and ladders on tracks are preferable to loose ladders.
- Spaces under stairs should be accessible for cleaning.
- It should be possible to reach open interior structures for cleaning since dust collects on them. This includes interior roofs over reception and similar areas.
- Cleaning is facilitated by open floor areas without pillars, furniture legs, piping or cables. Electric points and outlets for computer connections should be sited so that cables are short and not left lying on the floor, since that makes cleaning more difficult. Ceilings are the most suitable locations for electric points.
- Specifications should include requirements for a good working environment. Accessibility and adequate space to carry out maintenance should be considered, as well as ease of changing fan filters and other components that wear out. More than

- Locating load-bearing pillars near walls is important for ease of cleaning floors.
- For areas cleaned by machine, there should be access to as much of the floor area as possible.
- Window surfaces must be accessible from both inside and outside, which means that the ground outside a building must be suitable for a cradle installation. In addition, pillars and so on should not be located too close to inside windows, thereby preventing access to the entire window surface.

THE OFFICE OF A CLEANING COMPANY (ISS)

It is important to think things through right from the earliest design stage in order to avoid dust and dirt traps as well as providing easily cleaned spaces. Design detail is important.

The following example comes from the Stockholm office of the cleaning company ISS:

- Rounded corners make it easier for cleaning staff to move around and reach everywhere with their machines.
- There are no thresholds except where necessary, making it easier to navigate cleaning machines.
- Dust-gathering ledges have been avoided, so all floor mouldings are recessed into partition walls (this is more difficult on load-bearing walls, but there are places where this has been done), and glass sections have rounded mouldings.
- All electric and computer cables are laid in cable troughs and never on the floor.
- Cupboards and cabinets extend to the ceiling.
- Blinds are enclosed between window panes and collect little dirt or dust.
- There is an easily maintained marble floor in the entrance.

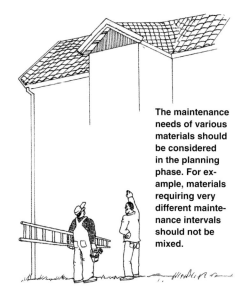

The maintenance needs of various materials should be considered in the planning phase. For example, materials requiring very different maintenance intervals should not be mixed.

On outward opening triple-light windows, one of the outer panes is always difficult to reach from the inside, as is the outer pane of outward opening single-light windows.

Source: *Bygg rätt för städning och fönsterputs*, Gudrun Linn, 1999

the minimum dimensions are needed for work to be carried out comfortably. Aids for heavy lifting should be available.

- High fixtures (including those in stairwells) are difficult to reach for servicing, e.g. changing light bulbs and for cleaning.

Factors to Consider

- Open-edged landings may allow dirty water to run over the edge and leave stains that are difficult to remove. A cleaning strip eliminates the problem.
- Doorstops should be installed high on walls and doors so that they do not get in the way of cleaning trolleys, vacuum cleaners, etc.
- Surfaces should be as smooth as possible for ease of cleaning. However, for flooring this requirement may not be compatible with skid-resistance.
- In areas to be cleaned by machine, durable flooring should be chosen, so thin ceramic tiles or plastic tiles may be unsuitable.
- Flooring in entrances should be tough, of good quality, and properly treated.
- Entrance walls may be exposed to wear, dirt and perhaps graffiti. Vehicle collision protection is often required.
- Good lighting levels are needed for cleaning.

Services

- Tiltable radiators that provide radiant heat simplify cleaning. Convectors (radiators with fins) that warm the air in a room are almost impossible to keep clean.
- Radiators placed close to the floor are a hindrance to floor cleaning and vacuuming of skirtings behind them.
- Suspended ceilings, fittings and ducting should be designed with cleaning in mind. Ceiling pipes should be positioned so that it is possible to clean on top of them.
- The backs of refrigerators usually are equipped with an evaporator that releases the heat taken out of the refrigerator. Some refrigerators have built-in evaporators, but on those that don't, it is worth pulling out the refrigerator regularly and vacuuming the coils. This can improve performance from 20–40 per cent.
- It is very important for a building to be clean during the entire building process.

So, for example, all ducts and spaces inside walls should be vacuumed out before they are closed off.

- It is a fact that intake air systems are not easily accessible for inspection and cleaning. So they are in fact never cleaned. Ventilation systems often pollute fresh air. It should be possible to clean them and especially the air intake ducts regularly.

Cleaning Equipment Storage Area

Cleaning equipment storage areas should be large enough so that cleaning equipment can be easily put away. There should be readily accessible electric points for recharging machines with rechargeable batteries. It should not be necessary to move anything, such as a machine, in order to reach something, e.g. a cleaning product, from a shelf.

Access to electricity and water is important for those who clean. Wall sockets, water taps, floor drains and outlets for central vacuum cleaning should be easily accessible. Access to water and electricity should sometimes be placed in areas where it otherwise wouldn't normally be, such as in stairwells. Water taps are needed on several floors in multi-storey buildings.

Rooms for Bicycles, Prams and Buggies

Rooms for bicycles, prams and buggies should be placed as close as possible to entrances since the wheels bring in dirt. In multiple-entry blocks of flats, small lockable rooms should be located adjacent to every stairwell so that fewer people have keys. Bicycle racks should be mounted on walls or ceilings to make it easier to clean the floors. It should be possible to lock bicycles to the bicycle racks.

Bathrooms and Toilets

Wall materials should be easy to clean, e.g. tiled. It stands to reason that bathroom floors should be made totally accessible by

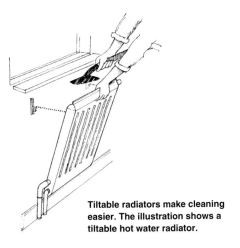

Tiltable radiators make cleaning easier. The illustration shows a tiltable hot water radiator.

wall-mounting toilets, basins, fixtures, piping and even toilet brushes. There shouldn't be any nooks or crannies. It is easier to clean the floor under bathtubs without fronts or with half-fronts. Built-in showers recessed into the floor are preferable to prefabricated units, which can be difficult to clean around. From a cleaning point of view, it is better if partitions are permanently attached to walls. Pipes should be accessible for repairs, but it is best if they are hidden behind a panel that opens so that dust doesn't collect on them. For ease of cleaning, 50cm is needed between walls, toilets and basins.

Furnishings

It is necessary to be able to get under furnishings everywhere in order to vacuum and clean. Furniture on castors is a good idea. There are many special types of castors on the market that can be used. Furniture on castors is easier to move, especially if it is heavy, and doesn't scratch the floor. Couches and armchairs can have protective covers that are easy to wash and replace.

Chemical-Free Cleaning Methods

A lot of chemicals are used for cleaning and surface treatments. Much of this chemical use is routine and could be replaced with dry methods or with soap and water. Steam cleaners may be used for professional clean-

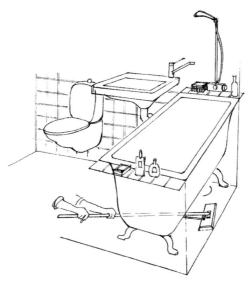

Bathrooms are often difficult to clean. Toilets, basins and toilet brushes should be wall-mounted. Avoid pipes that go down through the floor. Bathtubs without fronts are preferable so that it is easy to clean underneath. For ease of cleaning, fixtures should be at least 20cm above the floor.

ing. Windows do not normally need to be cleaned with a window cleaning solution. A microfibre cloth dampened with water often does the job just as well. Chemicals are seldom needed when a microfibre mop is used. If a surface is extra dirty, a little soap can be used in most cases. Microfibre cleaning cloths and mops should be cleaned often by boiling them on the stove or by being machine washed at 60 degrees, preferably in a net washing bag. Rinsing agents should never be used. They remove the static effect of the cloth that makes the dirt stick to it.

Cleaning Chemicals

Do not use cleaning agents unnecessarily, but if they must be used, choose environmentally certified ones. Not much planning is required for choosing environmentally friendly chemico-technical products. There are reliable lists of environmentally certified and environmentally hazardous products.

Bra Mijöval (Good Environmental Choice) or Swan-labelled products do not contain substances hazardous to health and have ingredients that are easily degradable.

When microfibre cleaning equipment is used chemicals are often unnecessary and use of water is adequate. However, if cleaning agents must be used, use milder ones such as soft soap, vinegar, citric and tartaric acid or environmentally labelled cleaning agents.

Soft soap is made of natural raw materials and can be used on most floors and surfaces.

Vinegar is good for disinfecting inside toilets.

Vinegar, tartaric and citric acid work well against rust and calcareous deposits in toilets, sinks and bathtubs.

Some cleaning agents are unnecessarily strong and can for example leave scratches on cleaned surfaces. They should only be used when really required. Chlorine should only be used in exceptional circumstances, for example to avoid transmission of infections via toilets.

Strongly alkaline substances should never be used to unplug toilets as they can damage piping. Instead, the stoppage may be able to be removed with lots of hot water or by dissolving a cup of bicarbonate and a cup of salt in a pot of boiling water and pouring it down the toilet.

Floor Cleaning Methods

About 70 per cent of all floor replacements are due to damage caused by improper cleaning methods and cleaning agents.

Dry-cleaning methods such as dry-mopping are becoming more common. Microorganisms can multiply in the moisture that remains after wet-mopping a floor. If a floor is first dry-mopped, then wet-mopped, the number of microorganisms doesn't increase.

For the everyday cleaning of linoleum floors, dry methods are used in combination with spot removal. For wet-mopping, wax or a neutral cleaning agent is used. In some locations, it is necessary to use polish in order to create an easily cleaned surface, e.g. entrances, dining areas as well as on some old floors.

As the contents and characteristics of plastic flooring can vary greatly, it is best to contact the supplier for proper maintenance instructions. However, in most cases dry-cleaning methods in combination with spot removal and wet-mopping are adequate. When moisture is added, a neutral cleaning agent prepared with water and/or a microfibre mop is used.

An impregnating agent is required both to finish concrete mosaic (terrazzo) floors as well as for the everyday cleaning process. The impregnating agent reacts with the lime in the terrazzo cement to create lime soap, which provides a dirt- and water-repellent surface. Other alternatives include special neutral soaps with a high fat content or soft wax. Polish or strong alkaline products (e.g. many common soaps) should not be used on concrete mosaic. Acidic substances should also be avoided as they corrode the flooring.

Stone such as marble, limestone, granite and gneiss that is sanded and polished should be cleaned with a neutral cleaning agent. Such stone surfaces are very dense and use of fatty products result in a sticky surface. When marble, limestone, granite and gneiss are sanded, the surface becomes porous. It is then suitable to use neutral soaps with a high fat content. These products fill and impregnate the pores, which results in a dirt- and water-repellent surface.

Ceramic flooring with tiles that aren't fully vitrified has a surface porosity that is comparable to a stone floor. Periodic maintenance with pore-filling expanded clay oil is recommended. Glazed and fully vitrified ceramic flooring has a density similar to polished stone flooring. Soap stays on the surface resulting in a sticky surface. Such flooring should be cleaned with neutral cleaning agents. There are, however, some exceptions and it is best to check with the flooring supplier.

Laminated flooring is sensitive to moisture and water and should be cleaned with neutral cleaning agents.

Wooden floors are also sensitive to moisture and water. Oiled floors should be treated one to four times per year with maintenance oil. For everyday cleaning, a product should be used that re-oils the floor and prevents drying. The product should be suitable for use with the appropriate maintenance oil. Lacquered wooden floors should be cleaned as dry as possible. If wet methods are required, neutral cleaning agents should be used.

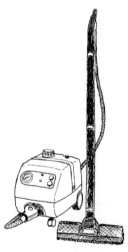

Steam cleaners are machines that can clean even extremely dirty surfaces without the use of strong chemicals.

Brown soap, made of vegetable oils, has been used for hundreds of years and is excellent for cleaning and surface treating wood, stone and textiles.

1.4 Implementation

Constructing an environmentally friendly building requires systems thinking that follows through the entire process. Construction should be monitored by an environmental manager, the architect or whoever is responsible for quality control, to ensure that work is carried out as agreed. The materials and systems used should be documented, and once the project is finished a follow-up evaluation should be carried out.

1.4.0 Environmental Management

If one is serious about working professionally and sustainably, environmental issues must be taken seriously. This means that a company first needs to make itself environmentally friendly as well as cooperating with other environmentally conscious enterprises. There are several different environmental management systems that can be applied. In general, modern environmental management encompasses many of the various activities carried on in enterprises and organizations, from market overviews, product planning, product development and product launching to market follow-up. In the planning and implementation of a construction project, many different actors are involved.

Environmental Management Systems

An environmental management system consists of several steps. First, the company management has to decide that they want an active environmental policy. Then a preliminary environmental review is carried out to identify the environmental impact of company activities. Once a company's environmental impact has been identified, various environmental performance goals can be set. It is common to speak of objectives, i.e. the focus of the work, and targets, i.e. the desired results to be achieved within a certain time period. With these as a basis, an environmental programme is written and an environmental management system is developed to implement the programme. The concept of continual improvement is included in environmental management systems. Therefore environmental audits are regularly carried out, and are presented in an environmental statement. From this, new improved environmental goals and an upgraded environmental programme are developed. One of the most commonly used environmental management systems in ISO 14001.

Environmental and Quality Management

A company can work internally with both quality management systems (ISO 9000 series) and environmental management systems (ISO 14000 series). Quality management systems deal with ensuring that the intentions expressed and the decisions made are carried out in a qualitatively correct manner. It can be advantageous to link environmental standards to quality management. In order to ensure that environmental ambitions aren't lost in the process, it is a good idea to work using a quality management system where all those involved are made aware of the environmental aspects to consider and are committed to working according to these guidelines. Throughout the process there should be regular evaluations of the established environmental requirements. A monitoring plan is developed. Those responsible for each requirement should be identified, and there should be a follow-up routine to ensure that the requirements are met. An environmental manager is appointed. The person responsible for quality management can also be responsible for environmental management.

The ISO 14000 Series

ISO is the acronym for the International Organization for Standardization. ISO is developing a number of different environmental standards in the ISO 14000 series. The most important is the ISO 14001 Environmental Management System standard. Absolute values for environmental performance are not given.

In the area of property management, environmental problems can be used as a basis, and an analysis can be made of the housing company's role in addressing these problems.

Source: Svenska Bostäder

EMAS

The Eco-Management and Audit Scheme (EMAS) is the EU's system for environmental management and auditing. EMAS is based on the ISO 14001 environmental management system, but also, among other things, requires companies to publish an annual public environmental report. EMAS is not used as widely as ISO 14001, but has an extra value since the system also provides verified data. Neither systems place limits on the extent of environmental impacts.

ISO 14001 – Environmental management
ISO 14006 – Guide on eco-design
ISO 14010 – Environmental auditing
ISO 14020 – Environmental labels and declarations
ISO 14030 – Environmental performance evaluation
ISO 14040 – Life-cycle assessment
ISO 14050 – Vocabulary
ISO 14062 – Integrating environmental aspects into product design and development
ISO 14063 – Environmental communication
ISO 14064–65 – Greenhouse gases
ISO Guide 64 – Environmental aspects in product standards

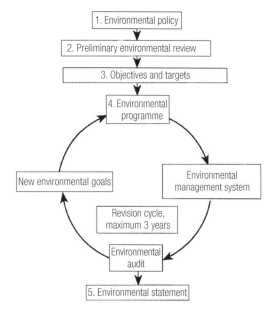

The EU has developed an environmental management system called EMAS.

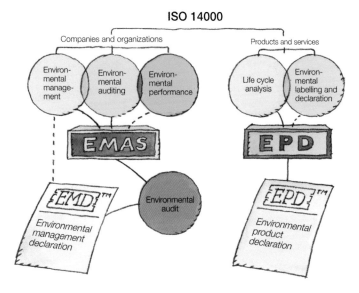

The ISO 14000 series was developed to facilitate environmental improvement and make it expedient and cost-effective. A guiding principle for all the standards is that overall effort should lead to continual improvement of environmental performance.

Source: The Swedish Environmental Management Council (SEMC)

1.4 The Maturing Process in a Project to Achieve Change

Leadership
— consistency
— perseverance
— systematic

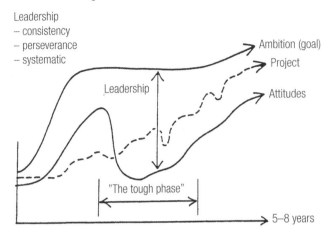

Getting new ideas to permeate through a large company can be a problem. The illustration shows the maturing process in a project to change to more environmentally friendly construction in the JM-bygg (JM Construction) company. Though a company may wish to become more environmentally friendly, it takes time for attitudes to change.

ENVIRONMENTAL POLICY
The environmental policy is here taken to be the organization's highest governing goal. In addition to the environmental policy there can be, e.g. a business concept, plans for the outdoor and indoor environment, a security policy and a quality policy. An environmental policy includes, amongst other things:
- That the organization intends to reduce and prevent environmental impacts within its activities,
- A promise of continual improvement,
- Reference to more detailed environmental goals.

The policy should be:
- Distributed, known and understood throughout the whole organization,
- Available to customers and the general public.

MANAGEMENT REVIEW OF THE ENVIRONMENTAL MANAGEMENT SYSTEM
- Periodic reveiw of the environmental management system, and
- The need for changes to any other elements of the environmental management system.

MONITORING AND CORRECTIVE MEASURES
- Continual monitoring and follow-up,
- Preventative and corrective measures,
- Regular environmental audits and auditing of the environmental management system.

PLANNING
- Inventory the main parts of the organization's environmental impact,
- Set objectives, i.e. long-term non-measurable goals,
- Set targets, i.e. short-term measurable goals,
- Implement an environmental management programme: when the targets should be met, where they should be carried out, and who is responsible.

INTRODUCTION AND OPERATION
- Internal training is required,
- Environmental performance should be documented,
- Internal and external communication is required, and
- Management is needed: new routines should be established and maintained.

Some construction companies have been certified according to ISO 14000 or EMAS. It may be easier to try to improve environmental performance in large companies that are already certified. The illustration shows a way of addressing environmental issues in a company.

AB Familjebostäder

Environmental Handbook for Planning and Construction
Work structure and organization

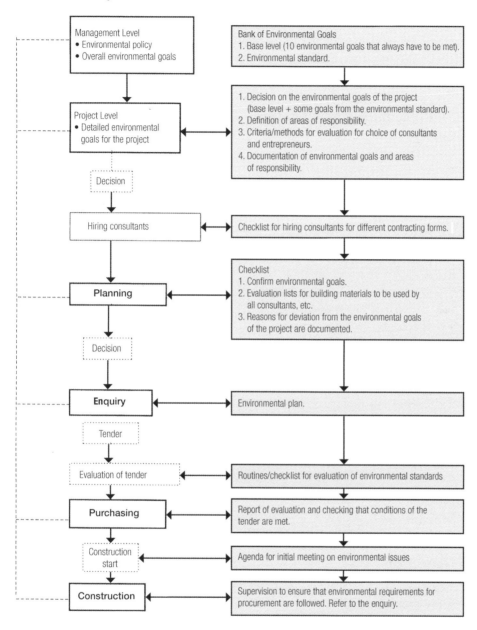

An important part of the work of the Stockholm company Tyréns is to prepare company-specific environmental handbooks for planning and construction. The illustration shows a chart that was commissioned by Familjebostäder AB (Family Housing Ltd).

1.4.1 Planning and Procurement

Those involved in the building process should be environmentally conscious and want to work environmentally both in their own organization and in carrying out the work. They should have knowledge of and a willingness to carry out eco-friendly projects. Planning with a long-term perspective is of the essence.

Environmental Management of Construction Projects

Contractors' experience of building in an environmentally friendly manner has shown that the following steps should serve as a guide.

1. There should be an explicit willingness. Otherwise, it is unlikely the building will be environmentally friendly.
2. Choose consultants and contractors who want to build in an environmentally responsible manner and know how to do so.
3. Detailed environmental specifications should be already worked out in the planning phase.
4. The original intentions must not be ignored during procurement.
5. There should be a functioning quality management system during the construction process.
6. Handover must be carried out with care. Operating and maintenance aspects should be included from the beginning, and it is at this point that they will be implemented.
7. There should be follow-up meetings where planners, builders and caretakers meet to exchange experience.

Guidelines in the Construction Sector According to The Environmental Code

The Ecocycle Council and the Swedish Environmental Management Council have formulated requirements that actors in the construction sector can demand of each other, and what others, including consumers, have the right to expect at reasonable cost. Project planning, production and management must be planned so that during its full life cycle a building will not influence its surroundings negatively and at the same time provide a good indoor environment during its whole period of use. This requires:

- environmentally friendly construction;
- energy-efficient operation and low carbon dioxide releases;
- good ventilation and low emissions from the construction materials used.

Proposals in the guidelines include:

- Clear environmental requirements when tendering for consultants and entrepreneurs.
- Greater energy-efficient requirements than specified in the National Board of Housing, Building and Planning regulations. It is proposed that energy consumption in homes that have other forms of heating and electricity should be 30 per cent lower than the 2009 construction regulations. For zone III (southern Sweden) this means 75kWh/m²/year (Atemp, or the heated area of abuilding that is used for energy calculations), compared to the 2009 construction regulations for southern Sweden of 110kWh/m²/year (Atemp).
- Requiring moisture-proof construction. An example is that moisture and

environmental checks should be carried out periodically during the whole production period.

- Greater sound quality requirements than specified in the National Board of Housing, Building and Planning regulations. It is proposed that at least sound class B should be met for constructions that separate apartments and for sound levels indoors from installations.

- Requirements for the chemical properties of construction materials and documentation of built-in materials.

- Rigorous requirements for the management of chemicals and construction materials in the building phase.

Environmental Plan Checklist

Proposed environmental plan checklist for construction entrepreneurs prepared by the Ecocycle Council and the Swedish Environmental Management Council:

- The on-site person responsible for environmental matters has been specified.

- The environmental plan is dated and signed by the entrepreneur's person responsible for environmental matters.

- The environmental plan refers to the client's environmental requirements.

- The environmental plan refers to an environmental inventory made for the project.

- The fractions to be separated out respectively during the construction demolition phases are documented.

- The on-site classification system for separating fractions is given. For example, containers (including number and size), smaller containers, shared containers, sacks, storing indoors until being removed, etc.

- A site plan during the construction phase of the project is included.

- The hazardous waste separated out and/or taken care of is given as well as how dealing with it will take place. For example, mercury-contaminated sewage pipes are disconnected carefully and cut into one metre lengths, plugged at both ends and placed in sealed containers for removal.

- The waste that any subcontractors themselves are responsible for is given as well as how the waste will be dealt with.

- There is a description of how the waste statistics will be compiled and reported to the client.

- There is a description of how the entrepreneur ensures that the Swedish Chemicals Inspectorate's phased-out and priority risk-reduction substances are not used. For example, who checks this, how the checks are made and they are documented.

- There is a description of how safety sheets are dealt with on site.

- There is a description of how communication about environmental plans with employees and subcontractor takes place.

- There is a description of how often environmental inspections are carried out. For example, inspections can include checking waste management, decontamination, approvals and licences, management of chemical products, how product and safety information sheets are dealt with, storage of materials, noise, dust protection, etc.

- There is a description of how the entrepreneur ensures that drying down to relative humidity (RF) <85 per cent takes place before surface protection is applied.

- The methods of storing and managing construction materials in order to avoid moisture damage is given.

- There is a description of how noise disturbance in minimized.

- There is a description of how spreading dust to surrounding activities and residents is avoided. For example, fans that create low pressure in the construction area, sealing windows and doors, use of mats that can be wiped off, etc.
- There is a description of how deviation from the established environmental plan is dealt with.
- There is a description of how quality assurance of sealing the weather-protecting layer around the building is carried out during the construction phase.

Environmental Programmes and Environmental Plans

The Swedish Federation of Consulting Engineers and Architects, which is the trade and employers' association for Sweden's architects, building and engineering consultancies, have defined the concept of environmental programme and environmental plan as follows.

The environmental programme is the part of the complete building programme that covers environmental requirements including owner priorities, setting environmental goals, and documented follow-up. The environmental programme should include, for example, who is responsible for the programme being followed for moisture protection and for dealing with construction waste. All involved should have knowledge about the programme. The programme should also provide a basis for future operation and maintenance instructions.

The environmental plan contains the planned measures to be taken in order to meet the requirements of the environmental programme. This is normally a part of the project's quality assurance plan. It should be remembered that environmental issues are inter-professional in nature, and therefore it is important to have careful control of who does what during the design stage. A system for evaluating building materials should be chosen. An issue that the owner should be involved with is if the so-called 'Environmental Handbook' (a tool for systematically applying an environmental perspective to a building or facility during its full life cycle) should be used. How verification and internal control will be carried out also needs to be decided.

Municipalities

Municipalities have a big responsibility regarding facilitation of achieving completion of environmentally friendly projects. An example where some municipalities provide dispensation, and some not, for building permits is regarding deviations in the number of floors and height. Further, an ecological building often has more insulation against the ground and in the walls and ceiling than normal. This means that outer walls and the roof can be 20 cm thicker than usual. If a building permit is used to the maximum it means that an environmentally friendly building gets much less activity space than a less energy-efficient building. Municipalities have various ways of looking at the issue of providing exemptions for this reason. Some municipalities choose to see building permit violations due to extra insulation as a small deviation in the local development plan, others not. In some cases, when the specifications of building permits are strictly applied, an unwillingness to provide dispensation can lead to a project not being completed.

Choosing Consultants and Contractors

Consultants and contractors should be chosen with adequately documented competence in the environmental field for carrying out the project according to specifications. They should familiarize themselves with the project's environmental requirements and goals. Large contractors should have an environmental policy, provide information about the person responsible for environmental matters, be able to give an account of their employees' environmental training and of the firm's experience with environmentally oriented projects. They should also use an environmental management system and a system for environmental evaluation of construction materials to help the purchaser make informed decisions.

Cooperation

Constructing buildings is an art and good results require the weaving together of many different skills. It is therefore important for good cooperation to be established between architects, builders, heating, ventilation, sanitation and electrical consultants. For many ecological buildings, additional specialists are involved, e.g. heating, ventilation and sanitation is often handled by a ventilation expert, a specialist in sewage management, and an energy expert who is familiar with heating and cooling. It is also important to involve landscape architects at an early stage, especially for large projects.

The Importance of the Project Manager

The project manager helps the contractor run the project by directing daily work during the planning process and by coordinating the different parties involved in the construction project. It is very important to have competent project management. Many people are involved in a building process and it is important that all are willing to work without prestige and share their knowledge. The aim is to have a project that is good in its entirety.

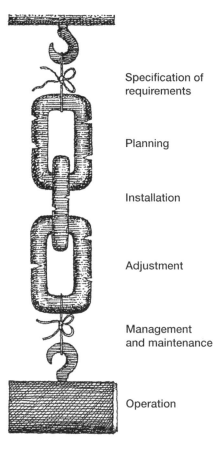

- Specification of requirements
- Planning
- Installation
- Adjustment
- Management and maintenance
- Operation

A chain is only as strong as its weakest link; an old saying which also applies to the construction process.

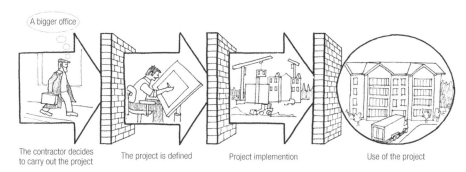

The contractor decides to carry out the project — The project is defined — Project implemention — Use of the project

One difficulty in carrying out environmentally responsible construction is that there is often poor communication between the different phases of the building process and a lack of knowledge in each phase about what is involved in building ecologically.

Actors in the Planning Process

In some projects there can be up to 200 parties involved. In ecologically oriented projects there is an attempt to increase co-operation between the actors. In order for the project idea to survive and thrive, good cooperation between those involved in the process is required.

Comparing Building Projects

The environmental impact of building projects varies. If the ambition is to construct buildings with minimal environmental impact, methods for comparing projects are needed. It is best to use these methods in the planning phase so that various projects can be compared. Such methods are often complicated and time-consuming; it is not possible to count on being able to apply them routinely. They are sometimes used to get an idea of how construction ought to take place, and ways of connecting environmental assessment tools to computer-aided design (CAD) programs are being developed.

Some of the many actors involved during the construction of the ecological community Understenshöjden in Björkhagen, Stockholm, mid 1990s.

Notes: * SMÅA = The Small Building Department of the Stockholm City Real-Estate Office.
** HSB = The Swedish National Association of Tenants, Savings and Building Societies.

Environmental Impact of a Building

A building impacts on many things. Within the construction sector, indicators and measurements to describe environmental impact are sought; such indicators facilitate the decision making that supports sustainable development.

Environmental Classification

Different environmental classification systems are used in different parts of the world. Among the most important are: LEED (North America), BREEAM (Great Britain), Code for Sustainable Homes (Great Britain), Green Star (Australia), CASBEE (Japan), DGNB (Germany), Minergie (Switzerland), and the Sustainable Building Tool (SBTool) of the International Initiative for a Sustainable Built Environment (IISBE). In Sweden there is EcoEffect and *Miljöklassad byggnad* (Environmentally Classed Building). All of these systems have been developed to make it possible to objectively evaluate resource consumption, environmental impact and indoor climate. Most of the systems can be used both for already constructed buildings and in the planning phase for new buildings. The purpose is to stimulate construction of environmentally friendly buildings. A classification system makes it possible for environmentally friendly buildings to have a higher market value.

In the beginning of the classification procedure a large number of properties are chosen to be given appropriate values for the chosen parameters. Some criteria must be met and the different properties are weighted to finally determine classification. The different systems are quite different from each other, both with regard to what is considered and how the different properties are evaluated. there is not yet any generally accepted environmental classification system. Rather, different systems are used in different countries. In Sweden, a Swedish system is used as well as LEED and BREEAM. Most important is that the results of a system are trusted and that the time and cost involved to carry out the evaluation is considered acceptable.

The main value of environmental classification systems is that they start an important discussion about how measuring and evaluation of environmental properties be carried out. Knowledge about the environmental impact of buildings is fundamental to being able to build more environmentally friendly buildings in the future. Work on environmental classification systems and concepts for buildings will hopefully lead to increasing knowledge in these areas. An important aspect of classification systems is that they are transparent so that it is easy to understand how different properties are evaluated.

The BREEAM system in Great Britain was the first environmental classification system and other countries followed after. There was dissatisfaction that the system did not cover a wide enough area. Conferences were organized called Green Building Challenge, of which the first took place in Canada in 1996. People learned from each other at the conferences and the environmental classification systems were a result. More and more countries became involved. The ambition was to create an international method and an organization called the International Initiative for a Sustainable Built Environment (IISBE) was created for this purpose. Regardless, countries have built up their own systems. The conferences continued to be held and contribute to the beneficial development of environmental classification systems. The name of the conferences was changed from Green Building to Sustainable Building. The purpose of this was to make requirements more rigorous. The green building concept compared a building to a standard building. Sustainable building measures impact on nature and climate.

SBTool (Sustainable Building Tool) is a computer program developed for the Green Building Challenge. The initiative for this environmental classification system came from Canada, but it has been developed by international cooperation within IISBE since 1996. There are three variations of the method for all types of buildings: one for schedules and analysis, one for design and one for operation. The method has been developed primarily for designers, but can also be used by buyers and users. The results are shown in bar charts with the contribution of various construction materials to various impact categories and an equivalent diagram for life-cycle costs.

The **LEED** Green Building Rating System was developed by the US Green Building Council. The first version came in 1999. The system is useful for schedules and analysis, design and operation. Homes, offices, public facilities and schools can be classified, as well as large renovation operations and neighbourhoods. Points for various criteria are added together and a construction project is classified as: certified, silver, gold or platinum.

BREEAM (Building Research Establishment's Environmental Assessment Method) is a British system developed in the early 1990s. The system can be used for homes, offices, public facilities and schools. It can be used for schedules and analysis, design and operation. Points are given for each criterion and the points are added up to a total. A summary evaluation is given as either: acceptable, good, very good or excellent.

The Swiss environmental classification system, **MINERGIE**, came in 1998. It is a voluntary system for new and renovated buildings. In 2002 a system was developed for passive buildings, called Minergie-P, and in 2006 came Minergie-Eco which also takes into account the environmental properties of construction materials. Minergie-P-Eco classifies passive buildings that use environmentally friendly materials. Some banks give reduced interest rates for Minergie buildings. The system also includes a computer program for calculating energy consumption. The system includes schedules and analysis, design and the building phase and can be used for homes, offices, public facilities and schools. The final result is a certificate that shows that the building has met Minergie requirements.

The **Code for Sustainable Homes** can be seen as part of the British government's climate goal that starting in 2016 new homes will not release any carbon dioxide at all. The method is a further development of the BREEAM version EcoHomes, and is a standard for new homes in England, Wales and Northern Ireland. The method can be used for schedules and analysis, design and the construction phase. Starting 1 May 2008 all homes being sold are required to be classified. Classification is done for each house, and the final result is given as one to six stars.

The **CASBEE** environmental assessment system in Japan receives federal government support. Development began in 2001 to meet the political and market demand for sustainable buildings The construction sector, universities and government agencies have taken part in the development process, which has been administrated by the Japan Sustainable Building Consortium (JSBC). The system includes many tools, from designing to demolition, and for all types of buildings. Results are shown in four different ways: 1) as roses with total values for each area, 2) as bar charts with total impact values for each category, 3) As a diagram showing eco-efficiency (quality/impact) and 4) as certain quantified indicators.

Green Star was developed by and is administrated by the Green Building Council of Australia (GBCA). The organization has technical committees that cooperate with trades people within the construction and environmental area to develop criteria. There are versions for schedules and analysis, design and operation. The system can classify many different types of buildings. The following final results and points summary are used: Best Practice, Australian Excellence and World Leadership.

DGNB (*Das Deutsche Gütersiegel Nachhaltiges Bauen*) is an environmental assessment system introduced in Germany in January 2009 by the German Sustainable Building Association. DGNB has active support from the German Ministry of Transport, Building and Urban Development. About 50 criteria are evaluated. Contrary to most other systems, the German system also takes into consideration economic and socio-functional qualities. The final grade is bronze, silver or gold.

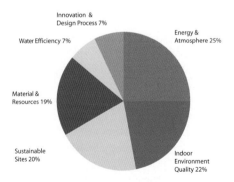 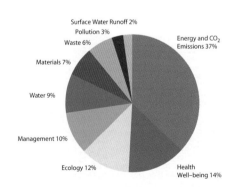

Pie charts that show how various criteria are weighted in relation to overall evaluation in the American environmental classification system LEED (the left circle) and the British Code for Sustainable Homes (CSH) (the right circle).

Source: Building Environmental Assessment Tools II – Detailed tool comparison, Marita Wallhagen, Mauritz Glaumann, Ulla Westerberg, 2009.

EVALUATION CRITERIA FOR ENVIRONMENTAL CLASSIFICATION SYSTEMS

There are many environmental classification systems. They differ between the properties they examine, how they are evaluated and how the final results are presented. A summary is given here of the properties taken into consideration by the most common systems. Environmental classification systems are important because they provide an instrument to market buildings and increase awareness of what needs to be considered to build environmentally.

ENERGY AND POLLUTANTS

Criterion	Systems
Energy Information	
Energy analysis	L
Output need	B D J L M
Output need 10W/m²	S
Output need 30KWh/m²	S
GWP (Global Warming Potential)	C D E J
CO_2-releases	B C S
ODP (Ozone Depletion Potential)	D E
SO_x	D E
Energy Efficiency	
Energy-eff. buildings (U-value)	L M S
Air density, insulation thickness, 3-glass windows	S
Area efficiency	B
Energy-efficient installations	J
Electrical Efficiency	
Energy-efficient devices	L M
Appliances	C S
Drying room	C
Energy-efficient lighting	B C L M
Outdoor lighting	C
Renewable Energy	
Portion of renewable energy	B C D E J
Local energy production	
District heating	M
Solar collectors	M S
Solar cells	
Wind, hydropower (green electr.)	M
Environmentally cert. biofuel heat	E M

MATERIALS AND WASTE

Criterion	Systems
Choice of Materials	J S
Ecological footprint	C E
Energy consumption, production	C E
Fair Trade cert. construction mtrl.	C
Fair Trade furnishings	C
Renewable raw materials	J L
Long lifetime, robust	B J
Energy efficient transportation	L
Hazardous Substances	
Environmentally friendly prod.	L
Hazardous substances (chemicals, metals, fibres)	B E M
Hazardous coolants	B
Emissions (TVOC)	B S
Emissions (formaldehyde)	S
Materials documentation	M
Household Waste	
Room for storing fractions	B C
Composting	C
Construction Waste	
Reduce waste during mtrl prod.	L
Minimize waste at constr. site	L
Sort construction waste fractions	C
Reuse products	L
Recycle material	L S
Dismantlability	D

INDOOR CLIMATE AND WELL-BEING

Criterion	Systems
Installations	
Ventilation	L
Air quality (NOx, air change rate)	B C D E J M
Reuse of heat	S
Electro-climate	E
Technical climate	E J
Adjustable temperature	B
Operative summer temp.	D M
Operative winter temperature	D M
Light	E J
Visual quality	D
Views	L
Adjustability	B
Daylight	B C E L M
Sunshine hours	E M
Technical Design	
Radon	E M S
Moisture problems	D M
Acoustics	B C D E J L S
Sound class	M
Sound insulation	C
Cleanability	D J
Easy to maintain	J

WATER AND SEWAGE

Criterion	Systems
Hydrology	
Surface water management	B C E L
Develop the water landscape	E L
Healthy watershed	E
Clean Water	
Groundwater protection	B
Water quality	B M
Tap water temp. (legionella)	M
Conservation	
Water consumption	B C D J
Irrigation	C
Water-efficient equipment	L
Measuring	B
Water leakage monitoring	B
Secondary water (e.g. rainwater)	B
Sewage	D
Sewage releases (biochemical oxygen demand, N, P)	M
Recycling of nutrients	D E
Over-fertilization	

CITY LIFE

Criterion	Systems
Social Life	D
Social openness	D
Public art	D
Homework places	C
Private sphere	C
Lifelong residency	C
Societal Structure	
Proximity to services	D
Use of existing server structure	L
Use of existing infrastructure	L
Use of existing transport system	L
Access to media, e.g. fast www	D
Leisure	
Access to green areas	L
Access to exercise facilities	L
Access to walking areas	L
Transportation	
Long-term sustainable solutions	L
Bicycle comfort	B D
Bicycle storage	C
Pedestrian security	B
Access to public transportation	B D
Comfortable public transport.	B

THE SITE

Criterion	Systems
Choice of Lot	
Site analysis	C D
Building on already used site	L
Density	L
Level of exploitation	C D
Neighbourhood influence	D J
Influence of surrounding area	J
Outdoor environment	B J
Consideration of local culture	J L
Consider. of region environment	D
Risks	
Flood and earthquake risk	B C D
Electro-climate, power lines	E
No contaminated land	B E L
NO_x from traffic in indoor air	E M
Surface ozone	D E
Noice	D E
Reduce soil erosion	L
Avoid disturbing sound & light	L
Microclimate	D
Heat islands/cold air sinks	L
Shade	E
Wind	E
Flora and Fauna	
Study the local flora and fauna	C
Min. impact on the ecosystem	L
Avoid using green areas	E L
Adapt to the cultural landscape	L
Green roofs	D
Preserve valuable ecolog. areas	B C J
Biological diversity	B C E J
Replace removed green areas	C
Establish gardens	E

Legend:
- B = BREEAM (GB)
- C = Code for Sustainable Homes (GB)
- D = DGNB (Germany)
- E = EcoEffect (Sweden)
- J = CASBEE (Japan)
- L = LEED (US)
- M = Miljöklassad Byggnad (Sweden)
- S = Minergie (Switzerland)

REALIZATION AND MANAGEMENT

Criterion	Systems
Economics	
Life-cycle costs	D E
Long-term value	D
Planning	
Environmental policy	B
Environmental quality of programme	D
Eco-integrated planning (prof.)	D
Hire prof., environmental issues	L
Optimization and holistic view	D
Procurement & sustainability	B D
Friendly environmental operation	D
Environmental training	
Owners	L
Planners	L
The Construction Site	
Reduce pollutants	L
Environmental considerations	D
Environment	C
Labour	C
The public	C
Environment. aware subcontractors	D
System. division of responsibility	D
Quality control	D
The Use Phase	
Accessibility	D
Flexibility	C D J
Robust technology	J
Safety	C D
Free safety	D
Security monitoring systems	B
Users	
Environmental training	B L
User handbook	B C
User behaviour	D
User questionnaire	M
Management	
Management handbook	B
Environmental management	B
Maintenance plan	B
Operation	B J

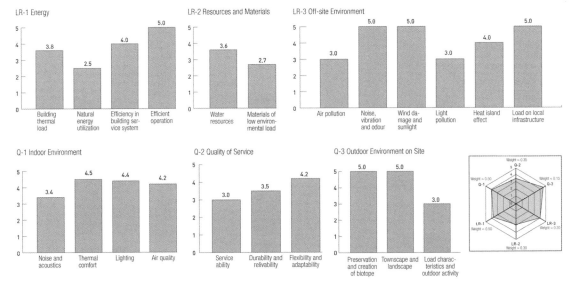

The Green Challenge is an international initiative that is being further developed at national level. The illustration shows the Japanese system, Comprehensive Assessment System for Building Environmental Efficiency (CASBEE). Summary of results of a building's impact on both the interior and outdoor environment.

EcoEffect is a Swedish method developed at the Royal Institute of Technology (KTH) in Gävle. EcoEffect takes a holistic view of environmental issues by in parallel dealing with: energy consumption, materials consumption, indoor environment, outdoor environment as well as life-cycle costs. An attempt is made to quantify environmental impacts as much as possible, which are reported in the form of environmental profiles in bar charts that show a building's contribution to various environmental impacts.

Miljöklassad Byggnad (Environmentally Classed Building) is a Swedish system that classifies in the areas of: energy, indoor environment, materials and chemicals. The system has been developed by the dialogue project 'Building-Living and Property Management for the Future', carried out together by construction and property companies, experts and researchers, as well as municipalities and the federal government. It is a simpler-to-use and cheaper system than EcoEffect. It can be used for all types of buildings, is easy to communicate and encourages continual improvement. The classification system has four levels: classified, bronze, silver and gold. 'Classified' means that basic requirements are not met. Bronze is for when only the basic requirements are met.

Building Information Modelling (BIM) with Environmental Weighting

Building Information Modelling (BIM) is a process for storing most types of data when a building is constructed and during its life cycle. Using BIM it is possible to measure geometry, geographic information, spatial relationships, quantities and properties of building components. Three-dimensional (3D) modelling is often used with time

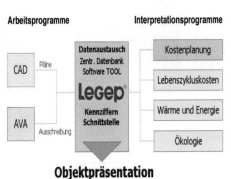

LEGEP, a tool for integrated life-cycle performance of buildings. The structure of the LEGEP software: plans and descriptions are put into the system in order to calculate costs, energy efficiency and environmental impact, and to carry out a life-cycle analysis.

Source: www.legep.de

aspects (4D) and calculations taken into account. Computer systems are available where building owners, architects, building consultants and entrepreneurs can input energy use, life-cycle analysis and ecological aspects as part of the planning process.

In Germany, the computer model **LEGEP** has been developed. This tool for integrated life-cycle analysis supports planning teams in the design, construction, quantity surveying and evaluation of new or existing buildings or building products. The LEGEP database contains a description of all elements of a building. All information is structured along life-cycle phases, construction, maintenance, operation (cleaning), refurbishment and demolition. LEGEP establishes: (i) building costs, (ii) life-cycle costs (construction, maintenance, refurbishment and demolition), (iii) energy needs for heating, hot water, electricity, etc., and (iv) environmental impact and resource consumption (detailed material input and waste). LEGEP uses four software tools, each with its own database that can simultaneously calculate energy and resource consumption during the building phase, and energy consumption during the use phase. There is an attempt to develop a model to measure the health effects of time spent in a building.

Impact Assessment

An impact assessment is when the consequences of one or more proposals are compared for various special interest groups. They can be used as a basis for making decisions.

In his book *Samråd* (Consultation), Örjan Wikforss (1984) compares two alternative proposals for a road crossing, one with a tunnel (see illustration below), as follows:

1 Consequences for special interests:

Environmental and ecological: noise, air quality, ground and building vibration, impact on nature (e.g. trees in the boulevard), energy consumption, medical and psychological effects on the individual.

Aesthetic: the townscape as a whole, local environment, design of the tunnel, design of the roadways, ramps, signs, railings, landscaping and trees.

Social: which groups are prioritized and for which groups is the proposal disadvantageous?

Demographic: will the area become more or less attractive to various groups and is the pattern of relocation affected?

Technical: road construction, tunnel, water and sewage, groundwater, lighting, and the connection in old buildings between vibration, groundwater and timber piles.

Economic: national economic costs; state, municipal (highways department, park administration) and individual investment costs; operational costs; capital costs (fixed and variable).

Implementation: what will happen during the construction phase, how will vehicle traffic function, pedestrian and bicycle traffic, commerce, etc.?

Organizational: general planning, project planning, construction, inspection, operation and maintenance, and supervision.

Legal: changed property borders and compensation.

Cultural: is the way of life affected?

Political: what are the implications of the proposal for different voter groups, is the proposal in agreement with overall planning goals, and what opinions are there?

2 Consequences for various groups:

Motorists: local traffic, through traffic, speed, safety, comfort, costs and parking.

Pedestrians and cyclists: cross-route to town centre, along roadways, speed, safety, comfort, costs, noise, air quality, journey experience, children, the elderly, people with disabilities and 'walking through tunnels at night'.

Housing: beside roadways, slightly removed from roadways, noise, air quality, accidents, accessibility, 'is it possible to have the windows open during the day and/or night?', 'will a person be disturbed when outside?', play opportunities for children, the local environment for adults, and the effect of distance (for bus access, parking, commerce, other services, playgrounds and parks).

Comparison between two renewal proposals.

Source: *Samråd*, Örjan Wikforss, 1984

Commerce: does the situation change for businesses (more or fewer customers)?, access to distributors, accessibility for customers arriving by car, and the work environment for employees.

Contracting Alternatives

It is important to know what division of responsibilities applies to different contracting alternatives. Otherwise, environmental goals can easily disappear during the building process. Traditional contracting forms are used less and less. It is important to know that written agreements always take legal precedence over drawings. Where there are **multiple contractors**, the purchaser negotiates various subcontracts for construction, heating, ventilation and sanitation, and electrical work. The purchaser is responsible for coordinating the various contractors, if this responsibility isn't delegated to one of them. This type of contracting requires knowledge of the construction industry and familiarity with leading construction projects.

When there is a **general contract**, the building contractor is often appointed as the general contractor. A contract is signed by the buyer and general contractor, who then puts the subcontracts out to tender. The buyer supplies drawings and instructions and is responsible for coordinating the consultants. The general contractor has responsibility for coordinating the construction.

A **turnkey contract** is when the buyer gives responsibility for both planning and construction to a single contractor, who is then called a turnkey contractor. Turnkey contractors either use subcontractors or have their own specialized departments, and have great flexibility in their decision making. A buyer needs to be well qualified to work with a turnkey contractor and be able to get what they want. In a guided turnkey contract, the buyer controls certain specified aspects of the contract.

Partnering is a relatively new form of planning and procurement. In partnering projects all involved (owners, clients, consultants, entrepreneurs, installers, etc.) cooperate flexibly together at an early stage. There is an openness towards each other and an attitude of trust. Risks, solutions and economics are presented in a transparent manner. The participants together set the goals that they will jointly work towards. Environmental issues are included at the outset. Partnering is best suited for complex projects.

Procurement for a particular function, as the term indicates, is a type of procurement where a particular function in the finished building is ensured. Environmental requirements can be set for procurement for a particular function if they are made at an early stage, and if there is follow-up to ensure they are met.

Tendering Procedure

An invitation to tender is issued; contractors then calculate their bid. During the tendering period, the buyer should be available to provide any clarification requested. This can be handled by the architect or consultant concerned. It is important to understand what a tender covers. A contractor may include some extra compensation clauses. Anything that is unclear should be clarified during negotiation with the contractor and this should be documented in a tendering record that serves as a basis for final choice of contractor and for commissioning the work. Changes should be incorporated into the document as otherwise problems can occur.

Soft parameters such as the ethics, morals and overall competence of a construction company may be taken into account in the tendering process. Questions to help make a decision can be put to the various contractors, e.g. concerning quality assurance systems, environmental management systems, references, implementation plans, company organization, competence and rating. Using this information, the tender amount can be multiplied by a factor, e.g. from 0.9 to 1.1, in order to fairly compare different tenders.

Inspection

Once a building is finished, a final inspection should take place and a request for final approval is made to the local building authority. The inspection document should include: date of inspection, name of inspector, names of those present, an attached list of any problems noted, a statement on whether or not the project is approved and if necessary, an agreement on rectifying problems.

Handover

Handover to an environmentally responsible building management should take place carefully: for example there should be simple and clear operational and maintenance instructions. Environmental certifications for materials and products used should be included.

Operating Instructions

Building management includes establishing and maintaining good contact and communications with the occupants. To help users become familiar with the materials and technical systems, all related information should be kept in a binder. If there are systems that are affected by the occupants, it is all the more important that they are familiar with the system and materials, and also the history of the building if it isn't a new construction.

Feedback

Follow-up from a building project, including occupants' experience, is an important aspect that is unfortunately often omitted. Follow-up should be included in quality control. Examples of issues that should be discussed at follow-up meetings are:

- Have the project's environmental requirements and goals been met?
- What changes/deviations have been made and why?
- What experiences should inform future projects?
- What experience was gained from working across specialist boundaries?
- How did handover to environmentally oriented management work out?

A method for comparing tenders based on multiplying the tender amount by a factor dependent on soft parameters.

Source: SIAB

When they move into a building, occupants receive a binder containing descriptions of materials and systems.

Source: JM Bygg

It is important to have follow-up and feedback so that construction develops in a positive direction. Sometime after everything is completed, all the parties involved should meet to discuss whether or not requirements were met and how the building functions.

Source: *Entreprenadupphandling inom byggsektorn. Grundläggande handbok om regler, förfarande, ansvar,* Eric Nytell, Hans Pedersen, 1995

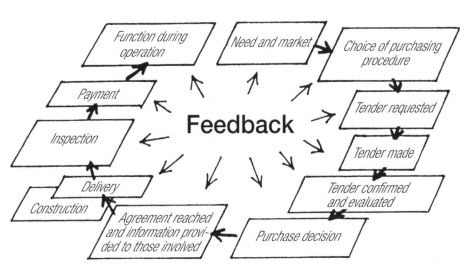

1.4.2 Economics

The economics of building are not simple. A high initial price may mean low long-term costs, or just the opposite. There are construction costs, operating costs, maintenance costs and life-cycle costs. Often it is simple and expedient to build with a smart layout that is space-efficient and compact. However, there is a difference between price and cost. It is best to build during a recession and to be a good negotiator.

Construction Costs

Construction costs include materials, salaries and other contracting costs such as transportation and use of machinery. Included in owner costs are the price of land, project design, municipal fees and insurance, as well as capital costs during the construction phase. Between 1995 and 2004, the cost for building materials increased by just over 17 per cent, which is more than for other industrial goods. Widening the market and investigating possibilities for importing building materials may be a way of keeping costs down. In the last 15 years, the salaries of construction workers have increased by more than inflation and the salaries of other groups. Why is that so? A lack of competition throughout the entire construction sector affects the cost of both work and materials. In order to build sustainable and healthy buildings, time to plan and organize an efficient construction process is necessary.

The Difference Between Price and Cost

An important aspect that must be considered is the difference between prices and costs. Costs can usually be calculated using methods based on statistics and personal experience. Prices are something totally different, and depend a lot on supply and demand, but also on whether there is, in practice, free competition. Price is something that can be negotiated until an agreement is signed and includes the contractor's profit. For an owner, it is most profitable to build during a recession.

Taxes and Fees

Half of costs are not really actual construction costs, but taxes and fees such as value added tax (VAT) and connection charges, as well as the cost of the site. Overall, taxes amount to about 40 per cent of total construction costs (including VAT, which accounts for one-fifth of total costs) depending on whether tax is levied on new builds or not. Municipal costs vary greatly between councils. Land prices in large urban centres in many countries have increased by 200 per cent in 17 years. Municipalities charge what they can for their land, which can make it difficult to build homes at a reasonable price.

Operating Costs

Operating costs cover heating, electricity, water and waste disposal, as well as caretaking, administration, staff and cleaning; in addition to taxes. Such costs are greatly reduced by constructing resource-efficient buildings.

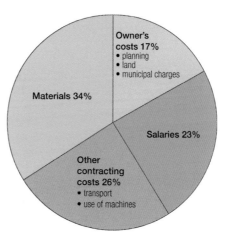

Division of construction costs according to the National Board of Housing's Construction Costs Forum Project Manager, Sonny Modig.

Source: *Arkitekten*, 4/2002

Maintenance Costs

Maintenance costs depend on how well a building is built, how durable the materials are and how long the life of the different parts is. A high-quality building has lower maintenance costs. Long-term costs are affected by the amount of maintenance required for a product to work in the long term and actual maintenance costs. It is important to choose products that have maintenance requirements as simple and inexpensive as possible, even though the price is often higher than for similar products intended for the same use. Investment costs usually comprise only about 10 per cent of the total costs of a building calculated over a life of about 50 years, and running costs make up the remaining 90 per cent. So there is great potential for reducing total costs over a 50-year period by using products of high quality and long-term durability, regardless of high initial cost.

Life-Cycle Costs

Choosing products with low life-cycle costs (LCC), instead of just choosing the cheapest, means that technical options with a higher purchase price can prove to be economically favourable and provide lower long-term operating and management costs. Life-cycle costs include not only construction costs but also heat, electricity, water and sewage, waste disposal, cleaning, maintenance and repairs, as well as management costs for staff, administration, etc. Real-estate taxes are also included in LCC.

Construction and running costs should not be reduced by lowering quality. During the Million Programme in Sweden, up to 100,000 new homes were built per year over a ten-year period from 1965 to 1975. It is estimated that 20–25 per cent of these homes were built at a low price and with a relatively low technical quality and standard. The life of these homes has been less than 30 years and a large proportion of them require extensive renovation and refurbishment, greater than would have been the case if long-term, durable options had been chosen at the start.

Building Efficiently

A building not designed for long-term use, fit for purpose, is expensive to manage in the long run, even if it was inexpensive to put up. Reducing investment costs for homes by cutting back on interior design

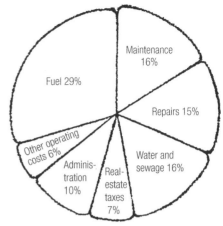

Division of average operating costs for all SABO homes in 1986.

Source: SABO

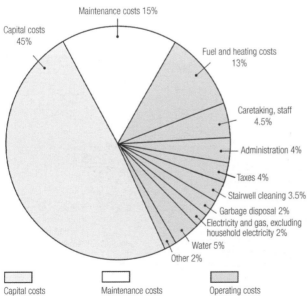

Total annual costs for a building, including capital costs, operational and maintenance costs.

Calculation of Total Cost

Requirements		Total costs during the building's life-time (30 years)				
Description	Quantity	Purchase price	Replacement costs	Maintenance costs	Cleaning costs	Total costs
Wall to wall carpet	4,000 m²	2,000,000	5,832,000	3,056,000	6,148,000	17,056,000
Linoleum	4,000 m²	2,400,000	2,644,000	2,460,000	5,536,000	13,040,000

Choice of flooring and cost determination should take into account total expenses during the lifetime of the building.

Source: Brochure from Forshaga linoleum

and equipment, or by reducing room size so that furniture placement, flexibility, accessibility and usability are compromised, limits possibilities for using the home for varied housing needs at different points in the future. Appropriate and flexible systems, well-designed housing environments and access to good services and good communications increase a home's longevity, reduce long-term costs, lower resource consumption and so reduce environmental impact.

Building efficiently often involves increased industrialization, which means standardization of technical building systems, greater efficiency on site, and an increased proportion of prefabricated elements. Use of prefabricated units can reduce construction costs by 10–25 per cent and shorten construction time by 20–50 per cent. It is easier to build healthy homes as the units are built indoors under dry conditions and everything can quickly be made weathertight. A constraint to consider is the size of the lorries used. For up to 24m long and 2.6m wide transport trucks no special measures are required. For longer and wider trucks, marking, and sometimes warning vehicles, manoeuvrable axles and an escort may be required. It is always the driver's responsibility to make sure that the vehicle can get through when the shipment is more than 4.5m tall.

Some walls and floor structures are thicker than when building on site. Risks with volume production are that the buildings can suit the site and the surrounding environment less well and that the designs become monotonous.

Self-Building and Partial Self-Building

Partial self-building means that the future users contribute their own labour in order to keep costs down. The amount of work is limited and is directed by the contractor with regard to both choice of materials and knowledge. The schedule incorporates the future occupant's labour so that it fits into the rest of the building process. What often happens is that an unfinished building is turned over to the owner, who chooses how much of the remaining work they will carry out themselves.

Self-building is a term used by SMÅA (the Small Building Department of the Stockholm City Real Estate Office). They have helped self-builders build their own homes since they opened in the 1920s. SMÅA has obtained land and provided prefabricated materials and instructors. Help has been available for planning and administration of e.g. building permits, loans, schedules and keeping the project within its budget. Using contractors, SMÅA has carried out all the landscaping and foundation work, installation of heating, electricity, water and sanitation systems, as well as delivery of outside framework. SMÅA is now a limited liability company owned by the JM company, the Swedish National Association of Tenants, Savings and Building Societies (HSB) and the staff. Up to €15,000 can be saved through the owner's labour contribution.

1.4

HOW BIG AND HOW MUCH?

Limiting size, compact building design, simple building structure and building efficiently are ways to reduce building costs.

Source: *Ökologische Baukompetenz*, H. R. Preisig, W. Dubach, U. Kasser and K. Viridén, Zürich, 1999

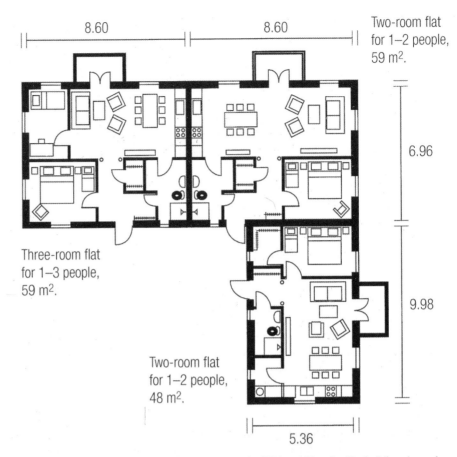

Design by *BoKlok-hus* (Live-Smart Flats), a concept run by IKEA and Skanska. The buildings have six flats on each of two floors. The idea is to provide a high standard at a low monthly cost. Both rental flats and cooperatively owned flats are being built. Part of the concept is to purchase land when prices are low. Residents must help take care of the buildings, courtyards and gardens.

Bank Loan

In some countries, such as Germany, there is a well-thought-out system for bank loans intended to help those that want to build in an environmentally friendly manner. Since an ecologically oriented building is very energy efficient and it is assumed that the building will be valued highly many years into the future, banks guarantee a low interest rate and it is possible to borrow up to 95 per cent of investment costs. In Germany there is a federal bank that guarantees a 10 year last mortgage loan (5 per cent of the total loan) at a fixed interest rate. Further, the ability to make monthly payments is not the only concern when considering borrowing capacity.

In Austria subsidies are generous. In the federal state of Voralberg, passive-house builders get one quarter of the loan without interest.

1.4.3 The Construction Site

It makes a big difference if construction workers are interested in and understand environmental building. It is important to avoid moisture problems by handling materials in the correct manner, storing them in a dry place, and adhering to drying times. Time estimates should be on the high side to avoid carelessness. Coordination between trades on site is important for keeping problems to a minimum. In addition, waste should be minimized and there should be on-site sorting of waste.

Quality Assurance

The proprietor of a building is responsible for the technical side of construction (and also for the impact of construction materials on people and the environment). It is the proprietor who appoints one or more people to be responsible for the quality of the construction project. Those responsible for quality assurance verify the quality of the work and collate checklists of work carried out. Information about who is responsible for quality assurance and a building application must be submitted to the municipality before work can begin. When working with an inspection plan and a person responsible for quality assurance, the final inspection is a formality where the inspection reports are submitted showing that inspection has taken place according to specification and work has been approved. The types of inspections required depend on the nature of the project and the type of quality assurance system used. The better the quality assurance routines, the fewer inspections required. The person responsible for quality assurance sends reports to the local housing authority and after all the reports have been received, final certification is issued and the building is approved.

Construction Site Meetings

Environmental issues should be raised at every construction site meeting so that there is a continual monitoring of environmental goals. If something changes, it is evaluated and decided upon. It is good practice to keep an environmental logbook to document decisions changed during the construction period that may have an environmental impact. The environmental logbook should be maintained throughout the life of the building.

Protection of Surroundings

In order to adapt buildings to their surroundings, decisions should be made during the planning phase about what should be protected and preserved. The site should be divided into a fenced-in area, where construction activity is not allowed, and an area where construction activity is allowed. The ground on a construction site often gets so compressed that it is difficult for vegetation to establish itself. It is possible to either lay a protective layer of gravel in advance, which is removed after the building is completed, or to treat the top layer of ground once the building is completed. It is also possible to mix gravel into the earth to restore the ground's original drainage characteristics. Individual trees earmarked for preservation should be well protected and penalties should be established so that the customer receives economic compensation if they are damaged.

Managing Materials

Materials management at the construction site must be examined. Materials should be stored in a dry place, preferably under cover. Even when materials without an environmental profile arrive during the building process, they should be examined from an environmental perspective. In some projects, all building takes place under a large tent.

Clean Building

It is important to keep the entire construction site clean and dry. It is also important to avoid building-in waste and materials into suspended foundations and under fitted cupboards, etc.

Construction Moisture

Buildings are often put up quickly and fans used to attempt to remove building moisture, which isn't always successful. Poorly constructed buildings often have damp and mould problems caused by covering ground slab foundations with insulation, inappropriate or faulty drainage, flat roofs that leak, built-in concrete that hasn't dried out, leaking water pipes, poorly ventilated suspended foundations, wooden exterior wall sills without moisture barriers, wetrooms with plastic wallpaper and painted fibreglass fabric, as well as faulty construction techniques and thermal bridges with moisture (condensation). Improperly stored insulation material exposed to snow and rain during construction without the opportunity to dry out can also cause damp problems.

Some materials, such as concrete, require long drying times. How fast concrete dries is affected by the thickness and quality of the concrete, the addition of drying agents, as well as whether one or both sides need to dry. The drying process involves a combination of ventilation, heat, dehumidification and air circulation. Drying times can be reduced by using prefabricated construction parts pre-dried under controlled factory conditions.

The moisture content of some building materials must be reduced to a certain moisture level before it is safe to continue. The maximum relative moisture content of underlying concrete prior to the application of outer coverings such as carpets, paint and waterproof layers should not exceed 85 per cent. This means that moisture content must be checked in both floor structures and walls. Furthermore, critical moisture limits should be given for moisture-sensitive materials, such as glue and sealing compounds. If these limits are less than 85 per cent, surface layers cannot be applied at a higher relative moisture content.

The moisture content of timber is commonly referred to as the moisture content ratio, and is the ratio between the weight of water in the wood and the dry weight of the wood itself. The moisture content is often given as a percentage. Air-dried timber should have a moisture content of 18 per cent. A solid wood floor should have a moisture content between 10 and 12 per cent, and dry furniture wood should have a moisture content of 8–9 per cent. A solid wood floor should be allowed to dry in situ before being fastened down. If removable skirting boards are fitted, it is easy to gain access to the floor and fill any gaps.

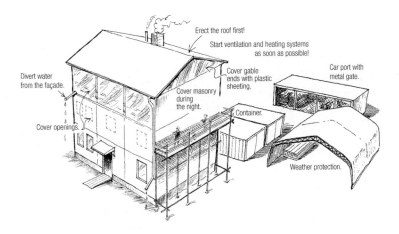

Moisture is a problem that must be attended to on site. Materials should be stored in the dry. The building should be protected from weather and wind. It is best to erect the roof first, then clad the walls. It is advantageous if exhaust air and heating systems are started as soon as possible.

Source: JM Bygg

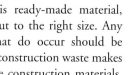

Examples of drying times.
Source: Håll torrt på bygget, BFR, 1987

Monitoring Moisture

A system for monitoring moisture should be prepared and included in the quality assurance plan. The moisture content in floors, walls and ceilings should be measured, especially if concrete and wood have been used. Two complementary methods of measuring concrete should be used: embedded monitoring points connected to computer equipment as well as extracted drill cores sent to a laboratory. When glues, varnishes and sealants are used, information should be provided by the supplier about the critical moisture level that should not be exceeded for construction to continue without risk.

Floor Laying

Floor should be laid as late as possible in the building process in order to minimize the risk of wear and damage. It is of the utmost importance to cover the floor carefully during the building phase. Before covering the floor, all dust, powder and refuse should be cleaned off. The protective material used to cover wood or laminate flooring must be breathable as there may still be moisture in the underlay. Inappropriate or faulty cleaning can damage a floor's appearance and function.

Construction Waste

Good planning for the use of materials prevents waste products. A method used more and more is ready-made material, e.g. flooring pre-cut to the right size. Any waste products that do occur should be sorted. Sorting of construction waste makes it possible to reuse construction materials, reduce disposal costs and deal with hazardous waste.

Waste Sorting

Every building site should have a system for waste sorting and a person responsible for implementing it. The aim is to minimize the amount of mixed waste and increase recycling. The waste should be at least sorted into the following fractions: wood, plastic, metal, gypsum, burnable, fillers and mixed waste. Hazardous, electric and electronic equipment waste should be handled separately. The contractor should ensure that waste sorting is implemented. Each site should have enough waste containers that can be moved to larger collection bins when full. Bins and skips should be sited so that transport is as short and uncomplicated as possible. No one should have to take a detour to sort waste. Construction sites can be cramped, e.g. in the town centre, and land for bins and skips can be expensive to rent. It can sometimes be more practical to use smaller bins for construction waste collection.

Paint Management

Cleaning paintbrushes and other painting tools is an environmental problem. Every year large amounts of paint from washing brushes end up in the sewage system. Sewage treatment plants have a difficult time dealing with waterborne paint, and there is a risk of polluting lakes and watercourses. Equipment for dealing with leftover paint is available, as are bins and containers for depositing paintbrush washing liquid. The polluted water can be mixed with separators

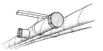

Service ducts can be plugged during transport so that they do not become dirty before use.

(bauxite), and after a few minutes the paint forms clumps that can be filtered out. About 80–90 per cent of paint can be removed using this simple method.

The Ragnsells company has developed a method called KRYO that can be used to recycle paint packaging and oil filters using four steps. First, the waste is collected in separate receptacles for hazardous waste. The waste is then transported to a KRYO facility where filters and paint containers are divided under extremely cold conditions (–180°C). Paint and oil are then sorted from metal and plastic and the fractions can be recovered for recycling. Paint, oil and filter wastes are treated for energy recovery, and metal is returned to the steel industry.

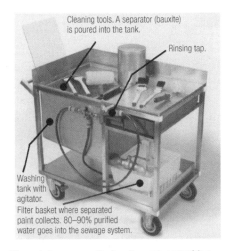

The paint company Beckers's wash cart with a paint separator provides an on-site solution for the environmental problems arising when paintbrushes are washed. The cart was developed by master painter Alf Karlsson.

Construction waste should be sorted on site. In SIAB's sorting system, waste is separated into the following categories: wood; appliances; metal; mineral wool; gypsum; concrete, brick, tile and mortar; mixed refuse; and hazardous waste.

1.4.4 Recycling Construction Materials

The aim should be to find a market for used construction materials and construction waste through reuse or material recovery, as well as to sort other material according to local regulations. Local requirements for recycling of construction materials must be determined at the time of demolition or construction. It is preferable to handle materials locally in order to minimize transport. There are companies that specialize in planning building demolitions so that as much as possible can be reused. It is of course best to plan for demolition during a building's planning phase.

Planning Demolition

Prior to demolition, an inventory of the materials in a building is made. An estimate is made of the volume of material, its composition and separability. This provides an estimate of the profitability of demolition as it is used as a basis for judging the marketability and price level of the material. Planning demolition involves developing a foundation upon which carrying out the work and managing the materials is based. An object's special qualities must serve as the starting point for determining the demolition method and sequence, deciding how to deal with the materials on site, organizing the site, as well as handling demolition materials.

For removal of floor boards, wallboards and tongue-and-groove ceiling boards, there is a special-purpose crowbar available that allows a lot of force to be used without damaging the boards. These crowbars are available in different sizes to suit various types of timber.

Source: *Rivningshandboken-Planering Demonteringsmetoder Verktyg*, Johanna Persson-Engberg, Lotta Sigfrid, Mats Torring, 1999

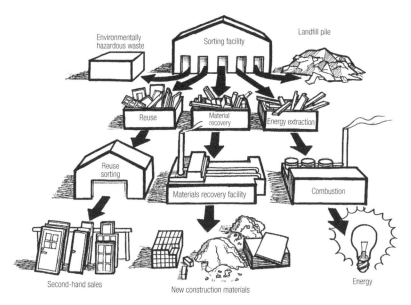

The fundamental idea is that construction waste should be sorted into five categories: hazardous waste that is destroyed, leftover construction material to be sent to scrapyards for reuse, construction material that is recycled and made into new construction material, material used for energy extraction and finally material that is disposed of in landfill sites.

Selective Dismantling

Development of selective dismantling is in its initial phase and so tested dismantling methods are not yet available for all types of building structures. The purpose of regulations for the inspection and control of demolition work is primarily to make sure that waste is handled and processed so as to protect the environment. Another purpose is to allow for reuse and recovery of demolition material. The laws involved when a building or facility is torn down include the Environmental Code, the Planning and Building Act, as well as the Occupational Safety and Health Act. There is also a specific ruling concerning hazardous waste with regulations for transport, intermediate storage and handling.

Detailed Demolition Plan

When an application for demolition is required, a detailed plan for dealing with the demolition waste must be included. According to the Planning and Building Act, when an owner draws up a demolition plan, an individual responsible for quality assurance must be appointed. The following information should be included in the demolition plan: (i) methods for identifying hazardous substances and hazardous waste; (ii) work methods and protective measures to be taken during the removal of hazardous materials; (iii) how to sort, handle and transport all hazardous materials and products as well as how to deposit the waste; (iv) demolition methods for and management of building elements that can be reclaimed, and for construction material that becomes waste and needs to be dealt with by materials recovery, incineration or delivery to a landfill site; (v) protective measures and methods for extermination of vermin, insects that destroy timber, and dry rot fungus of the Serpula family.

Reuse and Recycling of Construction Materials

County councils have information about which fill materials may be used where, as well as information about authorization for and management of various materials. Municipal waste advisers or local scrapyard staff have information on recycling. The following information comes primarily from 'Tips for Increased Recycling – Waste Products from Construction and Demolition', (in Swedish only, original title *Tips för Ökad återvinning – Restprodukter från bygg- och rivningsverksamhet*) by Lotta Sigfrid.

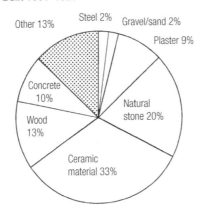

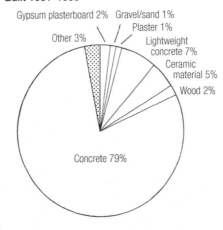

When demolition building material is to be recycled, it is important to know what materials are in the buildings that will be torn down.

Source: *Byggmaterial på 1900-talet*, RVF 92:12, Lotta Sigfrid

Concrete from whole concrete elements can be reused. The concrete can be crushed to separate fractionated ballast material, and can be used as filler once reinforcement is removed. The landowner can use the material as filler with permission from the county council. Reinforcement removed from concrete can be melted down and made into new reinforcement.

Cellular plastic in insulation that contains chlorofluorocarbons (CFCs) is classified as hazardous waste. (i) Entire construction elements or sheets can be reused. (ii) Cellular plastic can be recovered. Contact the appropriate plastic information authority regarding possibilities. (iii) Cellular plastic may be incinerated in licensed facilities. Expanded polystyrene (EPS) without CFCs, which is a totally white cellular plastic (styrofoam), can be recycled.

Kitchen sinks and taps should, if possible, be reused. Metal kitchen sinks can be taken to scrapyards for materials recovery.

Doors should be reused in their entirety complete with frames. Metal doors can be sent for materials recovery and wooden doors can be incinerated for energy extraction.

Windows (i) Used windows should be reused. (ii) There is a market for old window glass. (iii) Glass can be sent for materials recovery. (iv) Wooden window frames, including glass, can be incinerated in licensed facilities.

Gypsum (i) Plasterboard dismantled in whole sheets can be reused. (ii) Gypsum recycling takes place at gypsum production plants. (iii) Crushed gypsum may be used as agricultural sulphur-based fertilizer. (iv) Gypsum is sometimes mixed with mineral mass as fill material.

Whole **tiles and expanded clay slabs** have a second-hand value and should be reused. Prior to demolition, the possibility of removing tiles and expanded clay without breaking them should be assessed. Removal without breakage depends, among other things, on the kind of mortar used. Tile and expanded clay may be included in concrete mass when the material is recycled into fractionated ballast material, and they are sometimes permissible ingredients in ballast material.

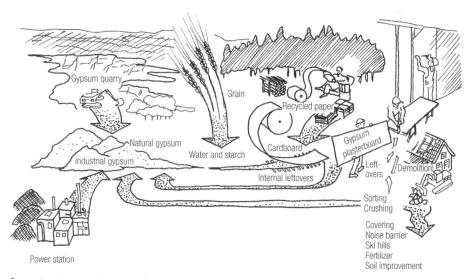

Several producers of construction material arrange recycling systems in order to make their production more environmentally acceptable. Pure gypsum plasterboard is made from gypsum recovered from leftovers and demolitions.

Source: adapted from a Gyproc brochure

Copper is found on roofs, façades, as sheet metal, water pipes, drainpipes and cables. Copper is also a component in the alloy brass. Copper has a high second-hand value and is sent for metal recovery.

Linoleum (i) Linoleum flooring that has not been glued down should be reused if possible. (ii) Linoleum mats can be incinerated in licensed facilities.

Lightweight concrete in whole block form can be reused if it does not emit radon. It can be made into fractionated ballast material. Lightweight concrete may be used as filler, but it is not suitable for use near harbours since it floats.

Used **mineral wool** should not contain mould spores, pollutants, allergens, infectious matter, etc. (i) Whole mineral wool bats can be reused. (ii) Some recycling of mineral wool takes place in factories that produce new mineral wool. (iii) Mineral wool can, if the quality is good, be recycled in bulk form.

Mineral wool in the form of fibreglass may be made from 70 per cent recycled glass. About 10 per cent of the glass wool is made from recycled leftovers. Less energy is required to melt crushed glass than to melt new raw material.

Plastic (i) Plastic mats that have not been glued down should be reused if possible. (ii) Plastic pipes and plastic mats (most often newer material) are in some cases taken for recycling by producers. (iii) Plastic material may be incinerated in licensed facilities.

Radiators (i) Sheet metal or iron radiators should be reused if they are deemed to be intact. (ii) Radiators can be taken to scrap dealers. Their value depends on prevailing metal prices.

Pipes and cables as well as scrap metal are taken to scrap dealers. Alternatively, cables are sent directly for cable recycling. County councils have information on facilities licensed to receive and deal with waste metals and cables.

Sanitary porcelain and fixtures should, if possible, be reused. Crushed sanitary porcelain is sometimes permissible for use as filler.

Whole **chipboards and veneered boards** can be reused, and chipboard can be incinerated in licensed facilities.

Whole **stone** blocks and slabs can be reused. The amount of natural stone usually comprises a small portion of total demolition material. Stone is salvaged when the second-hand price makes it worthwhile. It can also be used for fractionated ballast material or as filler.

Whole **bricks** can be reused. Whole roof tiles should be reused. Crushed brick can be used for fractionated ballast material, and brick can be used as filler. For brick to be reusable it has to be dismantled so that it isn't damaged, then sorted and packed so that it is transportable at the demolition site. Manual dismantling and cleaning of brick (with small hand-held machines) is technically possible even on a large scale, provided that the dismantled wall was made with lime mortar. Manual production of reusable brick, from dismantling to stacking on pallets, can be carried out at a rate of about 25–30 bricks per person-hour, and is done with a brick hammer, wire brush and metal screens. Brick cleaning machines, which can

The various fractions of crushed concrete can be sorted according to size.

save time, should be used if available. Bricks are sorted according to quality, colour, size and intended use (wall, façade or chimney brick). Even if the brick will be used as filler or gravel, the fractions should be kept uncontaminated by other materials. Landowners can use the material as filler with permission from the council. Used chimney brick cannot be used again in chimneys as the quality is usually reduced by wear.

Timber generates a large amount of wood-based waste during both new construction (packaging) and demolition (timber). If the frame of a building scheduled for demolition is made of wood, it pays to dismantle selectively. (i) Permanent and loose wood carpentry work should be reused if possible. (ii) Wooden floor boards and construction timber, e.g. roof trusses, can be reused once nails have been removed and they are sawn up. (iii) Untreated wood can be burned for energy extraction. (iv) Painted or impregnated wood can only be burned in licensed facilities. Timber tainted by vermin, dry rot or mould is not suitable for reuse.

Appliances should be reused if possible. It can, however, be better to scrap them than to reuse appliances that consume a lot of energy. Most appliances can be sent for materials recovery.

Zinc is found in zinc sheeting, zinc galvanized metal and older kitchen sinks. Zinc is sorted and sent for metal recovery.

Reuse of Building Products

There are various ways to acquire used building products. Second-hand dealers often buy and sell cookers, kitchen sinks, toilets, sinks, windows, doors, cupboards, etc. Appliance dealers handle refrigerators, freezers, washing machines and dishwashers. Demolition dealers may stock most products, but deal primarily with toilets, sinks, timber, roof tiles, stoves and wardrobes. Antique shops are beginning to buy and sell building products, usually older products for building restoration. Building products can also be found in the growing internet marketplace.

Used Building Materials

To make efficient use of used building materials, it is important to know which materials are available and which are in demand. Searchable databases are an efficient way for sellers and buyers to contact each other and nationwide databases may be set up, where information can be added before a building is dismantled.

The Recycled Building

The advantages of recycling building materials are that resources are conserved and waste is avoided. In the construction sector, the attitude towards recycling building materials is overwhelmingly positive. There is, however, scepticism about the economic feasibility of large-scale building of residences using recycled materials. Finding reusable material is seen as a problem and there is a lack of quality standards, which makes guarantee issues more difficult. In order for an increase in interest in using recycled materials, the following systems are required: development of selective dismantling systems, quality classification of reusable construction material, recycled building material depots with inventoried stock, as well as purchase systems. The time aspect is an important factor as time-consuming inventories of available reusable material are currently required, which lengthens the planning task.

Construction materials most desired by customers are brick, wooden window frames and doors. Reused brick is considered to be more 'alive' as new brick often has a more uniform colour and may be thought of as boring. Furthermore, many people feel that older, solid heartwood used for structural elements as well as for window frames and doors is superior to new wood in both quality and appearance.

The building on Turesen's Street is going to be dismantled but the materials will continue to be used.

Windows from the dismantled building waiting for renovation.

During selective dismantling some of the building materials are manually dismantled.

The brick has been cleaned and sorted and is ready to be used in a new building.

The floor boards that have been taken up are passed down and gathered together with the other dismantled timber.

Nails are removed from the dismantled timber with the help of a metal detector.

The building in the middle has been newly renovated. All the timber and windows come from dismantled buildings in Copenhagen.

In Denmark, they have come a long way with the reuse of building materials. They use selective dismantling and have developed special warehouses for sorted dismantled materials.

Source: Adapted from *Hus igen...*, Johanna Persson, 1993

The Filborna recycling facility in Helsingborg, Sweden is a good example of reusing building materials and building parts. Many materials have been reused there, including a large glass pavilion from a building exhibition that was taken down and moved to the facility.

Source: SWECO FFNS Architects through Per Lewis-Jonsson and Jonas P Berglund as well as Nordvästra Skånes Renhållnings AB. Glass pavillion designed by Kjellander och Sjöberg

Conservation

2

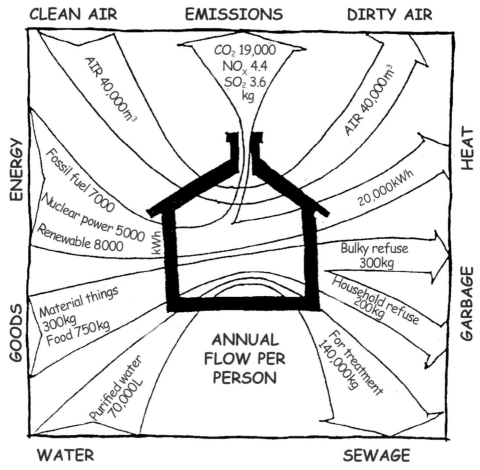

Various resources are required in order to use a building. A building can be regarded as a flow system. By studying the inward and outward flows and finding ways to reduce them, it is possible to create buildings that conserve resources.

Source: Byggd Miljö, Mauritz Glaumann, KTH Gävle

2 Conservation of resources

Conservation of resources when a building is used entails minimizing the resource flows, such as high-quality energy that is transformed to heat, clean water that becomes sewage and products that create waste.

Heating and Cooling –
How to Build an Energy-Efficient Building
Architecture affects a building's energy requirements. Its shape, type, zoning, heat storage, and passive heating and cooling all have an effect. The most important factor, however, is the building's shell, including the insulation, type of windows, and weathertightness. Another aspect to be addressed is heat recovery.

Efficient Use of Electricity –
How to Make Electrically Efficient Buildings
Electricity for heating should be avoided. Household appliances and lighting systems with the most efficient electrical technology should be used. Lighting should be controlled and regulated as required. Approaches that do not require electricity should be prioritized, e.g. daylight, cool larders and naturally ventilated areas for drying.

Clean Water –
How to Conserve Water
Water conservation measures include using water-conserving taps, showers and toilets, properly insulated hot water pipes and tanks, and individual metering. Another aspect of water conservation concerns local water resources, e.g. private wells and rainwater, and their purification and use.

Waste –
How to Conserve Materials
Waste can be minimized through the choice of products and packaging. Construction should be planned so that there is room to sort waste products into separate fractions for recycling. Organic waste can be composted, inorganic waste can be reused, and hazardous waste can be destroyed.

Energy – How Will it be Used and How Much?

In an energy study published in 1980 by Thomas B. Johansson and Peter Steen entitled, *'Energi – Till vad och hur mycket?'* ('Energy – How Will it be Used and How Much?' Swedish only), the following conclusion is drawn:

'We have shown that, despite a 50 per cent greater consumption of goods and services just after the turn of the century, it is possible to reduce energy consumption from the current 400TWh/year to between 200 and 250TWh/year. The reduction could be achieved by a much more efficient use of energy than is currently the case through using existing economical technology and technology under development.'

Reviewers of the study agreed that a 50 per cent reduction in energy use is possible even with a significant growth in consumption, but that the timeframe for this is more than 20–25 years, and is strongly dependent on future politics, both in the energy sector and in general.

Factor 4

In the book 'Factor Four – Doubling Wealth, Halving Resource Use' by Ernst von Weizsäcker, Amory B. Lovins and L. Hunter Lovins, 1997, the authors make a case for the following statement: 'Factor 4 means that productivity can and should increase

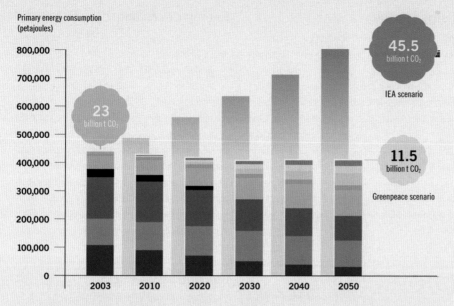

According to the Greenpeace study, by 2050 half of global primary energy demand could be supplied by renewable energy sources. See www.energyblueprint.info.

Source: *The Energy (R)evolution*, Greenpeace, 2005

The International Energy Agency (IEA) expects global energy demand to double by 2050. That means more and more fossil fuels, i.e. ■ coal, ■ oil and ■ natural gas, will be burned. If this happens, there would be a dramatic rise in CO_2 emissions.
We will have to cut our carbon emissions in half by the middle of this century to prevent the climate getting totally out of control.
The Greenpeace study Energy [R]evolution shows how this can be achieved. Saving and more efficient use of energy can cut back consumption – without endangering the global economy. ■ Nuclear power stations will be faced out in 2030 in this scenario. By the year 2050, half the world's primary energy requirements will be met by renewable energies, i.e. ■ biomass, ■ water power, ■ solar, ■ wind and ■ geo-heat.

fourfold, i.e. that the value created for every unit of resources can increase fourfold.'

In other words, we can live twice as well while simultaneously halving resource consumption. The book shows how this can be accomplished sector by sector. Existing buildings are described that require only one-quarter as much energy as current standard homes. There are cars, already in production, that consume 2.5L/100km, compared to cars currently in use that consume 10L/100km. The authors examine each industry in turn and show that Factor 4 is also applicable to industry. So it is not a question of whether or not the technology is available, but rather whether the will exists to change our society to a sustainable one.

Resource Consumption in the Construction Sector

The construction sector is one of the biggest resource consumers. It is sometimes called the 40 per cent sector because in many countries it consumes 40 per cent of total energy and 40 per cent of total materials. Globally, the construction sector consumes one-quarter of all timber and one-sixth of all clean water. As buildings have a long life, it is especially important to construct buildings that use resources efficiently in operation, and to renovate existing buildings to make them more efficient.

Energy Quality

Actually, the concept of energy consumption should not be discussed without mentioning energy quality. This means, for example, that electricity is valued about three times higher than space heat. A consequence of this is that when an electric heat pump is being considered for home heating, it should have a heating factor greater than three in order for it to be worth installing from an energy conservation point of view. The heating factor is the relationship between the amount of low-temperature heat that is utilized by the heat pump and the high-quality energy required to run the heat pump. The best home heat pumps are beginning to reach a heating factor of about three.

Entropy describes energy quality. Energy cannot be consumed; however, entropy can be lost. Entropy is lost during every process of conversion, i.e. the energy remains but it is of a lower quality. This means that during consumption of energy, an attempt should be made to have as few energy conversions as possible. For example, generating electricity with a diesel-fuelled generator heating water means that there is a large loss of entropy.

Exergy is a concept that connects the amount of energy with its quality. Electricity for example has a greater exergy value than the same amount of energy in the form of heat. This means that in principle high-quality energy sources should never be used for low-quality energy end uses, e.g. heating homes to 20°C with electricity.

Energy Cascading

Energy cascading means to use the same energy several times, or in other words, to make better use of exergy. An example is a power plant where fuel is burned to produce steam. The steam runs a turbine which produces electricity. The steam then cools to hot water that can be used to heat buildings in district heating. The paper industry provides another example where one fuel is burned to generate high-temperature process heat, and after the process there is enough residual heat with a sufficiently high temperature for it to be used for district heating.

Individual Metering

Household energy consumption varies considerably. It depends a lot on energy conservation awareness. Electricity consumption depends on many small things, e.g. turning lights off when leaving a room, how long the kitchen extractor fan is left on and using the right size pan on the right size cooker ring. Is the lid on the pan, is the ring turned off before cooking is finished in order to make use of residual heat, is the freezer defrosted, what are the temperatures inside fridges and freezers? These factors all affect energy consumption. A good way to increase energy consumption awareness is individual metering. The meter display should be located so that consumption is easily seen, and rental agreements should include payment for actual consumption. It has been shown that individual metering is the measure that results in the greatest reduction in household electricity consumption. Consumption habits are quite similar for electricity, heat and water – people either conserve or squander.

Energy Cascading

	Temperature	Process		Product
Energy in	1000°C	Electrolysis	Steam	Aluminium
	800°C	Smelting	Steam	Aluminium
	500°C	Back-pressure	Steam	Electricity
	180°C	Boiling	Steam	Paper pulp
	100°C	Drying	Water	Paper
	50°C	Ethanol distillation	Water	Ethanol
	30°C	Local heating	Excess	Heat
	10°C	Waste heat		

Note: This is a hypothetical example that shows the energy cascading principle. There is an almost 90 per cent saving compared to using 1000°C heat for local heating.

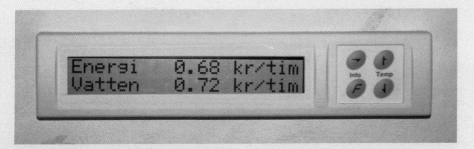

Individual metering is being introduced in new buildings. The photo shows the individual meter display, giving energy and water cost per hour, used by the housing company Svenska Bostäder in a building in Hammarby Sjöstad, Stockholm. The display is located so that personal consumption is easily visible.

Energy Declarations

The European Union decided in 2006 that an energy declaration carried out by certified energy experts should be made for all buildings; it should provide information about the building, the ventilation system, the hot water system, heat distribution system, heat production system, control and regulation system, household electricity, operating electricity, and show how much energy and electricity are used in the building. The aim of the declaration is to improve energy savings. So proposals on how to save energy in the building should also be included.

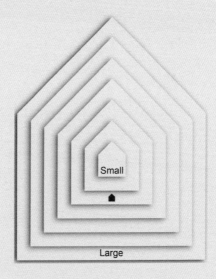

Once an energy declaration has been carried out, a label with a symbol and results is placed on the building in an easily seen location.

USER HABITS

For identical buildings with a mean consumption of 18,000kWh/year, the variation in energy consumption due to user habits is usually about 10,000kWh/year, with the lowest consumption at 15,000kWh/year and the highest at 25,000kWh/year.

Ways to try and change user habits include individual metering of heat, hot water, cold water and electricity, as well as a metering system where individual households pay for the amount they consume.

Heat

Letters relate to labels on the figure opposite.

- A good room temperature is 20°C, and 18°C in bedrooms. (E)
- Close curtains, blinds and venetian blinds at night, but maintain air circulation from radiators. (A)
- Don't place furniture in front of radiators and make sure that curtains don't block heaters. (J)
- Reduce ventilation in the winter but don't close it completely. (I)
- Air rooms quickly, preferably with cross-draught. (K)

Electricity

- Don't run washing machines with half loads.
- Don't leave the kitchen fan on unnecessarily. (G)
- It should not be colder than +6°C in refrigerators and −18°C in freezers. (H) Defrost regularly. Examine energy consumption when purchasing new household appliances.
- Use lids on saucepans.
- Use an electric kettle for boiling water.
- Choose a cooker ring that isn't larger than the bottom of the pan used.
- Vacuum regularly behind fridges and freezers.
- Thaw out frozen goods in the fridge, so that the cold can be used by the fridge.
- Don't leave TVs, radios, computers or other electrical devices on standby when no one is using them. Turn them off with the power button instead. (L)
- Turn off lights in rooms no one is using. M
- Change to LEDs (light emitting diodes).
- Disconnect battery chargers when not in use.

Water

- Seal or report leaking taps and running toilets. (B)
- Don't wash dishes under running water. (F)
- Run the dishwasher when it is full.
- Fully load the washing machine when washing clothes. (C)
- Avoid prewashing.
- Take quick showers more often than a bath. (D)
- Don't let water run unnecessarily.
- Don't flush unnecessarily. Don't flush waste down the toilet.

Waste

- Think about what you buy. Don't carry home unnecessary waste.
- Choose reusable packaging.
- Avoid disposable products.
- Refrain from accepting unnecessary packaging.
- Sort rubbish.
- Reuse as much as possible.
- Take as much as possible for recycling.
- Use non-poisonous products instead of poisonous products.
- Take environmentally hazardous waste to an approved facility.

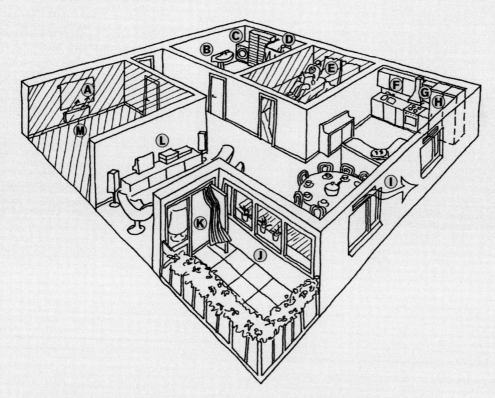

Resource conservation has a lot to do with user habits, i.e. being aware of how different actions affect living costs.

2.1 Heating and Cooling

Good architecture can be created for every climate by combining energy-efficient construction with passive heating and cooling. To reduce energy consumption and achieve a sustainable society, future buildings must be energy efficient, and energy conservation measures must be adopted in existing buildings.

2.1.0 Heating Efficiency

About 40 per cent of total energy consumption in cold northern climates is used to heat buildings and supply them with electricity. Most of this is for space heating and hot water.

Energy consumption standards for new buildings have been made more strict. There is a proposal that, by 31 December 2018 at the latest, EU Member States must ensure that all newly constructed buildings should be zero-energy houses that produce as much heat as they consume on site – e.g. via solar collectors and biomass, or heat pumps and photovoltaics.

The Energy Requirements of Buildings

To build energy-efficient buildings it is necessary to understand how buildings use energy.

Examination of the energy consumption of a building over a 50-year period shows that about 9 per cent of the total is used in the building phase (5 per cent for production of the materials plus 4 per cent to transport them), and over 90 per cent is used to heat and operate the building. The lesson learned here is that it is most important to reduce the energy required for heating and electricity supply. This means that if more insulation is used (investing more energy in the production of insulation), a large amount of energy for heating is saved.

In energy-efficient buildings, there is a radical reduction in total energy consumption. Furthermore, the division of energy use is different, so that about 20 per cent is used in the building phase and 80 per cent in operating costs.

Environmental Impact

A new way to show the environmental impact of two or more buildings is to compare heating in the form of energy consumption and related CO_2 emissions per year, as well as environmental profiles from the building phase, e.g. the environmental impact of consumption of materials (kg) and emissions of CO_x (carbon oxides), NO_x (nitrogen oxides) and SO_x (sulphur oxides)

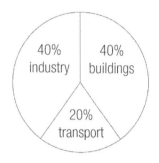

About 40 per cent of total energy consumption in Sweden is used in buildings, mainly for heating. About 40 per cent is used in industry and 20 per cent for transportation.

during production. All the data is presented in quantities per year of each material's expected life. The life of construction material is taken to be 100 years. Façade and sheeting material is considered to have a 50-year life. Certain sensitive materials are estimated to have a life of 20 years.

Energy Flows in Homes

Energy is lost from a house through the roof, walls, floors, windows, doors and from exhaust air and waste water. In addition, energy is needed to heat the cold water and fresh air taken into the house. A house is heated by several different heat sources: body warmth, solar heat through windows, and utilization of waste heat from electricity and hot water use. Other heating requirements are supplied by the building's heating system.

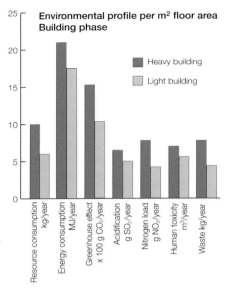

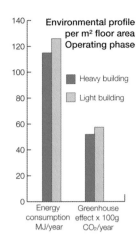

Environmental profiles comparison between two south-facing and well-insulated two-storey terraced houses. One is brick and the other is wood. It can be seen that in the building phase buildings made of heavy materials require more energy and result in a greater environmental impact than light buildings.

Heavy buildings require less energy for heating, and so the CO_2 emissions from heating are also less.

Source: *Arkitektur og miljø – form, konstruktion, materialer – og miljøpåvirkning*, Rob Marsh, Michael Lauring, Ebbe Holleris Petersen, Denmark, 2000

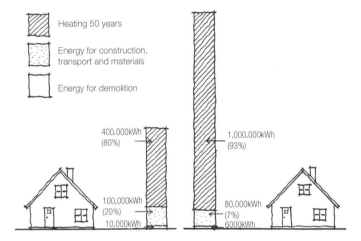

Energy consumption in a low-energy house and a conventional house over a 50-year period. The house on the left is a 150-m² low-energy house. The house on the right is a conventionally built house of the same size.

Source: *Mer bolig for pengene – Idéer til miljøvennlige boformer*, Nils Skaarer, Denmark, 2001

Energy Out

Ventilation is the biggest source of heat loss in a low-energy building. Fresh air has to be heated, exhaust air contains large quantities of heat that are lost and there is involuntary ventilation (draughts). The amount of heat lost depends on the building's airtightness, how much the building is ventilated and if heat is recovered from the ventilation air.

Windows are the most poorly insulated components of a building's shell and account for the second largest energy loss. The losses depend on how good (low) the windows' U-value is, how big the window area is, as well as whether insulating shutters or curtains are used.

Outer doors and especially balcony doors and their windows are usually poorly insulated, but there are usually relatively few outer doors, and so these losses are not so great.

The roof on a single-family home is the biggest insulated surface of the shell. Since heat rises, the temperature on the inside of the roof becomes relatively high. In addition, the roof faces the cold night sky. Losses through the roof are relatively high. The roof should therefore be the best-insulated part of the house.

The walls usually constitute the second largest surface on a single-family home after the roof. The walls should therefore be well insulated. It is more expensive to insulate walls than roofs since the entire building's area, roof and foundation increase in size as the thickness of the walls increases.

The amount of heat loss from **the foundation** depends on its construction. With a plinth foundation, the ground floor structure is in principle like an additional outer wall. A slab foundation is not exposed to outside temperatures but to the ground temperature, which is relatively constant all year round.

Cold water often has the same temperature as the ground (about 8°C), so when it comes into a building, it takes energy from it.

By using a simple heat exchanger, heat can be recovered from **waste water**. A tube and shell heat exchanger has an efficiency of approximately 50 per cent and can absorb part of the heat in the waste water, which can be used to preheat incoming water.

Energy In

Body heat is released as a result of people's metabolic processes. There are, however, various views on how much heat is given off. Besides the heat that can be felt, people give off just as much latent heat (through evaporation), which can be recovered with an exhaust-air heat pump.

Solar heat that comes in through south-facing windows and contributes to heating is referred to as passive solar heat.

Household electricity use contributes to heating. During the heating season, use of almost all electricity contributes to heating, while the heat from electricity use in the summer is lost since additional heat is not required at that time.

Operating electricity is the electricity used to run fans and pumps for a building's heating and ventilation systems. The amount of operating electricity required has had a tendency to increase due to the use of complicated mechanical ventilation systems, e.g. intake and exhaust systems with heat recovery.

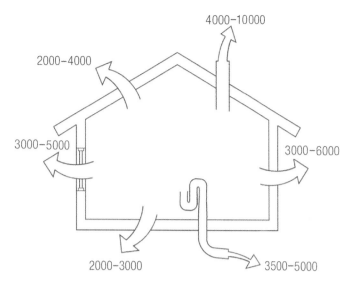

Energy is lost from a house through the roof, walls, floors, windows, doors and from exhaust air and waste water. In addition, energy is needed to heat the cold water and fresh air taken into the house.

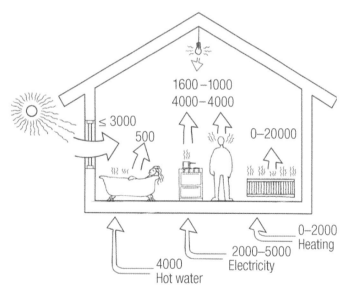

A house is heated by several different heat sources: body warmth, solar heat coming in through windows, and utilization of waste heat from electricity and hot water use. The rest of the heating requirements are supplied by the building's heating system.

Hot water contributes to heating. In a normal single-family home, about 4000kWh/year are used to heat hot water, of which approximately 500kWh contributes to space heating. If solar collectors are installed on a building, solar energy can provide about half the annual hot water requirement, or more, depending on the location.

The heating system supplies the remaining heating requirements.

Energy Balances

It is important to understand a building from an energy perspective. A good way to do this is to study the building's energy balance. An energy balance shows where energy comes into a building and where it goes out. In addition, the size of the different heat flows are determined.

Energy Balances for Different Single-family Homes

In an older single-family home built to earlier building regulations, about 200kWh purchased energy is used per m² per year. A newly built single-family home, built according to the current building code, uses about 100kWh purchased energy per m² per year. Newer homes may have thicker insulation, triple-glazed windows and heat recovery from ventilation. In an energy-efficient ecologically built house, the energy consumption can be further halved to 50kWh purchased energy per m² per year. Such houses are even better insulated, the windows have an even lower U-value, energy-efficient appliances and lighting are used, as well as water-saving technology. The amount of free energy from body heat and passive solar heat that can be taken advantage of decreases the more energy efficient a building is. The greater the energy efficiency the less heat is required and so there is less time when free heat can be used. Homes that use less than 50kWh/m²/year are not theoretical utopian visions. Such homes have already been built.

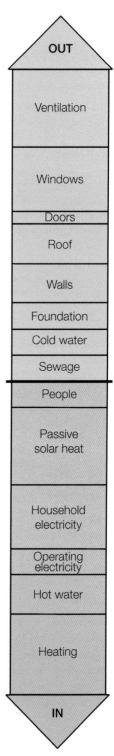

An energy balance shows how energy enters and leaves a building.

Energy Balance Calculations

There are a number of computer programs for calculating a building's energy balance. The calculations show, among other things, how a building can be improved. Relatively simple-to-use software is available that either does or does not take into consideration a building's heat storage capacity and gains from passive solar heat. Other programs that take these factors into consideration can be time-consuming to use.

It is possible to construct buildings that are much more energy efficient than the current average. Passive houses are now being built that are so well insulated that an extensive heating system is not required. It is this kind of energy-efficient building that should be the rule rather than the exception.

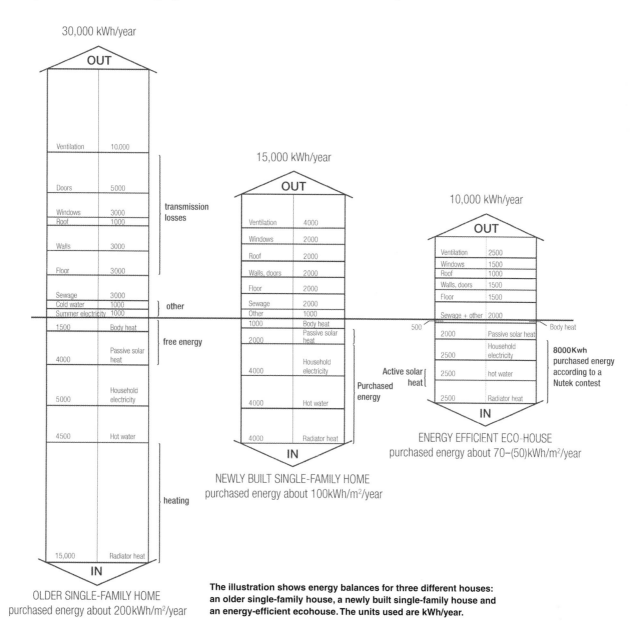

The illustration shows energy balances for three different houses: an older single-family house, a newly built single-family house and an energy-efficient ecohouse. The units used are kWh/year.

EXAMPLES OF LOW-ENERGY BUILDINGS IN SWEDEN

The Brandmästar Area in Karlstad, Sweden

In 1998, Karlstad municipality announced a contest to design and build energy-efficient, environmentally friendly blocks of flats. The buildings in the winning entry are super-insulated and built with healthy materials and solid wood foundations. The buildings are heated using solar and district heating, and use waste-water heat exchangers. The electric cables are shielded. The ventilation system is an adjustable exhaust air system. The buildings require about 60kWh/m²/year. There are garden plots, root cellars, composting facilities and water pumps in the courtyards. The storage buildings and carports have green roofs. There is a covered bicycle stand beside each building. The lobbies have space for special delivery boxes for internet purchases. Extra care was put into sound quality. Outside the clothes cupboards there are special spaces for airing and drying clothes. The area consists of 25 flats in a total of five blocks, which were designed to fit in with the existing surrounding buildings.

The Brandmäster area in Karlstad, Sweden. The winning entry in a contest to design and build energy-efficient and environmentally friendly blocks of flats.

Source: Architect Jonas Kjellander and Sören Sten, FFNS Örebro. Ecoarchitect Varis Bokalders. Builders Johnny Kellner and Sten Eriksson, JM Bygg

The Ekomer Concept of Karlson Hus, Sweden

This is the first prefabricated single-family house on the Swedish market that is energy efficient, heated with renewable energy and built with healthy materials. The outer walls are made of light beams and insulated with 300mm cellulose fibre (U = 0.13W/m²K). The roof is insulated with 450mm cellulose fibre (U = 0.09W/m²K). The foundation is insulated with 600mm cellular glass loose fill (U = 0.15W/m²K). The windows are triple-glazed with argon gas (U=1.2W/m²K). Special care has been taken to avoid cold bridges. The building has an air-to-air heat exchanger with 90 per cent efficiency. The total energy

The pellet stove with water jacket in the living room of the Karlson Hus. It's a 12KW Palazzetti where more than 80 per cent of the heat is stored in an accumulating tank.

Karlson Hus Ekomer concept house is the first prefabricated single-family house on the Swedish market that is energy-efficient, heated with renewable energy and built with healthy materials.

The 'House Without a Heating System' Project in Lindås – The First Passive House in Sweden

In the 'House Without a Heating System' project in Lindås (completed in 2002), 20km south of Gothenburg, Sweden, homes were designed where all the energy requirements are met without the use of traditional heating systems. The homes were designed by architects at EFEM Architecture Office with Hans Eek as the project manager. The technology used includes heat exchangers, extra-thick insulation, passive solar heating, and roof-mounted solar collectors that provide half the heat for hot tap water. In addition, triple-glazed windows with two metal layers and krypton between the panes ($U = 0.85 W/m^2 K$) were used.

needed for a house like this fulfils the passive-house standard. The heating system consists of a water-jacketed pellet stove in the living room, solar collectors on the roof and an accumulation tank in the basement. The pellet stove is a Palazzetti with an output of 12kW and more than 80 per cent of the heat is stored in the accumulation tank (500L). The stove has pellet storage of 20kg and has to be refilled twice a week during the winter. The flat plate solar collector is $7.5m^2$ and is sufficient to provide all hot water heating needs during the warmest six months of the year. There is an electric immersion heater in the accumulation tank that can be used if the house is vacant for a long period.

The terraced houses at Lindås are $120m^2$ and have in principle worked out as planned. Energy consumption was, however, a little larger than estimated. The measured energy consumption is about 6500kWh/year, i.e. about $55kWh/m^2/year$. There is a greater electricity consumption because the residents have more electrical devices than expected.

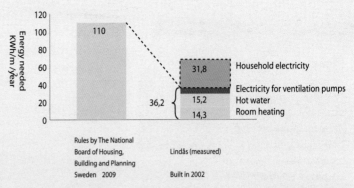

Energy consumption for the row house unit in Lindås. Energy consumption for hot water, space heating and electricity for ventilation and pumping is 36.2 kWh/m²/year. The following conclusions can thus be made for the future: the energy consumption for many passive buildings is totally provided by electricity (electricity is considered to be 2.5 times the energy quality as heat), which leads to the resulting costs. When buildings consume such a small amount of energy, use of household electricity should also be made more efficient. Further, it would be better to provide space and water heat by other means than direct electricity.

Energy loss via radiation, ventilation and sewage is minimized. All the electrical appliances are energy efficient. The total energy supplied is estimated to be 5400kWh/year, which is mostly household electricity for lights and cooking. A normal-sized single-family house heated by electricity requires about 20,000kWh/year. The 20 terraced houses are the result of a cooperative research project between Chalmers University of Technology, Gothenburg; Swedish Council for Building Research (now Formas, the Swedish Research Council for Environment, Agricultural Sciences and Spatial Planning); Lund Institute of Technology, Lund University; and the Swedish National Testing and Research Institute.

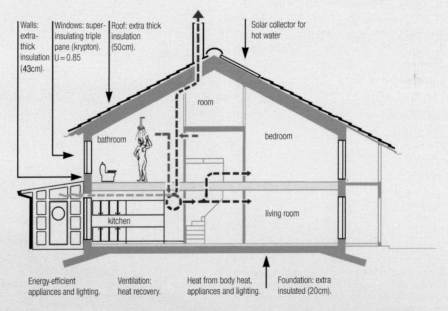

Cross-section of a terraced house unit at Lindås, Gothenburg, Sweden. The energy loss is so small that the heat inside the homes from body heat, passive solar heat, household electricity and hot water is enough to heat the buildings. These buildings do not have any traditional heating system at all. However, intake air can be preheated with an electric radiator when it is extremely cold or when people return from a trip during cold weather.

RULES FOR PASSIVE HOUSES

New housing in Sweden may not use more than 110–130kWh/m²/year for heating, ventilation and hot water. These figures do not include household electricity. Swedish passive houses may use 45–55kWh/m²/year, household electricity also included (not regulated by law). The lower figure refers to houses in southern Sweden and the higher figure refers to houses in northern Sweden.

If hot-water saving devices and efficient electric household appliances are used, the total energy needed in a passive house is 50–70kWh/m²/year. The annual use in a passive one-family dwelling 120m² in size is 3000kWh for hot water, 3000kWh for household electricity and 1000kWh for heating.

Energy-efficient houses are being built in many countries in Europe. There are many different terms used to describe and define these buildings. In Germany, the definitions do not cover hot water and household electricity, and are: low-energy houses (50–70kWh/m²/year); three-litre-houses that use maximum 30kWh/m²/year (10kWh is equal to 1 litre of oil), and passive houses that use 10–15kWh/m²/year. Plus energy houses (or zero-energy houses) are passive houses with e.g. solar cells that produce more energy than they use on a yearly basis. CO_2-neutral houses are buildings where all the energy needed is produced with renewable energy. The term 'Passive House' refers to a professional German construction standard set by the Passive House Institute (PHI). Similar standards are used in Austria. In Switzerland, the standards are defined by the organization Minergie.

There are about 15,000 passive buildings in the world, of which about 10,000 are in Germany. In Austria about 50 per cent of new buildings are passive. Government subsidy programmes are a main reason why there are so many passive buildings in Germany and Austria.

Passivhus (trade mark in Sweden by Hans Eek, Architect)

A passive house minimizes heat loss so much that the internal heat gains are adequate to provide good comfort all year round. Values that are in the definition of a passive house in Sweden in 2007:

Heat load at 20°C indoor temperature at lowest outdoor temperature:

- 10W/m² for multi-family houses;
- 11W/m² for attached terraced houses;
- 12W/m² for detached houses;
- 14W/m² in the northern climate zone.

Passive houses in solid wood construction in Potsdam, Berlin. Healthy materials and energy efficiency are combined.

Source: Joachim Eble Architektur

The first multi-family house built to the passive-house standard in Austria situated in Ölsbund, Dornbirn.

Source: Architect: Hermann Kaufman. Architectural photographer: Britt-Marie Jansson

Energy demand: Maximum 'bought energy' for the total energy supply of the building (heating, domestic hot water, operating electricity used for heating and ventilation systems, and household electricity):

- climate zone south: <45kWh/m²/year;
- climate zone north: <55kWh/m²/year.

Building requirements:

- air leakage through the climate shell: maximum 0.3L/sm² (at pressure +/−50pa)
- windows, doors, U-value <0.9W/m²Ka
- floor, walls, roof, U-value <0.1W/m²K
- Noise: At least sound classification B

Recommendation:

The use of household electricity should be minimized, e.g. standby functions for television sets, stereos and chargers should be switched off. Household appliances: A+ class is to be used, and low-energy light bulbs as well. Solar collectors for domestic hot water production during the summer are recommended.

To achieve this specified energy performance these houses are very well insulated, have super-insulated windows and use air-to-air heat exchangers (today the best ones can recycle 90 per cent of the heat). It is also important that the construction doesn't have any cold bridges and that the houses are extremely airtight. To achieve airtight seams in the windproof layers they have to be overlapped and compressed or connected with high-quality durable tape.

EXAMPLES OF PASSIVE HOUSES

Erik Hedenstedt's passive house in Trosa, built of healthy materials. The building has a foundation of foam glass and linen insulation in the walls. All the walls are clay plastered on the inside to provide a more even air humidity. Hot water and incidental heat gain is from solar collectors and a water-jacketed wood heater. Solar cells mounted on the building contribute to the electricity supply. Heat exchanging takes place both for ventilation and sewage. The building is super-insulated. The windows and doors are made of solid, oiled oak, with a U-value of 0.8W/m²K, are used throughout the building.

Source: Architect: Anna Webjörn.

Erik Hedenstedt at his water-jacketed wood heater.

In the bedroom, looking towards glass balcony.

A solid wood passive house with culvert ventilation, active solar water heating, clay plaster internal wall finishes, natural paints, etc – plus an innovative and economical textile façade.

Close up of the cloth façade

Openable windowdoors with a U-value of 0.7 W/m²K.

Architect: Walter Unterrainer, Vorarlberg, Austria.
Photos: Walter Unterrainer.

The first passive-house-standard municipal centre in Austria is situated in Ludesch, Voralberg.

Source: Architect: Hermann Kaufman. Architectural photographer: Britt-Marie Jansson

The first kindergarten in Sweden build according to passive construction standards, Stadsskogen kindergarten in Alingsås.

Source: Architect: Glantz Arkitektstudio AB, Alingsås.

INTERNATIONAL EXAMPLES OF PLUS ENERGY HOUSES

Marché's office building outside of Zurich

Marché's office building outside of Zurich is designed by architect Beat Kämpfen. It has been classified as a 'Minergie-P-Eco', which means that it is a low-energy building according to passive construction standards built with environmentally friendly materials. The surface of the roof is covered with amorphous solar cells which on an annual basis produce more electricity than consumed. It is a three-floor, solid wood construction with the only 35mm thick wood exposed to the interior. Other than the inside panels of upright studs, the load bearing structure includes an outside layer of horizontal studs made of the same panel material. The outermost stud keeps the airtight layer and façade boarding in place. This construction is filled with 35cm of insulation. The building has an FTX system where the supply air is preheated with heat from the ground via a heat pump connected to a water filled coil in the ground. The space heat and hot water needed by the building is produced by a geothermal heat pump. Sections of the façade are made up of transparent insulation filled with Glauber's salt that acts as heat storage with a phase change (using the GlassX product of GlassX AG Germany). The Glauber salt melts when the sun shines on it and hardens at sunset and the released stored heat contributes to heating the building. There is a modular pillar system in the building that contains all the installations (ventilation, electricity, telephone and computer cables). There are vertical green walls on every floor that moisturize the indoor air, so that it doesn't get too dry.

Award-winning office headquarters for the Marché company, Switzerland's first zero-energy commercial building, constructed at normal cost. It includes an integrated photovoltaic roof, salt-hydrate heat storage in translucent façade panels, solid wood, and standard passive construction.

Source: Architect: Beat Kämpfen, Zurich, Switzerland.

Sunny Woods is a four-floor apartment building with six two-floor apartments. The upper apartments have roof terraces and the lower ones have patios. It is a passively heated building made of solid wood and the roof is mounted with solar cell panels. The solar technology is attractively integrated into the architecture.

Source: Architect: Beat Kämpfen, Zürich, Switzerland.

Vacuum solar heaters for heating hot water decoratively cover all the fronts of the balconies on the Sunny Woods House.

Source: Architect: Beat Kämpfen, Zürich, Switzerland.

Plus-energy houses in Schlierberg, Freiburg, Germany. These passive house have photovoltaic roofs that produce more energy than they use on a yearly basis.

Source: Architect: Rolf Disch.

2.1.1 Insulation

One of the most important aspects of building energy-efficient buildings is to insulate them well. Different insulation materials vary in their insulating capacity and can be divided into those that are sensitive to moisture and those that are not. The insulating capacity of a construction is not only dependent on the type of insulation material used and how thick it is, but also on the construction's thermal inertia, moisture-buffering capacity and airtightness. The insulating capacity of a material or construction is given as a U-value (the unit used is W/m²K, watt per square metre and degree Kelvin).

Insulation Thickness

Historically, it can be seen that insulation thickness has increased as the price of oil has increased. In the 1970s 10cm-thick insulation in walls and roofs was still common. After the oil crisis of the 1970s, the thickness of insulation doubled. In the super-insulated buildings that should be built to achieve a sustainable society, an even higher level of insulation is required. The thickness of insulation in existing super-insulated buildings is 30–40cm in the walls

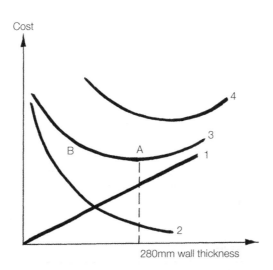

The optimum insulation thickness for walls can be calculated by overlaying the costs for building a wall of different thicknesses (curve 1), with the energy costs of heating the building (curve 2). During the first half of the 1990s, the optimum insulation thickness was about 28cm (curve 3, point A), but if energy prices increase the optimum insulation thickness also increases (curve 4).

Source: Professor Björn Karlsson, Linköping

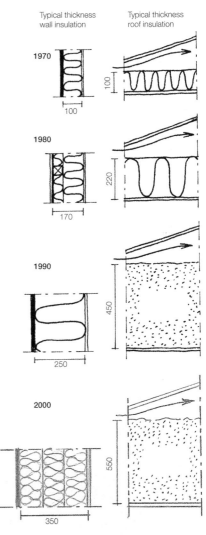

In Sweden, buildings are being better and better insulated. Building standards and methods have changed in pace with the increase in oil prices and environmental awareness.

Insulation Thicknesses and U-values for a Normal Terraced House and a Terraced House in Lindås (Built in 2002), which is a Passive House with No Traditional Heating System

	Insulation thickness (mm)		U-value (W/m²K)	
	Normal terraced house	Lindås	Normal terraced house	Lindås
Wall	240	430	0.17	0.10
Roof	320	480	0.12	0.08
Floor	150	250	0.20	0.09

and 50–60cm in the roof. Insulation in the foundations is about 20–30cm thick. There have been some attempts to calculate the optimum wall thickness from a life-cycle perspective. Results depend of course on the future price of energy. The cost curve is quite flat, which means that a somewhat better-insulated wall doesn't cost much more. It is, however, less expensive to have a roof that is better insulated since then the size of the building's foundation and roof doesn't increase. A problem in this context is that building codes stipulate how much of a building plot may be built on. This means that the thicker the walls (insulation) are, the smaller the surface living area there is to rent out. Building codes should be drawn up so that they stipulate the actual size of the living area in a town plan instead.

U-value

U-value is a theoretical value of a construction's insulating ability, and is the λ-value (lambda value) divided by the thickness. The best insulation material has a λ-value between 0.035 and 0.055W/mK. Insulation materials with extremely low (good) U-values have recently become available. These materials, which are based on the thermal principle, should make possible thinner, super-insulated walls. An example is the Vacupor brand (made by the German company Porextherm) with fumed silica in a vacuum between metal foils. The λ-value is as low as 0.005W/mK. They are, however, still expensive.

The table compares insulation thicknesses and U-values for a normal terraced house and a terraced house in Lindås (built in 2002), which is a passive house with no traditional heating system. All heating is provided by lights, appliances and people.

Thermal Bridges and Convection

The thicker the insulation, the greater the risk of convection in the insulation, and the more important it is to avoid thermal bridges in the construction.

Heat leaks out at thermal bridges. Condensation problems can occur at thermal bridges and it is often precisely in those places that dirtying of inner walls and inner roofs takes place. Thermal bridges are common where supporting constructions pass through the insulating layer, where holes are made in the building envelope (i.e. at windows, doors and chimneys), where architectural design has not provided enough insulation, and where insulation is poor due to technical building considerations. Convection can be avoided by insulating with several layers or choosing denser insulating material. Convection can more easily occur in mineral wool than cellulose fibre.

Weathertightness

For a building to be energy efficient, more than just good insulation is necessary. It must also be airtight, so that there isn't so much involuntary ventilation (draughts) that heat

Places where, in principle, thermal bridges can easily occur in a building if a structure is not well planned and built.

leaks out of the building. Leaks can occur where different construction parts meet, e.g. at windows, window frames, or in the seam between two construction elements. Leaks can also occur in constructions if they do not have good enough windproof layers.

Leaks

A building is made up of a large number of parts. There is a risk of leaks at the seams where parts meet. These leaks can occur due to carelessness during construction, or be caused by movement inside the walls, floor structure, etc. Buildings made of wood move all the time. Wood is a living material that changes in volume according to the relative humidity. When built-in wood dries, it shrinks, and leaks can occur with a risk of air leakage. It's important that the windproof layers are continuous and that all the seams are overlapped and compressed. There are a number of critical points in a building where seals and tapes may be inadequate:

- where outside walls connect to the floor;
- where window frames connect to walls;
- between window casements and frames;
- at the base of a roof;
- where inner walls connect to outer walls;
- at pipe penetration points; and
- at electrical boxes on outer walls.

Seams and Windproof Layers

Seams should be designed so that it is possible to caulk them properly. Windproof layers should be designed so that they can be laid in an overlapping fashion and clamped down with mouldings. Polyurethane foam is still commonly used to fill seams. In eco-buildings, flax or cellulose-fibre strips are preferred. Diffusion-proof plastic sheeting is often used as the windproof layer in constructions. In eco-buildings, windproof pasteboard, hard fibreboard or other windproof material (e.g. permeable plastic layers)

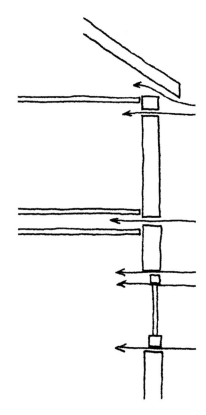

Buildings are constructed to be as windproof as possible to avoid draughts and save energy. Windproofing depends on the design details of connections between parts of a building. The illustration shows some of the weak points that must be carefully considered in order to achieve a 'tight' building.

are used instead, since the intention is to buffer moisture in the construction.

Moisture Damage and Excess Pressure

It is important for buildings to be weathertight so as to avoid moisture damage. If a ceiling isn't weathertight and allows warm, moist indoor air to go up into the loft and meet the cold outer roof, condensation and mould can occur. Another factor that increases the requirement for a weathertight building is a ventilation system that causes excess indoor air pressure.

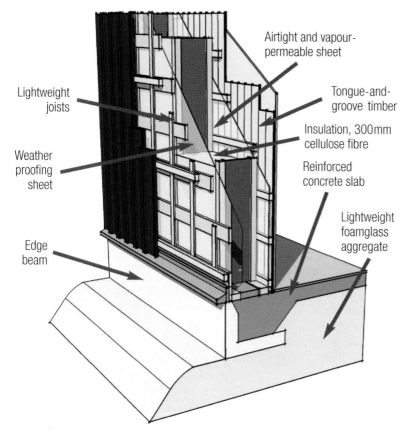

The wall and foundation insulation of the Karlson Hus Ekomer project. The airtight and vapour-permeable layer inside the wall is a Gore-tex-like material. The inner layer is less vapour-permeable than the outer sheet to avoid moisture problems, as vapour moves from the warmer inside of a building to the cooler outdoor space. This construction helps ensure that walls dry out rather than accumulate moisture.

Source: Karlson Husindustrier AB

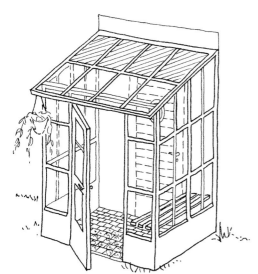

In order to avoid unnecessary heat losses, an eco-house should have a porch and well-insulated outer doors.

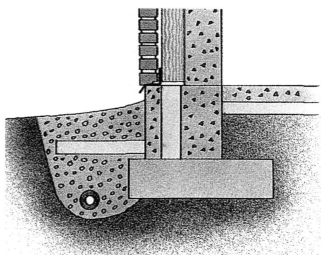

One of the critical thermal bridges is where the outer wall meets the foundation. This illustration shows way of avoiding that thermal bridge.

Source: Cementa

Weather stripping

Weather stripping that seals windows and doors well, e.g. made of textile or EPDM rubber, must be used. One of the most common recommendations made after energy crises was to seal windows and doors with weather stripping. It is of course good for buildings to be tight, but in many older buildings the leaks around windows and doors are part of the ventilation system. Sealing such buildings interferes with ventilation. In such cases, the weatherstripping at the top of windows can be removed, for example, so that air can enter (not in bathrooms and kitchens).

Test for Airtightness

As the standard for airtightness in buildings is currently quite high, it may be of interest to know that there are methods for measuring the airtightness of a building. These tests can be used to determine whether or not a contractor has fulfilled contract requirements. A building's airtightness is tested using a pressure method. With the help of fans, both excess and negative pressure can be created in a building. By establishing average values for flows during both excess and negative pressure, a standard value is arrived at for the building. A building's volume is confirmed at the same time. A building's airtightness is normally measured as the number of air exchanges per time unit.

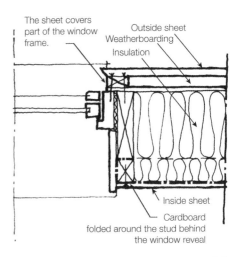

Energy-efficient buildings should be windproof. If pasteboard or hard fibreboard is used for weatherproofing, it is important to construct the seams and connections to windows, etc. very carefully.

It is especially important when windproofing a building to carefully plug the seams around windows and door frames, as well as between prefabricated sections (e.g. with packing of cellulose fibre, flax or coconut fibre)

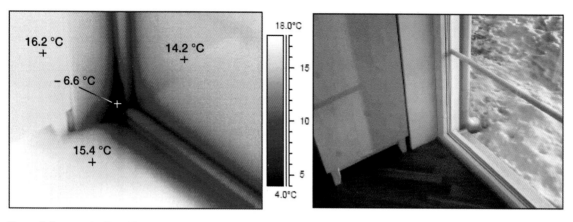

Thermal photography. The different colours show different surface temperatures. Thermal photography reveals faults in construction that result in energy leakage.

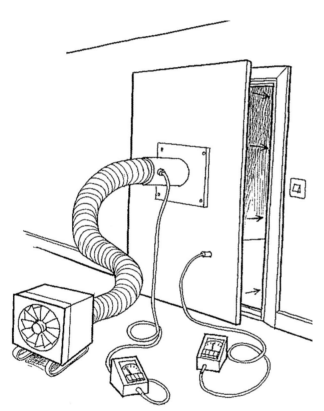

It is very important that a building be airtight in order to achieve low-energy consumption. The craftsmanship and the details of the airtight layer must be accurate. Airtightness can be tested by carrying out a pressure test of the building. The illustration shows the equipment required.

2.1.2 Windows

From an energy perspective, windows are the weakest link in a building, i.e. the part of the shell with the worst (highest) U-value. About 2500kWh per year leaks through the windows of a modern three-room flat. That is ten times the energy lost from an equivalent size wall area. So, for energy-efficient buildings it is especially important to install good windows with low U-values. With new well-insulated windows you avoid cold downdraughts.

U-values for Windows

What makes a window have a low U-value?

Radiant heat loss can be reduced with the help of:

- several window panes, e.g. triple-glazed (2+1), 4-glazed or 2+2-glazed (with two of the panes in one inner window);
- low-emission layers (thin layer of tin oxide or silver, which admits short-wave solar radiation from outside but doesn't let out long-wave heat radiation from inside)
- the space between the window panes influences the U-value; in a double-glazed window, a space of 2.5cm is best from an energy perspective;
- night insulation, window shutters, curtains or venetian blinds that are closed at night and open during the day.

Conductive heat loss can be reduced by using:

- insulated frames, or extra-thick, solid wood frames.

Convection heat loss can be reduced by using windows with:

- heavy gas between the panes (e.g. argon or krypton), which requires insulating glass.

Energy-efficient windows don't necessary cost more than conventional windows. In some cases, radiators are not needed under such windows. Better windows, e.g. triple-glazed, result in an increase in the temperature of the outside pane. So, there is a reduction in cold radiation and less energy is required to achieve the same indoor comfort.

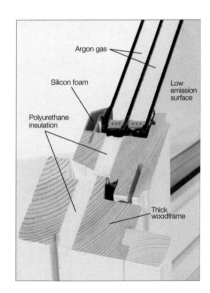

The company NorDan (NorLux) makes some of the most energy-efficient windows (NTech Passiv) in Scandinavia, with a U-value of 0.7W/m²K. It is a triple-glazed window with a low-emission surface on the glass, argon gas between the panes, and insulation in the frame. Other companies that make very energy-efficient windows are e.g. Häussler, Kneer and Variotec in Germany, and Galux in Poland.

Energy-efficient windows that do not leak too much heat are important for an ecohouse. Heat escapes from windows in three different ways: heat radiation through the panes, conduction through the frames and casements, and convection of the air between the panes.

Window Types and Approximate U-values. U-values Depend on the Whole Construction.

Number of panes	Design	U-value (W/m²K)
1-pane		5
	LE-layer (low-emission layer)	3.5
2-pane		3
	Gas-filled	1.9
	LE-layer	1.6
	Gas + LE-layer	1.4
	Gas + LE-layer + insulated frames	1.2
3-pane		2
	Gas-filled	1.3
	LE-layer	1.5
	Gas + LE-layer	1
	2-pane + plastic sheeting in the middle	1.1
	3-pane + 3 LE-layers	0.85
	3-pane + gas + 2 LE-layers + insulated frames	0.7
4-pane		1.2
	Gas + 2 LE-layers	0.7
2-pane	vacuum+LE-layer	0.5

Beautiful Windows

The primary purpose of windows is to permit contact between inside and outside. Sunshine, light, air and a view should be available both in dwellings and workplaces. It should be possible to open windows (including in bathrooms), and they should be easily accessible for cleaning. It should be possible to air dwellings with a cross-draught. The design of window details is especially important, not least when working with triple- and quadruple-glazed windows, in order to achieve an aesthetically pleasing impression. Indentation patterns on both frames and casements are very important.

Sound-Insulating Windows

There are windows that are especially designed to suppress outside noise. They are usually 2+1 windows with a divided frame and a large space between the panes. Special panes are often used. In the space on the sides between the panes, there is insulation to help suppress sound.

2+2-glazed Windows

The architect Bengt Hidemark has developed a way to obtain beautiful, elegant windows with a good U-value by using two double-glazed windows in each window opening. One of the double-glazed windows is placed on the outside of the window opening and the other on the inside in such a way that there is a small 'greenhouse' or 'display cabinet' between them. On cold winter nights, both windows are closed. When the sun starts to shine in, the inside window can be opened. During the summer, only the outside window is used. So, the number of panes in the window opening can be varied as desired and to suit the season. Another advantage of the window is that it is possible to hear birds sing in the spring through the outside window. Bengt Hidemark has designed several different options for how the inner double-glazed window, when open, can be integrated into walls or window embrasures. He has tested his own alternatives for many years in his architecture.

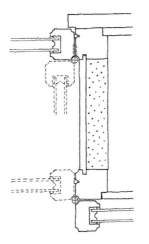

Some window producers are known for making beautiful, sleek, energy-efficient 2+1-glazed wooden windows with indented frames and glazing bars as well as transoms throughout. The Danes are well known for this design, but this window is from Allmogesnickerier in Leksand, Sweden.

Architect Bengt Hidemark has developed 2+2-glazed windows which are divided into two separate frames, one on the outside and one on the inside of the window opening. This results in a light and slim design. During the summer, only the outer double-glazed window is used, and the inside window is folded or slides out of the way.

2+2-glazed method where during the summer the inside window slides into the wall or behind screens on the wall.

2+2-glazed windows in the home of architects Karin and Gert Wingård, Tofta, Bohuslän, Sweden.

On certain windows in office buildings, use of sunscreen glass to reduce entry of unwanted solar radiation may be desirable. How this is carried out depends on the needs of those who will spend time in the offices. The photo shows the pharmaceutical company Novartis' office building in Täby, near Stockholm.

New Glass and Windows

Design for windows with increased energy efficiency has been developing fast. Work has been done with the glass, with various coatings on the glass, and with gas instead of air between the panes. Much research is taking place in this area. New window concepts open the door to new architectural possibilities. Energy-efficient glass will certainly become even better. Sunscreen glass makes it possible to use glass without letting too much heat into buildings with an oversupply of heat. Fire-protection glass allows the installation of windows in situations where a certain level of fire protection is required.

Insulating Glass

With triple- and quadruple-glazed windows, the resulting structures have been heavy and there are many panes to clean. Therefore insulating glass is often used, which consists of two or three panes with a spacer and an air gap between the panes. The air gap protects against influence from outside with the help of a seal. An absorbent material covers the small amount of moisture which, despite everything, may penetrate between the panes due to diffusion. The air in the spaces can also be replaced with gas, as its larger molecules and slower convection improve the U-value. The problem with insulating windows is their life; it is uncertain how long the seal between the panes lasts.

Ultra-Slim Vacuum Glazing for Optimum Thermal Insulation

Whereas the insulation effect at the transition from double- to triple-glazing is the outcome of greater system thicknesses, in the case of vacuum-glazing a genuine qualitative improvement is achieved by obviating the thermal conduction caused by the gas in the cavity between the panes. Using this form of construction at Bavarian Centre for Applied Energy Research (ZAE) in Würtzburg, even with double-glazing excellent insulation values of U_g = 0.5 W/m²K can be achieved with system thicknesses of less than 10mm. By 2011, suitable production techniques will also be available whereby the price of vacuum-insulating glazing should not be any higher than that for conventional triple-glazing.

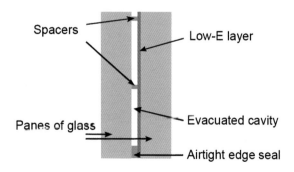

Diagram of vacuum-insulation glazing with virtually invisible spacers.

Source: Detail, English Edition 01/09

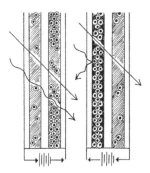

An electrochromic window can shield sunlight so it doesn't get too warm. It is regulated by passing voltage through an electrochromic layer which darkens the pane.

Glass with daylight screens.

Electrochromic Glass

In electrochromic windows the colour and characteristics of the windows are changed by a voltage source. There is 'electrochromic foil' inside the window. A low-voltage source is connected and by varying the voltage, the electrochromic layer darkens and changes colour, thereby altering the transparency of the window. The transparency of the window can range between 7 and 75 per cent by changing the voltage. The idea is to reduce the need for cooling in, e.g. office buildings, while retaining the view.

Thermochrome Glass

This technology is already in use for some sunglasses and car windows. The colour of these windows cannot be changed at will, it is automatic.

The windows consist of two sheets of glass with a material between that is affected by temperature. At low temperatures the material is transparent, and at higher temperatures it becomes opaque.

Glass with a Daylight Screen

During the winter, when the sun is low on the horizon, people want to let in the sunlight, but in the summer when the sun is high, people want to block it out. This

Principles of daylight screens

can be achieved by using a daylight screen built into windows. The screen, which consists of glass and metal, is like a third pane between two glass panes in an insulating window.

Prism Glass

This glass lets in direct sunlight from some angles, while light from other angles is reflected away. Daylight, however, always comes through. Such glass can also be used as an external Venetian blind, and can be adjusted as required for daylight and to block the sun.

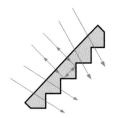

Prism glass.

Source: Siteco

Prism glass at Sparkasse FFB.

Source: Siteco

As it is not possible to see through transparent insulation material (TIM), it is alternated with glass.

Transparent Insulation

An interesting material has been developed that may improve the possibility of using passive solar heat. It is called transparent insulation, is transparent and lets in daylight while at the same time insulating and preventing heat loss from a building. Transparent insulation can, for example, be used to insulate Tromb walls so that they can be used to provide indirect solar heat to a building, or be used as insulation in south walls and so let heat and light directly into a building. It may be necessary to screen transparent insulation during the summer to avoid high temperatures. The U-values of transparent insulations are between 0.3 and 0.6W/m²K.

Silica Aerogel

Silica aerogel is a material that is placed between two glass panes. The material is fairly brittle and opaque. It lets in 90 per cent of the light. Windows using silica aerogel cannot be used as regular windows as it is not possible to see clearly through them. They can function as a new architectural element, a well-insulated opaque glass surface that lets in light yet keeps in heat.

Ecotopia educational centre is located in Aneby, Småland, Sweden. Transparent insulation is used there, as well as used brick and timber from various deciduous trees.

Marché International Support Office, Kemptal, Switzerland. The building has 90m² of GlassX-modules. The building won the 2007 Swiss Solar prize.

Source: Beat Kämpfen, büro für architektur/Architecture office

Nanogel® is Cabot corporation's trademark for its silica aerogel products. If the space between the glass in skylights is filled with aerogel a diffuse general light is obtained and a better U-value than normal windows. Similar products are marketed by other companies. An example is Kalwall+Nanogel. which is made up of panels with transparent insulation filled with nanogel. The U-values is 0.28 W/m²K.

GlassX

GlassXcrystal integrates four system components in a functional unit: transparent heat insulation, protection from overheating, energy conversion and thermal storage.

A three-ply insulating glass construction provides heat insulation with a U-value of less than 0.5W/m²K. A prismatic glass implemented in the space between the panes reflects sun rays with an angle of incidence of more than 40°C (in summer, when the sun is high in the sky). On the other hand, the winter sun passes through the sun protection at full intensity. The central element of GlassXcrystal is a heat storage module that receives and stores the solar energy and, after a time, releases it again as pleasant radiant heat. PCM (phase change material) in the form of a salt hydrate is used as the storage material. The heat is stored by melting the PCM; the stored heat is released again when the PCM cools. The salt hydrate is hermetically sealed in polycarbonate containers that are painted gray to improve the absorption efficiency. On the interior side, the element is sealed by 6mm tempered safety glass that can be printed with any ceramic silk-screen print.

Heated Glass

Heated glass does not save energy but is a way to increase the level of comfort in areas where people spend a lot of time near large glass surfaces. This type of glass has a metal coating that electricity is passed through. The coating is invisible, as are the positive and negative poles, positioned underneath mouldings. A thermostat can be used to keep the temperature of the glass the same as the room temperature.

Self-Cleaning Glass

The following takes place when the outside surface of glass is coated with titanium dioxide. Oxygen and water vapour in the air come in contact with the titanium dioxide and are transformed into free radicals with

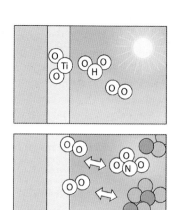

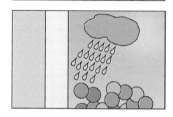

The principle for how a self-cleaning window works. Glass coated with titanium dioxide transforms oxygen and water vapour into free radicals with the help of sunlight.

Free radicals, such as O_2^- and OH^-, are highly reactive and easily combine with air pollutants such as nitrous oxides and organic particles to form relatively harmless substances that are washed away by rain.

Source: CSTB, adapted from *Ny Teknik*

the help of ultraviolet radiation from sunlight. The free radicals are very reactive and combine with air pollutants, such as organic particles and nitrous oxides, to form relatively harmless substances that do not affect the glass, and are easily washed away by rain.

Window Features

A window fulfils many different functions. It should let in daylight and solar heat when required. It should provide a view and in cold climates it should keep the heat in. Sometimes it gets too hot when solar radiation enters and something is required to screen out the sun. Sometimes more daylight and solar heat are required. In such cases, a reflector can reflect the sun to the inside. When it is very cold, an insulated window shutter can transform a window into a wall. In other words, there are many different ways of using devices other than the window itself.

Sunscreening

During the summer, screening out the sun can be necessary to prevent buildings from getting too hot inside, especially when passive solar heat is being utilized. Screens can be designed as permanent parts of a building, e.g. roof canopies or porch roofs, that shade windows during certain times of the year. Permanent sunscreens also result in a reduction in the amount of daylight entering a building. It is also possible to have movable screens, like awnings, roller blinds and venetian blinds. To be efficient, movable sunscreens should be placed outside windows. They require operation and maintenance. They need to be cleaned periodically and awnings are especially vulnerable to stress from the wind. Curtains and venetian blinds between window panes or indoors are also useful for screening out the sun. They protect against direct sunlight, but still let in heat. Another method to screen out the sun is to plant deciduous trees that lose their leaves in the winter and let in light, and shade the building during the summer when they are covered with leaves.

Sunlight Reflector

When more daylight is wanted, outdoor reflectors can be used, e.g. hatches made of reflective material that can be adjusted to the time of year and need for solar heat.

Reflective hatches can also be used to provide a view in a particular direction. Reflective venetian blinds can be used to block light or reflect direct sunlight towards the ceiling and so provide light without glare. A light shelf is a reflector that is placed high up on a window. It shades workplaces at a window at the same time as it reflects daylight towards the ceiling.

Insulation

Since the U-value of windows is inferior to the U-value of walls, it is possible to use movable insulation to improve a window's U-value, e.g. on cold winter nights. Options include insulating window shutters, insulating curtains and insulating roller blinds. Insulating shutters should be placed outside windows to avoid frost and condensation on the glass. Insulating curtains should be placed between the panes or inside the window on tracks to avoid convection. Insulating roller blinds are placed outside the window and also act as protection against burglary.

Regulation of Daylight

When sunlight is too strong, it is sometimes necessary to reduce the amount of light that comes in without totally blocking the view. This can be done with venetian blinds, roller blinds that are pulled down or a system of shading screens. At the Arab Institute in Paris, there is an advanced design by architect Jean Nouvel, where every pane in the south-facing glass façade is equipped with a diaphragm whose opening varies according to the brightness of the sunlight. The various methods can fulfil several functions, namely shading, insulating, reflecting and regulating.

Residential building in Nieuwland, Amersfoort, Holland. Sunscreening using adjustable sunscreens comprised of solar cells.

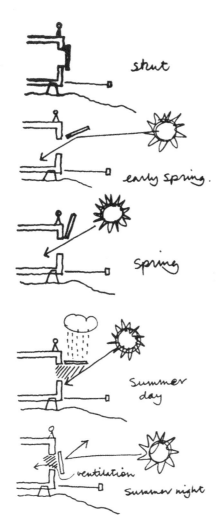

Different ways to use reflecting and insulating window shutters.

Source: *Ralph Erskine, Architect,* Egelius, Mats, 1988

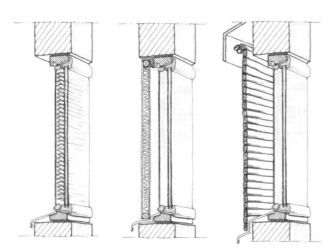

Window shutters, venetian blinds, curtains and roller blinds can regulate both solar radiation in and heat loss out through a window. Insulating shutters and roller blinds should be placed outside windows to avoid condensation.

2.1.3 Heat Recovery

Buildings lose heat through the shell, ventilation and waste water. Besides building a shell with a low U-value, there are several different ways to recover heat. Heat exchangers and heat pumps can be used to recover heat from ventilation and waste water. An air-to-air heat exchanger in the ventilation system was the first, more widely installed, heat recovery technology. This technology is now being replaced with exhaust-air heat pumps. Heat exchangers for waste water were first used in public swimming pools and laundries, but since the technology is both inexpensive and simple, it has started to be used in blocks of flats. Dynamic insulation is a method that has fascinated many architects and engineers, but has proved difficult to master. The best results have been achieved in roofs of indoor public swimming pools.

Air-to-Air Heat Exchangers

Air-to-air heat exchangers are used to recover heat from exhaust air to heat intake air. They require mechanical exhaust and intake air systems, which are both expensive and technically complicated. If a building is not airtight and if the system is adjusted improperly, excess pressure inside the building can result, which in turn can lead to moisture damage. Proper use of the system requires a very tight building, in principle a building built after 1980. There are several types of heat exchangers for ventilation systems.

Operating Experience, Historical

Experiences of the use of exhaust-air heat exchangers have been bad. The high heat recovery levels promised in marketing were not achieved. Heat recovery with air-to-air systems has often been an unprofitable investment. Problems have included freezing with ice in the winter, noise, leaking and poorly insulated ducts and exchangers. Systems have been difficult to access for service and cleaning, and operation and maintenance instructions have been insufficient. Difficult-to-handle approaches have resulted in neglected maintenance and cleaning, leading to dirt in the system as well as impaired heat recovery. Dirt in the duct system and clogged filters are big problems resulting in imbalances and reduced flow in the system. Installations have been done carelessly and adjustments have been incomplete. This is

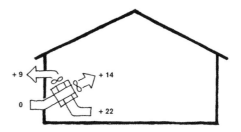

Heat from exhaust air can be recovered with an air-to-air heat exchanger. The exhaust air heats the intake air in the heat exchanger.

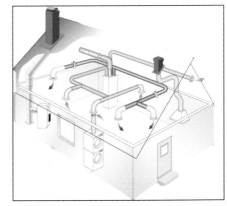

An air-to-air system sucks out air from the kitchen, bathroom and utility area and uses it to heat intake air. The heated fresh air is blown into bedrooms and the living room.

Source: REC Indovent AB

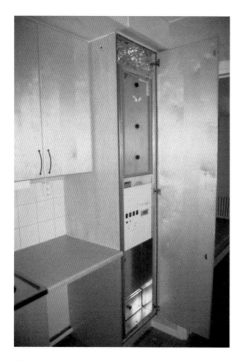

This is how a heat exchanger in a single-family house can be fitted in a kitchen cupboard. This house in Lindås, Gothenburg, Sweden has no heating system.

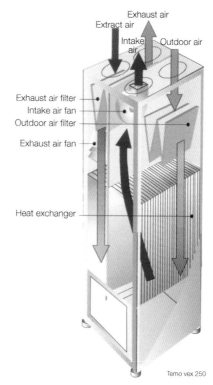

Temo vex 250

One of the winners in The Swedish Energy Agency's technology competition for better exhaust air heat exchangers. It is a Temo Vex 250 made by Temo Vex Svenska Ltd. A high level of temperature efficiency is achieved by connecting two counter-flow heat exchangers in a series. The other winning entry is called HERU 50 and is made by C. A. Östberg. It is equipped with a rotating heat exchanger that does not require defrosting.

Source: www.stem.se, 'Värmeåtervinning av ventilationsluft – Förbättra inomhusklimatet och minska energikostnaderna'

New Heat Exchangers

Better air-to-air heat exchangers have been developed in recent years. A competition was held for better systems for heat recovery from ventilation air. The winning systems have a high level of efficiency (85 per cent), are easy to clean, and have easy-to-change filters. They are airtight and quiet, easy to maintain, and come with clearly written instructions. They have indicators that show air flow and when the filter should be changed, and defrost automatically. The fans are adjustable, so the air flow can be reduced if the building is not being used; heat recovery can be disconnected during the summer. Electricity consumption is about 95W.

one reason why an obligatory ventilation inspection was introduced.

Exhaust Air Heat Pumps

An exhaust air heat pump removes heat from exhaust air on its way out of the building. The heat is used to heat hot water and sometimes also for space heating. The advantage of this option is that an intake and exhaust air system is not required. Instead, it can be installed where there is a normal exhaust air system. A problem is that heat pumps have often been subject to breakdowns. In addition, it must be pointed out that an exhaust air heat pump is quite expensive, and it does not pay to install one in an extremely

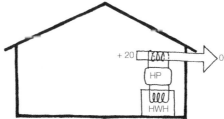

Heat from exhaust air can be recovered with a heat pump (HP). An exhaust air heat pump takes heat from exhaust air which cools down. This energy is used by the heat pump to heat hot water in a hot water heater (HWH). The heat pump can also make use of the condensation heat in air moisture.

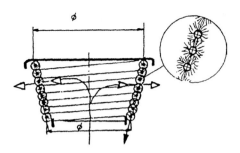

SPAR-VEN Ltd has developed a heat exchanger with such a low pressure drop that it can be used with natural ventilation. This heat exchanger has small metal fringes that absorb heat from the exhaust air and is placed up in a ventilation chimney where it is used together with an exhaust-air heat pump. The heat is often used to heat a building's hot water.

energy-efficient single-family house. Exhaust air heat pumps are often used in blocks of flats and other large buildings.

How a Heat Pump Works

A heat transfer medium is heated by exhaust air, which in heated rooms is at least +20°C. The temperature of the heat transfer medium is increased by compressing it in the heat pump, and the higher temperature can be used to heat hot water. In the past, most heat pumps contained freons. Currently, only 'soft freons' are used, but the goal is to use heat transfer mediums that do not affect the ozone layer or climate, such as butane and pentane in small heat pumps and ammonia in large ones.

Operating Experience

In recent years, many exhaust-air heat pumps have broken down. It was always the compressor that broke down. Some models have suffered from an unusual number of problems. According to the producers, problems were caused by the transition to more environmentally friendly cooling agents in the pumps. Some heat pumps have required a lot of maintenance and made too much noise. Insurance company websites sometimes provide damage lists for heat pumps and other devices. An exhaust-air heat pump requires

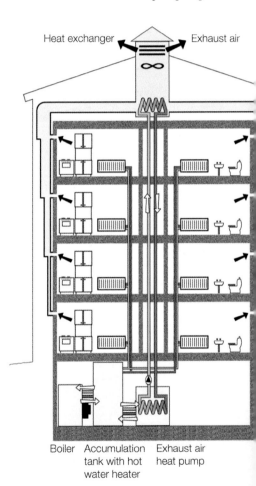

An exhaust-air heat pump can recover heat from a number ventilation chimneys in multi-storey buildings that are ve with fan-reinforced natural ventilation. The recovered hea to heat water in the hot water heater (accumulation tank).

Source: Illustration: Leif Kindgren

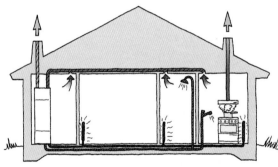

An exhaust-air heat pump in a single-family house. Exhaust-air heat pumps require some maintenance, have contained freon and been quite expensive. However, they are improving, becoming more dependable and cheaper, and there are freon-free alternatives.

a greater investment than a heat recovery unit, but provides more energy in return. It cannot be assumed however that the investment always pays dividends in the form of reduced heating costs. Use of exhaust-air heat pumps has been questioned. They are relatively expensive, and in super-insulated buildings with small heating requirements they are scarcely worthwhile.

Waste-water Heat Exchangers

Use of hot water releases some heat into the house but most of the heat exits along with the waste water. The heat in waste water can be recovered using a heat exchanger that preheats cold water brought into the house. This type of equipment is used where a lot of hot water is used, e.g. in public swimming pools, and has also been tested in energy-efficient single-family homes.

Design

Waste-water heat exchangers can be designed in a number of ways. The most important element of the design is ease of cleaning, since waste water contains a lot of dirt particles. It is possible, for example, to design a heat exchanger using the pipe-in-pipe principle where the waste water flows in the inner pipe and heats the clean water that flows in the outer pipe. Since the clean water is under pressure, as opposed to the sewage water, which isn't under pressure, there is no danger of contaminating clean water with waste water. If a leak should occur, the clean water flows into the waste water, and not vice versa.

SplitBox

SplitBox is a complete unit. It is a combined energy and sewage system that manages the total energy needs of the building, including systems for ventilation, heat recycling and management of water and sewage. Heat is recycled from drying processes (sewage + waste), sanitary waste-water and the building's ventilation. Nutrients are removed in solid form for fertilization. The system is available for single and multiple family homes.

Large Facilities

Ready-built components are available for large facilities. About 50 per cent of the heat in waste water can be recovered and transferred to fresh water with such heat exchangers. The best result is achieved if all the water goes through the heat exchanger, not just the hot water. To get really cold drinking water, the kitchen cold water pipe can bypass the heat exchanger's cold flow. In most heat exchangers, toilet waste water is excluded from the heat exchanger to reduce sludge problems.

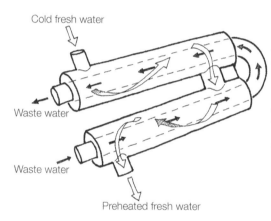

A waste-water heat exchanger in buildings with large waste-water flows can in principle consist of double pipes, with the waste water in the inner pipe and the fresh water preheated in the outer pipe.

Source: Power products AB, Åkersberga

Sewage heat exchanger in Erik Hedenstedts project-house in Trosa (www.ekologiskabyggvaruhuset.se). This type has to be mounted vertical.

Source: GAIA Scotland

Dynamic Insulation

Dynamic insulation is a construction method where intake air is sucked in through insulation material. There is heat loss from a heated building, i.e. heat escapes through the insulation in the roof and walls. If the construction is made so that it is possible to suck air in through the insulation, the sucked-in air is preheated by the air that is on its way out through the insulation.

Dynamic Roofs

'Air open' walls, which are built to allow air to be sucked or pushed through them, have big problems with wind since wind pressure affects walls differently depending on the wind direction. Therefore, dynamic insulation is used on loft floors where wind pressure does not play a decisive roll. A number of buildings have been made with dynamic insulation, but it has been difficult to determine whether there have been any energy savings in ordinary, single-family homes. However, it is a promising option for public indoor swimming pools where a lot of ventilation is required and the air is moist.

Dynamic insulation was first used in cowsheds where intake air was brought in through the hay in haylofts. The idea was further developed using mineral wool as the insulation material. For example, several sports centres have been built with dynamic insulation in the ceilings. In recent years, there have been constructions using wood-wool cement or cellulose fibre. The GAIA Group in Norway has experimented a lot with dynamic insulation.

Insulation Material

There is great interest in dynamic insulation among environmentalists, but some are concerned about contamination of the insulation material since it acts as a filter (albeit a very large filter). Mineral wool is made up of small fibres that are bound together by urea resin, which contains formaldehyde that is continually emitted. If for some reason the insulation becomes moist (which can of course happen in ceilings), large amounts of formaldehyde can be emitted. Cellulose fibre insulation can therefore be considered a better choice. Regardless of the type of insulation used, the insulation acts as a filter for intake air during the entire life of the building, without being cleaned or replaced. This is a source of doubt in the technology.

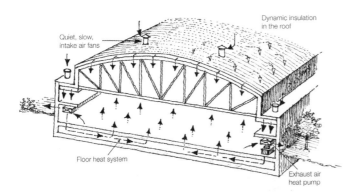

Architects Howard Little and Dag Roalkvam of GAIA Architects design indoor swimming pools and sport halls with dynamic insulation in the ceilings. Fresh air is brought in through the ceiling and the moist exhaust air from the swimming pool area is dehumidified with a heat pump before it is blown out. The condensation heat obtained from dehumidification is used to heat the facility via a floor heating system.

Indoor swimming pool with dynamic insulation, McLaren Community Leisure Centre, Callander, Scotland.

Source: Architect: Howard Little

2.1.4 Architecture

The approach of the architect greatly influences energy consumption. It has to do not only with insulation and technology, but also with an understanding of passive techniques. The goal of every architect should be to use their skills to contribute to energy efficiency.

Building Design

Smaller, well-planned buildings save energy. Architectural devices that influence energy efficiency include a building's shape and type, temperature zones, how much it is dug into the earth, as well as options that make it possible to use passive solar heat.

Building Shape

The objective for an energy- and materials-efficient construction is to encompass the largest possible volume with the smallest expanse of outer wall. Theoretically, the best shape is a sphere, but it is difficult to make efficient use of the space inside a sphere. So when it comes to single-family houses, the preferred shapes are a two-storey cube, a one-and-a-half-storey square shape with a hip roof and a finished attic, or an eight-sided two-storey house with a sloping roof to minimize the total volume of the building.

Building Type

The type of building is very significant. The number of storeys and how the various parts are put together can make it possible to reduce the surface area of the outer walls and roof. In semi-detached houses, there is one less outer wall and so heat loss is reduced. In terraced housing, there are two fewer outer walls, and in blocks of flats, there are three or four fewer outer walls as well as a reduced roof area.

Temperature Zones

One way to reduce heat loss is to reduce the volume heated. This can be done by dividing a building into different temperature zones, e.g. a cool larder, wood shed and storage area that are not heated on the north side of the building, and a veranda and porch, both glassed-in, on the south side. Different temperatures in the living areas can also be considered, but this is often difficult to achieve in a compact and well-insulated house.

Dug-out Buildings

Heat loss through the foundation varies depending on the type of foundation. In a plinth foundation, the floor is in principle like an additional outer wall. A slab foundation on the ground is not exposed to outside temperatures but to the ground temperature, which is relatively constant year-round and is equal to the average annual temperature. In a hillside or semi-subterranean house, some of the walls are built into the ground. In these 'earth-sheltered' buildings, only the windows and part of the walls come into contact with outside temperatures. The more a building is dug into the earth, the less heat is lost.

Heavy Mass

Experience has shown that heavy buildings often require less energy than light constructions. This is because heavy materials have the ability to store heat until the excess heat is required, just as the heavy mass functions in a tiled stove. According to a former building norm, U-values are adjusted by a certain factor for heavy walls. To reduce energy requirements in super-insulated buildings, it is possible to deliberately add thin layers of heat-storing material.

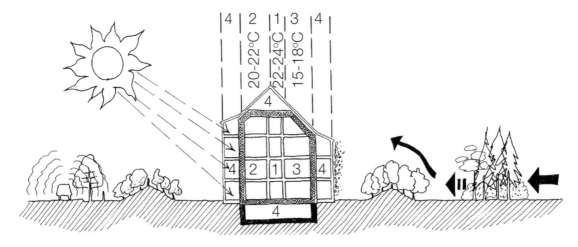

To save energy, a building can be divided into different temperature zones according to the purpose of the zone. Architect Joachim Eble divided his house in Schafbrühl in Tübingen, Germany into four temperature zones.

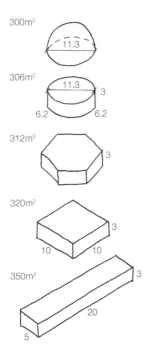

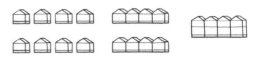

Eight homes as:	separate buildings	Terraced houses	2-storey block
Ground area	100%	70%	34%
Exterior surface area	100%	74%	35%
Heating requirements	100%	89%	68%
Costs	100%	87%	58%

This is a comparison of ground floor area, exterior surface area, heating requirements and construction costs for eight living units in different types of buildings. By building the units together in terraced houses or multiple-storey buildings, the outside wall area is reduced and thus also heat loss.

Source: *Ökologische Baukompetenz*, H. R. Preisig et al, Zürich, 1999

The area of the outer walls of a building depends on the building shape. The most energy-efficient shape is a hemisphere. The example shows the area of the outer shell (walls, roof and floor) for five differently shaped houses, each with a 100m² floor area.

ARCHITECTURAL EXAMPLES OF ENERGY-EFFICIENT BUILDINGS

The Tuskö House

This two-storey, 144m² house in Tuskö, near Östhammar, Sweden, is super-insulated and built using healthy materials. On the second floor, the ceiling height is lower at the outer walls in order to reduce the surface area of the outside walls. The stud frame is made of wood and the insulation material used is cellulose fibre. The house is divided into zones. On the south side there is an insulated and unheated glassed-in veranda that can take advantage of passive solar heat. On the north side, an extension with a storage area, wood shed and larder serves as extra insulation. There is an open stairway in the house that distributes excess heat from the veranda and kitchen to the other rooms. A soapstone stove in the ground floor kitchen is the main heat source. As the house is well insulated and has good windows, it has a small heat requirement. When passive solar heat and wood heat from the soapstone stove are insufficient, there are small electric radiators as a reserve. There are two urine-separating dry toilets in the house. There is a suspended foundation that is easy to enter from the outside, where the small faeces container for the dry toilets is located. Urine is collected in an underground tank beside the building.

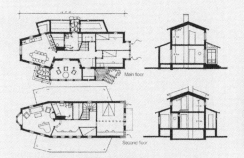

In a house in Tuskö, near Östhammar, Sweden, there is a larder, wood shed and storage area in an unheated zone on the north side. The veranda and porch are solar-heated zones on the south side. The suspended foundation under the house is a frost-free zone where all the services are located.

Source: Architect Lollo Riemer von Platen, Stockholm. Eco-architect Varis Bokalders, Stockholm

The Torrång Eight-sided House

This three-storey house is a good example of how architectural techniques can be used to reduce outside wall area. The house is also well planned with regards to temperature zones – the fireplace, stoves and accumulation tank are centrally located, and a winter garden and drying cabinet are unheated outer zones. The winter garden faces south to take advantage of passive solar heat. Doors to the winter garden can be opened to allow solar-heated air to flow into other parts of the house. The open floor plan makes it possible for the heat to spread. Heavy material in the house stores the passive heat.

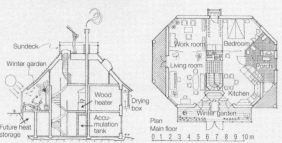

Architect Ola Torrång designs eight-sided houses with a sloping roof and sloping ceilings at the edges, large glassed-in verandas facing south, and a warm zone in the middle with a chimney stack, stoves and heaters.

Passive Heat

Buildings gain heat from occupants, household electricity and solar radiation. In order to make use of this heat, the heating system has to turn off when there is excess heat, the floor plan design should allow the heat to spread, and the materials used in the roof, walls and floor should have a large heat storage capacity.

A heavy mass in a building frame can store solar heat from day to night. Only the outermost centimetres of the heavy mass are involved in 24-hour heat storage. Therefore, it is more important that the mass exposes a large surface area towards the room than to have a heavy, thick structure.

The Heating System

It is important for the thermostat to be located where it can detect excess heat. An example of an improper location for a thermostat is on the radiator under a window. When it gets too hot and the window is opened, cold air falls down on the thermostat, which responds by increasing the temperature of the radiator instead of turning it off. It is also important that the heating system is easy to adjust, and that it has a small heat storage capacity. A floor heating system set in concrete has a large heat storage capacity, i.e. the whole floor is warm and continues to heat even when the floor heating is turned off.

Floor Plans

With an open floor plan, heat is distributed in a building through an internal stairwell or room that connects two or more storeys. It is common today for kitchens and living rooms to be joined together, but some people prefer more demarcated rooms. It is also possible to actively spread and distribute heat in a building, i.e. with fans and ducts. Ducts can be built-in or in the form of an open textile sock that runs from floor to ceiling. Hot air rises and collects under ceiling ridges. Heat is distributed better if this warm air is brought down, which can be done with a ceiling fan. It is important for such fans to be quiet.

A living room with an interior brick wall. The purpose is to obtain a relatively thin but heavy wall section that can store heat throughout the day.

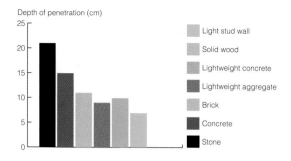

Penetration depth during the 24-hour temperature fluctuation in a building. The depth of penetration shows how much wall thickness contributes to a building's heat storage during a 24-hour period.

Source: Cementa, 2001

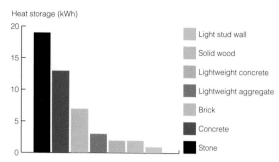

The storage potential for different materials. The maximum amount of heat that can be stored in a 100m² wall with an indoor temperature variation of ±1°C over a 24-hour period. A light stud wall is a compound construction with double gypsum plasterboard, mineral wool and light studs.

Source: Cementa, 2001

Heat-Storing Material

In order to distribute excess heat over time, the heat storage capacity of heavy materials can be used, i.e. the material's capacity to absorb, store and release heat. Important aspects are how much heat the material can store (heat capacity), and how far the heat penetrates into the material (heat conductivity). In 24-hour heat storage in solid material, heat manages to reach a maximum depth of about 10cm. During shorter periods between a heat surplus and a heat deficit, the depth of penetration is only a few centimetres. Water has a better storage capacity since convection takes place in a water tank (accumulation tank). Another way to store heat in material is to build-in substances with a melting point just above room temperature, e.g. Glauber's salt. It is possible to store larger amounts of heat from a change of state (e.g. the transition between solid and liquid states) than from a temperature increase.

Passive Solar Heat

It is possible to conserve heat by making use of free heat from the sun. All buildings with south-facing windows already do this. There are, however, other architectural methods that can be used to maximize use of passive solar heat such as: using glass on the south façade, regulating the solar radiation (with roof hatches, deciduous trees, awnings or venetian blinds), installing a heavy heat-regulating mass, designing so that heat can easily be distributed (e.g. open floor plan), and using a heating system that can be adjusted to make use of passive solar heat.

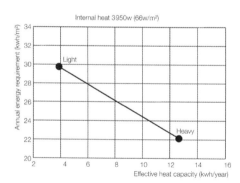

A classroom with an internal heat load from occupants, lighting and computers, area 66W/m², was simulated in order to study how the weight of a construction affects the annual energy requirement. The internal heat load is equivalent to 26 students, eight computers and normal lighting.

Source: Cementa 2001; adapted from Andersson and Isfält, 2000

The following can be concluded about methods that are suited to the Scandinavian climate:

- South-facing windows are suitable (i.e. the total window surface area is not increased, but more and larger windows face south than north).

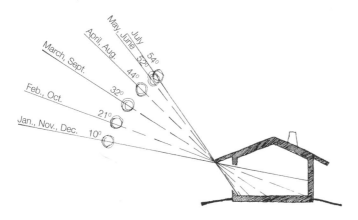

In order to make use of passive solar heat, it must be possible to regulate the solar gain so that it doesn't get too hot in the summer. One way to do this is to take into consideration the varying angle of the sun throughout the year and a building's heating requirements when designing roof overhangs and windows. The illustration shows conditions in Gothenburg, Sweden.

- Glassed-in spaces with double glass are also suitable (e.g. a glassed-in veranda or attached greenhouse. The heat is brought into the building by opening vents, a window or a door).

For a single-family house in Sweden, passive solar energy can contribute a maximum of 1500–2000kWh/year. It is thus especially important that systems are simple and cheap.

Solar Screening

When the south face of a building is glassed in, it can get too hot in the building at certain times of the year. A method of reducing excess heat is to screen the glass surface, which can be done in a number of ways. Since the sun is higher in the summer than in the winter, the summer sun can be blocked out with permanent constructions such as overhangs, balconies and porch roofs.

More flexible options include awnings, roller blinds and venetian blinds. Another way to block summer sun is to plant deciduous trees in front of the south façade that lose their leaves in the winter. The foliage is most dense when the need for screening is greatest.

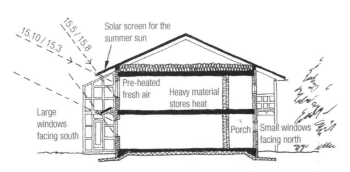

Passive solar heating involves designing a house to take advantage of much solar radiation, which reduces the house's heating requirements. Passive solar heat contributes much less energy in the Nordic climate than further south.

Source: 'Passive Solar Heating in a Scandinavian Climate,' Hans Ek, BfrD3:1987

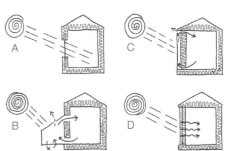

There are four ways to take advantage of passive solar heat: (A) direct gain via south-facing windows, (B) indirect gain via attached greenhouses or verandas, (C) solar walls (which are in principle solar air collectors), and (D) heavy wall construction with glass on the outside (a Tromb wall). The latter works poorly in Nordic climates.

MODERN GLASS ARCHITECTURE

Glassed-in Courtyards

After the energy crises in the 1970s, glassed-in courtyards were built with the intention of saving energy. However, studies have shown that in most cases, glassed-in courtyards do not save energy. It is possible to save energy with them if they are properly built: to save energy and be economical, the courtyard must be designed so that the façades that face the courtyard are relatively large compared to the area of the glass. The glassed-in area becomes a buffer zone (extra insulation) between the inside and outside, and can result in energy saving if it is not heated, which would result instead in an energy loss.

During the 1990s, courtyards in large ecobuildings were glassed in for other reasons as well. The high ceiling makes them suitable for use as part of a natural ventilation system. They are often used for the purpose of blocking traffic noise, while simultaneously being used as communication areas with galleries, lifts and stairwells. They are, however, most often designed as glassed-in gardens with plants and running water, restaurants, sitting areas and other recreational uses.

All the glass sections in the roof can be opened to avoid overheating in Prismahuset, Nuremberg, Germany.

Source: Joachim Eble Architecture

1. Transparency/view

Natural light

View

2. User controlled natural ventilation

3. Noise reduction

4. Reduced energy consumption

Pre-heated air

Transport away solar heat

Heat contribution in the winter

Protect the solar screen

Double glass façades – a step on the road to a sustainable society?

Source: *Dubbla glasfasader – Image eller ett steg på vägen mot ett uthålligt samhälle?* Anders Svensson, Pontus Åqvist, 2001

Double Glass Façades

Glass buildings have become a popular form of architecture, stemming from the dream of a transparent building. One of the main points of a transparent building that allows daylight to reach far into the building is the visibility and the view. The desire to see into the building, which varies according to the time of day and the weather, is another purpose of glass buildings. Glass buildings, however, experience problems with energy loss and excess temperatures. Double glass façades are not ecological. This expensive type of façade does, however, offer a technical solution to the difficulties of building large, tall glass buildings.

A double glass façade can be defined as a fully glazed façade with two layers of glass separated by an air gap of at least 50cm. There can be one pane in the outer layer and two on the inside or the other way around. Energy loss from a glassed-in building can be reduced by the buffer zone created by a double glass façade, since it improves insulation a little. If a double glass façade is used as an intake air route, it is possible to take advantage of some solar heat. Excess temperatures can be reduced if the space between the glass layers is used as a chimney, an exhaust air channel that ventilates away some of the excess heat by natural ventilation. Glass buildings require sunscreening which can be installed within the double glass façade. Double glass façades can block noise from the outside but, on the other hand, require much more window cleaning. Double glass façades make it possible to open windows even in tall buildings without experiencing unacceptably strong draughts, but can also cause problems with glare.

The large area of glass in double glass façades makes the design of the sunscreen very important since it must be lowered more often than for conventional façades.

It is often too hot in a double glass façade on the south side and too cold on the north side. Corner fans can even out the air temperature.

Source: The headquarter of Martela in Helsingfors, Tommila Architects

EXAMPLES OF PASSIVE SOLAR HOUSES

In the first eco-village in Sweden (finished in 1984), Tuggelite, near Karlstad, there are good examples of super-insulated passive solar houses. They are two-storey terraced houses, so the outer wall area is minimized. The buildings are unusually well insulated. There is insulation under and around the edges of the foundation, which prevents heat leaking out of the foundation and cold from penetrating under the building. The windows are triple-glazed and placed so that there are large windows facing south and only small window openings on the north side. There is a porch that faces north and a greenhouse on the south-facing side where intake air is preheated. The frame is heavy and there is a concrete foundation which is good for storing heat. Energy conservation has been very successful. Purchased energy amounts to about 6000kWh/year per residence (120m^2). As a comparison, according to the 1980 Swedish building code, a dwelling of the same size consumes about 15,000kWh/year. The Tuggelite buildings are heated with pellets from a shared central boiler. The roof of the boiler facility is covered with solar collectors that provide hot water during the summer.

Heat loss can be reduced by raising the level of the ground around outer walls or by building semi-subterranean houses. Some people have dug their houses into the earth, placing windows facing an atrium courtyard or on a façade that sticks out of the ground, but experience has shown that this often costs more than it is worth. It is expensive to build walls that can withstand both pressure and moisture from the earth.

The roof overhangs are shaped so that the low winter sun comes in, but the high summer sun is blocked.

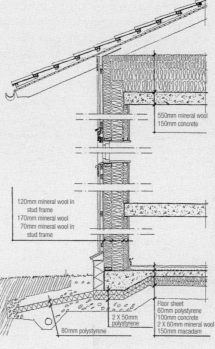

Cross-section of a super-insulated building in the Tuggelite eco-village. The insulation is 36cm thick in the walls and 55cm thick in the roof. Thermal bridges have been avoided and there are insulating sheets in the ground around the building to prevent cold from penetrating under the building.

Source: EFEM Arkitektkontor, Gothenburg

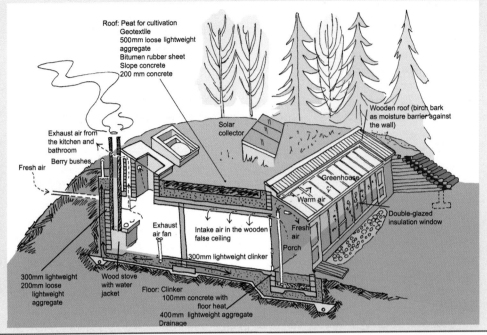

Architect Anders Nyquist's home (built 1991–1993) in Rumpan, near Sundsvall, Sweden, is extremely energy efficient. It is not only well insulated but also dug into the ground. The only façade, facing south, is glassed in and used as a greenhouse.

Passive Cooling

In hot climates it is important to design buildings so that excess temperatures are minimized and then to work with passive methods of cooling. In order to minimize excess temperatures, there are many architectural design factors to consider. The building itself can be protected from the sun and the indoor heat load can be reduced. Among other things, all windows should be shaded so that direct sunlight is blocked while indirect daylight can illuminate rooms.

Shape and Colour

How much a building is heated by the sun depends, among other things, on its shape and colour. A large, south-facing area gets very hot from the sun, while a dome only has a small surface area facing the sun's rays. A light coloured surface does not become as hot as a dark, absorbent surface. An insulated wall does not let as much heat into a building as an uninsulated wall.

Heavy Mass

In many climates it is hot during the day and cold at night. If a building is constructed of heavy material, the material's ability to store heat or cold can help provide a more pleasant indoor climate.

Shading a Building

There are various ways of shading a building so that it isn't heated by the sun. Buildings can be placed close together so that they shade each other, and double roofs or other construction details can be used to shade a building's shell.

Window Shading

It is especially important to shade windows in order to reduce internal heat load. This may be done in a number of ways, contributing to the overall design.

Zoning

In extreme climates, it is possible to build houses in which occupants move according to season. People in the Nordic countries

Zoning.

used to move into the kitchen for the winter. In hot climates, people can move out into the courtyard to get some shade, live on the roof to enjoy the evening breeze, or move into the basement on extremely hot days to be cooled by the earth.

Indirect Daylight

Though direct sunlight isn't desirable, letting in daylight is. This can be achieved by locating windows so that they aren't facing in the sun's direction, by shading windows with roof overhangs or by letting daylight filter in through a screen.

Sunlight and heat go together. In hot climates, screens are often used in windows for shading. They let some daylight in to illuminate the room while simultaneously keeping out most of the sunlight and heat. In addition, screens allow visibility outward without allowing visibility inward. The illustration shows window openings with screens (*Mashrabiya*) in a traditional Egyptian building in Cairo.

Source: *Natural Energy and Vernacular Architecture*, Hassan Fathy, 1986

Ventilation

In some climates, ventilation is the only thing that can provide a little cooling. Therefore buildings are designed so that they are easy to ventilate. Floor plans are used that allow cross-draughts. Spaces with high ceilings are included in buildings, such as stairwells, and inside courtyards capped with ventilation hoods (cupolas).

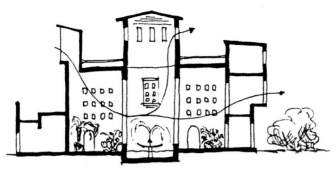

An Egyptian building with natural ventilation via a *malkaf*.

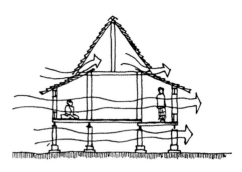

Cross-draught in a Malaysian house.

Advanced Ventilation

In order to increase ventilation, more extreme architectural elements can be used, such as ventilation chimneys and solar chimneys that increase natural ventilation, and wind-catcher vents (*malkafs*) that direct wind down into buildings. Walls and roofs can also be ventilated in order to cool down buildings.

Evaporation

Evaporation cools. There are many ways to make use of this fact to cool buildings. Indoor fountains can be used or vegetation. Water can be sprinkled on hot roofs, or running or evaporating water (a humidifier) can be placed in the fresh air intake.

Dug-out Style Buildings

By digging a building into the earth, it is possible to take advantage of the earth's cooling effect. A short way underground, the annual average temperature prevails. Another way to take advantage of the cool ground temperature is to use an underground air intake pipe that brings in air that is cooled in the process.

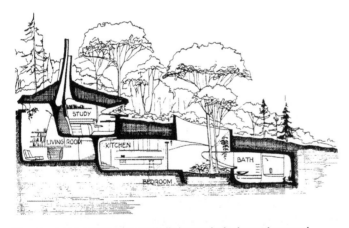

The American architect Malcolm Wells began designing underground earth-covered homes. He felt that vegetation was so beautiful and valuable that it shouldn't be built upon, and that any lost natural landscape should be replaced with a new version on the roof. In addition, such houses are protected from the weather. They can be well insulated and designed to admit a lot of daylight.

Source: *Underground Designs*, Malcolm Wells, 1977

There is often a '*salsabil*' in traditional Egyptian houses. A *salsabil* is a stone that is sculpted with an attractive relief pattern on the front. Water runs slowly over the stone while intake air blows over it. A cooling system is thus created in a simple and beautiful way.

Dehumidification

In climates where it is both hot and moist, dehumidification can provide cooling. This is because comfort does not have to do only with temperature but with air movement and relative humidity as well. Dehumidification can occur passively using salts that absorb moisture and are then dried using solar heat.

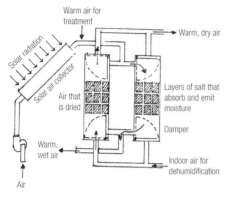

Cooling through dehumidification.

Green walls and roofs

In hot and humid climates experience shows that green plants on walls and roofs can lower the temperature in the wall by 3–4°C. In Singapore special green walls have been developed with vertical elements including integrated pots and irrigation systems. In Switzerland, vertical green walls are used indoors to aesthetically moisturize offices.

Radiation

On clear nights, a strong heat radiates towards the cold, night sky. In the Sahara desert, it is actually possible to make ice in shallow pools that are insulated from the ground. There are various architectural methods for cooling a building with night-time heat radiation, e.g. with the help of a shallow roof pool that is protected (covered over) from the sun during the day. There is also a special type of metal coating that has a great capacity for reflecting and emitting heat. This type of coated metal helps to cool buildings.

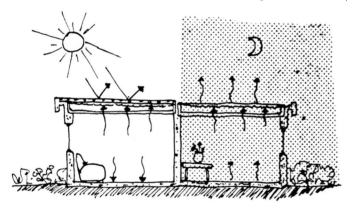

House with a roof pool.

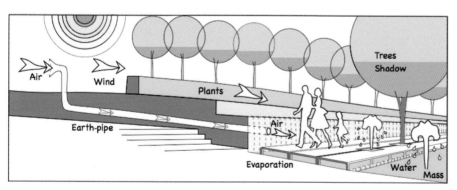

At the International Exhibition in Seville in 1992, work was undertaken to temper the hot climate with shadow from trees and plants, wind, and cooling air using earth-pipes, evaporation, water and mass.

2.1 MODERN EXAMPLES OF PASSIVE COOLING

Even in buildings in cold climates there can be problems with excessive indoor temperatures. This occurs primarily in buildings where there are many people and there is a lot of heat emitted from lights and electronic devices, such as office buildings and schools. The problem mostly occurs in the summer. Such buildings can thus require both heating and cooling.

Principles

If buildings are constructed in a way appropriate to the climate using heavy materials that are exposed indoors, the heat storage capacity of the material can balance out heating and cooling needs. Buildings can also be passively cooled with the help of underground pipes and night cooling. In this case, buildings are cooled during the night, the cold is stored in the heavy materials, which later help to cool the building during the day.

An excessive temperature problem may occur during the summer in glassed-in courtyards. Shading the glass during the summer is a possibility. Another is to make sure the glassed-in courtyards are properly ventilated, e.g. with the option of opening the entire glass ceiling. It is also possible to passively cool the intake air to the glassed-in area with the help of evaporation. One way to do this is for intake air to enter through a small waterfall.

Tegelviks School, Kvicksund, Sweden

Tegelviks School in Kvicksund (on Lake Mälaren, just north of Eskilstuna) is build of brick, which is a heavy, heat-storing material, and is ventilated with underground pipes that run in a culvert system under the buildings. Each classroom also has a ventilation chimney and windows that open so that ventilation can be forced.

Tegelviks School, Kvicksund.

Source: Architect: Bengt Strandberg

The Novartis Office Building, Täby, Sweden

The pharmaceutical company Novartis office building in Täby, just north of Stockholm, uses only 30kWh/m²/year for heating and cooling. The low energy requirement is due to: heavy, heat-storing material indoors; sunscreening glass and screening from the sun; underground ducts for intake air; and a capacity to force ventilation using skylights in the large, shared indoor courtyard that contains stairwells, lifts and sitting areas.

The Prisma House, Office Building, Nuremburg, Germany

The German architect Joachim Eble has designed an office building in Nuremburg with a wall containing a built-in waterfall. Intake air is brought in via the waterfall to the large, glassed-in courtyard. During the summer, the intake air is cooled inside the wall by evaporation of the water. The offices are supplied with intake air from the glassed-in courtyard.

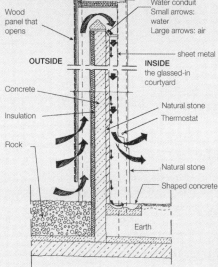

The Novartis office building in Täby, just north of Stockholm.

Source: Arkotek Architects Ltd

The Prisma House, Office Building, Nuremburg, Germany

2.2 Efficient Use of Electricity

Reducing electricity consumption in a building has to do in part with choosing efficient equipment available on the market and in part with minimizing the need for electricity by using architectural design that provides plenty of daylight. It is of course also important not to use electricity unnecessarily.

2.2.0 Use of Electricity

Today's high electricity consumption makes us dependent on nuclear energy and coal. Conservation of electricity and expansion of electricity production from renewable energy sources are the cornerstones of sustainable electricity use. Therefore, in a sustainable society, electricity should only be used where it is necessary (for electricity-specific needs). This means, for example, that buildings should not be heated with direct electric heating and that efficient electric products should always be chosen.

Electricity Use in Sweden

Large amounts of electricity are used in Sweden. In 2008, 144TWh (terawatt-hours) of electricity were consumed. Electricity is used in buildings, in industry and in the transport sector. In addition, about 10 per cent disappears in distribution losses from electricity lines.

Most of the electricity, 69TWh, is used in buildings – homes, offices and service buildings. This means that people working within the building sector, such as architects, engineers and builders play an important role in future electricity needs. With the help of energy-efficient technology it is possible to satisfy the demand for convenience with a significantly reduced consumption of electricity.

An International Comparison

In comparison with households internationally, households in Sweden use a lot of electricity. In Norway, where electricity is cheaper, there is the highest consumption of household electricity in the world. Denmark uses only half as much electricity as Sweden. The conclusion to be drawn from this is that if electricity is inexpensive, as in Norway, electricity consumption is high, but if electricity is expensive, as in Denmark, less electricity is consumed and it is almost never used for space heating. In Denmark, buildings and water are heated with fuel (fossil or biomass) instead of electricity. Electricity prices are however on the increase in Norway.

Electricity Specific

Electricity is a high-quality and expensive energy source and should therefore only be used where it cannot be replaced with a fuel, e.g. for running motors and for lighting. A sustainable society should try to limit the use of electricity and in principle only use it for electricity-specific needs. How important this is of course depends on how the electricity is produced. Today a lot of electricity is still produced with nuclear power and in power stations that burn fossil fuels.

When energy use is divided according to energy quality, it can be seen that electricity is required for only 10 per cent of energy end uses (lighting and operation of motors). The remaining end-use requirements can be met from lower-quality energy sources, such as fuel or heat.

Electricity consumption in Sweden is mainly by the housing sector, services, etc., amounting to 55 per cent of the total. Industry consumes 42 per cent of the total and 3 per cent is consumed by the transportation sector for trains, streetcars and subways.

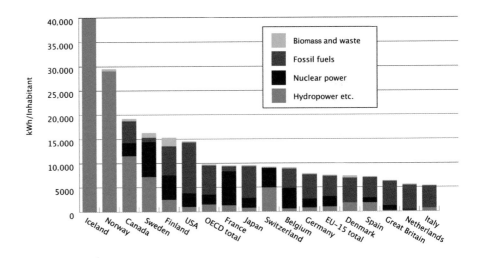

Electricity production per capita in 2006 by energy source. Iceland produces a lot of electricity from hydropower. This electricity is used for electricity-intensive industry, primarily aluminium smelting.

Source: 'Energiläget', The Swedish Energy Agency, 2008.

Household Energy Consumption

Electricity in homes is used for appliances, lighting, electrical devices and for pumps and fans (operating systems). Appliances are used for storing food (e.g. refrigerators and freezers), cooking (e.g. cookers and ovens), clothing care (washing machines and dryers), and cleaning dishes (dishwashers).

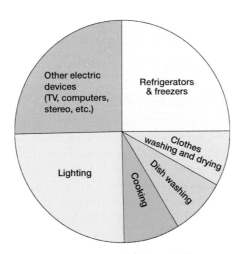

The electricity consumption of a typical household in Sweden is about 6000kWh/year and is distributed as shown in the diagram.

The appliances that use the most electricity are refrigerators and freezers since they are on around-the-clock, year-round.

Increasing Energy Efficiency

Technology is currently available to reduce electricity use without lowering the standard of living. For example, energy-efficient appliances and lighting should consistently be used. Furthermore it should be ascertained whether the desired end use can be met without electricity. Is it possible to reduce lighting needs with the help of more and better daylight? Is it possible to use smaller refrigerators and freezers with the help of larders and cellars? Can electricity use be replaced with a fuel source? Is it possible, for example, to cook with gas or use a gas-powered refrigerator? Is it possible to use a wood-burning baking oven or a wood-burning sauna? Do we need all our electrical devices, or could life be better if we did without some gadgets?

Electric Current Limiters

The cost of electricity consumption depends not only on the amount of energy used but also on how high the peak current is. The higher the peak current, the larger the electrical system and the fuses required. Electricity rates are based, among other things, on fuse strength, or how much current is required. It is self-evident that if all a household's electrical devices are used simultaneously, an oversized system is required. A current limiter cuts off the peaks since the load sources are given a priority order. For example, a current limiter can turn off the hot water heater while the electric cooker is being used.

Another way to keep electricity costs down is to use a cheaper night rate. Automatic controllers are available for this.

Four Levels of Household Electricity Consumption for a Four-Person Single-Family Home

The first column shows average consumption. The second column shows that total electricity consumption is halved if the best available technology on the market is used. The third column shows that electricity consumption can be reduced by a factor of four if better devices and more energy-efficient technology are used. The fourth column shows the electricity requirement where electricity is only used for electricity-specific end uses, e.g. in a house where electricity comes from a solar cell system.

Household electricity	Average	Best available technology	Greater efficiency	Solar cell electricity
Food storage (refrigerator and freezer)	1400kWh/year	415kWh/year	300kWh/year (refrigerator and freezer)	200kWh/year (larder and refrigerator/freezer) *
Cooking (cooker and oven)	1000kWh/year	650kWh/year (+kettle)	200kWh/year (+ gas range)	0 (only gas)
Clothes care (washing and drying)	1000kWh/year	365kWh/year (condensing dryer)	250kWh/year (connected to hot water)	50kWh/year (outside drying and cold water wash)
Dishwashing (dishwasher)	500kWh/year	220kWh/year	50kWh/year (connected to hot water)	0 (hand dishwashing)
Lighting	900kWh/year	550kWh/year (50% light bulbs)	300kWh/year (few light bulbs)	250kWh/year (energy-saving mode)
Electrical devices	1000kWh/year (many devices)	700kWh/year (normal amount)	350kWh/year (few)	225kWh/year (energy-saving mode)
Total	5800kWh/year	2900kWh/year	1450kWh/year	725kWh/year

Note: * In the solar cell alternative, to further reduce the need for electricity, a liquefied-gas-powered refrigerator/freezer could be used.

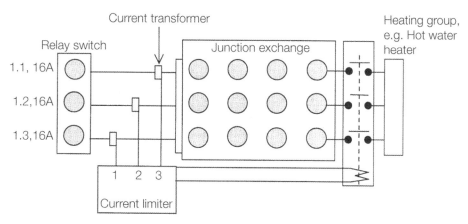

Schematic diagram of a single-stage current limiter connection. The current limiter can, for example, turn off a heating group when there is a current overload on one of the fuses L1, L2 or L3.

Source: Vattenfall Utveckling AB

2.2.1 Appliances

Appliances, especially refrigerators and freezers, use a relatively large amount of electricity in a home – about 25 per cent of total household electricity. When it is time to buy a new appliance, a model that uses the least amount of electricity should be chosen. It is also important to choose appliances that have additional environmentally friendly aspects, e.g. refrigerators and freezers that use environmentally friendly coolants and appliances that don't make too much noise.

More Efficient Appliances

Electricity consumption could be reduced by 50 per cent if the most energy-efficient products on the market were consistently chosen, and electricity consumption could be reduced to one-quarter of its current level if alternative methods that meet the same end-use needs were chosen. There is continual progress made in technology, and products in the prototype stage are more energy efficient than equivalent products on the market. It should be noted that functioning appliances should not be thrown away since it took a lot of energy and raw materials to produce them. It is when new appliances are going to be bought or when old ones that

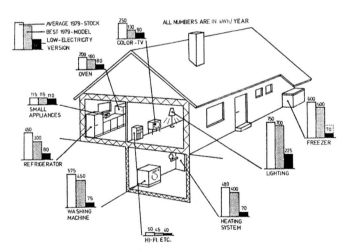

Technical development has made it possible to meet the same household end uses with less electricity. The bars show from left to right: (1) average stock, (2) the best model on the market, and (3) technically and economically possible low-electricity versions.

Source: Dr Jørgen Nørgård, Danmarks Tekniska Högskola, Lyngby, near Copenhagen

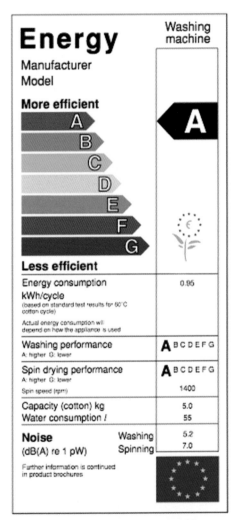

The energy label provides information about energy consumption and other important characteristics which makes it easier to choose energy-efficient household appliances. The labels are uniform within the EU and obligatory.

have stopped working are to be replaced that it is important to consider energy efficiency. In some cases, however, energy efficiency has improved so much that it can be worthwhile to replace appliances before they are completely worn out, e.g. in the utility room in a block of flats where replacing washing machines and dryers may be worth considering.

Energy Labelling

A large amount of electricity could be saved if everyone used energy-efficient appliances.

Energy labelling divides up products into categories from A to G, where A is the most energy efficient and those in category G use the most energy. From the perspective of sustainable development, only products in category A should be chosen. The newest equipment is more efficient than can be shown using this scale, today there are A+, A++ and A+++ products on the market. The machines have become so efficient that they have grown out of the energy labelling system that was introduced in the mid-1990s.

In Norway, larders with built-in condensing units are common. In the winter, it is cool enough in the larder. In the summer, when the larder gets too warm and the price of electricity is low, the condensing unit is turned on.

Refrigerators/Freezers

Refrigerators and freezers are one of the greatest electricity consumers in a household. Separate refrigerators and freezers are more energy efficient than combination models. However, if there is a larder, a single combination refrigerator and freezer may be sufficient and use less energy than two units. A chest freezer is more energy efficient than an upright freezer since cold air is heavier than hot air and 'falls out' of an upright freezer every time it is opened. The electricity consumption of refrigerators and freezers is also affected by the room temperature. So a kitchen shouldn't be unnecessarily warm, and it is a good idea to place chest freezers in a cool spot. One possibility is to place the condenser outdoors, but then the excess heat cannot be used to help heat the building.

In a cold climate a refrigerator is not needed during the winter. A larder is adequate, which can be cooled with a condensing unit in the summer when cooling is required and electricity is cheap.

The most efficient small refrigerator, with a volume of 150 litres, uses 135kWh/year. The most efficient large refrigerator, with a volume of 307 litres, uses a similarly small amount. The most energy-efficient combination refrigerator/freezers, with a 189 litre refrigerator and 96 litre freezer, consume about 192kWh/year. The most energy-efficient upright freezers, with a volume greater than 220 litres, use about 280kWh/year. Energy-efficient chest freezers, with a volume greater than 220 litres, use about 190kWh/year.

Washing Machines

The amount of electricity consumed by a washing machine depends on the temperature of the wash water and the amount of water used. In recent years, new models of washing machines have been developed that use much less water and are much more energy efficient. Washing machines should be loaded with the maximum possible load. Lower wash temperatures should be used and prewash cycles excluded except for very dirty items. A 90°C wash uses about twice as much electricity as a 60°C wash and four times as much as a 40°C delicate wash.

People wash about 150–200kg of washing per person per year. The portion of 90°C white colour wash has decreased to about one-tenth of the total. The rest is quite evenly divided between 60°C and 40°C washes.

Older washing machines may have a water consumption of about 67L/kg. In 2001, the best washing machines used about 40 litres of water for a 5kg wash and consumed about 180kWh/year. New washing machines often have a shower system that makes them much more water efficient. A pump continually supplies water to a shower that sprays water over the laundry. In the washing phase there is only a small amount of water in the bottom of the drum. An alternative to the shower system is to equip the inner perforated drum with scoops that pour water over the textiles. There are also washing machines with a built-in scale that weighs the wash and adjusts the amount of water and energy use accordingly. The water in clothes washers and dishwashers is usually heated by electricity in the washers. Central heating systems save electricity but require machines

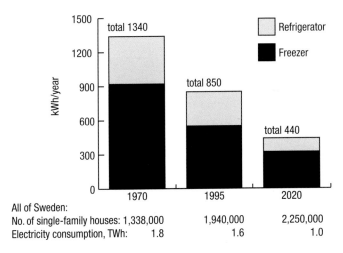

New energy-efficient refrigerators and freezers mean that the energy consumption for food storage can be greatly reduced. Appliances are commonly divided into three classes: the ones currently being used, the best available technology on the market, and future technology currently in the prototype stage.

Source: Gullfibers tidning *Isolerat*, 1996

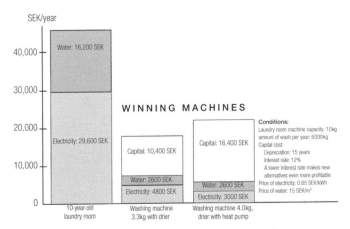

The winning machine uses so little electricity and water that it is economically advantageous to buy new machines immediately. In spite of the new capital cost of new equipment, the total cost per year is much smaller than before.

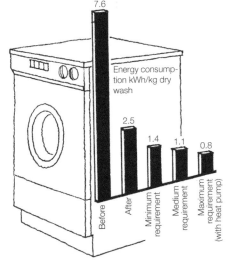

Nutek, the Swedish Agency for Economic and Regional Growth, asked the washing machine industry for help to make old utility rooms in multi-family houses more energy efficient. The winning technical solution for washing and drying was almost ten times more energy-efficient than the earlier ones.

that can be connected to both cold and hot water. Such machines are unusual in Sweden but can be purchased from Germany.

Utility Rooms

In utility rooms in the early 1990s, it was common to use 8kWh for each kg of dry washing. In 2003, the same amount of washing could be done with 0.8kWh. This means that the technology improved by a factor of 10. With such an improved energy efficiency, it even pays to replace functioning utility room equipment. The electricity and water consumption is so much less with new machines that the total cost is less despite new capital costs.

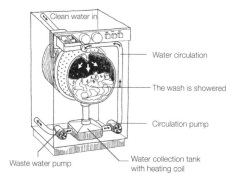

Energy- and water-efficient washing machines have been developed. Among others, Electrolux has developed a machine that recirculates the wash water and showers the water over the wash. There are also washing machines with a built-in scale that adjusts the amount of water according to weight of the load.

Drying Cabinets

Condensing drying cabinets use about 80 per cent less electricity than drying cabinets with heating elements. The drying cabinets that consume the most energy are those that are connected to an exhaust air duct and ventilate warm, moist air out of the building. A condensing drying cabinet is available on the market sized 60 × 60cm. The better the spin-dry cycle, the faster the laundry will dry. The centrifuge should spin at a minimum of 1000 revolutions per minute.

Clothes Dryers

Clothes dryers are more energy-efficient than drying cabinets but are harder on clothes. In a condensing clothes dryer, the condensed water is collected and pumped to a drain. A condensing dryer with a class A energy efficiency uses about 190kWh/year.

Dishwashers

Dishwashers use energy to heat water, run electric motors and dry dishes. They usually have a cold water feed. In this case, the water is heated only for washing and the last rinse. So a dishwasher connected to cold water uses 30 per cent less energy than one with a hot water feed, but takes longer to complete a wash cycle since the water must be heated in the machine. It is best of course if a machine has both hot and cold water feeds, especially if the building's hot water is heated with wood or solar energy. Electricity use comprises only 15 per cent of the total energy required to run a dishwasher – the remaining 85 per cent is used to heat water. Many dishwashers have an energy-saving cycle, where washing is done at 55°C instead of 65°C. This is perfectly adequate for normally dirty dishes and results in a saving of about 25 per cent.

Most dishwashers have an electric element for drying, but the dishes dry

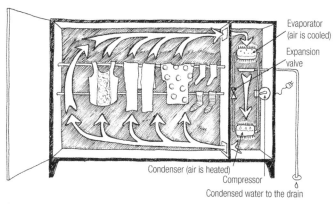

A condensing drying cabinet, where air from the cabinet is cooled and dehumidified, then heated and blown back into the cabinet, is much more energy efficient than other drying cabinets. The condensed water runs out into a drain.

almost as fast if the door is opened instead, which can result in energy savings of about 20 per cent. So it is best to choose a dishwasher with a drying phase that can be cancelled.

New models are often equipped with timers which makes it possible to start them during the night if there is a lower electricity tariff then. Most dishwashers have a built-in water softener, to which salt must be added regularly. A dishwasher with a water softener uses 3 litres more water per load than a machine without a water softener. Current eco-friendly dishwashing detergents require use of a water softener at a lower degree of hardness than before.

Cooking

Old cast-iron hot plates are being phased out. It is difficult to make electric hot plates more efficient. One way is for the producer to equip them with thermostats which make sure the hot plates maintain an even temperature. Cookers with ceramic tops are low energy consumers. New induction tops are the most energy efficient, but require special cookware and may emit a strong electromagnetic field.

Cookers have been energy-labelled since 2003. Gas stoves are more energy efficient than electric cookers. The energy consumption of an oven depends on how well insulated it is and how closely the size of the oven matches the amount of food to be cooked.

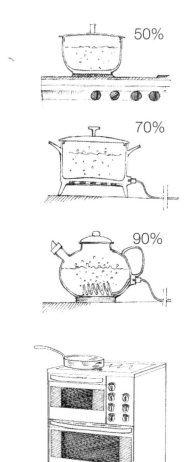

Different devices used to cook food require different amounts of energy. Electric rings are less than 50 per cent efficient, a pot with an element built into the bottom is about 70 per cent efficient, and a kettle with an immersion heater is 90 per cent efficient. A gas stove is about 70 per cent efficient, and a modern woodburning stove about 50 per cent efficient.
Ovens are usually oversized and use more energy than is necessary, since they are made to hold a large turkey. So, some new cookers have both a small and a large oven. Energy-efficient ovens should be well insulated.

There are dishwashers that have heat exchangers in a water reservoir, where water is preheated for use in the next phase.

NEW APPLIANCE TECHNOLOGY

Household technology will of course be more energy efficient in the future. For refrigerators and freezers, new development is focused on improving insulation with the help of a vacuum (the thermos principle), using two compressors instead of one in combined refrigerator-freezers, as well as running refrigerators using heat instead of electricity. Stirling engines combined with heat pumps, absorption cabinets with zeolites and magnets instead of compressors are possible future technological developments.

In order to develop washing machines that don't use as much electricity, several new technologies are being experimented with, including cold water washing, washing machines that use ultrasound and electrolysis to remove stains, as well as washing machines that use liquid CO_2 under high pressure at room temperature instead of washing detergent and hot water.

Using zeolites for cooling in refrigerators is a new type of cooling technology. Zeolites are a group of porous aluminium silicate minerals that have a great capacity to absorb and release water. Cooling is achieved using two containers that are connected to each other. The container above the refrigerator is filled with zeolite, and the one inside the refrigerator is filled with water. The containers are connected to each other by a pipe with a valve. The whole system is enclosed in a vacuum, which makes the water evaporate at room temperature, so cooling the refrigerator. The water vapour is absorbed by the zeolite which after a few days is saturated with water, and the cooling stops. If the zeolite is then heated for a short while, the water is again released and runs down into the container in the refrigerator where the process can begin again.

Magnetic refrigerators will use less energy, be more environmentally friendly than today's refrigerators, and will also be totally silent.

Kitchen in a residential area in Vadstena built in 2002. In addition to a refrigerator, there is a larder that is ventilated through an earth duct. It is a good example of how former practices can be combined with new technology in a beautiful, functional way.

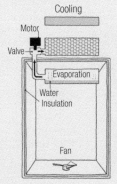 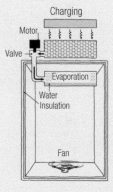

Refrigerator with zeolites.

A method of producing even more efficient refrigerators is to insulate the walls with a vacuum using the same method used in thermos flasks.

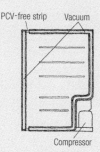

In the bottom of the refrigerator is a simple little device with a spinning disc (the size of a standard CD) that contains some of the element gadolinium, which is a strong permanent magnet, and a little water. Gadolinium has a unique characteristic: it becomes warm in a magnetic field and cold outside it (it is magnetocaloric). A cold segment on the rotating disk is heated by heat from within the refrigerator via a liquid-filled heat exchanger. Then the disk rotates so that the gadolinium segment is in the strong field created by the permanent magnets. The magnetocaloric effect further increases the temperature of the segment, and then the heat is taken away via another heat exchanger to a radiator outside the refrigerator. The segment is then rinsed with cold water and cools down. When the disk rotates further and the magnetic field disappears, the alloy becomes even colder and reaches temperatures below the normal refrigerator temperature. The ice-cold alloy is then once again warmed by the heat from the food inside the refrigerator, which cools the food. The only energy required is the energy needed to rotate the metal disk and to pump the liquid through the cooling coils.

Washing with Steam

LG Steam Direct Drive is the first washing machine in the world that washes with steam. It is extremely energy-efficient and is classified as A++, which means that it's 20 per cent more efficient than a washing machine in class A. It also has a bigger capacity than an ordinary washing machine, 8kg instead of 5kg.

Dry-Cleaning with CO_2

The company AGA has developed a dry-cleaning machine that uses CO_2 instead of perchlorethylene, which is dangerous for people and can pollute water. The electricity consumed by the new process is only one-tenth of the electricity used for an ordinary dry-cleaning system. The process is twice as fast, because drying is not needed since the CO_2 evaporates. The CO_2 used is collected from industrial processes, where it would otherwise have been released to the atmosphere.

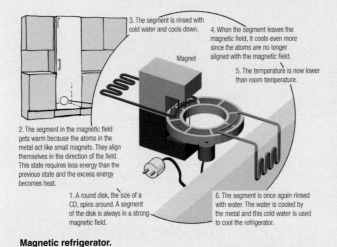

Magnetic refrigerator.

Modern cooker with both gas and electric burners.

To save electricity, much cooking in future will be done in insulated pans with built-in heating coils and thermostats. Such a prototype was developed by the Danish design group 02, using a design by Niels Peter Flint, for their exhibition on the future kitchen at NordForm 90 in Malmö, Sweden (among other places) in 1990.

KITCHEN FOR SUSTAINABLE LIFESTYLE

In a sustainable dwelling, kitchens undergo a marked renaissance. No more functionalist cocktail kitchens – make way for 'eco-functional'. In an eco-kitchen, food is cooked in an energy-efficient way in well-insulated pans and ovens, and on a gas stove; taps and dish sprayers are water efficient and practical, rubbish is sorted into compostable and other fractions, the larder is back, and in addition there is a greenhouse or window box for growing herbs and vegetables. The kitchen is a natural place for people to gather.

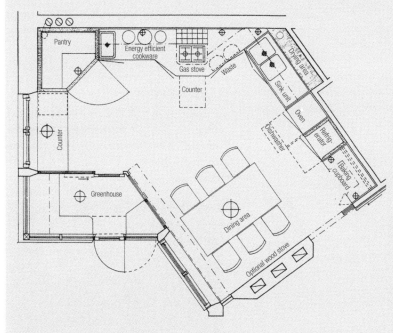

In 1993, interior architect Cecilia von Zweigberg Wike designed a kitchen for her final project at The National College of Arts, Crafts and Design in Stockholm. Her design incorporates a number of ecological features.

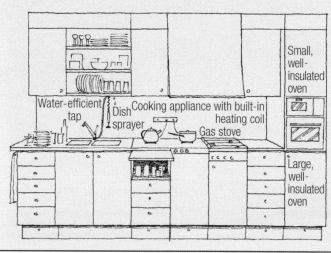

In tomorrow's kitchen, cooking will probably be done in insulated pans with built-in heating coils, on gas stoves and in well-insulated ovens of different sizes.

2.2.2 Lighting

Achieving energy-efficient lighting requires a thought-through system where electronic controls, light sources and fixtures all work well together. However, energy efficiency is not the only consideration. Other important concerns are light requirements and light quality. At workplaces it is good to be able to control brightness. In recent years there have been advances in the development of energy-efficient electronic controls and light sources. Completely new lighting technology is being tested as well.

Planning Lighting

Planning lighting means taking into consideration the whole chain from how many watts are put into a room to the intensity of illumination and the quality of lighting at a particular spot. The chain is made up of electronic controls, light sources, fixtures, the room and the conditions surrounding the place to be lit.

Light Fixtures

The purpose of a light fixture is to direct light where it is needed. A fixture usually has four components: frame with switch, reflector, bulb and shade.

Light Sources

The quality of light, i.e. its distribution and colour, is difficult to describe. Daylight is unequalled and should be the standard used to judge lamps by.

Plug	Electronic control device	Bulb	Fixture	Room	Lighting Task Adjustable
Electricity	Losses in the electronic control device	Heat loss the from the bulb	The fixture's efficiency	The room's influence	
Watts	Watts out/watts in	Lumen/watt	Specified lumen/ generated lumen	Lumen/watt, m²	Lux

In an ecological building, the amount of electricity used for lighting is kept to a minimum, so bringing daylight into a building is once again, as in the past, a priority. At the same time, options for energy-efficient lights and fixtures with good reflectors are considered. Good lighting considers the following four criteria: quantity, direction, distribution and spectrum.

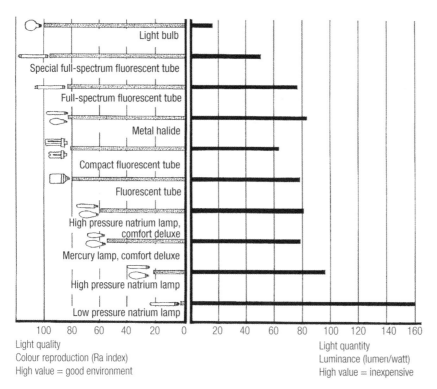

Light quantity versus light quality for some light sources.
The Ra index is a way to classify colour reproduction in lights. Unfortunately, some light sources with the highest luminance have a low Ra index. So lighting chosen for energy efficiency, e.g. low-pressure sodium motorway lighting, may have an unusual or poor colour. In places where people spend a lot of time, it is desirable to have lighting with good colour reproduction, so lights with an Ra index over 80 should be chosen.

Source: 'Belysning inne och ute, Watt med vett', Malmö Energi AB

Incandescent Light Bulbs

Incandescent light bulbs have long been popular due to their good ability to show colours. However, they have two problems: they use a lot of energy (only 8 lumens/watt) and they contain lead. For these reasons, in October 2008 the EU Energy Ministers gave the final approval to phase out all incandescent light bulbs by 2012.

Halogen Light Bulbs

Normal light bulbs become halogen bulbs when halogen is added. They provide significantly more light (up to 100 per cent more) and the light is much more brilliant as well. Energy consumption can be reduced by up to 30 per cent for every light source changed from incandescent to halogen bulbs. This is because the glass has an infrared coating that reflects heat. Infrared coating (IRC) technology saves energy by containing the radiant heat and reusing it inside the bulb. They also have a longer lifetime. EU will sharpen the regulations in year 2016.

Dimmable compact fluorescent tube with a screw-in fitting.

Source: Govena.

Low-Energy Light Bulbs

Low-energy light bulbs are compact fluorescent tubes and other mini fluorescent tubes. The mounting determines which term is used. Low-energy light bulbs contain mercury and are therefore not the best choice from an environmental perspective. Every bulb contains 1–5mg of mercury, which can not be recycled and must be put in a waste storage facility.

For general light fittings, low-energy light bulbs are still the best alternative from the perspective of functionality. The latest developments have made dimmable compact fluorescent tubes possible with such extremely compact design that normal fittings can be used, and the light is similar to that from incandescent light bulbs. Compact fluorescent light bulbs can be turned on and off many times and have a long lifetime. They have become 10 per cent more energy efficient then earlier and can run for 16,000 hours. EU will sharpen the regulations in year 2013.

Fluorescent Tubes

The lighting company Aura has introduced a model called 'Eco-Saver', which has been described as the world's first long-life energy-saving fluorescent tube. It uses up to 12 per cent less energy and has three times the lifetime of standard fluorescent tubes.

GE produces the T5 WattMiser, a fluorescent tube that uses 5 per cent less energy than standard fluorescent tubes.

Energy-Efficient Lighting

Office Lighting

Energy-efficient lighting from light sources that have high luminescence and long lives, in modern fixtures with good reflectors, together with daylight and more efficient control systems, can provide good lighting while saving electricity and reducing environmental impacts. The most important sustainability aspect of lighting is reducing electricity consumption. There is often a 30–80 per cent potential for savings with energy-efficient lighting. The lighting system is a chain from the socket to the object to be lit, and losses at every link in the chain can be minimized. The chain is made up of transformers, electronic switches, lamps, fixtures, rooms and objects to be illuminated.

45 W/m²

15 W/m²

7.5 W/m²

Development of energy-efficient lighting has progressed quickly. Energy-efficient light sources, electronic control devices, fixtures and better planning for lighting are now available. It was formerly common for offices to use 45W/m². Today, 15W/m² is used to manage the same light conditions. Today's technology can satisfy the requirements with 7.5W/m².

Source: Vattenfalls projekt 2000; illustration: Leif Kindgren

Staircase Lighting

Örebrobostäder, the housing agency run by the City of Örebro, is working on the development of LED fittings for staircase lighting since the fittings on the market did not meet their specifications. They want robust fittings that use little energy with diodes that last decades and with three levels of lighting: fully lit, resting and totally off. In addition, cost should be low.

Street Lighting

Use of LEDs save a lot of energy compared to current technology. They are 60 to 70 per cent more efficient than high-pressure natrium lights. Newly installed systems require smaller-sized cables as less electricity is required. The whole electricity installation is thus much cheaper. The high light quality with a Ra value of about 80 provides better visibility and colour reproduction. The very long lifetime of LEDs, up to 50,000 hours, minimizes maintenance needs and thus saves money. Their low heat release means the lights do not require ventilation and can be made totally enclosed.

They are ideal for operation with solar cells and wind turbines. The lamps are made to provide maximum cooling to the diode, which guarantees a long lifetime.

LED (Light Emitting Diode)

LEDs are small light sources that emit light different from that of other light sources. Diode light is produced by a process called electro-luminescence, which takes place in a semiconductor chip not more than one quarter of a square millimetre in size. The structure of the semiconductor material determines the colour of the light. LEDs produce a focused beam within a scattering angle of 160–180 degrees. Diode light as a completely different character to what people are used to. It is monochromatic within a narrow spectral width. The use of LEDs for lighting is relatively new. It wasn't until 1997 that active development of 'white', high-efficiency diodes began. Warm-white light emitting diodes are still not equivalent to cold-white versions with regards to colour stability and luminous efficiency. Colour reproduction (Ra), which is rated according to an index with 100 being optimal, has increased over time to Ra 80–85. Cold-white diodes are already available with a Ra index of 90.

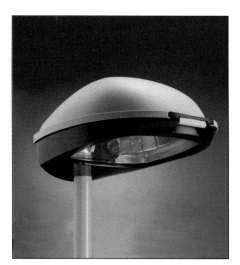

Street lighting with LED technology that is attractively designed, fits in well and does not collect dirt and water.

Source: AEC Illuminazione, an Italian company, and a Swedish reseller Tivalux AB.

Transformer for dimmable LED-lighting.

Source: Philips

Common high-efficiency diodes (high-power LEDs) are millimetres in size, about the size of a pinhead. For lighting purposes diodes are complemented with a reflector or a plastic prism optical lens with a high reflective index. A disadvantage to be aware of is that uncovered high power LEDs are often extremely bright and very uncomfortable to look at. The design and placing of fittings must thus be carefully chosen. Incandescent bulbs and fluorescent tubes spread light in all directions while LEDs have a focused beam, and thus reflectors or lenses must be use to spread the light. Much design work is under way to develop new fittings appropriate for LED technology. The light loss of a diode during its lifetime is not linear but rather accelerates towards the end of its lifetime. LEDs must therefore be replaced before they go out, which can take a long time. they should be replaced at the latest when the total light from the fitting is less than 70–80 per cent of what it was to begin with. Lifetime is often given as over 50,000 hours, but in reality with a surrounding temperature of 25°C the service life of a diode is between 20,000 and 40,000 hours.

Light emitting diodes produce light without thermal radiation (IR), though the energy input creates excess heat that must be properly reduced as in all electronics. The flow of light from a diode can be halved and its lifetime is greatly reduced if cooling of the fitting is inadequate. LEDs are powered by direct current and all light diodes require a ballast. The ballast can be built into the fitting so that 230V alternating current (AC) can be directly connected. There are some LED lights with a built-in transformer that can directly replace an incandescent bulb. There are also light emitting diode tubes that can replace fluorescent tubes in a fitting. The ballast can be mounted externally and used for one or more if fittings. LED lights are also suitable for use with a dimmer switch. High-frequency diodes are used for lighting purposes. The output of light diodes ranges up to 3.6W for the largest high-output diodes, which is equivalent to a 35W halogen bulb. As the light from a LED is more focused than that from an incandescent bulb it has become common to measure the amount of light from the fitting instead of the lumens for individual diodes. Currently, even in well-designed fittings, 10 to 20 per cent less light can be expected than what is promised.

Advantages of LEDs

- highly efficient and bright with a very low energy consumption, produce 10 to 25 per cent light from the energy supplied and development is rapidly progressing;
- the smallest light source, and make small and easily placed fittings possible;
- light instantly with full luminance flux and are suitable for use with a dimmer switch;
- insensitive to repeated turning on and off as well as bumps and vibration;
- have a much longer lifetime than incandescent light bulbs and fluorescent tubes.

Disadvantages of LED

- relatively expensive to purchase;
- very sensitive to heat and require fittings with good heat dissipation;
- light emission is reduced over time and they should be replaced before they go out;
- may be experienced as blinding and difficult to look at;
- bulbs with warm-white light have less stable colour than incandescent bulbs and 30 per cent less luminous efficiency.

One of the biggest advantages of incandescent bulbs has been the high quality of light. LEDs are availablse in a variety of shades of light. Intensive development is under way to produce LEDs that provide as good light as from incandescent bulbs. There are LED lights that contain green, red and blue diodes that can be adjusted to produce the colour desired.

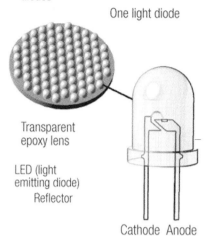

Light diode.

Source: Illustration: Leif Kindgren

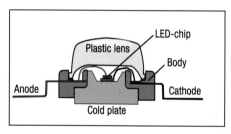

Cross-section of a light diode.

Source: Det nya LED-ljuset - Ett kompendium från Annell Ljus+Form AB, 2009

LED lamps with varying fittings that can replace both incandescent light bulbs and low-energy light bulbs (that contain lead and mercury, respectively).

Source: Kloka Hem 2008 & www.varuhuset.etc.se

Control Systems

In this context the computer bus system distributes information between devices. There is a small communication unit in each fitting that 'talks' with a computer. Such a system provides full control of how every single bulb works and how much energy it consumes. It is also possible to adjust whether or not the bulb is on or off and how bright it is. The system is unique in that all functions are controlled electronically. The lighting can be controlled from a keypad or a

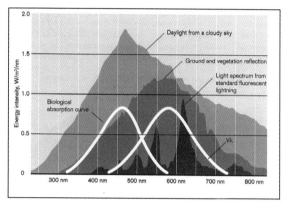

The diagram shows different types of light spectra and absorption spectra for biological and visual receptors. The so-called $V\lambda$ curve, showing the visual sensitivity of the eye, correlates well with the curve for light reflected from vegetation. The biological absorption curve correlates well with the diffuse solar radiation spectrum, daylight from a cloudy sky.

Source: People and Buildings – Both Need Light, Hans Arvidsson and Lars Bylund, ÅF-infrastruktur AB, Stockholm

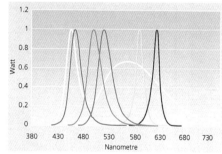

Spectra of white and coloured LEDs. The spectra for white LEDs reminds of the daylight spectra and the reflected light from vegetation. This may be the reason why white LEDs are considered to give a high-quality light.

Source: LED – Ljus ur lysdioden 17, published by Belysningsbranschen in cooperation with Ljuskultur, 2008

computer. Experience has shown that so-called lighting on demand can reduce energy consumption in street lighting by half. Bus systems are also being used in energy efficient buildings. The software allows automatic adjustment of lighting in rooms according to the current level of activity. The system can be programmed so that standby mode is turned off when not needed.

Efficiency of Various Light Sources

Light source	Luminous efficiency (lm/w)
Incandescent light bulbs	8–15
Halogen light bulbs	9–25
Compact fluorescent tubes	50–88
Straight fluorescent tubes	75–104
Metal halogen	80–120
Cold-white led	47–70*
Warm-white led	25–50*

Note: approximate values during 2008. The numbers can in reality be higher or lower depending on the fitting and location. New values are normally published several times a year. Note that the luminous flux (lm) from LEDs is directed in a sector of only 160° to 180° and thus lights a surface or object much more efficiently than other light sources.
Source: Det nya LED-ljuset – Ett kompendium från Annell Ljus + Form AB, 2009.

Photo from the Stockholm City Museum where they have tried a LED light prototype armature. The employees that use this room are satisfied with the quality of light and don't suffer from as much headache as before.

Source: Lars Bylund, Professor TTA, Bergen Architect School, Norway

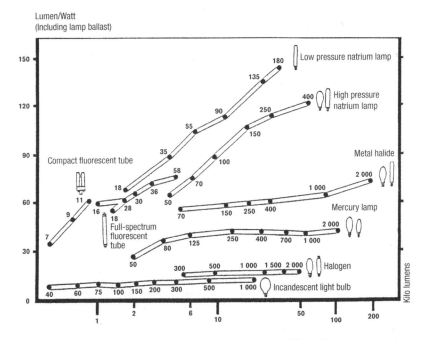

The diagram shows how light exchange (Lumen/Watt) increases for different light sources and different sizes of lamps. The numbers in the diagram represent watts for the respective light source.

Source: 'Belysning inne och ute, Watt med Vett', Malmö Energi AB

Lamp Ballasts

Discharge lamps such as fluorescent tubes must be connected to a lamp ballast. The lamp ballast has two functions: it provides the necessary starting voltage and it limits the current to the right level. A distinction is made between conventional (magnetic) and modern (electronic) lamp ballasts. Magnetic lamp ballasts contain a heavy iron core. They are cheap to produce but use more current.

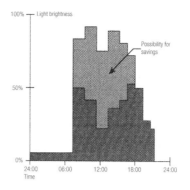

The diagram shows how much energy can be saved by using a daylight sensor connected to a dimmer.

Source: *Advanced Lighting, Guidelines: 1993*, US Department of Energy, California Energy Commission; Electric Power Research Institute

Electronic high-frequency (HF) ballasts use less electricity, weigh less and reduce the power requirement. HF ballasts should be used in order to obtain electric efficiency, long life and flicker-free light. HF ballasts also allow brightness control, daylight control, movement detection, as well as connection to various types of control systems. This means that HF ballasts allow the regulation of light sources as required, which is an excellent way of saving electricity.

Adjustment

Large amounts of electricity for lighting can be saved with the help of control equipment, which has become cheaper and more sophisticated.

Motion detectors can be used to turn off lights in rooms that are not being used. Modern motion detectors sense heat and movement from people in a room. The latest ones react to sound/infrasound and the light dims on and off according to those. 75–95 per cent of the electricity can be saved in staircase and garage lighting etc.

Motion detectors can turn off lights when no one has been in a room for a certain length of time or turn on a light in a stairwell when someone enters. With the help of this technology, lights are not on unnecessarily. Motion detectors can easily be installed in existing electric systems. The motion detector can turn on lights automatically and be connected to control ventilation, heating and cooling.

Daylight sensors dim (adjust) electric lighting according to daylight brightness. Daylight sensors can save a lot of electricity used for lighting during the lighter times of the day.

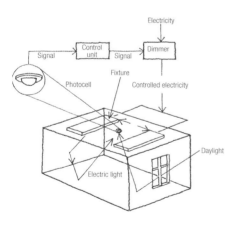

A daylight sensor connected to a dimmer can decrease the brightness of electric light as daylight strength increases, and turn off the lights when there is enough daylight.

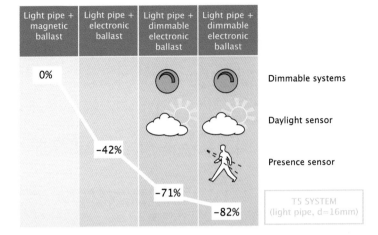

How much electricity you can save by using different systems of detectors, sensors and control devices.

Source: Peter Pertola, WSP

NEW LIGHT SOURCES

Improvements to lighting include attempts to develop energy-efficient discharge lamps, e.g. without mercury, that provide good quality light. Alternatives to mercury for discharge lamps include plasma lamps, lamps with electron cannons, and lamps with microscopic fibres (nanofibres).

New fixtures are being developed that provide several lighting functions. A light pipe in the ceiling can light a whole corridor from one source; with a bundle of fibre optics, one light source can create a whole starry sky; and with directed and concentrated reflectors, fixtures that provide general lighting can also be used for workplace lighting.

Light Pipes

Light pipes are a type of opaque fixture. A strong, efficient light source is directed into the opaque pipe. On its way through the pipe, every reflection lets through 2 per cent of the amount of light, and a long, even stream of light enters the room.

Mercury-Free Compact Fluorescent Bulbs

Mercury-free compact fluorescent bulbs are a promising development. Ordinary fluorescent tubes and compact fluorescent bulbs sold contain a considerable amount

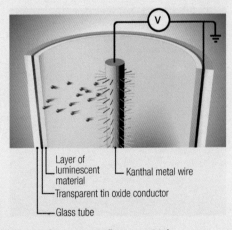

Mercury-free compact fluorescent tube.

Source: Illustration: Leif Kindgren

Light pipe.

of mercury. About two-thirds of this dangerous heavy metal is lost every year. There is now a mercury-free fluorescent tube in the prototype stage that uses the same principle as the electron cannon in a TV picture tube. The fluorescent tube provides new types of more daylight-like white light. A metal wire sits in the middle of the fluorescent tube. The wire is made of kanthal alloy covered with a layer of microscopic carbon nanofibres. Electric voltage is put through the wire and a thin conductive layer of tin oxide (luminescent material) on the inside of the airless glass tube. The electric voltage pulls electrons from the layer of nanofibres, and light is produced when the electrons hit the luminescent material on the inside of the glass tube.

OLED (Organic Light Emitting Diode)

Organic LEDs (OLEDs) are a new type of LED. They are soon going to make new opportunities possible for lighting as they can be used for lighting of flat surfaces. Currently known organic materials can generate all colours within the visible light spectrum, even white. White can be produced in the organic layer by mixing colours. Using this method of mixing colours, white and coloured OLEDs can be made that are totally transparent when turned off.

The Development Potential of LEDs

Several exciting research projects are under way to try to make LEDs cheaper. Examples are Qunano in Lund and Nanosys in the US where work is being done with LED technology using silicone nanofilaments.

Researchers at Purdue University in Indiana, USA, are working on new technology that will make bright, white LEDs cheaper to produce by substituting expensive sapphire substrate with silicone.

In Nyköping, TD Light Sweden is working on producing a diode light tube. The product looks like a light tube but a row of white LEDs is inside the glass. Their lifetime is 10–15 years, energy consumption is less than half of that for normal light tubes, the light is similar to natural daylight and the tubes do not contain mercury. They are, however, more expensive to purchase.

The company Aluwave is developing a circuit board material with a ceramic material instead of a polymer. Such light diodes would be cooler and thus last longer and have a higher light quality for a longer time.

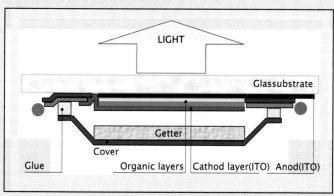

Cross-section of an organic light diode.

Source: LED – Ljus ur lysdioden 17, Utgiven av LED-gruppen i Belysningsbranschen i samarbete med Ljuskultur

2.2.3 Electrical Devices

We surround ourselves with more and more electrical gadgets. Many use electricity even when they are not in use, e.g. TVs, stereos, video machines, photocopiers, water beds and computer printers. These are in reality wasted kilowatt-hours of high-quality energy, benefitting no one. With regard to computers, Energy Star or TCO-certified computers automatically go into standby mode after a certain amount of time.

Electrical Equipment

As lighting and white goods have become more energy efficient, so the development of electrical equipment, pumps and fans is moving in the same direction.

Office Equipment

Photocopiers, computers, computer printers and faxes are used in offices. Copiers use the most energy. Energy-efficient equipment should be chosen. Equipment should be turned off at night and at holiday time. Energy-efficient certification can promote visual quality on screens, low-energy consumption, and minimize electric and electromagnetic fields, use of heavy metals and dangerous flame retardants.

The appropriate size copier, with power control (goes into standby mode when not being used) but without a hot-drum, should be chosen. Choose a machine that can do double-sided copying. Choose a computer with standby mode. Don't buy a bigger monitor than necessary and make sure it has a screen-saver mode. Consider whether or not a laptop computer is adequate; they almost always use less energy than desktop models.

Buy a computer printer with standby mode and double-sided print facility. Consider whether or not an inkjet printer is adequate, as they are much more energy efficient than laser printers.

Faxes should have low-energy consumption in standby mode. It is possible to receive faxes via a computer as email messages.

Today, there are all-in-one devices that copy, fax, print and scan. Such multifunctional machines are often adequate for small offices and home use.

Household Appliances

Many electrical appliances are used for such short time periods that they don't make much difference from an energy perspective, e.g. vacuum cleaners, hair dryers, shavers and electric mixers. Appliances that are turned on for long periods or that use a lot of power should be avoided.

Standby Power

Despite many household appliances becoming more energy efficient, Swedish households use more and more electricity. An explanation is that more and more household electric devices are purchased that consume electricity even when they are not being used. 10 per cent of household electricity consumption is made up of standby power. The three worst environmental villains, according to the magazine *PC för alla*, are Sony's TV game system Playstation 3, plasma TVs, and in third place is stationary computers. Multifunctional computer machines (printer/scanner/copier), speakers and TV digital boxes are other products that use a lot of electricity in standby mode. In 2008 it cost about €200 per year to have a computer with old monitor and printer turned on all the time. There are now products available that can be plugged into a normal electric socket and make it possible to turn off standby electricity with a remote control.

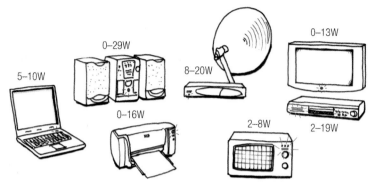

Avoid standby mode. To save electricity it ought to be possible to turn off any devices not being used. Many devices, such as TVs, satellite dishes, stereos, videos, etc., are equipped with a standby mode so that they can be turned on with a remote control. This means that they use electricity continually, up to 5–10 per cent of the total household electricity. Many of these devices actually use more electricity when not being used than when they are being used. The standby function also constitutes a fire hazard. There are devices that can be plugged into the wall sockets so that you can switch off the electricity with a remote control.

An electric sauna uses a lot of electricity due to the high power requirement. Wood-fired saunas are better for the environment.

Waterbeds consume a lot of energy, and the actual need for one can be questioned.

Aquariums use pumps and lighting around the clock, all year round, which means that they use quite a lot of electricity.

It is still common for vehicle engine preheaters not to be connected to a timer but they use a lot of electricity if they are on all night. Cars should be parked in a cold garage and engine preheaters controlled with a timer. A timer can turn on the heat 30 minutes before the car is going to be used, saving energy. Electrically heated seats can eliminate the need for a car heater.

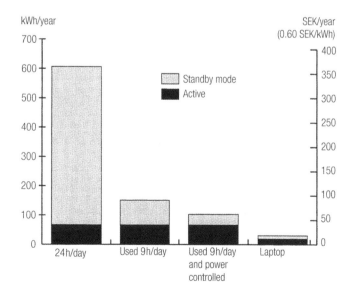

With regard to computers, there is a lot of energy that can be saved, e.g. by turning off the computer at night, by using energy-saving functions that turn off the monitor if the computer isn't used for a few minutes, and by using computers that have energy-efficient components.

Source: NUTEK

Operational Electricity

New technology for pumps and fans has also developed in the direction of energy efficiency. This has to do with reducing flow resistance in pipe and duct systems, and working with and not against natural forces (such as natural ventilation and gravity systems). In addition, frequency controls for pump and fan motors are becoming more common; these allow speed adjustments instead of running them at full speed and braking them with a damper. (See Section 1.2 on ventilation and heating systems.)

Shared Electricity Use

The electricity used by buildings varies of course, and includes electricity for laundry, lighting, fans, pumps, as well as any engine preheaters and lifts. All these devices except lifts have already been discussed. The electricity consumption of lifts has improved with the development of new technology. Lifts that move a little slower use less energy. Energy is also saved when more people often choose to use stairs. More energy-efficient hydraulic lifts, where the machinery is located under the lift, are available.

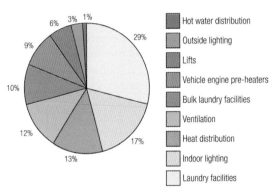

Breakdown of shared electricity use in a block of flats

Source: Uppdrag 2000, Huge Bostäder, Huddinge

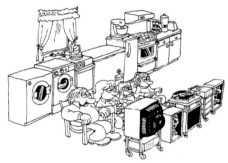

Appliances that save time have driven away shared activities from homes.

Source: Illustration: Claus Deleuran

This table shows four different cases of operational electricity use in a single-family dwelling. The first column shows normal use for a single-family dwelling. The second column shows use in a single-family dwelling with the most energy-efficient pumps and fans on the market. The third column shows how little electricity is needed in a well-planned energy-efficient home. The fourth column shows that natural ventilation and heating with natural circulation can be used in a home.

Operational electricity (single-family dwelling)	Normal kWh/year	Best available technology kWh/year	Energy efficient kWh/year	Without electricity
Pumps	420	230	100 (small pump)	0 (natural circulation)
Fans	1070	750	500 (pressure controlled)	0 (natural ventilation)
Total	1490	980	600	0

2.2.4 Getting Things Done Without Electricity

In the past when there was no electricity, lighting, food storage and clothes drying were done differently. Many of these methods are still practicable and can be combined with energy-efficient technology to create good holistic approaches that integrate high and low technology.

Avoiding Electricity Use

Since electricity is a high-quality and expensive energy source, electricity should always be replaced where possible with fuels (e.g. gas), heat (e.g. solar collectors), or old-established methods (e.g. cellars).

Daylight

Good access to daylight reduces the need for electric lights, and exposure to daylight and the circadian rhythm is beneficial for people's well-being.

Windows with attractively profiled glazing bars and with white, angled window recesses provide good conditions for letting in daylight. Windows high up on a wall and skylights allow in more daylight than windows lower down. In anthroposophical architecture, daylight entering a room from at least two directions is preferred. This creates a more modulating light.

One method for bringing daylight far into a room is to use light shelves.

Advanced Daylight

In recent decades, work with sophisticated ways of bringing daylight further into buildings has been under way. Reflectors, prisms and diffusers are several of the tools employed. Methods using fibre optics and light pipes have also been investigated.

An angular niche makes a window seem larger and reduces the risk of glare.

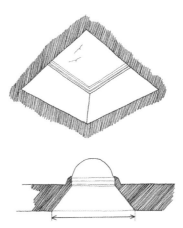

An angular space between the skylight opening and the ceiling provides desirable contrast balance.

Good daylight conditions reduce the need for electric light. Daylight access is affected by the placement of windows and skylights, the design of window recesses, the colour and material used for the recesses, mouldings of the glazing bars and frames, as well as the design of the room.

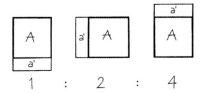

From a lighting point of view, light entering high up is much more effective than light entering low down. For a normal window, the light that enters near the top edge is four times more effective than the light that enters at the bottom edge.

Source: 'Sol energi form', Adamson, Hidemark, BFR T2:1986

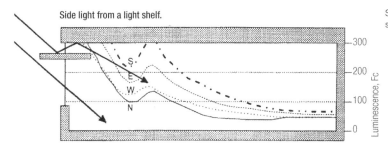

To bring daylight further into a room, a reflecting light shelf can be built into a window. The light shelf reflects light up to the ceiling, which casts light further into the room. At the same time, the light shelf shades the area near the window.

Daylight via fibre optics is a new invention. One of the finalists in the Environmental Innovation 2002 competition was Parans Daylight. This company has developed technology to transport daylight in fibre optic cables and release it indoors.

The Parans Solar Panels are 1 m² modules that are mounted on roofs or façades. Inside the panel, 64 Fresnel lenses move uniformly around their axis, tracking and concentrating sunlight. The Parans SP2 employs an active tracking system, guiding the Fresnel lenses so that they are always orientated towards the sun. This movement is achieved with three motors, consuming on average under 2W.

The Parans L3:s are spotlights leaving great freedom to the user to design the light experience. The Parans L3:s have adjustable focal ranges and are easy to direct in different angles.

From each Parans Solar Panel come four optical cables. These are 6mm in diameter, and can be ordered up to 20m long. The bending radius can be as small as 50mm, making it easy to turn tight corners. The light transmission is 95.6 per cent per metre. Using special fibres, it will be possible to transport light up to 70m.

Advanced daylight is used in the GreenZone in Umeå, Sweden. It is an ecologically designed building, designed by architect Anders Nyquist, that houses a car dealership, petrol station and hamburger restaurant. There are skylights on the roof with upside-down cones of reflecting material inside them. The cones lead light down a shaft lined with reflecting material that opens out into a fixture made of opaque material that diffuses the light. From the inside it looks like regular lighting, but the light comes from the sun. The daylight fixtures are supplemented by fluorescent tube fixtures.

Delivery of daylight via fibre optics. The patented Swedish System developed by Parans Daylight AB.

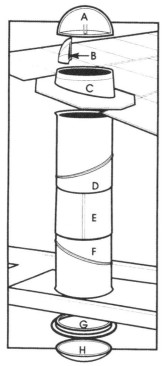

Delivery of daylight via fibre optics inside houses.

A UV durable acrylic plastic dome
B Reflector
C Roof sleeve
D Reflecting luminous tube
E Telescope to adjust length
F Possibility to adjust the angle from 0 to 45°
G Roof ring
H Diffuser (light distributor)

Daylight fixture. Daylight is taken down a 'solar tube' via a pipe and can light rooms inside a building.

The 'daylight catchers' are put up on roofs and façades. Dimensions:
980 x 980 x 180mm
Weight: 30kg
**Number of Fresnel Lenses: 64
Fiber Optic Cable Quantity: 4
Power Supply: AC 220–250V
Average Power
Consumption 2W
Shell Material:
Eloxated Aluminium
Glass Surface:
Hardened Glass**

Source: Parans Daylight AB

DAYLIGHT AND ARCHITECTURE

Use of daylight in buildings gives beautiful indoor light, saves electricity and provides opportunities for exciting architectural designs. There are many different ways to let in daylight: through windows, skylights, lanterns and glassed-in spaces. Daylight can come into a building along walls, and along shafts and light courtyards.

Rönninge school near Stockholm. A school with good daylight, using lanterns on the roof.

Source: Tallius-Myhrman Arkitekter AB

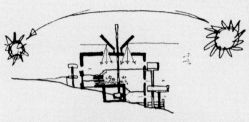

In the Ström family home in Stocksund, architect Ralph Erskine has brought in daylight and a view of outdoors through a periscope. Upon entering the dark hall, the whole archipelago can be seen in the mirror of the periscope.

Daylight enters via lanterns in the exhibition hall of the National Park building, Tyresta, near Stockholm.

Source: Per Liedner, Formverkstan arkitekter AB

There is a holographic heliostat in a multiple family building in Hammarby Sjöstad, Stockholm that directs light from the sun down into a stairway so that it is lit up better.

Source: Deltalux

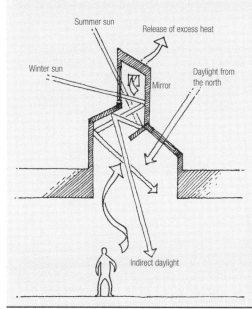

Lanterns designed to let in daylight can also be used as ventilation chimneys for exhaust air in natural ventilation systems.

Different principles for making use of daylight.

Source: 'A Green Vitruvius – Principles and Practice of Sustainable Architectural Design', ACE (Architects' Council of Europe), ERG (Energy Research Group, University College Dublin), SAFA och SOFTECH, 2001

Drying Boxes

Clothes can of course be dried without using an electric drying cabinet or dryer. In the past, it was more common to hang clothes outside to dry. These days there can be problems with theft and air pollutants. A drying box is a type of crate fitted with louvres that is placed outside a wardrobe window. Clothes can be hung up inside to air and dry. While the wind can move freely through it, the construction offers protection from precipitation, dirt and theft.

Gas

Gas appliances can be used for cooking (stoves and ovens), food storage (refrigerators and freezers), heating hot water and lighting. Use of gas radically reduces dependence on electricity. The most common use of gas is for cooking, since stoves require a high power output. People often choose an electric oven even though they have a gas stove since gas ovens are more difficult to use. If a house is equipped with solar cell electricity for lighting, it is common for the stove and refrigerator to be run with bottled gas. Bottled gas is a fossil fuel, but perhaps the least environmentally harmful one. It is also possible to use environmentally friendly biogas.

A drying box on a block of flats in the Brandmästar district of Karlstad, Sweden. The drying boxes are very popular among the residents.

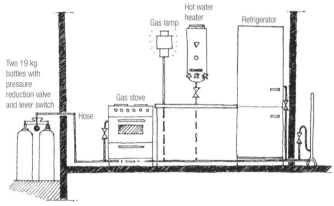

This is how a gas-powered kitchen can look. A gas stove provides security in the case of electric power cuts and many people think they are better to cook on. The gas bottles are placed outdoors for safety reasons.

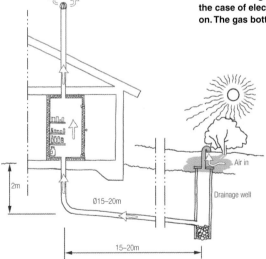

It often gets too warm in larders in the summer. Ventilating with an underground duct avoids this problem and in addition allows freedom in locating the larder (doesn't necessarily have to be placed on a north-facing outside wall) as well as with floor-plan design.

Larders

In the past, it was common to store food in a larder. It worked perfectly in the winter, but larders sometimes got too warm in the summer. It is best to locate a larder on the north-facing side and by an outside wall so that the temperature can be regulated with an air vent. Larders are experiencing a renaissance in eco-houses. These modern larders are insulated and have an insulated, tightly fitting door. If they are equipped with an underground duct that supplies intake air and a ventilation chimney for exhaust air, they work better during the summer as it is possible to take advantage of the cool temperature of the earth. Such a larder can be put anywhere in a house as the vent on an outside wall is no longer needed. A fan can be placed in the exhaust air duct so that ventilation via the underground duct works better.

Food Storage

Today, it is common to store most food either cold or frozen so that it won't go bad. However, this isn't entirely necessary. Instead, food can be preserved, dried or lacto-fermented. The food can then be stored at room temperature or in a cool place. For goods preserved in metal cans, 10 per cent of vitamins are retained. However, drying retains 80 per cent of the vitamin content. The lacto-fermentation process also releases vitamins that our bodies could not otherwise utilize.

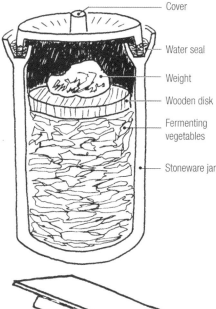

A ceramic jug for lacto-fermentation, with an airtight water seal lid.

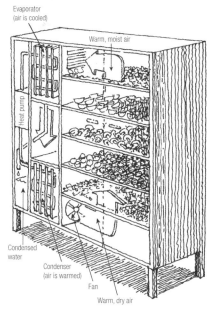

Condensation dryers dry food quickly and aromas are preserved as well.

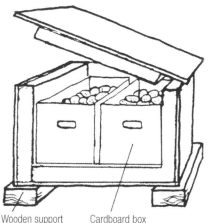

A simple insulated box for storing potatoes, apples and root vegetables on a balcony. The top does not have to be insulated. It's better to be able to add thick layers of insulating material on top inside the box when required.

Source: 'Jordkällare och skafferi', Kerstin Holmberg

2.2 ROOT CELLARS

Root cellars are currently available as prefabricated concrete or plastic modules. Instructions are also available on how to construct one. Traditionally, the entrance to a root cellar faces north so the sun won't shine in, and there is a small entry room to keep out the cold. A root cellar shouldn't get too damp, and so it must be carefully drained and be protected from rain with a small roof or a waterproof layer of clay on the top. It should not get too dry, either, as then the cold can get in. A root cellar must have good thermal contact with the ground so that it doesn't get too warm and good outer insulation so that it doesn't get too cold. Therefore, they often have a dirt floor and stone or concrete walls. It's common for root cellars to be dug into the ground, with a thick enough layer of earth on top, but a root cellar may also be built above ground if it is covered with enough earth. Root cellars must be ventilated. By regulating the ventilation the relative humidity can be kept at a suitable level. The intake duct can be placed in the door and the exhaust duct can be located in the roof. Some foodstuffs cannot be stored in the same space. Fruit and root vegetables should be kept separate. Apples, for example, release a gas that accelerates the aging process in other fruits and root vegetables.

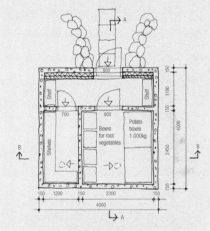

It is best if the entrance to a root cellar faces north and has a small entry room. The storage area should be divided into two areas, one for root vegetables and the other for fruit, jam and juice.

Source: *Bygga jordkällare*, Urpo Nurmisto

Root cellars should be built so that they 'breathe'. The structure should protect the cellar from direct water ingress from the ground while allowing just enough moisture to enter so that a suitable humidity level is maintained. Excess moisture is ventilated away. Due to seasonal changes in outdoor temperatures, a root cellar should keep out heat during the summer and keep warmth in during the winter.

Source: *Bygga jordkällare*, Urpo Nurmisto

Root cellars on the island of Öland, Sweden are often built above ground due to the surface bedrock.

Root cellars are a traditional way of storing root vegetables and fruit. These days, root cellars are also built adjacent to blocks of flats and other urban settlements.

The Sankt Botvid spring near Bornsjön in Salem Municipality, Sweden.
Source: Photo: Maria Block

2.3 Clean Water

Water is the requirement for all life, therefore we have to keep our waters clean through careful maintenance.

Clean water is provided by municipalities or private wells, depending on the location. Water conservation and use of water-saving technology should be a matter of course.

2.3.0 Water Use

The world's water resources are used first and foremost for agriculture (70 per cent), followed by industry (20 per cent), while households account for only 10 per cent of the usage. One illustration of agricultural water consumption is as follows: 1 litre of milk requires about 800 litres of water to produce and 1 kilogram of beef requires 6000 litres of water to produce. In the US, half of all water consumed is used for irrigation. In Israel, more than 75 per cent of all water consumed is used for irrigation. In Sweden, industry accounts for the majority of water use.

Water Needs

A human being needs at least 3.5 litres of water per day to survive. We need at least 25 litres per day to maintain hygiene and avoid disease. The United Nations policy guidelines are that 50 litres of clean water per person per day is reasonable. Household water consumption varies from country to country. In Sweden it is 215 litres per person per day, in Denmark 110 litres, and in the US 450 litres.

Water-Related Diseases

For household water consumption, the main concern is not lack of water but lack of clean water. In many developing countries, 70 per cent of all diseases are related to water being polluted by bacteria, viruses and parasites. A different pattern is seen in industrialized countries. The World Health Organization (WHO) believes that in industrialized countries, 70 per cent of disease is water-related due to the insidious development of disease caused by long-term exposure to small doses of contaminants. The main challenge is thus to supply the world's population with clean water that doesn't contain excessive levels of bacteria, viruses, parasites, and metals and chemicals that are health hazards.

Simple Water Cleaning

An invention by Petra Wadström called *Solvatten* (Solar water) is a simple patented method for purifying water with sunlight only. The product is a black, portable plastic container with a filling hole for dirty water

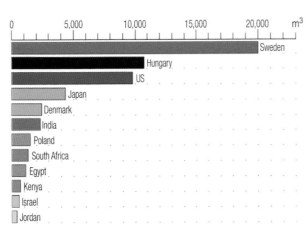

Access to water per capita per year in various countries. Access to less than 1700m³ per capita per year is considered to be water scarcity, and per capita access to less than 1000m³ per year, chronic water scarcity.

Source: VVS Företagens Teknikhandbok, 2009

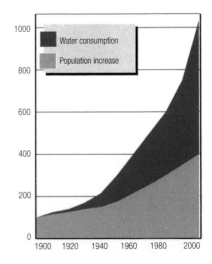

During the 1990s, the world's population increased almost fourfold, while water use increased by about 10 times.

Source: *Omvärlden*, no 4, 1998

The *Solvatten* tank holds 10 litres of dirty water that takes five hours to clean.

and a drainage hole for water that has been solar heated and made free from bacteria. Purification is by heat and ultraviolet rays. The container can be folded out so that a transparent exposed surface is presented that will kill hazardous organisms. A finely-meshed filter in the filling pipe reduces the amount of suspended matter and separates out unwanted organisms such as amoeba. A mechanical temperature indicator shows when the water has reached 55°C and it's clean.

Virtual Water

The amount of fresh water needed to produce a product at the location where it is produced is called virtual water. The total amount of water required to produce the goods an individual consumes is called a water footprint.

The average water footprint for a person in Sweden is 2150m³ per year. According to the World Wide Fund for Nature every individual in the country uses almost 6000 litres per day if all the virtual water concealed in the production of goods such as clothes and food is included. Regarding food, a vegetarian consumes half as much water per day (2500 litres) as a person who eats meat (5000 litres).

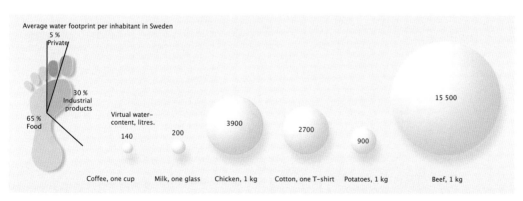

Examples of water footprint and virtual water contents.

Source: Water Footprint. www.waterfootprint.org

Water Access in Sweden

Sweden is rich in water compared to many other countries. The south-west Skåne region of Sweden has the lowest per capita water supply while the Norrland region has the highest. Water use for the whole country is about 3600 million m³ per year. The slightly more than 7 million urban inhabitants get more than 90 per cent of their water from municipal water supply plants. Rural households (about 1.2 million people) and holiday cottage households (about 1.3 million people) usually use groundwater from private wells.

Water Use in Sweden

Water use in Sweden increased greatly in the post-World War period, but has decreased since 1960. The reduction is primarily due to the more strict environmental requirements placed on industry. When the environmental requirements became more strict, industry made a concentrated effort to reduce water use and discharge of water-borne pollutants. Household water use also increased in the post-war period, but the increase ceased in the 1970s and currently, a certain reduction due to the introduction of water-saving technology can be seen.

Water use in Sweden has decreased greatly since the 1960s. One of the most important reasons is the concentrated effort made by industry to reduce water use and polluted discharges.

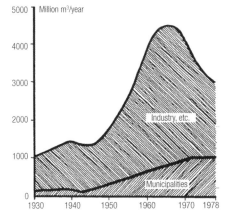

Water Use Distribution in Sweden

Industry	2540 million m³	(70%)
Households	575 million m³	(16%)
Other	320 million m³	(9%)
Agriculture	171 million m³	(5%)

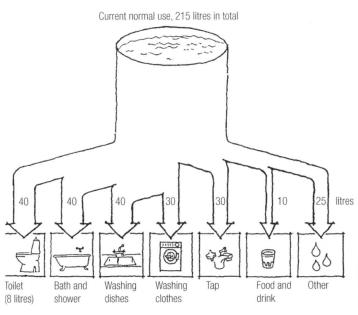

Current daily household water use in Sweden, litres per person per day.

2.3.1 Water-Saving Technology

A normal household in a developed country uses about 215 litres of water per person per day. With the new water-saving technology now available on the market for showers, taps, toilets, as well as clothes and dishwashing machines, it is possible to reduce water consumption by 50 per cent without difficulty and without lowering the standard of living or of hygiene.

Taps

There has been a revolutionary change with regard to water fixtures, especially taps. In the past, there was one tap for cold water and another for hot, then came the mixer with one spout and two taps, and now there

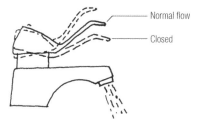

Normal flow
Closed

Full flow: the lever is pushed up and returns to normal flow when released

are single-lever mixer taps. The single-lever mixer tap has increased convenience while at the same time reducing water consumption by about 30 per cent. Ceramic washers are used to reduce leakage, increase life and make maintenance easier.

Single-lever Mixer Taps

Single-lever mixer taps regulate both water flow and temperature with a single lever. Such taps save water since they are easier

Single-lever mixers save water since they can be quickly adjusted to the right temperature. On several models the normal flow can be set as required, e.g. 18L/min in a bathtub, 12L/min in a kitchen sink and 6L/min in a bathroom basin.

A water-saving tap is a single-lever tap with low water flow in the normal position. It is possible to force the water flow by pushing the lever upwards, but as soon as it is released it automatically returns to the normal flow position.

The figure above shows temperature zones for the different positions of a tap pointing sideways, and where the water is warm in the central position. The bottom figure shows the difference in temperature zones for a tap that runs cold instead of warm water when the lever is in the central and up position.

The Grohe company sells a stainless steel tap. Ordinary taps are made of brass and contain lead and tin. Many fittings have surface coatings that contain nickel and chrome. Lead is a health hazard and should be phased out; chrome and tin are metals toxic to aquatic organisms. Nickel is an allergen.

to quickly adjust to the right temperature. Additional developments for single-lever mixer taps include the capacity to set maximum flow at different tap locations and to set maximum temperature. There are also water-saving mixers that reduce the flow to a normal level as soon as the grip on them is released. In addition, there are mixers that run cold instead of warm water when the lever is opened in the central position. These mixers save up to 50 per cent of total hot water. In public places there are taps controlled by photocells that save even more water. Tap quality has also improved over the years. It is important to turn off single-lever taps gently in order to avoid water hammer. The costs for converting to single-lever water mixers can be recovered in 5–10 years through savings in energy and water costs.

The company Gustavsberg has taken a holistic approach to environmental requirements for water taps. Their e-TAP environmental label covers specifications for production, recycling and releases to water as well as energy efficiency aspects. The connecting pipes used are made in an environmentally friendly manner of corrosion-resistant and recyclable material. The Nautic brand water tap has been approved by the Water Research Center (WRC), which means that all components are 'food approved'. It also has soft closing which protects from water leaks and extends the lifetime. It meets the Nordic Quality standard, which means that industry high-quality and service standards are met according to the INSTA-CERT's specifications. Discharges in the production process are minimal, consumption of materials are kept at a low level and no environmentally hazardous materials are used.

Perlator

A Perlator, or aerator, is mounted at the end of the spout. Air is mixed with the water and

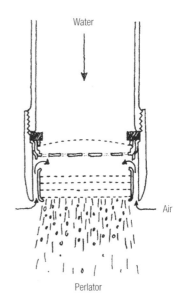

The figure shows an aeration device, a Perlator. It distributes the water flow so the same soaking effect is achieved with a reduced water flow. They can be mounted onto most old and new taps.

despite a reduced water flow, the soaking effect is good and the water stream ample. There are different types of aerators: those without extra features, those with a ball and socket joint (where the ball and socket can be purchased separately), and those that can be adjusted from jet to spray.

Flow Restrictor

The simplest type of flow limiter on an old tap is a flow restrictor, i.e. a plastic washer with a slot that limits the water flow.

Leakage

Dripping taps and leaking toilets waste large amounts of water. Gaskets, O-ring washers in traditional taps, wear out after a certain amount of use. Ceramic seals have a much longer life, at least five years or 100,000 closures, and are now common in single-lever mixers. 100 litres a day can be lost from a fast dripping tap. Leaking toilets are just as bad of course.

Slow dripping tap. Loss of 20 litres water per day or 7.3m³/year.

Fast dripping tap. Loss of 100 litres water per day or 37m³/year.

Running tap with a thin flow (1.5mm). Loss of 380 litres water per day or 140 m³/year.

Make sure that no taps or toilets drip or leak, which can quickly lead to large water losses. If it is hot water that is dripping, energy is also lost.

Shower Heads

Improvements have been made in water-saving shower heads. Some old shower heads had a flow of 20–25L/min, and most of the water never hit the person showering. By designing a shower head that distributes water better and mixes the water with air, it has been possible to reduce water flow. Today there are water-efficient showers that use 6L/min. It is possible to make showers even more water efficient, but experience has shown that if the amount of water is reduced to less than 6L/min, many people complain that there isn't enough water.

Thermostatic Mixers

The advantage of a thermostatic mixer is that the water quickly reaches the desired temperature and remains constant, which saves both water and energy. Thermostatic mixers are usually delivered with the maximum temperature set at 45°C, but the setting can be changed to increase energy savings.

Shower Stop Button

For temporarily stopping water flow while showering, e.g. to apply soap or shampoo, a shower stop button is practical. They make it possible to turn off the water while preserving the hot-cold adjustment. A shower stop button is installed between the mixer and the shower hose.

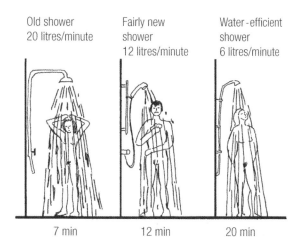

It is possible to shower for three times as long with the same quantity of water in new showers as in old ones.

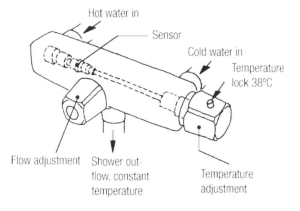

Thermostatic mixers save water as the temperature can be set in advance.

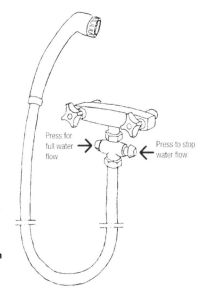

A shower stop button makes it easy to turn off the water while soaping down and then to turn it back on at the same pressure and temperature.

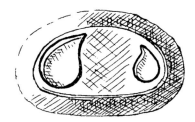

There are now flush toilets with a water-saving flush.

Toilets

Flush toilets are sometimes used as a symbol of a resource-squandering society. The 'waste' is flushed away and what happens to it later is of no concern. In a sustainable society, dry toilets or toilets that are as water efficient as possible are used. There are many different types of toilets and consumption of flush water varies by a factor of 100, which is one indication of great water-saving potential. At the same time, flush toilets have become a symbol of prosperity and many people do not like the thought of having a dry toilet.

Water-Efficient Toilets

The first flush toilets had a cistern up near the ceiling and used 12 litres of water per flush. There were even toilets that used 25 litres of water per flush. When the cistern was moved down behind the seat, flush volume decreased to 9 litres. During the 1960s and 1970s many homes had toilets that used 6 litres of water. Since 2000 most toilets use 4 litres. Toilets with two-button systems are also common, one button for a small 2-litre flush and another button for a large 4-litre flush. It is possible to achieve the same result with considerably less water. There is also equipment on the market that can easily be put into the cistern to reduce the flush water by approximately 2 litres per flush.

Some separating toilets, e.g. urine-sorting flush toilets, flush faeces with 5 litres of water and urine with 0.2 litres. Statistics show that people go to the toilet five times per day (of which four times are to urinate). Total water use per person per day would then be 4 × 0.2 + 5 = 5.8 litres. A common water-saving toilet would use 5 × 4 = 20 litres per person per day. Another way to separate is to use a double sewage system, a black-water system for toilet waste and a grey-water system for

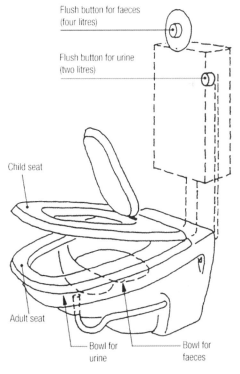

The Dubblett urine-sorting flush toilet.

Water Consumption of Toilets

Water consumption of different models of flush toilets varies greatly. Over the years development has moved in the direction of water efficiency.

Type of toilet	Litres/flush
US (older models)	25
Ceiling-level cistern	12
Seat-level cistern	9
1970s	6
1980s	4
1990s (two-button system)	4 or 2
urine-sorting toilet	4 or 0.2
boat toilet	1
vacuum toilet	0.5
aeroplane toilet (low-vacuum)	0.1

other waste water (dish, shower, sink and laundry). It is common to use vacuum toilets in such systems. Vacuum toilets use 0.5–1.2 litres per flush. Such toilets are common on boats and trains, but are starting to be installed in blocks of flats. One disadvantage of the systems is the loud noise from each flush. Research is under way to make quieter low-vacuum toilets.

On some aeroplanes (e.g. jumbo jets) there are extremely water-efficient toilets. The inside of the toilet bowl is Teflon-coated so that nothing sticks to the sides. When flushed, 0.1 litres of water or oil are released, a low vacuum is created, and 'swish', the contents are carried away to a collection tank. Such methods require a short and steep waste pipe.

Flow Increaser

When the flush volume is reduced in toilets, problems can occur with transporting faeces and toilet paper through sewage pipes. This

In Japan there are toilets that, when flushed, refill the cistern with water from a small sink where hands can be washed. So the water is used twice.

Flow increaser

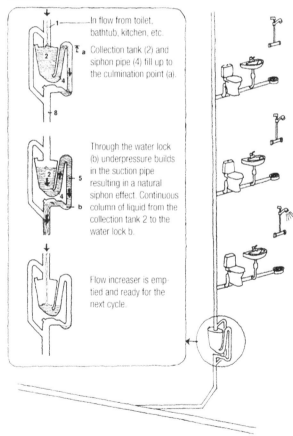

1 — In flow from toilet, bathtub, kitchen, etc.

a — Collection tank (2) and siphon pipe (4) fill up to the culmination point (a).

Through the water lock (b) underpressure builds in the suction pipe resulting in a natural siphon effect. Continuous column of liquid from the collection tank 2 to the water lock b.

Flow increaser is emptied and ready for the next cycle.

If there is a risk that the water flow in sewage pipes will be too slow after installation of water-saving technology, a flow increaser can be installed.

problem can be solved by installing a flow increaser under the toilet on the vertical soil pipe. It consists of a collection tank that collects three or four flushes before emptying. The resulting flush volume is about four times as large as a single flush and the water flow is therefore temporarily large enough to be able to carry the contents through sewage pipes with a gentle slope. Emptying of the collection tank takes place according to the siphon principle. When the container is full, the flush water exceeds a threshold and draws all the water out of the flow increaser.

Water-Efficient Appliances

Some appliances, such as washing machines and dishwashers, use both electricity and water. It turns out that energy-efficient appliances are also water efficient, since the electricity is primarily used to pump and heat water in the machines. (See 2.2.1 Appliances.)

LG Electronics Nordic sells a washing machine that uses both steam (that easily penetrates into textiles) and water in order to reduce water consumption. A built-in scale measures the weight of the wash in order to calculate the exact amount of water that should be used. The steam washing machine has a special energy-saving programme that is 20 minutes long, and that uses steam to flatten wrinkles and remove odours.

According to the Swedish Consumer Agency a family of two adults and two children can save 25,000 litres of water per year by using a dishwasher instead of hand washing. A Swan-labelled dishwasher is energy and water efficient, quiet, should easily be recycled, should not contain environmentally hazardous materials, e.g. that cause cancer or have plastic parts that contain flame retardants harmful to reproduction. The dishwashers should also meet strict requirements for washing and drying efficiency. Some Swan-labelled dishwashers use 30% less water than most others on the market, and are also quieter. The Siemens speedMatic model uses only 10 litres of water per wash. In the early 1990s, dishwashers required between 30 and 40 litres per wash. Whirlpool has introduced an energy-efficient steam dishwasher that utilizes the steam always created by dishwashers.

2.3.2 Hot Water

Of about 200 litres of water used per person per day, about 70 litres are hot. Hot water use varies from household to household. An energy-conscious family uses about 3500kWh/year to heat hot water while a wasteful family uses 6500kWh/year. To reduce use of hot water from taps, short, well-insulated water pipes are needed as well as hot water circulation that functions well. In order to economize on hot water it is more important to meter hot water than cold.

Saving Energy and Hot Water

Use of hot water makes up a major part of energy use in energy-efficient houses. There are several different ways to make hot water use as energy efficient as possible.

Temperature

The temperature of hot water at the supply point should not be below 50°C (due to the risk of legionnaire's disease) or over 65°C (due the risk of scalding). To be 50°C at the supply point, the temperature of the hot water must be higher when it comes out of the boiler. Inside the boiler, where the water may be stationary, the water temperature should not fall below 60°C. If the heat from the boiler heat source has a lower temperature (e.g. heat from a heat pump or solar heat), it should be possible to raise the water temperature to 60°C with the help of an electric heater.

Legionnaire's Disease

Legionnaire's disease (a type of pneumonia) infects via small airborne water droplets (water mist) that enter the lungs when showering and from aeration devices on taps (perlator) and whirlpool baths among other things. The disease is not caused by drinking the bacteria. Most documented cases of legionnaire's disease are related to large facilities where water stood still for a long time in pipes (hospitals, swimming pools and schools). The mortality rate is about 10 per cent and the sickness strikes primarily older people with a poor immune system.

The larger and more complicated the tap-water supply system, the greater the risk of growth of legionnaire's disease bacteria. The bacteria survive best in stationary water at a temperature of about 40°C. At 60°C the bacteria die within about 10 minutes, and at 70°C they die in less than 1 minute. To avoid the risk of legionnaire's disease, the water system should be kept clean, especially the boiler. In addition, the cold water should be really cold.

Pipe Insulation

Hot water pipes should be insulated to a level equivalent to 30mm mineral wool. Suitable insulation materials are cork, cellular glass, EPDM rubber and in some cases expanded rubber. Flax fibre and wool are good insulation materials from an environmental perspective. Increased insulation of hot water pipes in new buildings saves only a small amount of energy. However, it can be worthwhile to improve the insulation in existing buildings. The profitability of doing so depends on the condition of existing insulation and the possibility of efficiently carrying out the work. Insulation of hot water pipes in blocks of flats results in only marginal energy savings. However, it is important that hot water pipes in apartments do not have a larger diameter than necessary. Large diameter pipes result in large losses due to cooling and long waiting periods when a tap is turned on. In blocks of flats, it is preferable to install hot tap-water pipes with a small heat storage capacity, e.g. cross-linked polyethylene.

Hot Water Circulation

In large buildings a hot water circulation system is usually needed in order to avoid long waiting times. At a normal flow of 12 litres/minute, the waiting time should not exceed 20 seconds in a multi storey building, or 40 seconds in a single-family house. A circulation system is made up of a well-insulated water pipe with a smaller diameter than the hot water pipe lying parallel to it, as well as a circulation pump. In offices and other workplaces where there is no activity at night or weekends hot water circulation can be turned off to save energy. Electric on-demand heaters can be installed in places where hot water is rarely used, like caretaker's rooms.

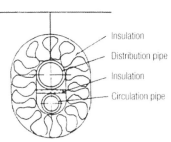

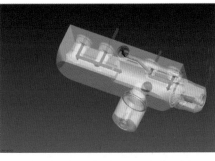

The heart of Zeonda™ Cirkulationsmetod (circulation method) is a mixer that makes it possible for all the water to circulate. A diagrammatic sketch of the Gustavsberg Zeonda™ Germedic shower mixer with recirculation of both cold and hot water and automatic emptying.

Source: Zeonda - Cirkulationsmetod som marknadsförs i Sverige avis marketed in Sweden by AB Gustavsberg AB.

An energy-efficient design for hot water pipes where shared insulation is used for both the tap water and central heating pipes.

Viega Smartloop In Line System, a pipe-in-pipe system for shaft installations with the tap water circulation inside the hot water intake line using polybutylene piping. This results in less use of materials, less work hours, 20–30 per cent less heat loss, a smaller hot water pump, fewer tenders, and less installation.

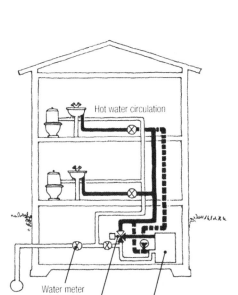

To eliminate the wait for hot water when the hot water is turned on, hot water circulation in a special pipe loop is used. It should be well insulated to reduce heat loss.

Hot Water Boilers

There are three types of boiler: (1) on-demand or instantaneous, (2) storage, and (3) plate heat exchanger. The size of a hot water heater should be determined by the amount of hot water used. A hot water heater for a four-person household should have a capacity of about 200 litres.

Problems with Hot Water Heaters

There are two main problems with many older hot water heaters: large heat losses due to poor insulation and poor durability (they begin to leak). Some older models can lose 1200kWh/year, of which half is completely lost and cannot be made use of for space heating. A new hot water heater should have at least 6cm insulation. One of the most expensive repairs for home owners is fixing a damaged hot water heater. Enamel-coated hot water heaters built into combination boilers account for 95 per cent of all hot water heater breakdowns. Enamel hot water heaters contain a very important galvanic anode. It should be checked regularly and replaced about every third year, since it is meant to corrode instead of the hot water tank. When purchasing a new hot water heater there are a number of things to consider and insist upon. The heater should be able to supply an adequate amount of hot water. It should be well insulated and have low energy consumption. It should have easily accessible controls and connections that simplify operation and maintenance. It should be possible to connect alternative energy sources to it, such as solar collectors, wood boilers and heat pumps. If an enamel-coated metal hot water heater is chosen, it should be easy to access and replace the galvanic anode.

Insulation of Hot Water Heaters

Older hot water heaters are often poorly insulated. If an old hot water heater is replaced with a new one, it is certain to be more energy efficient, even if there are some shortcomings with regard to the insulation. Hot water heaters can be insulated with cellulose fibre or flax fibre. They are usually insulated with mineral wool or urethane foam. The least heat is lost from hot water heaters insulated with polyurethane foam. From an ecological perspective, hot water heaters should be insulated better than is currently the practice and with cellulose fibre. Hot water pipes should be kept short and also be well insulated. About 45kWh is needed to heat $1m^3$ of hot water. Heat losses from a hot water heater and pipe system are about $20kWh/m^3$. Therefore, normally about 65kWh are needed per $1m^3$ water. An important task for researchers is to reduce heat emissions from storage hot water heaters for single-family homes, e.g. by supplementing them with a jacket where hot water is preheated.

Choosing a New Hot Water Heater

Before a new hot water heater is purchased it is important to find out about the quality of the water. This is especially important for people with private wells, who should obtain a water analysis before choosing a hot water heater. Hot water heaters are still flawed in several respects with regards to corrosion. The weak points of hot water heaters are at their welded joints. Three options for hot water heaters according to water quality are: (1) If the water is acidic as well as aggressive, a stainless steel hot water heater should be chosen. Copper hot water heaters are inappropriate

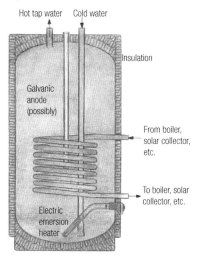

Cross-section of a storage-type hot water heater. The coil in the hot water heater heats the tap water in the tank. In an on-demand hot water heater, e.g. an accumulation tank where hot water is stored, the hot tap water is heated inside the coil instead.

Hot water at a distant supply point can be provided by a separate electric on-demand hot water heater supplied by the cold water pipe. This method eliminates the need to lay both hot and cold water pipes a long distance as well as eliminating the large amount of heat that would be lost from a long hot water pipe.

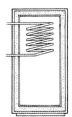  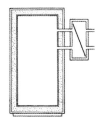

On-demand boiler (left), storage boiler (middle), and a plate heat exchanger (right).

Source: *Värmeboken – 20°C till lägsta kostnad*, Anders Axelsson och Lars Andrén, 2000

as copper is dissoluble. (2) If the water has a high chloride content, an enamelled hot water heater should be chosen. Stainless steel hot water heaters can be sensitive to water that is chalky and contains high levels of chloride. (3) Hot water heaters containing electric immersion heaters are unsuitable for hard water. They can be damaged by lime deposits that build up on the warm immersion heater. To deal with this problem, hot water heaters with electric immersion heaters that aren't in direct contact with the water are available.

On-Demand Hot Water Heaters

On-demand hot water heaters consist of one or two heating coils, usually of finned copper pipe, inside an accumulation tank or boiler. Thermal stratification is created in the tank with this type of hot water heater, but they are not suitable for hard water. There are also on-demand hot water heaters where the water is heated with electricity directly at the supply point, which can be a good approach for single, distant water supply points. Since the heater is attached directly to the cold water line, supply is by only one water pipe. In addition, the hot water arrives very quickly. A disadvantage is the high power requirement, which is usually 3–9kWh for a single-family house.

Storage Hot Water Heaters

A storage hot water heater heats water in a water reservoir inside an accumulation tank or boiler. The hot water capacity is large, but it is important to ensure proper temperature levels to avoid legionnaire's disease. This type of hot water heater is recommended for hard water.

Plate Heat Exchanger Hot Water Heaters

Plate heat exchangers are a relatively new way of heating tap water. They have a good hot water capacity and can be adjusted to suit the water quality. Plate heat exchangers for heating tap water in accumulation tanks provide the best conditions for thermal stratification.

Solar-Heated Hot Water

Enamelled hot water tanks should be avoided in systems with solar heat, because of the high temperatures produced by solar heat. Further if the water contains a high level of chalk then the heat exchanger in the hot water heater can get furred up. Pipes without cooling fins should be chosen.

2.3.3 Water Supply

About half of all our drinking water comes from surface water, a quarter from natural groundwater and a quarter from artificial infiltration. Surface water comes from lakes and rivers. Groundwater is found under the Earth's surface and is accessed via wells. Artificial infiltration means that the amount of groundwater is increased by pumping surface water to reservoirs on top of gravel and letting the water slowly run down (infiltrate) through the ground. The raw material for drinking water is called raw water. Both surface water and groundwater can be sources of good drinking water, but groundwater is preferred as it contains fewer organic substances and fewer bacteria, and is thus easier to purify. It is important to preserve gravel for future drinking water supplies.

Municipal Purified Water

Municipal purified water from groundwater catchments is generally of good quality. In large urban areas, surface water is used as a source of purified water. Before it is delivered to consumers, it requires purification in a water purification plant. The same high standards for purified water apply to both large urban waterworks and small municipal ones.

There are physical, chemical and biological factors that influence water quality. Water should in practice be free from bacteria. It should also have a satisfactory appearance, i.e. have low colour and turbidity values, and be without odour and taste. The levels of iron and manganese must be low in order to avoid spots on laundered clothes. The hardness values, i.e. the quantity of calcium and magnesium, should be low so that the water doesn't fur up pipes. The level of chloride, which gives the water a salt taste, should not be too high. In addition, the level of nitrogen compounds should be low for hygienic reasons. A pH that is too low and a CO_2 level that is too high makes the water corrosive and corrodes water mains.

The biggest drinking water problem is elevated levels of undesirable micro-organisms. When people get sick from drinking water it is often due to a leakage of sewage water into drinking water mains. This can be due to temporary overloads where overflow systems don't work. Old water mains that allow in undesired water can also be a cause. There are also problems with other types of substances, e.g. pharmaceuticals, that are not removed in sewage treatment facilities. The levels of such substances are currently low in drinking water, but it is difficult to estimate the long-term consequences.

From an environmental perspective, which chemicals are used in chemical purification and how the chemical sludge is managed are important issues. From a health perspective, important considerations are whether or not the water is chlorinated, the type of water mains used to distribute the water, and the levels of minerals, purification chemicals and chlorine found in the water.

Water Purification Plants

The poorer the quality of the raw water, the more treatment it requires. Good groundwater, for example, only has to be oxygenated and filtered through gravel and sand, while surface water intended for drinking water usually has to go through a number of different processes.

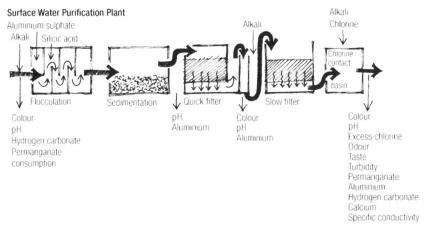

Purification principles for a surface water treatment plant.

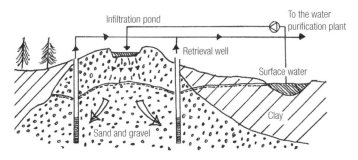

Many water purification plants use artificial infiltration. Surface water is pumped up to a reservoir on top of gravel and filters through the gravel. The gravel purifies the water and the groundwater supply is augmented. The retention time is normally two–three months.

Wells

It is possible to find water-bearing gravel or sand layers between layers of earth or rock, and under clay deposits. Groundwater, as a rule, is not found in clay. The best water is found in gravel or sand layers, especially in boulder ridges. There water is often abundant.

Drilled Wells

The most common way to obtain a private water supply is to drill a deep well through rock. With the highly developed technology currently available, great depths can be reached in a few days. However, there are problems. If the drill hits brittle rock it can get stuck, and primary rock can be difficult to drill in and in some places contains inadequate amounts of water. Porous sand and limestone formations are most suitable for wells, as they are full of cracks and can transport large amounts of water.

How deep it is necessary to drill varies of course from place to place, but as water use needs have increased and the level of the water table has gone down, it has become necessary to drill deeper. Where it once was sufficient to drill 50m, it may currently be necessary to drill to a depth of 70 or 100m. If enough water isn't found at a depth of 100m, it probably isn't worth continuing to drill in the same place. Once a certain depth has been reached, drilling is stopped and water is pumped down the hole under high pressure. This widens the cracks and water begins to flow into the well. There are also methods for detonating explosives in boreholes drilled through rock that have not produced water.

It is important to have a written agreement about how many holes will be drilled and how deep the holes will be. Some drilling companies provide a cost guarantee, i.e. they guarantee that they will find water.

Driven-Point Wells

These days when people talk about 'digging a well' they are usually referring to driven wells, which involves forcing a perforated tube down into the earth to a water-bearing layer. A driven-point well consists of a special drive point well screen and a sturdy casing. This type of well is used primarily when the water-bearing layer is located under a layer of clay. A filter well is a further development of the driven-point well where the drive-point well screen is adapted to the characteristics of the water-bearing layer. In terms of both water quality and quantity, driven wells are comparable to drilled wells, and are often cheaper for the consumer.

If the overburden is thick enough (8–10m) it is often worthwhile to try to make use of the groundwater. A method for ascertaining the depth of the overburden is to carry out a sounding using a sounding rod in the ground, which helps to determine the soil type and depth. Driven wells are relatively cheap compared to drilled wells.

Dug Wells

The old way of making a well is to dig and blast, and then remove rock and earth. This method can unfortunately turn out to be expensive if water isn't found as quickly as expected. A dug well requires water at a depth of not more than 5–6m. If an old dug well doesn't provide enough water, it can be worthwhile to deepen it with a driven well. It depends on how deep the well is and what type of soil is underneath. It is possible to make soundings and investigate whether or not there is water-bearing material under the bottom of the well, i.e. gravel, sand or moraine. If such a layer is found, it is probably worthwhile drilling.

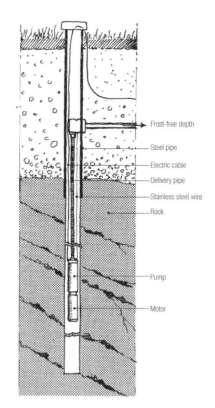

Drilled well with a submersible pump. A drilling depth of greater than 60m is common.

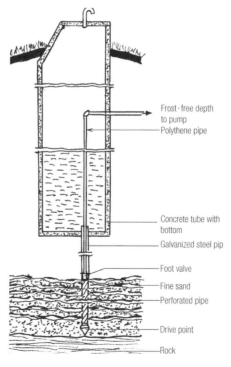

A driven well, i.e. a well made by driving a perforated pipe with a drill point down into the earth, can also be used to deepen an old well that has run dry.

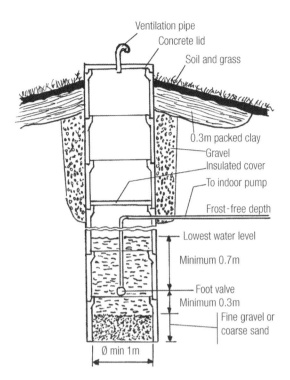

A dug well is made with concrete rings, preferably 90–100cm in diameter and 60cm high. It is important to have a tight seal between the rings and ground that slopes away from the well so that surface water doesn't run into the well.

Threats to Wells

A well's location and construction can be crucial to the quality of the water. So, it is important to know something about how groundwater behaves in the ground and how the groundwater and well can be protected from contaminants.

Sources of Contamination

Sources of contamination include sewage drains, sewage pipes, manure piles, timber and bark piles, arable land, road ditches, spillage from farm tanks, road salt, etc. Chemicals used in gardens and agriculture can also pose a risk. A rule of thumb in this context is that the source of contamination should be at least 50m away.

The further the groundwater is from the surface, the more protected the water is and the better the water quality. In general, deep driven wells usually provide better water than shallow driven wells, and a deep well drilled through rock usually provides better quality water than a driven well.

If rock outcrops near a well, there is a risk that contaminants will penetrate down cracks and get into the well without having been purified along the way. How contaminants reach a well has to do with the direction of the groundwater flow. A rule of thumb is that groundwater moves in the same direction as the ground slopes.

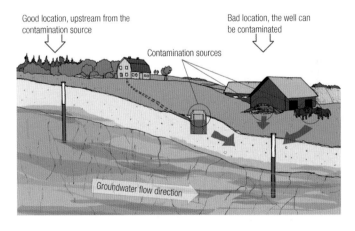

The direction of the groundwater flow is crucial when it comes to a well's contamination risk. A downstream location can imply a large risk even if the source of contamination is far away.

Source: 'Ditt viktigaste livsmedel', Information om brunnar och grundvatten från Sveriges geologiska undersökning (SGU) och Livsmedelsverket. Illustration: Leif Kindgren

Penetrability

To avoid contamination, it is of course important for a well to be impermeable. It is essential to check all places where any form of contamination might get in.

Heat Pumps

Ground-source heat pumps with a vertical borehole are becoming more common. The holes are drilled so deep that there is an increased risk of salt groundwater penetrating up and spreading to drinking water wells. If the water in a drinking water well flows in from a single fracture, there is a risk of increased chloride levels since salt water penetrates when fresh water runs out. A well with fresh water flowing into it from several fractures has a reduced risk of salt water penetration since the fresh water flows in from several directions. Salt groundwater is mostly found near coasts and in areas that were once covered by sea water. The risk can be eliminated by filling the heat pump holes with a sealant such as bentonite clay.

Roads

Road salt runs as a brine beside roads and seeps downwards to groundwater reservoirs. The result is an increase in the chloride level of the groundwater. Drinking water with elevated chloride levels can corrode and rust water mains, washing machines and dishwashers. When road salting stops, the chloride level decreases slowly. Road salt contamination in a drinking water well is an indication that other more serious contaminants are getting in. Along strips of road that pass over groundwater reservoirs, drainage systems for carriageway runoff can be built that direct the salt water via watertight ditches to other, less sensitive areas.

Private Water

If water is obtained from a private water source, more than just a well is required. A whole system is necessary, with a pump, pressure tank and purification system. Pumps are either placed down in the well (a pressure pump), which is common for deep wells, or in a pump house (suction pumps) if the well is shallow.

Pumps

These days, electric pumps are most common. They are both cheap and practical if electricity is available. Many holiday home wells and wells used for grazing animals don't have access to electricity. In such situations it is often possible to operate a pump with solar energy, wind power or water power.

If a well is deep, the pump must be placed down in the well (pressure pump). If the well is shallow, a suction pump can be

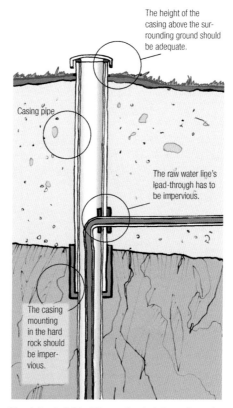

Check the height of the casing in tube wells and ensure that the raw water pipe penetration and casing mounting are watertight.

Source: 'Ditt viktigaste livsmedel', Information om brunnar och grundvatten från Sveriges geologiska undersökning (SGU) och Livsmedelsverket

used (they can lift water 7–15m). There are also systems where the pump is in a pump house, usually for medium-deep wells, where air is pumped down into the well and water is forced up.

Pressure Tanks

When clean water pipes are installed in a building, in order to have running water, pressure is required. The force of gravity can be used to provide running water by installing a water tower or tank in the loft. These days, a pressure tank (hydrophore) is usually used and the same pump that pumps water up from the well is also used to create pressure in the tank. The slower the water flows, the bigger the reservoir that is required. Slow flow can be dealt with by installing either a large hydrophore or two parallel connected hydrophores.

Hydrophores must be placed in frost-free locations such as a cellars, suspended foundations or pump houses. Two factors require consideration: there is noise from the pump and the hydrophore gets so cold from the cold clean water that there can be a lot of condensation which can give rise to damp problems. So it is necessary to eliminate the risk of damage due to damp.

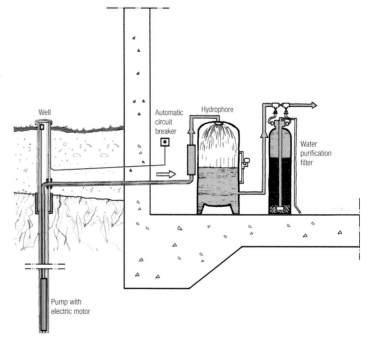

The water in wells drilled through rock is usually good. There can, however, be radon in the water or an unsuitable chemical composition. When a radon filter is required, other filters usually become necessary as well. The oxygen that radon removal adds can change the chemical composition of the water. The filters should be placed indoors so that they can be checked and maintained.

Source: illustration adapted from Hus&Hem 1998

Water filter (blue cylinder) for purification of fresh water and a hydrophore (red in colour).

DIRECT DRIVE PUMPS

Wind Pumps

Using the wind to power pumps is an old technology that is back in favour again. With modern technology, pumps are now made simpler, cheaper and more durable than they used to be. In particular, much less maintenance is now required. There are three types of wind pumps. The model selected depends on how far under the surface the water is located: there are pumps to pump water less than 7m, pumps that pump about 25m, and pumps for deep drill holes 60–100m deep. In Denmark, there are several producers of the two smaller types. Good wind pumps for deep holes are made in the US, Australia, South Africa and Kenya.

Wind pumps require considerably less wind speed than wind turbines. A speed of 2.5–3.5m/s is adequate. One limitation of wind pumps is that they must be placed directly above the well, which requires quite open terrain. So, a wind pump can't be sited up on a hill if the well is down in a valley. Wind pumps are used primarily to pump drinking water for people and animals, as well as to circulate water in purification facilities.

Hand Pumps

Hand pumps were common in the past and are still available. A hand pump can be placed over a well or indoors over a kitchen sink. There are several models available. Very durable hand pumps have been developed for Third World applications (by the World Bank and the United Nations Development Programme, UNDP) which are also suitable for use in industrialized countries.

Solar Pumps

Electric pumps can be powered by solar cells. Such systems don't require batteries. The pumps run when the sun shines and store water in cisterns or water towers. Battery-run pumps where the batteries are charged by solar cells are also common, especially for summer cottages and boats. In these cases, it is a question of small amounts of water in places where electricity is unavailable.

Modern wind pump in Kenya, the Kijito pump, with a latticework tower, direct drive without transmission box and self-lubricating bearings.

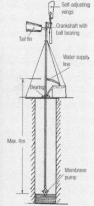

Wind pumps are usually divided into different categories depending on how far the water needs to be raised. The illustration shows a small wind pump for short lifts.

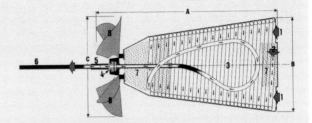

1. Bottom cover with screened openings. 2. Rear end of the hose/water intake with filter. 3. Spiral wound pump hose. 4. Swivel coupling. 5. Hose connection for feeder hose and attachment for the anchor wire. 6. Feeder hose for outgoing water. 7. Flotation device. 8. Propeller blade.

Source: JTM Invest AB, Jukkasjärvi

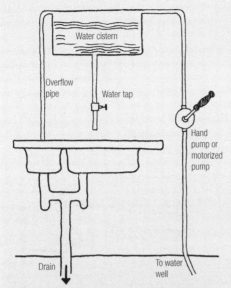

With a small cistern and a hand pump it is simple to have running water. This type of system is good to have in case of power cuts or in buildings where electricity is produced by solar cells and electricity is only used when absolutely necessary.

Sling Pump

A sling pump is a mechanical water pump driven by running water. The pump is anchored in a watercourse. The water speed in the watercourse must be at least 0.5m/s. The construction is designed so that the whole pump rotates in the water. This is accomplished with the help of a propeller at one end of the cone-shaped pump body. A polyethylene hose is wrapped around the inside surface of the pump body in a spiral. The hose is open on the downstream end and connected to a swivel coupling on the upstream end. With each revolution of the pump, water is taken in and is pushed forward and out a feeder hose by the rotating movement. The pump can manage a pressure head of 25m and has proven to be durable.

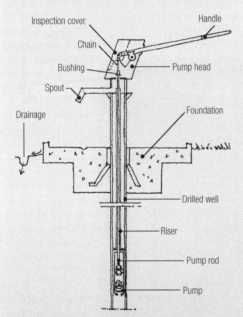

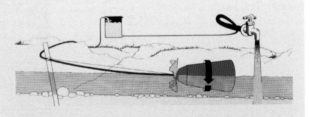

A sling pump is anchored with a cable fastened to a pipe stuck in the bed of a watercourse. The flow of the water makes the pump rotate and the water is forced through the pump via a swivel coupling into the feed hose and to an intermediate storage tank or directly to the point of use.

Source: JTM Invest AB, Jukkasjärvi

Hand pumps are being manufactured in many parts of the world. The India Mark II pump is a very durable hand pump that was developed by the UNDP and World Bank for use in developing countries. It can be used and maintained by people with no technical training. Such pumps would also be suitable for use in industrialized countries.

Hydraulic Ram

A hydraulic ram pumps water using the power of the moving water itself. The technology is an ancient one and requires neither electricity nor fuel. A hydraulic ram consists of a large air chamber, a spring operated valve and two check valves. If there is flowing water in a valley and water is to be raised to a building located at a higher point, a hydraulic ram is placed in the flowing water and a pipe is run from the ram up to the building. The ram stays in the stream and ticks pleasantly away, day in and day out, and with every tick, a little water is pumped up the pipe. Hydraulic rams are very durable. The only wear that occurs is that the spring becomes weaker and must be replaced after several years.

Solar cells can be used to run water pumps. An advantage of this technology is that batteries are not needed. The water is pumped when the sun shines and is stored in a reservoir.

A hydraulic ram, a pump that is driven by the power of moving water.

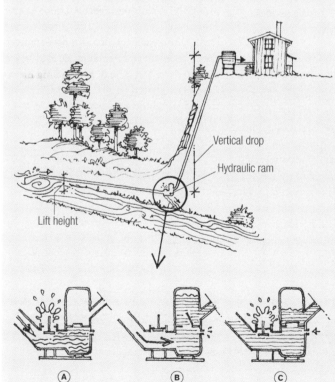

A hydraulic ram is a water-powered pump that harnesses the energy of moving water to raise water from creeks and streams. A hydraulic ram consists of an air chamber and two valves, one of which is spring operated.

(A) Water flows into the hydraulic ram and out through the spring-operated valve until the water pressure becomes greater than the spring pressure, whereby the valve closes.

(B) A check valve then opens and the water flows up into the air chamber. The air in the chamber is compressed and a recoil effect occurs.

(C) The recoil forces water up the water line. The pressure in the air chamber then falls, the valve in the air chamber closes and the spring operated valve opens once again, and the process begins anew.

A hydraulic ram pumps water year in and year out with a pleasant ticking sound until the spring wears out and needs to be replaced.

Water Quality

The water quality in private wells may be periodically poor; this is particularly common in summer cottage wells in the spring.

The Environmental Situation

Pollution in lakes and watercourses is widespread. The main problems are acidification and excessively high mercury levels in fish, as well as eutrophication where surface water is over-fertilized, which leads to abundant algal blooms, lack of oxygen and increased plant life. There are also problems with organic environmental contaminants, such as DDT, PCB, dioxins, etc., which do not break down naturally. Many are fat-soluble and therefore concentrate in living organisms. Heavy metals that leach from mines and metal industries are another problem. Outside older forestry operations there are fibre embankments that often contain mercury. Poor quality groundwater may be found in regions where acidification and nitrogen leakage from agriculture affect the groundwater with nitrates, etc.

Acidification

Sulphur and nitrogen fallout during the last half of the 1990s has resulted in radically acidified groundwater in some regions. Shallow water is most acidified, which primarily impacts private water supplies from dug wells, but even larger groundwater reservoirs for municipal water supplies have been affected. Reduction in pH causes corrosion of water lines and release of aluminium, which can have consequences for health.

Nitrogen Leakage

Use of nitrogenous inorganic fertilizers in agriculture has increased dramatically since the 1950s. Nitrates from fertilizers leak into groundwater. This is a slow process, so current nitrate levels do not yet fully reflect the consequences of modern agriculture.

Salt water

Penetration of salt water into wells is primarily a problem for wells drilled into rock in coastal areas. The problem of salt water in wells can be caused by improper extraction of water, e.g. withdrawal of too much fresh water and wells that are too deep.

Fluoride

Some substances that occur naturally in groundwater can be present in levels that are unacceptable for drinking water. Fluoride is one such substance. Elevated levels are often found in groundwater and prevent it from being used for drinking.

Radioactivity

All radioactive substances are included in the total ionizing dose (TID). It is primarily uranium and radium that contribute to increased TID. Uranium can cause kidney damage. Radium is a uranium decay product and radon is formed when radium decays. Radon decays into the radon daughters. Radon levels >100Bq/L are considered suitable with risk and radon levels >1000 Bq/L are considered unsuitable.

Arsenic

In several areas of Sweden there are high levels of the carcinogen arsenic in water from bored wells. Geologists have found levels over the threshold limits for potable water (10 micrograms/litre) in Skellefteå Field, Västernorrland, as well as in the city of Södertälje. Arsenic may also be found in other areas.

Boron

Many wells on the island off Gotland exceed WHO's recommended acceptable level of 500 micrograms/litre.

2.3.4 Water Purification

Very little water is suitable to drink just the way it is. Most water requires some kind of treatment. Water pollutants can be divided into two categories, those that affect health and those that affect the water in other ways, e.g. odour, taste, appearance or hardness (pH value). As water quality varies greatly from one source to another, purification methods must be adapted to local conditions. So, water must be analysed before a suitable purification method can be determined.

Purification Methods

There are four main methods for purifying water: filtration, reverse osmosis, distillation and ultraviolet (UV) light. Filtration is the most common method. There are many varieties of filters. Availability of increasingly fine filters makes it is possible to remove more and more undesirable substances. Different filters are often combined in a purification process. In principle, everything but viruses can be removed. Reverse osmosis removes minerals and chemicals, but can't remove flavours and micro-organisms. Reverse osmosis is primarily used to purify salt water. Distillation removes everything but volatile chemicals. Distillation is energy-intensive and provides tasteless water without salts. UV light kills micro-organisms. UV filters are often used as a last stage in purification to kill any remaining micro-organisms. Acidic water can be neutralized by passing it through a reservoir tank with lime in it. Aerating water can remove sulphur odours and radon gas.

Unsuitable Contents

Pollutants that are health hazards are categorized into five groups:

1 micro-organisms (bacteria, viruses and parasites);

2 poisonous metals and minerals (e.g. heavy metals, nitrates and asbestos fibres);

3 organic pollutants (stable organic, bio-accumulative, chemical substances, e.g. pesticides and herbicides);

4 radioactive particles or gas (e.g. radon); and

5 additives used in the water purification process (e.g. chlorine, fluoride and chemical precipitates).

Brown water can be caused by organic substances (dirt particles) or iron. Odour can be caused by sulphur compounds, e.g. a rotten egg odour can be caused by hydrogen sulphide in groundwater. Taste can be affected by surface water or metals, e.g. a high level of iron in groundwater. The pH level is important. Acidic water corrodes water mains and water supply and sewage systems, and lowering the pH can result in the release of minerals from soil, rock and pipes. High pH values do not affect health.

Filtering

Filtration often begins with a sediment filter (a sieve) that catches particles. The finer the filter, the smaller the particles removed. The next step is a carbon filter, which removes chemical substances, additives, taste and odour substances, and radon. Membrane or ceramic filters remove micro-organisms. They are sometimes called bacterial filters. Redox filters (reduction-oxidation) remove metals and minerals. Most filters eventually get dirty and must be cleaned or replaced.

Nanofilters, which filter out micro-organisms, can be used instead of chlorine to disinfect drinking water. After going through conventional purification stages, water is filtered through a membrane that is pierced

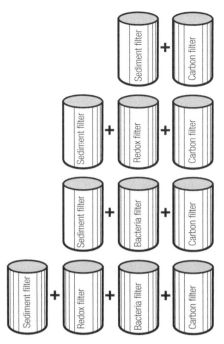

Different filters can be combined in various ways depending on purification requirements.

Source: *The Drinking Water Book*, Colin Ingram, 1991

with microscopic holes with a diameter of 10 Ångström, i.e. 1 nanometre. The procedure removes bacteria and viruses, and other organic materials; so avoiding the addition of chlorine to disinfect water. Water is brought into the system under high pressure and is passed through a number of filters so that it is completely clean. A disadvantage is that a lot of energy is needed to force the water through the membrane.

UV Light

All micro-organisms, bacteria and fungi die when exposed to sufficiently strong ultraviolet light, and it is thus possible to achieve complete disinfection. However, any particles in the water must first be filtered out since they block UV light. Chlorine and UV radiation are competitors when it comes to drinking water purification. The advantage of UV light is that no chemical additives are needed. One disadvantage compared to chlorine is that there is no long-lasting effect, so a UV system must be located near the point of use.

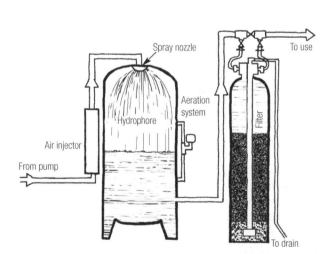

Private water supplies may require purification, e.g. with a filter. The type of filter depends of course on the nature of pollutants in the water. The illustration shows a system that purifies using circulation and aeration to remove iron and manganese, and to improve smell, taste and colour.

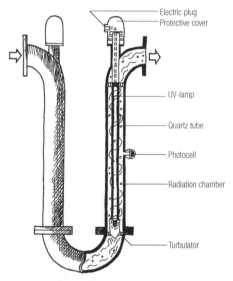

UV filter. A device that forces the water to swirl (turbulator) is used to push the water into a radiation chamber past UV lamps. A photocell senses if the water lets through too little light or if the UV lamps are too dirty to carry out adequate disinfection.

Source: *Ny Teknik*, no 40, 1987

Radon

A country may set maximum allowable levels of radon in drinking water, for adults and primarily to protect children. These levels may apply to both municipal and private well water. High levels of radon in water are found most often in wells drilled through rock, in cold springs and in dug wells where the water comes from fissures in rock. However, it is almost exclusively in water from wells drilled through rock that radon levels are problematic. Radon can be removed from water through aeration.

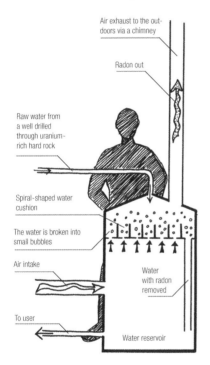

Radon can be removed from drinking water using aeration. If aeration equipment is not available, the water can either be boiled or stirred vigorously for at least three minutes.

Ozone

There are systems that use relatively low voltage to produce ozone directly from the air. The system is simple, uses a small amount of energy and requires little maintenance. The extracted ozone molecule is made of three oxygen atoms and next to fluorine is the strongest oxidizing agent. It oxidizes organic and inorganic poisonous compounds and destroys viruses, bacteria, fungi and micro-organic parasites. During the process, the ozone itself is reduced to natural oxygen.

Research into purification methods that combine UV light and ozone is under way. Most toxic substances seem to be broken down by ozone and UV light.

Membrane Filtering

Membrane filtering involves using a semipermeable membrane to separate solutions into different components. The liquid passes over the membrane and is divided into two parts, one part that passes through the membrane and another part that is a concentrate containing the contaminants. Which substances and how much passes through depends on the characteristics of the membrane, such as the type of material and its density. In addition, the chemical potential difference of the membrane is important, and this can be achieved in different ways. There can be a concentration difference, a pressure or temperature difference, or an electric field. Reverse osmosis and nanofiltering use added pressure. Membrane distillation uses a temperature difference.

Reverse osmosis is mainly used to desalinate sea water. The reverse osmosis process uses a membrane that only allows water molecules through. The water to be purified is put under pressure and the very small water molecules are pushed through the membrane. Sodium and chloride ions are removed by the membrane. Ultrafiltering takes place. The method also removes metal ions, organic substances, bacteria and sludge. The only substance that isn't more than 97 per cent removed is radon. Reverse osmosis is an expensive way to purify water and it is also difficult to achieve a sufficient capacity to purify all water. However, the 20–30 litres per day needed for drinking and cooking can easily be produced. Osmosis purifiers require daily rinsing and cleaning.

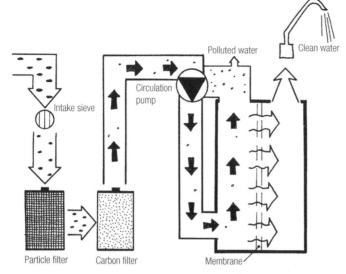

Purification of saline water is considered energy-intensive and expensive. Electrolux has developed a small-scale purification apparatus (RO 400) for saline water that uses osmosis technology. There are three pre-filters to protect the membrane and a circulation pump that recirculates the water at a high speed across the surface of the membrane. The pollutants are automatically flushed away.

Water at 95°C flows in one channel, and cold water in the other. Between them is a Gore-tex membrane. The hot water releases steam that goes through the membrane; however, water molecules cannot get through. It doesn't matter how polluted the raw water is.

In-Situ Treatment

Vyrmetoder Ltd has developed an interesting water treatment method. The Vyredox process removes iron and manganese from water while the water is still in the ground. This type of purification system consists of several satellite wells placed around each pumped well. Degassed and oxygen-enriched water is injected into the satellite wells. A 'natural' oxidation zone is created in the aquifer. Iron and manganese oxidize and remain in the overburden, and only purified water free from these minerals reaches the main well. The nitredox process is a further development that also removes nitrate and nitrite. In the nitredox process, water containing some form of carbon nutrient is injected into an outer ring of satellite wells. Satellite wells in the inner ring push away nitrogen gas as well as oxidize iron and manganese. Purified water is then pumped out of the central well.

Osmosis purifiers are so efficient that they remove the natural minerals in water. This means that if water purified by osmosis is used for drinking, it is important to eat a well-balanced diet or to take nutritional supplements.

Membrane distillation is a method that requires two channels and one membrane.

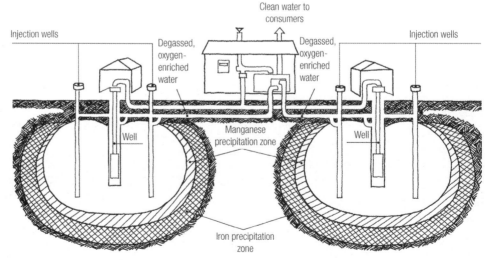

Vyredox is a method where iron and manganese are removed while the water is still in the ground. This is accomplished by pumping oxygen-enriched water down into the ground around the well, which results in the oxidation of mineral ions in a precipitation zone.

The photo shows an ecocycle house in Borlänge, Sweden.

Source: Architect Bertil Thermaenius in cooperation with Professor Nils Tiberg

2.4 *Waste*

Our society produces too much waste. It is one symptom of our faulty production and consumption patterns. The waste causes environmental problems, takes up space and results in high costs. Actually there is no such thing as waste. Rubbish is material in the wrong place at the wrong time.

For construction material waste see 1.4 Implementation.

2.4.0 Waste from Human Activity

Waste can divided up into the categories of organic, inorganic, liquid and airborne (i.e. molecular waste that cannot be seen with the naked eye). Waste doesn't disappear when it is flushed down a drain. Sooner or later it emerges in lakes or oceans. Waste doesn't disappear if it is incinerated either. It becomes airborne waste that falls to the ground or into water (or remains in the ashes).

- Environmentally hazardous waste
- Sewage sludge
- Household waste
- Industrial waste
- Mine waste

The total waste in Sweden by origin.

Source: Swedish Environmental Protection Agency

Waste Management Today

Some countries have recently banned the disposal of combustible rubbish or organic waste in landfills. Household rubbish may be incinerated for energy use, or recycled. Recycled rubbish may include organic waste composted or treated in biogas plants. So only a small percentage of rubbish needs to be disposed of in rubbish dumps. Hazardous waste must be taken care of separately.

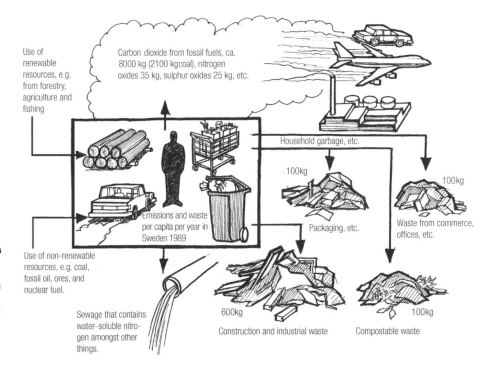

Volume of resources consumed in Sweden from resource to waste, per person per year. The data is from 1989 but is still of interest.

Source: Professor Nils Tiberg, LUTH

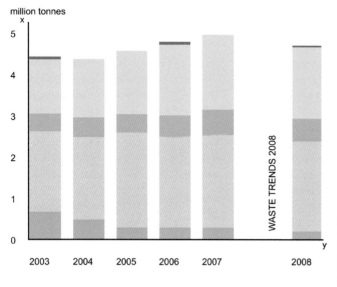

WASTE TRENDS 2008

Total amount of treated household waste	Tonnes	%
Hazardous waste	40,890	0.9
Material recycling	1,737,720	36.8
Biological treatment	561,300	11.9
Incineration, with energy extraction	2,190,980	46.4
Deposition	186,490	4.0
Total amount treated	4,717,390	100.0

Recycling continues to increase and 48.7 per cent of household waste is recycled, including treatment of organic materials. One of Sweden's environmental goals is that at least 50 per cent of household waste will be recycled, including treatment of organic materials, by the year 2010. If all of Europe would incinerate as much garbage as is done in Sweden and Denmark, that could supply about 20 per cent of the energy needed and result in 20 per cent less carbon dioxide emissions. But it might even be better to recycle more and burn less, as they do in Germany, the Netherlands and Belgium.

Source: Swedish Waste Management 2008 (www.avfallsverige.se)

Waste Incineration

The great challenge we are faced with is to make our use of materials and management of waste sustainable.

When we sort, compost and reuse our waste, a combustible fraction will still remain that will require incineration in appropriate facilities. Therefore, rubbish should be sorted into combustible and non-combustible fractions. Waste incineration causes air pollution. Emissions to the air contain acidifying substances, heavy metals and dioxins. In order to reduce the quantity of these pollutants, it is important to have good incineration facilities and flue-gas filtration. In new incineration plants, the use of fluidized beds is becoming more common. This technology combusts waste while it is suspended, resulting in a more complete

The waste incineration plant in Vienna. The artist Friedrich Hundertwasser was one of those responsible for the design. The flue gas filter on the chimney is painted in gold as this is the most important part of the process.

combustion and a smaller quantity of ash. As good combustion is hard to achieve when incinerating mixed waste, dioxins are sometimes created. It is preferable to design incineration plants as district-heating power plants to produce both electricity and district heat.

Some of the dioxins are the most dangerous environmental toxins of all the chlorinated hydrocarbons. They can damage the gene pool and cause cancer in animals and presumably in humans. Dioxin emissions have decreased radically in recent years to one-hundredth of the 1980 level. Dioxins may be released by the pulp industry, vehicle traffic, waste incineration heating plants, and the metal industry.

The company Götaverken Miljö has specialized in separation of environmentally hazardous pollutants from flue gases as well as energy recovery. Their patented method, called ADIOX, uses plastic tower packings (6cm in diameter and 3cm in height) that bind dioxins. Once enough dioxins are accumulated the tower packings are combusted in a controlled manner so that the dioxins are

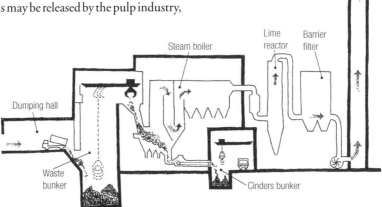

Simplified description of a waste incineration plant. One problem with waste incineration is the smoke emissions which occur despite flue-gas cleaning. When people's rubbish sorting practices improve, only the combustible fraction will be incinerated, which will produce cleaner smoke.

Annual Emissions to the Air from Waste Incineration in Sweden

Substance	1985	1991	1996	Change 1985–1996 as percentage
Flue dust	420 tonnes	45 tonnes	33 tonnes	–92
Hydrogen chloride	8400 tonnes	410 tonnes	412 tonnes	–95
Sulphur oxides	3400 tonnes	700 tonnes	1121 tonnes	–67
Nitrogen oxides	3400 tonnes	3200 tonnes	1463 tonnes	–57
Mercury	3300 tonnes	170 tonnes	77 tonnes	–98
Cadmium	400 tonnes	35 tonnes	8 tonnes	–98
Lead	25,000 tonnes	720 tonnes	214 tonnes	–99
Dioxins	90 grams	8 grams	2 grams	–98

Source: SVEBIO 7/98

destroyed and thus do not enter into ecological cycles.

At the company Cementa's facility in the town of Slite great accomplishments have been made in environmental efficiency and they are the best in Europe at cleaning flue gases. The efficient purification plants operate at high temperatures that allow combustion of wastes that are difficult to deal with, such as tyres, dried digested sludge, solvents and paint, and meat and bone meal.

Landfills

Landfill sites are on their way to being phased out. They take up space, leach out hazardous substances, they smell and attract large birds and rats. There are two strategies for solving the problem of large, smelly landfills. One is to try to seal them off so they don't leak, purify the leach water, cover them, etc. The other strategy is to try to decrease the amount of waste deposited in landfills.

They occupy increasingly larger surface areas and it is difficult for many municipalities to find locations for new landfill sites. However, the greatest environmental problem is leaching. When rain falls on landfills, hazardous substances leach out and pollute the environment. Practically all landfills contain slowly leaching contaminants, primarily heavy metals. At modern landfills there is an attempt to collect leach water, which is purified, and the remaining sludge is dumped back in the landfill.

Sooner or later the environmentally hazardous substances will spread in the ecosystem. Instead of putting mixed waste in landfills, dangerous waste must be removed by minimizing use of such materials, isolating them and reusing them, or enclosing or destroying them. In many landfills a decomposition process takes place that releases landfill gas (methane gas). Pipes may be placed in landfills to collect the gas and use it as an energy source.

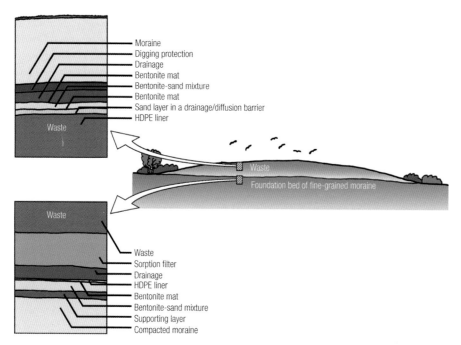

In Swedish landfills there is an attempt to limit leaching by using different layers.

Source: SAKAB

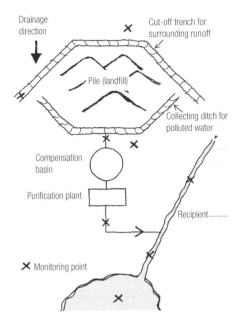

Outline of the basic arrangement of a landfill. A problem with landfills is the leach water that is created. It can contain large amounts of pollutants and must be purified. After purification, the sludge is returned to the landfill. Better rubbish sorting would result in reduction of environmentally hazardous waste reaching the landfill.

Producer Responsibility

In order to reduce the quantity of waste that ends up in landfills, producer responsibility for waste has been introduced. This means that the producers of certain types of products have the responsibility for ensuring that their product does not end up in a landfill but is reused instead. The cost for this is added to the product in question. So far producer responsibility applies to packaging, tyres, recyclable paper, cars, and electrical and electronic products. Product responsibility is an area which is developing all the time. The intensification of measures for recycling plastic and aluminium is desirable. The long-term goal of product responsibility is for it to lead to more environmentally friendly product development.

Tomorrow's Waste Management

The solution to the problem of waste management is conservation of resources, responsible production, environmentally conscious consumption, sorting rubbish at source, composting, reusing, recycling and energy extraction. To encourage such development calls for knowledgeable consumers and responsible producers combined with political initiative, laws and environmental fees.

Strategy

Right from the start in the production phase, it is important to prioritize useful products made from high-quality ecological materials that can be maintained, repaired and renovated. Of course, production should take place in an environmentally friendly manner and product wrapping and packaging should be well thought through. Transport distances should be kept as short as possible.

Conscious consumption involves being able to choose the right products and perhaps most importantly to not buy unnecessary items. Here, environmental labelling and comprehensive content declarations play an important role. A good approach is, 'Don't buy what you need, only buy what you can't do without.'

In order to solve the problem of increasing quantities of rubbish, it is good to have an overview of the materials concerned and how they can be managed. The materials can be categorized as follows: dry products which can be reused (as products), recycled (as materials), incinerated, sent to landfill or destroyed as environmentally hazardous waste; wet products including liquids such as oils, solvents and grease which can be recycled, and sewage which can be sent to treatment plants; organic material which can be composted and made into new soil; sewage sludge which can be used in the composting process; and cinders and ashes from waste incineration which can be sent to landfill.

In order to increase rubbish sorting and reuse, some municipalities have introduced higher fees for unsorted waste. The aim is to increase people's motivation to sort their waste. Several waste management companies have differentiated waste charges. Many construction projects sort their waste on site, and so cut down significantly on waste charges.

2.4

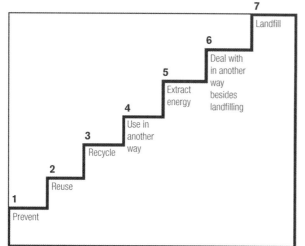

The most important measure is to prevent the production of waste. It is important to try to stay as close to the foot of this staircase as possible. The figure shows the Töpfer scale, introduced by one of Germany's Ministers for the Environment.

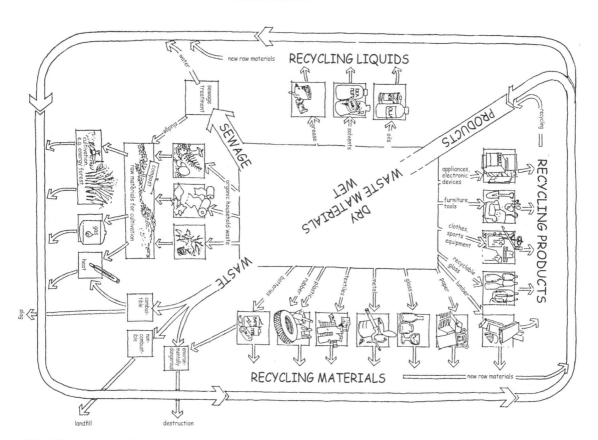

All the different waste fractions, wet as well as dry.

Source: *Tänk om – hjälpmedel för kommunal restmaterialhantering*, Birgitta Jerer, Olof Stenlund och Conny Jerer

2.4.1 Waste Sorting

Sorting waste requires carefully considered systems for homes and workplaces, as well as refuse storage rooms and recycling centres. The system must fit in with the municipal waste management system. Since it is important to have enough space in order to manage waste in a practical manner, architecture and planning are affected. How waste management is worked out differs of course from place to place.

Household Waste

In 2007 within the EU, an average of 522kg of household waste was generated per person. The amount ranged from 294kg in the Czech Republic to 800kg in Denmark. The amount in Sweden was about 515kg per person, just under the average.

There have been major changes in waste management in Sweden since the 1990s. From 1998 to 2007 the amount of household waste increased 24 per cent to a total of 4.7 million tonnes. Recycling, including biological treatment, increased from 35 to 49 per cent. Combustion of waste with energy extraction increased from 38 to 46 per cent. In 1998, one million tonnes of household waste was dumped. The amount dropped to 0.2 million tonnes in 2007. Landfilling continues to go down and was at 4 per cent in 2008.

Waste is a resource. About half of the household waste in Sweden is combusted. Combustion converts the waste into energy. In Sweden both heat and waste are produced from combustion. The separated food waste from a household of four can produce enough biogas to drive a car 7.2km per week.

Waste Statistics

Detailed studies have been done of the contents of household waste and these serve as the basis for the organization of waste sorting. The paper fraction is greater in large urban areas due to thick daily newspapers and the quantity of advertising.

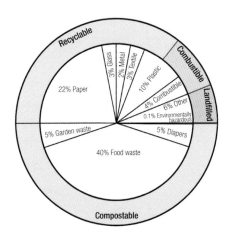

The outer ring shows the fractions that rubbish should be sorted into: compostable, recyclable, combustible, and material to be sent to landfill or destroyed. The inside ring shows statistics (per cent of weight) for household waste.

Compostable waste constitutes half of some countries' household waste. Composting should take place as locally as possible to avoid unnecessary transportation. Nappies account for 5 per cent of compostable waste. Unfortunately they still contain plastics that do not break down easily so it is problematic. Product development is required. An alternative to composting is rotting in biogas facilities which results in lower nitrogen emissions.

Recyclable waste accounts for about one-third of what we want to throw away. Paper (22 per cent), glass (3 per cent), metal (2 per cent), textiles (3 per cent) and plastic

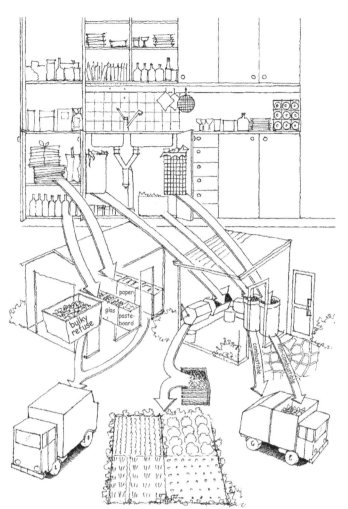

A waste-sorting system is based on a chain where all the links are present and functioning: space in flats for sorting and a place for collecting recyclable paper and glass; a waste storage room in the building for storing compostables and other waste; local waste-sorting stations equipped with suitable containers; space for maturing compost and land to use the composted material on; and established organizations for waste management and reuse.

(10 per cent) can be separated out for recycling. Plastics are a problem. They should be sorted and recycled but rubbish contains so many different kinds of plastic with many different characteristics that in practice recycling them does not work well.

There are hard plastics and soft plastics, ecologically harmful plastics and plastics with a low environmental impact. Cardboard packaging is often difficult to recycle because it may have a plastic or aluminium coating and because packages may not be completely empty. It is especially important to recycle aluminium as it is so energy-intensive to produce. Recycling glass, steel and corrugated cardboard works relatively well. Glass is separated into coloured and clear. Local collection points need to be aesthetically pleasing or at least not ugly.

Combustible waste can be used for energy recovery. Wood makes up 4 per cent of combustible waste. As long as the recycling of plastic and cardboard packaging doesn't work very well, a portion of these fractions can also be incinerated. Considering the large expansion of waste incineration facilities, the combustible fraction is extra-important since sorted combustible waste produces much cleaner flue gases.

In Hammarby Sjöstad, Stockholm, there is an automated waste collection system from Envac. It consists of a fully enclosed vacuum system, doing away with foul-smelling, dirty refuse collection rooms and containers in the streets. No one needs to come into contact with refuse sacks or containers. Sorting at source is handled by using one inlet for each fraction: organic waste, paper and the rest.

Source: Envac

Landfills are the final destination for waste that nothing else can be done with. About 6 per cent of total waste still goes to landfill, but much is being done to reduce this figure.

Environmentally hazardous waste must be handled carefully. The portion of household waste that is environmentally hazardous waste has dropped to 0.1 per cent. However little, since it is hazardous, it is important that it be dealt with and recycled or destroyed.

Levels of Sorting

It is important to sort waste as close to the source as possible. By establishing logistics at different levels, waste sorting can become an automatic part of everyday life.

In Homes

There are under-the-sink units on the market designed for kitchen waste sorting. These primarily have space for two fractions: compostable and other. There are also units for three fractions: compostable, combustible and other. Some of these units have a hatch on the countertop which serves as a waste drop. There are also inserts available for kitchen sinks that act as a sieve to collect peelings from root vegetables, fruit, etc.

Special cupboards for sorting recyclable materials are also on the market. There are closet-type cabinets, under-the-sink kitchen cupboards and other specially designed furniture.

There is not enough room under the kitchen sink for all recyclable material. Therefore, homes should have a place for collecting paper, metal, glass, plastic and returnable packaging.

Source: Leif Kindgren

Sorting often starts in the kitchen under the sink, so that is where suitable receptacles should be placed. In the illustration, there is a container for compostable waste on the left door. It has a perforated cover which can, when required, be placed in the sink to collect fruit and vegetable peelings. On the right side there are two containers for other waste, one for combustibles and the other for non-combustibles.

Source: adapted from Gun och Jan Hallbergs Skulmodul

In Residential Areas

Many waste-sorting stations have been considered ugly and untidy. An alternative is to locate waste-sorting areas in blocks of flats or to place containers for recycling in small buildings specially designed as recycling stations. In this way, ugly containers are removed from public areas and placed in semi-public areas.

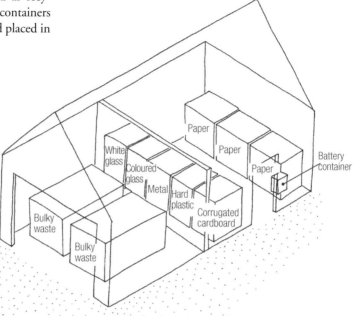

To make it practical to recycle waste, environmental stations equipped with containers for waste sorting are built in residential areas. There is one room for paper, glass, cardboard, etc., and one room for bulky waste. All the containers are on wheels and the building is built so it is easy to pull the containers to the collection trucks.

A waste-sorting building constructed for the housing exhibition in Staffanstorp in 1997. The building has a section with room for 11 large containers and a room for warm compost. It is 48m² in size and designed for use by 60 households (46 flats and 14 single-family homes). The building was the result of a student competition in the Architecture Department, Lund Institute of Technology. It was designed by Maria Dagås and developed together with White Architects and Svenska Landskap in Malmö.

Recycling Centres

In every municipality there should be a number of recycling centres where hazardous and bulky waste can be left. The recycling centres should be staffed, have ample business hours, and be able to receive fractions such as garden waste, concrete, glass, wood, electronics, metals, bulky waste, cardboard and corrugated cardboard and environmentally hazardous waste.

Second-Hand

Second-hand centres can be located in residential areas and organized so that everything is free, is traded, or is sold in a second-hand store. Second-hand shops and barter websites like Freecycle are popular and becoming more common. Trading centres often specialize in particular products, e.g. ski equipment or items for small children. Another possibility is to arrange an exchange room in a residential area where gadgets and clothes that are no longer wanted are placed, and can be picked up by others who have a use for them.

Rural Areas

In certain rural municipalities far from recycling centres, various fractions are collected from households. Household waste-sorting containers have been developed for this purpose, as well as refuse collection vehicles designed for picking up the various fractions at the same time. A refuse collection vehicle with two compartments can be used to pick up newspapers and other rubbish on one trip and combustible waste and glass on a second trip.

In sparsely populated areas, there are containers for sorting waste adjacent to each house. Compostable waste is composted on site by residents.

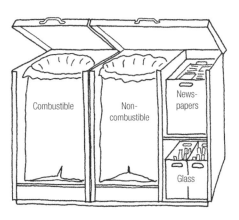

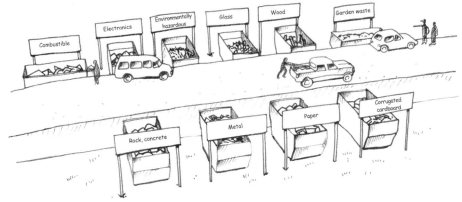

A recycling centre.

Source: Leif Kindgren from a brochure from SRV och Huddinge Council

Offices

Office waste should be sorted into more fractions than household waste. Offices use more fluorescent tubes and light sources than are used in homes, and these should be dealt with in an environmentally suitable way when they wear out. There are companies that recycle mercury. Used toner cassettes are taken back by suppliers and can be refilled and reused. Worn out electronic equipment may contain environmentally hazardous substances such as PCBs, brominated flame retardants, mercury and cadmium. Environmentally toxic materials are dealt with by special companies. It is a good idea to have plastic containers to collect certain wastes, e.g. all sharp objects.

Several different fractions are sorted at environmental stations. In offices, the largest fractions may be white paper, newspapers and brochures, soft plastic, corrugated cardboard and other combustible waste.

2.4.2 Composting

Composting can be done using garden waste composting, cold composting, hot composting, wormeries, a large composting container, drum or tunnel composting or by windrowing. In addition, ecocycle buildings containing chickens, a wormery, and a greenhouse can be constructed. The compost-maturing stage and use of the finished compost requires it to be near cultivation areas, e.g. flower beds, private gardens or nearby agriculture.

Hot Compost

There is a great variety of compost containers on the market, e.g. stationary, rotating, sunk into the ground and large drum composts. Compost containers should be designed so that pests such as seagulls, crows, rats and mice are not attracted. Mice can squeeze through holes more than 6mm across. After about two months of composting, the mass no longer attracts such pests.

Maintenance

Composting requires attention. Containers need to be checked and sawdust or peat must be added as required to maintain a proper carbon–nitrogen balance. Compost must be stirred and clods need to be broken up with a compost rake. It is important to be tidy and clean up spillage. Water leachate should be returned as necessary. When the containers are full, they are closed to allow time for decomposition. When the compost is ready, the container is emptied and the compost taken to a suitable location for maturing.

One way to make use of organic waste is to feed it to pigs.

The praised hot compost 'Biokuben' from SanSac.

A place for composting doesn't have to be anything special. The most important requirement is to have it close by but not too close.

Source: *Lottas kompostråd*, Lotta Lanne.

Large Composters

Large composters are designed either as rotating drum composters, tunnel composters or long compost rows. Compost processors that break up the waste, mix it with straw and then distribute it to several smaller containers for composting are available. A drum composter can handle waste from 50 to 80 households. A tunnel composter is a long tunnel where waste is put in smaller containers and moved through the tunnel as it composts. Tunnel composters may be able to handle waste from up to 300 households. Compost processors can process organic waste from 50 to 200 households.

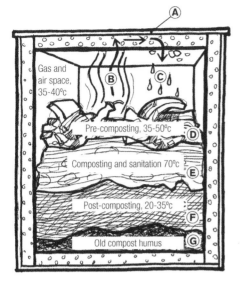

The following takes place in a hot compost: **(A)** Water vapour condenses on the inside cold metal surface and runs through the insulation. **(B)** The nitrogen and ammonia that evaporate in the decomposition process bind to the water vapour and heat causes it to rise. **(C)** The nitrogen and water drip back into the compost mass. **(D)** Due to the high temperature, pre-composting begins right away in the surface layer. **(E)** The +70°C composting temperature shortens the decomposition phase to 2–4 weeks. **(F)** After rapid decomposition, the temperature goes down and post-composting begins. **(G)** A little old compost humus with micro-organisms can be left in the container to get the new compost going.

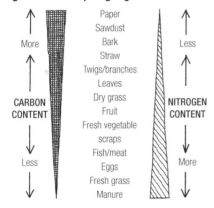

The relationship between carbon content and nitrogen content in a compost is important. One part nitrogen requires 25–30 parts carbon. Carbon-rich material usually needs to be added to a household composter. To reach the right balance, layers of nitrogen-rich household waste are alternated with paper, sawdust, leaves, peat, etc., i.e. material with a high carbon content.

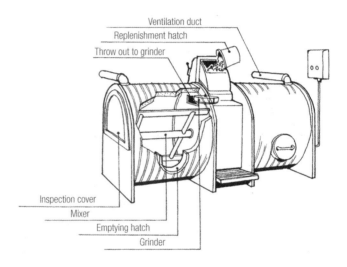

In a drum composter, waste is put into one end of a drum and eventually comes out the other end as composted waste. The drum inclines and slowly rotates, or is equipped with a mixer that pushes the material forward in the drum.

2.4 To close the ecocycle for compostable household waste, there should be space close by for the maturing phase and cultivation.

Source: EFEM architects office, Göteborg

Ecocycle Buildings

An ecocycle building consists of a greenhouse, a henhouse and a hayloft. The chickens are given organic waste and eat what they can. The remains and chicken manure are then placed in a wormery, and finally in a post-composter. The chickens can go into the greenhouse or a fenced-in area outside the building. In an ecocycle building, organic waste is disposed of and at the same time, eggs and tomatoes are produced.

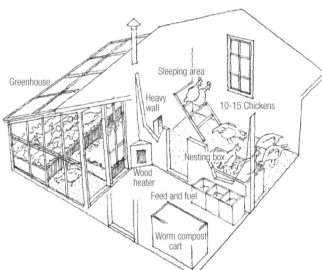

Design for the ecocycle building at the 1992 housing exhibition in Örebro, Sweden. The actual building is located in the courtyard of a block of flats. It was designed by architect Bertil Thermaenius in cooperation with Professor Nils Tiberg.

Centralized Composters

Most hot composters are located so that compostable waste must be carried to them. There are also centralized systems that use vacuum systems with waste grinders or grinder pumps to convey the waste to treatment, either anaerobic (biogas) or aerobic (wet reactor). These composting systems often mix organic household waste with other organic waste.

Waste Grinder

An alternative way to deal with organic kitchen waste is to grind the waste using a waste disposal grinder down into the sewage system. It is possible to have a sewage system that separates black water (toilet and kitchen waste) and grey water (bath, dish and washing water). The black water, where the goal is to maintain as high a level of dry matter as possible, can be pumped to a biogas facility for decomposition.

Biogas

Biogas can be extracted from compostable waste via anaerobic decomposition and may be a solution to many current waste management problems. It is also very advisable to combine waste management and sewage digestion. In addition, it is advantageous to mix in other products such as night soil, sewage sludge and organic industrial waste, e.g. slaughterhouse remains. The mass is moistened until it can be pumped (8–10 per cent dry matter) and is then transported to a dissolving facility where the fibres and small particles not removed in the first phase are separated to remove as much of the heavy metal as possible. Then the sludge is moved to airtight decomposition chambers where an anaerobic breakdown process takes place.

Black-Water Systems

Black-water systems where the sewage is divided up into two components: sanitary waste-water that is treated locally, and the organic waste from toilets and kitchen waste mills (black water), will become more common. The black water is piped to anaerobic biogas facilities or aerobic composts. Black-water systems are in operation at Tegelvik school in Kvicksund in Sweden, the agricultural university at Ås south of Oslo in Norway, and in the residential area Lübeck-Flinten-Breite in Germany.

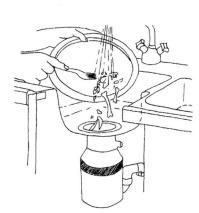

This is how a waste disposal grinder can be fitted under a kitchen sink.

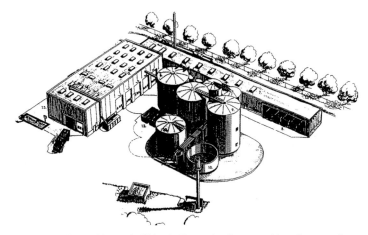

One of the biggest biogas facilities in Europe for decomposition of unsorted organic waste from households, restaurants, schools, etc. is located in Helsingør, Denmark.

2.4.3 Recycling

Worn-out products and packaging taken for recycling must be transported long distances because specialist recycling plants are few and far between.

Packaging

Packaging and containers make up a large portion of all waste, so it is important for people to choose packaging with care. According to the industry's own calculations, packaging constitutes about 10 per cent of the total cost of food. In Sweden, companies can join **REPA**, the service organization of Sweden's business and industry sectors for the collection and recycling of packaging. Membership certifies that a company takes responsibility for their packaging. Via contractors, the companies collect, recycle or extract energy from packaging material. The customers provide the contractors with specifications about the material to be picked up at their company. Collection and recycling of packaging can not cover its own costs. Therefore, in order to finance the recycling system, the member companies pay fees. In the end, however, it is consumers who finance producer responsibility by paying a little more for packaging.

Hazardous Waste

It is especially important to take care of hazardous waste. The largest quantities of environmentally hazardous waste are, in terms of annual production: oil wastes, waste containing heavy metals, solvent wastes, and paint and lacquer waste. In addition, there are acidic and alkaline wastes, wastes containing PVC, wastes containing mercury, lime wastes, waste containing cyanide, biocide wastes, as well as waste containing cadmium. Other environmentally hazardous waste includes special waste from laboratories and hospitals.

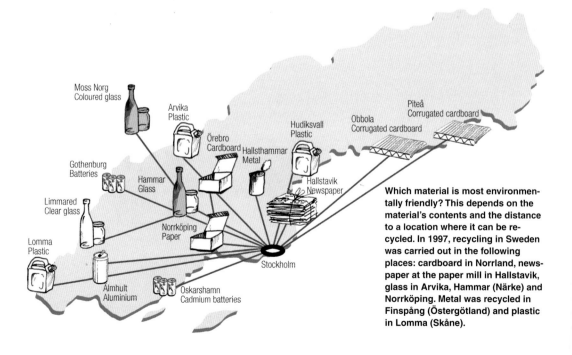

Which material is most environmentally friendly? This depends on the material's contents and the distance to a location where it can be recycled. In 1997, recycling in Sweden was carried out in the following places: cardboard in Norrland, newspaper at the paper mill in Hallstavik, glass in Arvika, Hammar (Närke) and Norrköping. Metal was recycled in Finspång (Östergötland) and plastic in Lomma (Skåne).

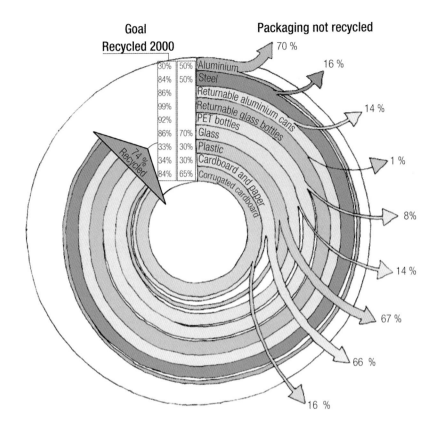

The amount of consumer packaging of different materials that was recycled in the year 2000, the amount wasted in each category, as well as the goals set by the government.

Source: *Ny Teknik*, no 35, 2001

Recycled materials from Swedish households in 2007

	How much (%)	Goal of government (%)
Newspapers	85	75
Officepapers	61.5	
Cardboards	72.6	65
Metal packages	67	70
Plastic packages	30.1*	70**
Glass packages	95	70
Electronic waste	80	
Refrigerators etc.	95	
Metal from household waste	95	

* 34.5 per cent were energy regained – total per centage of recycling 64.6.
** From this energy regain 30 per cent.

An ingenious approach using trays instead of crates provides room for more bottles per unit of volume. This results in a reduction of shipping packaging waste.

The packaging industry is trying to find environmentally friendly alternatives. One alternative is starch-based compostable packaging made from potato flour.

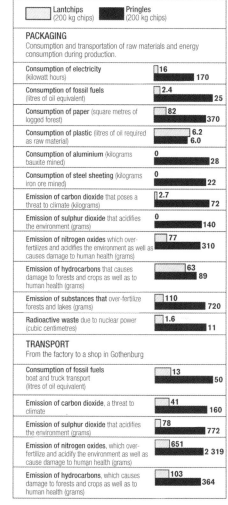

	Lantchips (200 kg chips)	Pringles (200 kg chips)
PACKAGING Consumption and transportation of raw materials and energy consumption during production.		
Consumption of electricity (kilowatt hours)	16	170
Consumption of fossil fuels (litres of oil equivalent)	2.4	25
Consumption of paper (square metres of logged forest)	82	370
Consumption of plastic (litres of oil required as raw material)	6.2	6.0
Consumption of aluminium (kilograms bauxite mined)	0	28
Consumption of steel sheeting (kilograms iron ore mined)	0	22
Emission of carbon dioxide that poses a threat to climate (kilograms)	2.7	72
Emission of sulphur dioxide that acidifies the environment (grams)	0	140
Emission of nitrogen oxides which over-fertilizes and acidifies the environment as well as causes damage to human health (grams)	77	310
Emission of hydrocarbons that causes damage to forests and crops as well as to human health (grams)	63	89
Emission of substances that over-fertilize forests and lakes (grams)	110	720
Radioactive waste due to nuclear power (cubic centimetres)	1.6	11
TRANSPORT From the factory to a shop in Gothenburg		
Consumption of fossil fuels boat and truck transport (litres of oil equivalent)	13	50
Emission of carbon dioxide, a threat to climate	41	160
Emission of sulphur dioxide that acidifies the environment (grams)	78	772
Emission of nitrogen oxides, which over-fertilize and acidify the environment as well as cause damage to human health (grams)	651	2 319
Emission of hydrocarbons, which causes damage to forests and crops as well as to human health (grams)	103	364

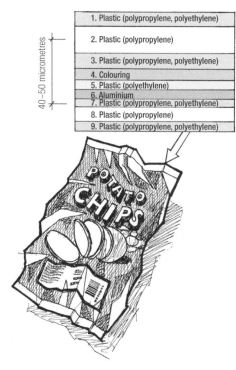

Some packages are impossible to recycle. The illustration shows a crisp bag with nine layers: seven layers of plastic, one of aluminium and one of colouring.

A comparison of the environmental impact of two types of packaging for crisps.

Source: *Dagens Nyheter*, 27 March 1996

Environmental Stations

Hazardous waste can be left at environmental stations. Every municipality should have a sufficient number of these so that people feel it is convenient to dispose of. Petrol stations already take care of hazardous waste such as waste oil, car wash detergent, etc. So, it would be relatively easy to expand these to lockable environmental stations where hazardous and environmentally dangerous waste could be left.

Environmental stations can accept the following fractions, which are stored in a locked area:
- environmentally hazardous liquids;
- environmentally hazardous solids;
- car batteries;
- waste oil;
- home electronics.

Examples of common products that should be turned in at an environmental stations are acetone, biocides, photographic fluids, thinners, leftover paint, white spirit, chlorine, vinyl polish, Tippex and oil residues.

Shops that Accept Returns

A service provided by many shops is to accept the return of products they have sold that have worn out and may constitute an environmental danger or damage the environment. Camera shops take back batteries and photographic fluids, and pharmacies take back medicines and mercury thermometers, etc.

Destruction Facilities

Every country should have a special facility where domestic environmentally hazardous and/or damaging waste can be taken care of or destroyed. An example of environmentally damaging waste is material that is dangerous when combusted.

Recycling Hazardous Waste

In today's society, some products that are difficult to do without contain so much hazardous waste that special recycling systems are needed to deal with them, if they are to be used at all. Electronics, batteries, light sources and tyres are examples of such products.

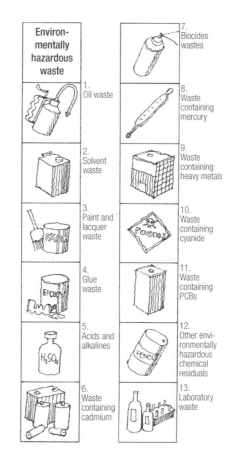

Hazardous waste must be handled with special care. The illustration shows the fractions that are considered hazardous.

Source: Naturvårdsverkets förteckning, SNV 690

Light Sources

Fluorescent tubes and compact fluorescent bulbs contain mercury. About 50kg mercury is lost each year in the form of discarded light sources (according to a study by *Ny Teknik*, no 12, 2001). This cannot continue. Companies that can take care of the mercury in fluorescent tubes are beginning to appear. A fluorescent tube contains 5–30mg mercury. Compact fluorescent bulbs and tubes contain 5–10mg mercury, and metal halogen lamps contain 45mg. Mercury lamps contain 20–50mg, and high-pressure sodium lamps 20mg. Neon tubes contain 0.5–2.0g.

Incandescent light bulbs do not contain mercury; however they do contain lead in the base. Up to 40 tonnes of lead is lost per year from incandescent light bulbs disposed of in household waste. This is the largest known source for the spread of lead. There are technical developments under way to reduce the amount of heavy metals in light sources. Fluorescent tubes are now being produced that contain less mercury, attempts to make incandescent light bulbs without lead in the base are under way, and work is being done to develop completely new types of light sources that do not contain any heavy metals at all.

Electronics

There are companies that dismantle computers so that various parts can be reused. About 97 per cent of a computer can be recycled. The average age of a computer is seven years. Iron, aluminium and copper are the metals most commonly used in computers, along with lead, silver and gold. Many of the plastics are treated with flame retardants and must be combusted at high temperature (1300–1400°C). Conventional computer monitors contain three types of glass, each containing a different proportion of lead: from 5 per cent at the front of the screen to 25 per cent in the rear parts. If a computer is TCO-certified, it should not contain lead or cadmium. Monitors also contain many uncommon substances such as beryllium, palladium, strontium, etc. It is difficult to recycle glass from computer monitors, but new technology is being developed that can do this.

Greenpeace's 'Guide to Greener Electronics' ranks the best electronic companies according to how environmentally friendly they are.

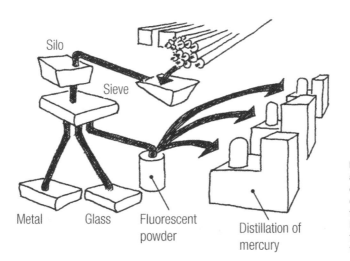

If hazardous materials are used they should be carefully collected in a controlled manner so that they can be reused. The illustration shows a system for recovering mercury from fluorescent tubes.

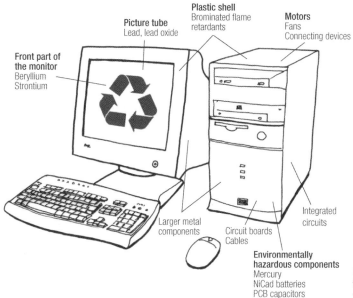

Materials recovered from computers have a variety of uses.

Tyres

One problem with tyres is that the softeners used in the rubber mixture are highly aromatic oils (HA oils) that are highly carcinogenic. Many tonnes per year of poisonous tyre substances are released from the wear of tyres on roads. Some tyre producers choose to replace HA oils with other mineral oils. A worn-out car tyre still contains 85 per cent of its total rubber content. Tyres can be retreaded. There are several companies that recycle worn out rubber tyres by producing rubber granules and powders from them that serve as new raw material for various rubber products. For example, recycled rubber can be used as an ingredient in new asphalt surfaces as it lowers the noise level. Most recycled rubber comes in a granular form and is used to make sports tracks. Some is mixed with new rubber. New products made from recycled rubber are being developed e.g. elastic ground slabs.

Freon

New technology exists to remove freons from old refrigerators and freezers, from both cooling coils and insulation.

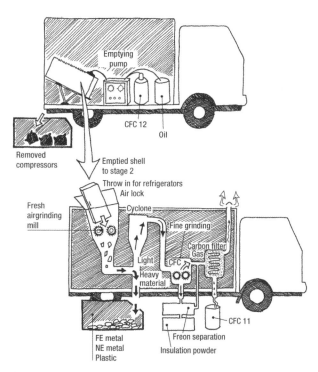

Systems for recovery of freon from cooling devices have improved. In step one, the cooling circuit is emptied and 95 per cent of the freon is recovered. In step two, the casing and insulation are ground down and 80–90 per cent of the freon in the insulation is recovered.

Source: Svensk Freonåtervinning AB

2.4.4 Ecological Design

Ecological design involves incorporating a life-cycle perspective into the development of a product right from the beginning. How can the life of a product be extended? How can the product be repaired? Can the product be reconditioned (by making a greater effort to make it function again)? What happens to the product when it can't be used anymore?

Dangerous Products

It is important to remember that the biggest waste problems today come not from the production and packaging of products but from the products themselves which sooner or later become rubbish.

Unwanted Substances

Many of the goods that we surround ourselves with contain hazardous substances. Often we don't think about that fact. It is justifiable to ask whether or not the goods are better because they contain these substances or if there are production methods that don't include hazardous substances. Two of the most common heavy metals, chromium and cadmium, are found in many everyday objects and will be an environmental problem in the future. Does my sofa have to contain hydrocyanic acid? When telephones burn, must they contain brominated fire retardants? A well-known car manufacturer uses the sales argument that whoever buys a car from them gets the car with the most chromium plating on the market. At the same time, we know that chromium is a heavy metal that should not be spread. A lot of waste is not visible but surrounds us in the form of molecular waste. For example, most leather shoes are tanned with chromium. Every step taken with tanned leather shoes spreads chromium molecules around us.

Life

A life-cycle analysis can be done by comparing the environmental impact of various products. The analysis examines what takes place during the entire life of a product. How much raw materials and energy are consumed? What is the environmental load from emissions to the ground, air and water? When these questions are answered, the various data are divided by the product's life. The life is thus highly significant. We should get away from being a throwaway society and instead produce products that last a long time.

Ease of Repair

When a product stops functioning it is often the case that one of the product's parts has worn out and broken. If it is possible to simply and practically replace that part, it is worthwhile repairing the product rather than purchasing a completely new one.

Illustration from cover of *Dangerous Products*, a report from the Swedish Ministry of Environment (Ds 1992:82).

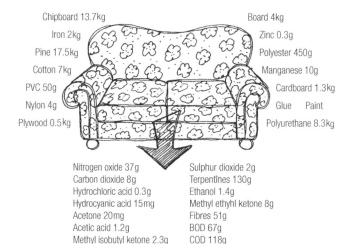

Materials and substances in a common sofa. Some materials emit gases that can affect human health.

Source: Miljö- och naturresursdepartementet, 1992

Reconditioning

Some products wear down more evenly than others, e.g. tyres, wood floors and shoe soles. By retreading tyres, sanding down wood floors, and resoling shoes, product function is regained. Such products should be designed so that reconditioning is possible.

Reuse

If a product of sufficiently good quality is produced, with respect to both aesthetics and materials, it can be used for a long time. Many clothes, cast aside when fashion changes, often become fashionable again after a couple of decades and are sold in second-hand shops. As life changes, an item can lose its value for one person, yet be desired by another. Especially nice things become antiques and their value increases with time. A fundamental difference emerges here. A poor quality piece of furniture made of chipboard has a short life and must be thrown out after about ten years. A high-quality piece of furniture made in the 1700s is an object of great value.

For more complicated products, some parts may break down while others may last longer. It should be possible to dismantle products easily, and to acquire spare parts. Systems for this have been developed in the vehicle sector and a similar development is taking place in the building sector. There are databases shared by car breakers and similar databases for reusable building products.

Material Recovery

In a sustainable society, is should be possible to recover materials from all products. This places two requirements on the product. It should be possible to break it down into its components and to sort the pieces. The vehicle sector as been in the forefront of this process. SAAB was one of the first manufacturers to content-label every part. One interesting product is a British toaster that is equipped with a red recycling button. If the toaster breaks down, then the red button is pushed and the toaster falls apart into metal, plastic and other components. A big problem today in recycling plastic products is that it can be difficult to know what kind of plastic material is involved and how to deal with it.

Choice of Material

In ecological design, substances and materials that pose an environmental and/or health hazard should be avoided. Materials with as small an environmental load as

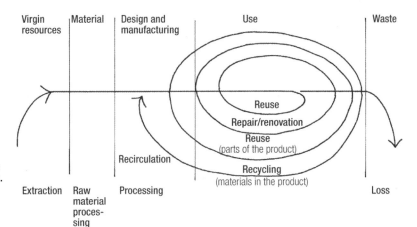

In a sustainable society products are made to be used over and over again. It should be possible to repair or recondition them, and their contents should be recyclable or suitable for energy extraction.

possible should be chosen. This information is found in life-cycle analyses, which show how much energy, raw materials and pollution of the land, air and water an individual product accounts for. Life-cycle analysis can be done in many different ways and results vary, depending on the terms of reference. Equivalent analysis is needed in order to compare different materials. One way to try to achieve this is to use environmental product declarations (EPDs), where products are delineated according to a particular model, and inspected and certified by an accredited third party.

Multipurpose

In a consumer society, more and more special purpose products are developed, which leads to a materialistic society where people surround themselves with a large number of things. One way to counteract this is to design products that can satisfy several different needs. An example could be to buy a single, multifunctional fax, computer printer, scanner and photocopier device instead of buying four separate devices. Instead of having a set of dishes that includes a dessert plate, soup bowl, tea cup and dinner plate, it is possible to have bowls that can be used both for dessert and tea, and plates that are both flat and deep that can be used for both soup and other courses. The new generation of mobile phones contain a telephone, camera, music system and a hand-held computer that can be connected to the internet, etc.

Leasing Agreements

Many products have looked basically the same for decades, e.g. kitchen and laundry appliances and computers. Some components break down; others go out of date. There has been discussion about whether a person needs to own a computer or refrigerator or whether instead they could enter into a leasing agreement where a producer commits themselves to providing a function rather than a product. This could lead to a different kind of design where a refrigerator is of high standard both aesthetically and functionally, but where the producer takes on the responsibility for replacing old components when they wear out. Instead of replacing a refrigerator after 15 years at the most, a refrigerator could be made that has a longer total life and parts that are replaced over time.

Ecocycles

3

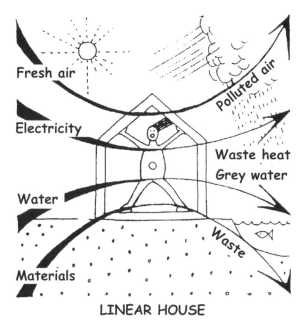

LINEAR HOUSE

A linear house where large amounts of resources flow through and become problematic waste.

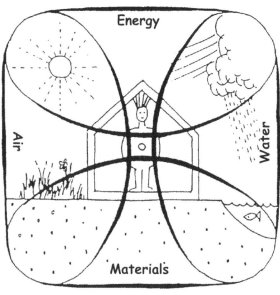

ECOCYCLE HOUSE

An ecocycle house where the resource requirement is reduced and resources move in a cycle.

Source: *De siste syke hus*, Architect Björn Berge, 1990

Closing Ecological Cycles

The resources used in a sustainable construction should circulate in an ecological cycle. Energy systems based on renewable energy sources are used. Sewage systems are chosen that recycle nutrients. Organic material is returned to the earth to provide nutrients for vegetation and cultivation.

Renewable Heat:
Heating and Cooling with Renewable Energy Sources
Renewable heat involves using biomass in combination with solar heat. Such systems use an accumulation tank to store heat. Heat pumps are also used. They make use of low-temperature environmental heat to heat and cool.

Renewable Electricity:
Producing Electricity with Renewable Energy Sources
This refers to generating electricity from hydroelectric power. Additional renewable electricity may come from biomass power plants and wind power. In the long run, solar cells will become cheaper and be integrated into roofs and façades.

Sewage:
Sewage Systems that Recycle Nutrients
Sorting sewage is a way to minimize pollution. Sewage can be separated into runoff water, traffic water, grey water, urine and black water. Mechanical and chemical purification can be complemented with natural purification methods. It should be easy to return the final products to agriculture in a hygienic manner.

Vegetation and Cultivation:
Recirculating Organic Material
Vegetation and cultivation fulfil several functions: they produce food and energy and return organic waste and sewage sludge to ecological cycles. The biological diversity, beauty and well-being they provide are also important. Wild areas, parks and gardens, vegetation around buildings, on buildings and in buildings are important features.

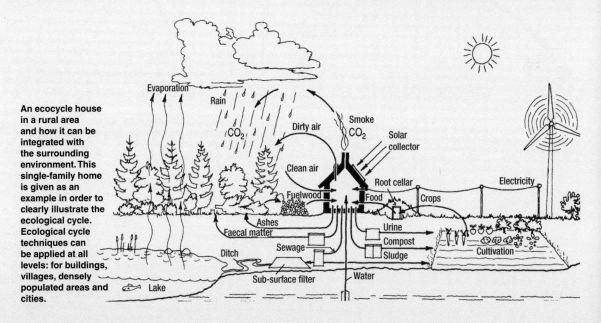

An ecocycle house in a rural area and how it can be integrated with the surrounding environment. This single-family home is given as an example in order to clearly illustrate the ecological cycle. Ecological cycle techniques can be applied at all levels: for buildings, villages, densely populated areas and cities.

Ecocycle Buildings and Sustainable Society

In an ecocycle building the flows used are investigated to determine their origins, their destinations and whether they are in harmony with nature. The same flows described in the section on conservation of resources are dealt with here, but here we go further and look at the integration of these flows into ecological cycles.

Energy supply today

80 per cent of the energy used in the world is produced with fossil fuels, and 80 per cent of the greenhouse gases come from the use of fossil fuels. To create a sustainable energy future there are two strategies: to use energy more efficiently and to develop renewable energy sources. It's about solar energy, biomass, hydropower, wind energy and geothermal energy. Solar energy has the biggest potential of them all and it can supply us with all the energy needed. Photovolatics can give us electricity and solar collectors can produce heat. Heat pumps can help us to utilize the warmth in air, water and soil heated by the sun. Wind energy is used to produce electricity and to pump water. From the seas we can get wave energy, tidal energy, current energy and OTEC (ocean thermal energy conversion). Biomass and organic waste can be used as fuels. Geothermal energy can give us both heat and electricity. Hydropower is an old technology that can be expanded. There are many possibilities.

Energy supply in Sweden

Sweden's dependence on fossil fuels is still quite large. The current annual consumption is slightly greater than 200TWh. Even in an economic growth situation it is possible to use less energy than is currently used. Annual energy consumption in Sweden is about 450TWh, of which 150TWh is electricity. Future scenarios, including one by the Swedish Environmental Protection Agency's Climate Commission in 1998, show that by applying conservation and

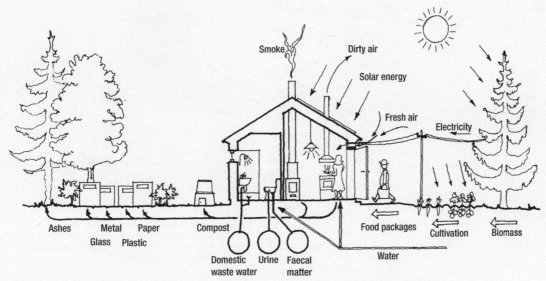

A schematic representation of a single-family ecocycle house that uses the latest available technology.

efficiency measures it is possible to get by very well on 300TWh per year, of which 100TWh are electricity.

In Sweden the potential for meeting energy requirements with renewable energy resources is great. Today 100TWh of biomass are used, which can be expanded to 200 with existing resources. Within the near future, all of Sweden's electricity could be provided by hydroelectric power together with biomass-fuelled district heating and wind power. About half of the electricity consumed in Sweden is currently produced by hydroelectric power (65TWh). The potential for small-scale hydropower, without too large a negative environmental impact can be acheived by making existing hydropower plants more efficient. The estimated potential for wind energy is 30TWh/year. With new improved technology and more cogeneration plants, district heating can contribute up to 33TWh electricity/year.

The goal proposed for Sweden by the European Commission of 49 per cent of energy supply being met by renewable energy sources in the year 2020 can be reached by biomass-based combined power and heating plants according to construction plans in 2009 as well as increasing the construction goal for wind power to 25TWh.

Global Energy Supplies

It is not difficult to meet the energy needs of large countries with small populations with renewable energy sources. Nor is this difficult in countries with large sources of biomass, hydropower and wind power. The big challenge is meeting the entire world's energy requirements with renewable energy sources. Investigations of the possibilities show that supplies of biomass, hydroelectric power and wind power are limited. Geothermal energy is found only in certain locations. The greatest potential lies in the use of solar energy.

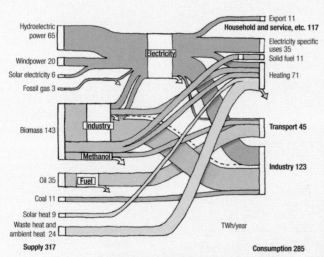

Energy flows in the post-materialistic energy scenario for Sweden in the year 2050. CO_2 emissions have been limited to 4 Mtonnes C/year, a 75 per cent reduction compared to the level in the early 2000s.

Source: *Energiläget År 2050*, Naturvårdsverket 1998

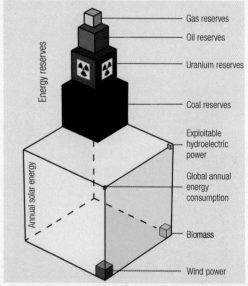

This illustration shows the relationship between annual solar insolation, potential wind power resources, hydroelectric power and biomass, as well as total reserves of non-renewable energy sources.

Source: *Nye fornybare energikilder*, Norges vassdrags- og energidirektorat, Norges forskningsråd, 2001

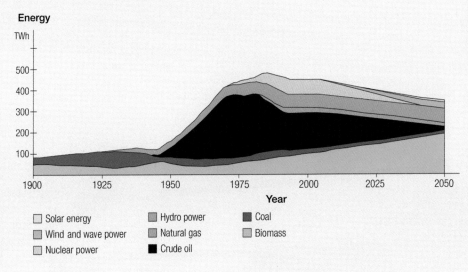

Energy consumption was low up until the end of the Second World War. After the war, energy consumption increased, especially of oil. In the 1970s, energy use stagnated and nuclear power came into the picture. The challenge now is to phase out nuclear power and fossil fuels, and instead use solar, wind, biofuel and hydroelectric power.

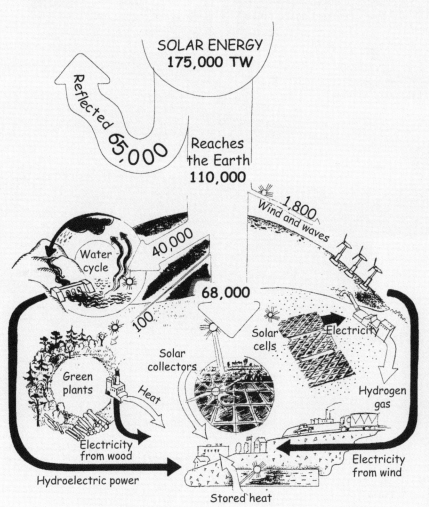

If we could only learn to make use of a fraction of the energy flow from the sun, the energy problem would be solved. Solar energy drives the winds, the water cycle, and makes green plants grow. Most of the sun's energy, however, comes to Earth as direct sunlight.

Source: Illustration: Erik Sandegård

The district heating facility in the Tuggelite eco-village, Karlstad. The energy sources used are biomass (pellets) and solar heat.

3.1 Renewable Heat

In order to achieve a comfortable indoor climate, buildings must be climatized, i.e. heated or cooled. In northern Europe, most buildings need to be heated during the winter. Some buildings, particularly offices, need to be cooled during the hot time of the year. In a sustainable society, efforts are made to use renewable energy for heating and cooling. There are a number of ways to heat buildings and tap water with renewable energy sources: heating with biomass, solar heat, heat pumps that use low-temperature heat from the environment, and waste heat.

3.1.0 Biomass, Solar Energy and Accumulation Tanks

A common way to heat buildings using renewable energy sources is to use a combination of biomass and solar energy together with an accumulation tank to store the heat. Solar energy is used primarily during the summer to heat water, and biomass is used mostly during the winter when both space heat and hot water are needed. This system is called the 'energy trio'.

Size of the Heating System

The energy trio is suitable for both small- and large-scale applications. The system can be used in single-family homes. A local heating plant can supply a group of buildings while a district heating system can supply the heating requirements of entire urban districts.

Domestic Heating

It is common to use the energy trio in single-family homes. An accumulation tank is connected to a roof-mounted solar panel and a combustion system fitted with a water jacket. A wood-burning stove, tiled oven, kitchen combination wood stove and boiler, or a biomass boiler in a special boiler room can all serve as combustion systems. It is also possible to have a combustion system without a water jacket complemented by another heating system. In this case warm water is heated separately.

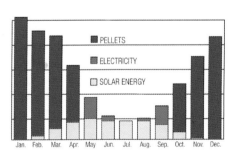

There are many advantages to using solar energy to complement biomass. In Sweden, lack of winter sunlight means that solar energy needs to be supplemented with a fuel. During the summer, however, solar energy can be used to heat hot water and the biomass boiler can be turned off. During low-load periods, all forms of combustion are inefficient and thus result in an increased environmental impact. Electricity is therefore used to supplement solar energy during the summer.

Source: Jan Olof Dalenbäck, Chalmers University of Technology, Gothenburg

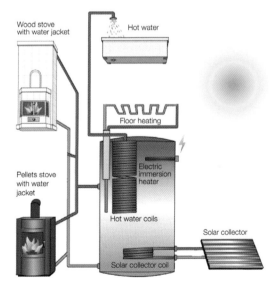

Heating system using renewable energy in an energy-efficient house. Stoves with water jacket and solar collector charge the heat storage tank. Hot water and heating are supplied from the storage tank.

Source: Nordic Värmesystem

3.1

An example of an energy system in an ecological home. The house is highly insulated and is designed to use passive solar heat. The building is equipped with a solar panel on the south side of the roof and a tiled oven with a water jacket. The heat is stored in a tank which supplies the building with space heat and hot water.

Source: John Karlsson

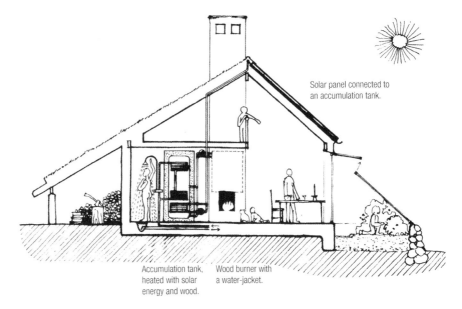

Solar panel connected to an accumulation tank.

Accumulation tank, heated with solar energy and wood.

Wood burner with a water-jacket.

Local Heating

Small-scale, pellet-burning heating plants can heat buildings via a local heating system. Such systems are used to heat small residential areas, eco-villages, large farms, blocks of flats, schools and hospitals. Use of pellets makes it possible to fully automate the heating system with no extra work for the users. Both two- and four- pipe systems can be used to distribute heat. A two-pipe system supplies space heating from a central heating plant, and hot water is produced locally in each unit. A four-pipe system supplies both hot water and space heating from a central heating plant using two parallel systems.

Solar panels can be located at the central heating plant or on the roof of each building. The main accumulation tank is located at the central plant, but can be supplemented with small accumulation tanks in each unit.

An example of the 'energy trio' in a home heating system. The system is composed of a solar panel, a wood boiler and an electric heating coil connected to two accumulation tanks. One of the tanks is used to store heat from the solar panel during the summer, and both tanks store heat from the wood-burning boiler during the heating season. Electricity is used as a back-up.

Source: Gunnar Lennermo, Energianalys AB, Alingsås, Sweden

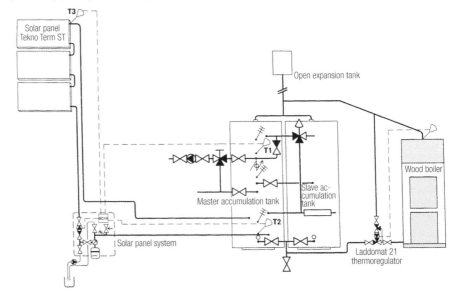

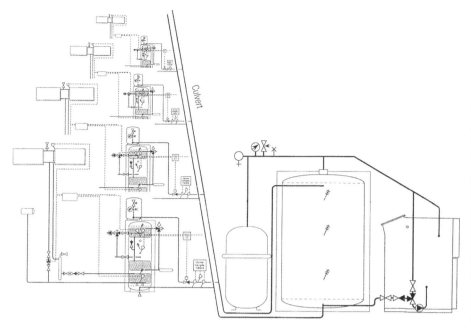

Heating system in the Understenshöjden eco-village. There is an accumulation tank in each unit even though there is a local, central boiler. Energy loss from the distribution of solar heat is thus reduced and each building has its own reserve system – an electric heating coil in the accumulation tank. An accumulation tank is still required for the central boiler in order to obtain efficient heating.

Source: Gunnar Lennermo, Energianalys AB, Alingsås, Sweden

A distribution system for local heating is relatively simple and inexpensive to install. The distribution pipes and insulation form a unit (a culvert) that is 'ploughed' down into the soil. Outermost on the culvert is a pliable corrugated protective casing filled with insulation. In the centre is an empty plastic distribution pipe. The culverts are made in fixed lengths of tens of metres that are easy to join together. They are easy to handle and can be laid in narrower trenches than conventional ones. Once the culvert has been laid, the water supply pipe is pulled through. This method means that there are no underground joints.

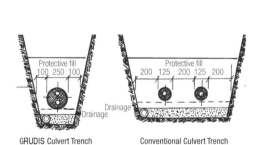

A 'GRUDIS' culvert trench requires less space than a conventional trench.

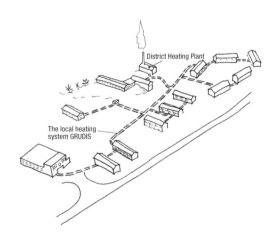

GRUDIS is a local district heating system with many advantages. The prefabricated culvert is an insulated unit, easy to lay, takes up little space and is relatively cheap.

3.1

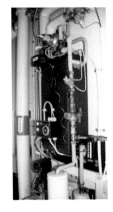

Connection box to the central heating system in a private home.

District Heating

Municipal district heating is used in densely built-up areas. The heat can be produced in biomass-burning heating plants or waste heat can be taken from industry. District heating plants use advanced flue-gas cleaning systems. This combined with fewer chimneys results in a reduction of air pollution from combustion. The choice of location for district heating plants should be determined by building density. The denser the built-up area, the lower the distribution losses and the greater the efficiency of the system. The savings of a municipal district heating plant decrease if it is extended to low-density areas.

Passive solar blocks of flats require so little heat that the return water to district heating plants can be used as a heat source (via a heat pump). This method is used at the Kvarteret Seglet block of flats in Karlstad. The negative contribution from the heat pump is compensated for by investments that reduce total electricity consumption (household and building electricity).

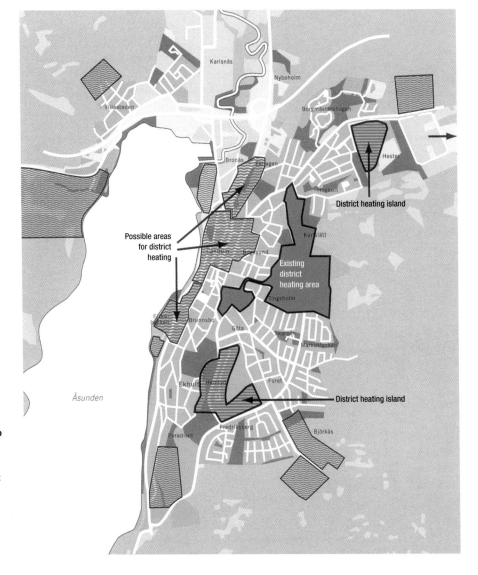

A piece of a map of the heating plan for Ulricehamn, Sweden. It would be quite simple to extend the network from the 'existing district heating area' to the 'possible areas for district heating'. In the areas marked 'district heating island' it would be possible to set up small local district heating plants. The map also shows areas that may be developed in the future where heating needs could be met in a variety of ways.

Source: *Att bygga ett hållbart samhälle* (Building a sustainable society), Birgitta Johansson and Lars Orrskog, 2002

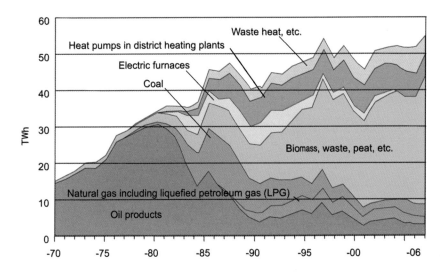

Supplied energy by district heating in Sweden 1970–2006.

Source: Statistics Sweden and The Swedish Energy Agency.

Development is currently moving in the direction of improving insulation of district heating pipes and making distribution technology more efficient. First, the maximum distribution temperature can be reduced from 120°C to 100°C and sometimes even lower. Second, new technology for laying district heating pipes has been developed. Third, work is being done to adjust and replace local receiver stations with the aim of lowering the return temperature.

The most efficient way to use biomass is to burn it in a power station that produces both electricity and heating (see Combined Heat and Power with Biomass in Section 3.2 Renewable Electricity).

Waste Heat

Industrial waste heat may be used for heating in district heating systems. In the steel industry, the huge amounts of energy released when steel cools could be used as a source of heat. Heat pumps can be used if the temperature of waste heat is too low, for example in a sugar refinery. In some industries, the temperature of the waste heat is so high that it is inefficient to use it for space heating. Instead, it should be used to provide secondary heat to industries that require such heat, and the waste heat from that process could in turn be used for district heating (this is called cascading).

A research group in chemical engineering at the Royal Institute of Technology (KTH) has recommended that surplus heat from the steel plant in Oxelösund be transferred to the airport at Skavsta by chemically storing the energy in containers transported by train, ship or truck. The method uses sorption technology that stores heat in containers filled with zeolite.

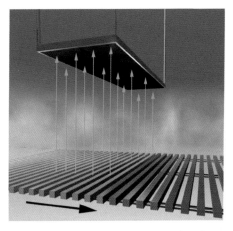

A solar panel can be used as a tool to recover waste heat in a steelworks. This method was tested in the Hofors steelworks in Sweden and found to be 100 times more efficient than conventional solar heat. Theoretically, the entire Swedish steel industry could heat a total of 6000 single-family homes by using waste heat from the cooling beds.

Source: Illustration: Leif Kindgren

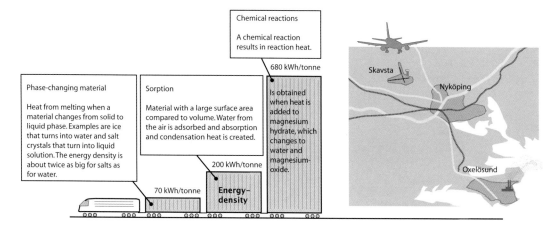

Waste heat from the steel works in Oxelösund can supply the airport in Skavsta with energy. The illustration shows three different methods to chemically store heat for later transport by train to Skavsta. The heat storage methods are: phase change of salt, sorption in zeolite, and chemical reaction of magnesium hydrate.

Source: *Ny Teknik*, no 21, 2007

Flue-gas Cleaning

The combustion efficiency of biomass has developed to the extent that high levels of efficiency and relatively clean flue gases are now common. To reduce air pollution even more, various methods of cleaning flue gases are used, especially in large facilities. This is particularly important when incinerating refuse that may contain contaminants. At large plants, several different methods of cleaning flue gases are combined. Small facilities use catalytic flue-gas cleaning.

Flue-gas Cooler

A flue-gas cooler is a form of heat exchanger that can be installed in large plants to extract energy from the flue gases. However, enough heat must remain in the gases for them to rise up and exit the chimney without condensing and damaging the chimney.

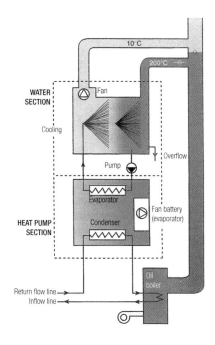

This flue-gas cooler cools and cleans the gas by spraying water. Water, which is a condensate for the flue gases, is heated and pumped away to be used as an energy source.

Source: Renergi, Svenska Rökgasenergi AB, Tranås, Sweden

3.1.1 Bioenergy

An advantage of biomass is that burning it does not affect climate change because the CO_2 released during combustion is absorbed when new trees grow. Combustion technology has improved and resulted in reduced air pollution. Returning the ashes to the soil is important for completing the ecological cycle and preventing soil depletion.

Biomass

Firewood comes from the forest's wood raw materials and agricultural fuel from agriculture. Peat fuel is made of a biological material found in bogs that is not completely decomposed. Spent liquor and tall oil are by-products of the pulp industry. It is also possible to use combustible waste as fuel.

More biomass could be used if the ashes were returned to woodlands. This means that the use of biomass could be doubled without threatening the long-term sustainability of the forestry industry.

Wood for Fuel

Wood for fuel consists of tree tops, branches, stumps, deciduous trees with no industrial use, forest thinnings and trees left after felling. Various types of biomass are made from firewood including wood, pellets, briquettes, clogs, wood chips, bark and shavings. Wood is suitable for limited heating needs, especially if a person has access to a forest. Pellets are also suitable for limited heating requirements, and have the advantage of being homogeneous and fed automatically. Briquettes are larger than pellets, can be fed via an automatic delivery system, but are seldom used in single-family homes. Wood chips, bark and shavings are often used in large heating plants as they are cheaper, but they require continuous supervision.

Peat

As peat is a fuel that is renewed very slowly, combustion of peat is considered to increase the amount of CO_2 in the atmosphere. Peat fuel is produced commercially in two forms: peat briquettes and milled peat. How harvesting and burning peat affect climate change is unclear. Methane, which is a powerful greenhouse gas, is emitted from virgin bogs, while burning of peat emits a

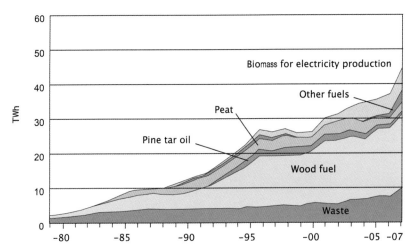

The use of biomass, peat, etc., in district heating plants, 1980–2007.

Source: Statistics Sweden and Swedish Energy Agency

peat are less than from the combustion of coal, but more than from natural gas. The peat industry is relatively well developed with smoothly functioning technology for extraction, transport and use. The industry plays an important role in providing rural employment locally.

Energy Forests

Fast-growing species such as willow (*Salix*) can be grown specifically as an energy crop. This has great potential and could generate a great quantity of energy. In the upper north of Scandinavia reed canary grass (*Phalaris arundinacea*) has better chances to grow.

Energy Crops

Energy crops are grown on agricultural land and are used for energy purposes. Straw fuel is the term for straw and energy grass. Straw is agricultural waste, and reed canary grass is the most suitable energy grass species in Sweden. Today, straw is used in limited quantities for energy production. Oil-yielding plants can be processed into fuels that can replace diesel. A common example is rapeseed/canola. Grain can be fermented into ethanol and used as vehicle fuel instead of, or to mix with, petrol and as a substitute for diesel.

The fermentation of pasture crops and manure produces biogas which can be used for heating, electricity and liquid fuel. The fermentation wastes can be returned to agriculture as fertilizer. Alfalfa and reed

The use of biomass and potentials according to SVEBIO (TWh). In 2005 Sweden used 112TWh of biomass. According to Naturvardsverket (Swedish Environmental Protection Agency) This can be increased to 200TWh in 2050. The actual potential is even greater.

considerable amount of CO_2 to the atmosphere. According to recent research, the net emissions of greenhouse gases from peat in its natural state and from the burning of

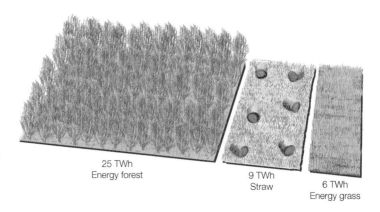

In Sweden in the year 2020 agricultural energy crops could produce 40TWh. The energy potential within agriculture, together with the additional 20–50TWh that can be harvested from forests, is equivalent to, at the very least, the total energy production from nuclear power.

25 TWh Energy forest

9 TWh Straw

6 TWh Energy grass

Fuel chips	Forestry waste	Bark and sawdust
Cut chips from deciduous trees. Net calorific value at 45% moisture content: 9.0MJ/kg Moisture content: 30–50% Ash content: about 1% Size: 1% > 100mm, 10% < 5mm Oil comparison: 10–12m^3s per 1m^3 Fo 1	Broken-up logging wastes, clearing and thinning wood and stumps Net calorific value at 50% moisture content: 8.4MJ/kg Moisture content: 45–55% Ash content: 2–5% Size: 1–2% > 150mm, 20–30% < 5mm Oil comparison: 12–14m^3s per 1m^3 Fo 1	Waste from the sawmill industry. Net calorific value: 55% moisture content: 7.3MJ/kg Moisture content: 45–60% Ash content: 1–3% Size: bark 0–100mm. 20% < 5mm Oil comparison: 18–20m^3s per 1m^3 Fo 1
Briquettes, clogs	**Pellets**	**Peat bricks**
Industrially produced solid fuel. Pulverized and compressed dry wood waste. Net calorific value: 17.0MJ/kg Moisture content: 8–12% Ash content: <1% Size: depends on the production machine Oil comparison: about 3.5m^3s per 1m^3 Fo 1	Industrially produced solid fuel, e.g. Dried, pulverized and then compressed wood waste. Net calorific value: 15.9MJ/kg Moisture content: 10–15% Ash content: 1–2% Size: depends on the production machine Oil comparison: about 3.5m^3s per 1m^3 Fo 1	Net calorific value at 35% moisture content: 13.4MJ/kg Moisture content: 35–50% Ash content: 1–10% Size: cross-section 50 x 100mm depending on the production machine Oil comparison: 6–7m^3s per 1m^3 Fo 1

There are many different biomass fuels on the market. Wood chips should be combusted in large facilities managed by trained personnel. Pellets and briquettes work well in automatic, small-scale systems. Fo 1 = fuel oil quality 1.

canary grass are pasture crops that are suitable for fermentation.

Bark, Lyes and Oils

Black liquor and tall oil are by-products of the pulp industry. Black liquor is already used as an energy source but it is possible to recover much more energy from it. Most is utilized internally within the pulp industry as a heat source. In addition, some electricity is produced (back-pressure power). Black liquor is created during production of sulphate pulp, a type of paper pulp. The black liquor is separated from the sulphate pulp after boiling, and it can be combusted to run a steam turbine. If the black liquor is gasified instead, electricity production can double. The gas is purified and combusted in a gas turbine. The steam produced during combustion can run a steam turbine. Gasification makes it possible to use the energy in two steps. Tall oil is a product created from the pith of pine trees. The more heartwood there is, the more oil can be extracted. Tall oil is used as fuel in district heating plants.

Combustible Waste

Household and industrial waste can be combusted to produce electricity and heat, primarily for district heating. Concurrent with an increase in waste sorting, two interesting fractions will become available: degradable organic waste that can be fermented into biogas, and combustible waste that currently goes to landfill. The biogas can be used to produce electricity, space heat and fuel. The combustible waste can contribute to energy supplies.

Suitable times of the year and moisture content for wood-fuel management.

Source: illustration Leif Kindgren

Biomass

Biomass is processed into a homogeneous form so that it can be used effectively and automatically. This homogeneity improves its combustion characteristics. Processed biomass comes as pellets or briquettes. These fuels are produced commercially and are purchased ready to use from a distributor.

Trees destined for use as **firewood** should be cut down in the late fall or early winter, sawed and split in the spring and dried under cover (preferably for two years).

It is good practice to bring the wood into a heated area two or three weeks prior to burning it. Wood must be dry in order to achieve good combustion and low emissions. Wood with a moisture content of 20 per cent releases two to three times as much heat as newly cut wood. It is of course possible to use a sawhorse and chopping block, but practical equipment is available such as circular saws and hydraulic wood splitters as well as machines that both saw and split. Electric equipment is available, and equipment that may be connected to a tractor's hydraulic system.

Wood chips require a storage area, a drying system, a delivery pipe to feed the chips into the boiler, and in some cases a pre-oven where the material is gasified and ignited. Forest owners need a chopper that can be connected to a tractor and a suitable truck to transport the chips. Electric chip choppers made primarily for small-scale use are also available. It is important for the chips to be dry both during storage and combustion. Moist chips can easily become a breeding ground for mould, which constitutes a health risk for users. Material should be stored unchipped until it is dry. Trees should be felled when they are leafing so that the leaves draw out the moisture. The

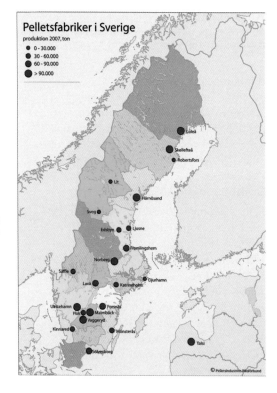

Use of pellets is on the rise. This illustration shows production locations in Sweden and how many tonnes they generated in 2007.

Source: Pelletsindustrins riksförbund

trees should dry in the sun for some months until the moisture content goes down to 30–35 per cent. The branches and tops left over from logged forest produce good chips. Logging remains should be chipped during the summer or covered to be kept dry until they can be chipped.

Briquettes (clogs) can be used instead of solid wood in all types of combustion systems. However they are most suitable for use in district heating plants specifically designed to burn briquettes. Such heating plants are equipped with a briquette storeroom and delivery pipe that delivers the briquettes to the boiler. Dry content: about 10 per cent.

Pellets are growing in importance as an environmentally friendly and economic biomass that produces minimal emissions and can be fed and combusted using automation. They are suitable for use in combustion systems of all sizes, from a boiler in a single-family home to the largest district heating plants. A warehouse or silo for storing pellets is needed for large-scale plants, as well as a delivery pipe to feed the pellets to the boiler. Pellets can be stored indoors or outdoors. Heating with pellets on a small scale requires suitable storage space and a pellet-fuelled boiler. Dry content: about 10 per cent.

Wood powder is a dry and homogeneous biomass fuel made from forest wastes, primarily used in large-scale thermal heating plants, but there are also attempts to use wood powder as a motor fuel.

Ashes

When forest fuels are removed from forests, large amounts of plant nutrients disappear with them. It is therefore important to return the nutrients to forest lands by spreading the ashes left after combustion. The ashes contain almost all the minerals and nutrients found in the original fuel, except nitrogen. The ashes also contain most of the heavy metals that were in the tree, e.g. cadmium and lead, but on the other hand,

Pellet boiler with silo and chimney at Tegelviks school in Kvicksund, near Eskilstuna.

do not add any new metals to the soil. The characteristics of the ashes vary according to the fuel they originate from. Ashes are basic, which means that they counteract soil acidification. If the soil is highly acidified, combining ashes with lime provides an even better antidote to acidification.

Due to their alkalinity, dry untreated ashes can be corrosive. Ashes need to be treated before being spread in order to facilitate ease of handling and return them to the environment in an acceptable form. They are hardened and made into balls or grains that are simple to deal with and slowly disintegrate into the soil.

Combustion Facilities

There is a big difference between old combustion facilities and modern, environmentally certified ones. Modern ones are both more efficient and have lower emission levels. An environmentally certified combustion facility is designed so that combustion is good and so that emission levels of

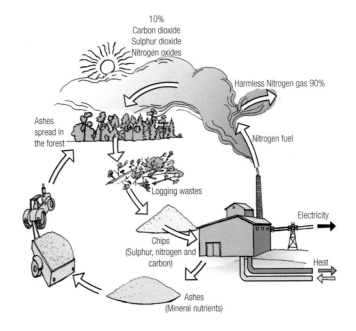

Returning ashes. With the exception of nitrogen, an equivalent amount of the substances released during biomass combustion are returned to the environment and then eventually go up in smoke again when new biomass is combusted. Most of the nitrogen in the biomass becomes nitrogen gas, the most common component of air, which does not have an impact on the environment. Use of biomass counteracts excessive accumulation of nitrogen in soil and water.

Source: illustration Leif Kindgren

organically bound carbon (OGC) and carbon monoxide (CO) are not high.

In large combustion facilities, such as district heating plants, the level of efficiency reaches over 100 per cent, which may seem impossible. This 100 per cent plus efficiency is reached due to the method used to calculate efficiency. These facilities make use of flue-gas condensation, and percentage efficiency includes utilization of energy that is actually outside the efficiency calculation model.

Boilers for Local Heating (500kW)

There are many boilers with auxiliary equipment in this size class. They have a greater output than home boilers (which are available up to 40kW). The largest have an output of up to 800kW. A biomass central heating plant consists of various

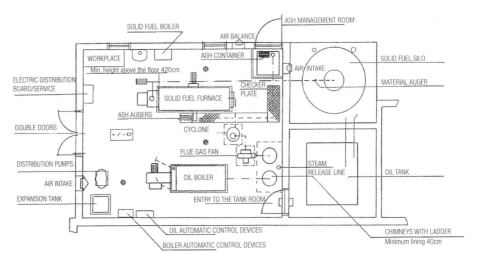

The plan shows a boiler room for biomass at Tolita school in Kils Municipality. It clearly shows how much space everything needs: furnace, fuel silo, flue-gas cleaning, ash management, etc. There is a reserve oil boiler.

Source: Chiron Energi och Fastighetsekonomi AB, Gothenburg

components such as a fuel storeroom, a feeder (e.g. augers and stokers), a pre-oven, a burner, a boiler, ash removal and accumulation tanks. Companies often custom-build central heating systems to suit local conditions. The boilers can be fuelled with different types of biomass, e.g. straw, chips, sawdust, briquettes, wood or pellets. Some systems can also use other fuels.

Boilers for District Heating (3MW)

Many types of biomass can be used: firewood, peat and combustible waste. District heating plants try to achieve the best possible combustion efficiency and flue-gas cleaning. Two kinds of new combustion technologies are used: fluidized beds and gasification.

Fluidized bed combustion involves the suspension of a mixture of fuel, sand and sometimes lime within a rising column of air inside a boiler. Fluidized bed technology makes it possible to use a broad spectrum of fuels. The technology also provides the necessary conditions for limiting sulphur and nitrogen oxide emissions.

The main difference between gasification and combustion is that a smaller amount of air is added for gasification. Through gasification, a fuel is converted to a combustible gas that can be used in a gas turbine to generate electricity. To increase efficiency, the energy in flue gases is used, both to preheat the combustion air and to heat hot water. In such large facilities, it is possible to use advanced flue-gas cleaning, e.g. with electrostatic filters and textile filters.

Household Combustion Devices

Biomass combustion devices that can be used in living rooms include tiled ovens, hot air stoves, wood-burning stoves and pellets stoves. Wood-burning stoves are used in kitchens.

They release heat into the room, but can also be equipped with water jackets and connected to an accumulation tank and waterborne heating system. In weatherproof and energy-efficient buildings it is important to bring in intake air through a special air intake duct. This is especially important when natural ventilation is being used to avoid unwanted reverse flow.

Heating source	Efficiency
Old wood-burning boiler	40–70%
New wood-burning boiler	80–90%
Open fireplace	5–15%
Fireplace insert	50–70%
Tiled oven	50–70%
Modern tiled oven	70–90%
Hot-air stove	50–80%
Pellet burner	80%
Pellet stove	70–90%
Pellet boiler	90%

Wood boilers burn wood to heat the water used to heat a building using its waterborne heating system. Modern wood boilers have a boiler efficiency of 80–90 per cent. This means that if an old boiler is replaced with a new, highly efficient one, it is possible to save 7–8m^3 of wood per year for an ordinary single-family house. Emissions of organically bound carbon (OGC), carbon monoxide (CO) and particles are reduced, e.g. emission of tar substances can be significantly lower (by a factor of 100). Boilers should be connected to an accumulation tank which makes it possible for the boiler to work only at optimum combustion speed and thus maximum system efficiency is achieved.

Pellet burners can be installed on thermal boilers (wood, oil or combination boilers). Replacing an oil burner with a pellet burner is the cheapest way to convert to pellets. The boiler is supplied with pellets from a storage space via an auger or vacuum system. Buildings with a traditional boiler room are most suitable as the pellets require space. Sometimes the space issue can be solved by rearranging use so that a garage or storage area is used. Aside from the need for soot and ash

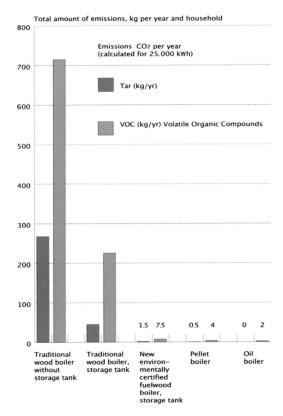

Total release of hydrocarbons in kg per year per single-family house (based on 25,000 kWh) for various heating systems.
Biomass should be burned in a modern, environmentally certified boiler connected to an accumulation tank. There are also environmentally certified tiled ovens, airtight heaters and wood-burning stoves.

Source: Swedish Energy Agency

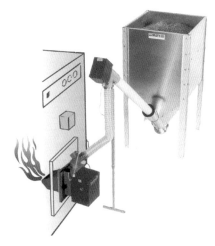

In small properties a pellet burner often replaces an oil burner. The burner is integrated with a storage silo filled with pellets. The silo can be placed 5–10m from the burner if necessary.

Source: Bioenergi-Novator

removal, the equipment is generally fully automatic. A signal can indicate if something goes wrong. As a rule, a pellet burner has an automatic ignition system (electric coil or hot air) and is adjusted with the boiler thermostat.

Pellet boilers are available in sizes from just over 20kW and up. The fuel storage space in boilers can cover requirements for a few days during the winter. An alternative is to feed pellets from a separate indoor or outdoor storage space. The boiler should be connected to an accumulation tank with a backup electric immersion heater. It is important that ash removal is easy as this often has to be done once a week in winter.

A kitchen boiler is a wood boiler surrounded by a water jacket intended for installation in kitchens. They are usually equipped with burners and sometimes with a baking oven. A distinguishing feature of kitchen boilers is that they provide facilities for both

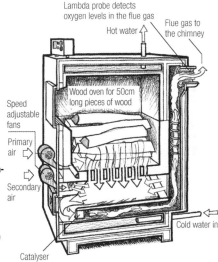

An environmentally friendly wood boiler should have a properly designed combustion chamber, controlled combustion with a lambda probe, speed-adjustable fans and a catalyser. It has a ceramic combustion chamber and uses reverse combustion with primary and secondary air.

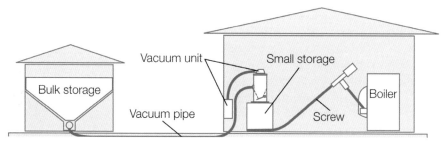

Transport of pellets from bulk storage to smallstorage next to the boiler can also be done with a vacuum system. This is a good solution if it is far from the bulk storage, and if the bulk storage is situated so that screw transportation is difficult. This also makes it possible to have a large pellet storage that can be filled once a year.

Source: Eurovac

cooking food and heating. They can be designed as wood stoves with cooking as the primary function or as small boilers with heating as the main function. Kitchen boilers have become more and more popular in recent years, especially in energy-efficient houses with small energy requirements. The expense of a separate boiler room is thus saved. It has been difficult for kitchen boiler producers to meet environmental specifications, but there are currently a few environmentally certified kitchen boilers on the market.

Wood-burning stoves are traditionally made of cast iron. Modern wood-burning stoves are often made of sheet iron. A wood-burning stove isn't meant just for cooking, but is also a heat source for the whole house. Metal wood-burning stoves (cast iron or sheet iron) must be kept going continually while wood-burning stoves made of soapstone and brick store heat in their mass and can heat a house with a couple of firings per day. Cooking on a wood-burning stove is an environmentally friendly way to cook, providing a modern, environmentally certified stove and dry wood are used. Wood-burning stoves are often supplemented with a gas or electric stove for use in the summer. Brick wood-burning stoves often come as do-it-yourself kits. Drawings and cast iron details can be purchased, primarily from Finland.

Kitchen furnace, with pellets module, that can burn both wood and pellets. The output can be set continuously between 2.5kW and 7kW. The kitchen furnace is programmable and can run automatically. In Källby, outside Lidköping in Sweden, there is a showroom where systems can be seen and tested connected to different types of furnace, accumulation tanks, solar collectors, control units, etc.

Source: Johan Walther, www.lohberger.se.

Baking ovens are often built into wood-burning stoves and kitchen boilers, but separate baking ovens are also available. In Finland, modern baking ovens have been developed that incorporate the tiled oven principle with the addition of an afterburn chamber for secondary combustion above the oven. These ovens are often made of soapstone but are also available as brick do-it-yourself kits. Sometimes they are included in larger units combined with a wood-burning stove. Such a unit placed centrally in a home can constitute the main heating source and provide warmth to all the rooms.

Tiled stoves have two main characteristics: they are built of heavy, heat-storing material and the flue-gas channels go up, then down and then up again to the chimney. The long flue-gas channels make up a large heat-transfer surface area that transfers heat from the flue gases to the heat-storing material. This means that when a fire is lit in a tiled stove, a large part of the heat is stored for later release into the room over a relatively long period of time.

Tiled stoves are usually covered with tiles, but there are also stoves without tiles. The latter were made as early as the 1700s and were called pipe stoves or 'the poor man's tiled stove'.

Old, poorly insulated buildings needed a tiled stove in every room. In new, well-insulated buildings, one stove is adequate for a whole building. New stoves are made according to the old principle, but it is also possible to dismantle and rebuild an old tiled stove. Some county museums that have taken care of old tiled stoves lend them out so they can be of use.

There is also an extensive second-hand market. Traditional tiled stoves were built of brick and clay mortar. New ones can be made of brick, soapstone sections, olivine stone or fireproof casting compound. The latter has a greater heat storage capacity than brick and therefore releases heat over a longer period of time. The combustion

The AGA stove, a famous Swedish invention by Gustaf Dalén. It burns wood and has a water jacket.

A Finnish bake oven built according to the tiled stove principle in a kitchen, Alma farm, near Arboga, Sweden.

chamber in modern stoves can also be better designed from a combustion perspective, and there are some (especially Finnish models) that have secondary air supply and an afterburn chamber above the combustion chamber. Another method of increasing the efficiency of stoves is to install an air intake duct that brings air from the outside into the combustion chamber so that already heated indoor air isn't used. Tiled stoves are heavy and require a mounting base or reinforced floor structure.

Air-jacketed stoves heat a cold house faster. These stoves are equipped with two air ducts that take cold air from the floor, warm it inside the stove and expel it from openings in the upper part of the tile oven. It is often possible to then close the openings so that the mass of the stove is heated instead.

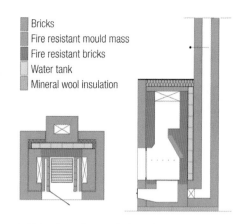

In Finland, architect Heikki Hyytiäinen has developed an improved stove. It is built in three layers: the innermost layer is a combustion chamber and an afterburn chamber made of fireproof bricks and cast fireproof compound elements, the middle layer consists of flue-gas channels made of fireproof brick, and the outer layer can be designed in various ways. Secondary air supply is distributed half way up the combustion chamber. The chimney reaches to the bottom, which reduces temperature stress and increases its life. This kind of stove is called a masonry heater.

Source: Drawing of a mass oven made in consultation with Johannes Riesterer, The Swedish Earth House Association (*Svenska Jordhus ek. for.*).

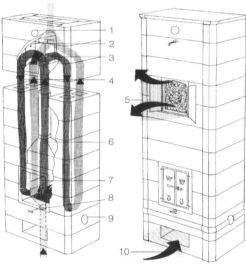

1. Inspection hatch
2. Flue gas damper
3. Separating wall between the front and back channels
4. Flue gases
5. Heated room air
6. Combustion chamber insert
7. Combustion chamber
8. Combustion air duct
9. Soot hatch
10. Unheated room air

The Wasa stove is a modern stove built of olivine stone. The combustion chamber was developed at a university (the Royal Institute of Technology, Stockholm), and combustion air is taken from outdoors. The built-in channels can quickly heat the air in the room when necessary.

This beautiful masonry heater in Sweden is designed by Vera Billing and built by Johannes Riesterer, Svenska Jordhus. Johannes has built most of the masonry heaters in Sweden.

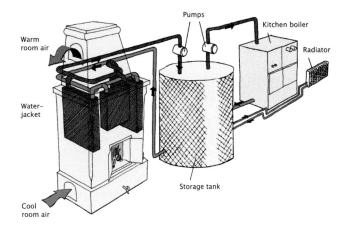

Water-jacketed stove in operation with a storage tank, a kitchen boiler and radiators.

Water-jacketed stoves are equipped with a heat exchanger where some of the heat (about 30 per cent) is transferred to a water coil in the stove. This heat can be stored in an accumulation tank and used to heat hot water and/or radiators. A stove with a water jacket can serve as the main heat source for a small, very well-insulated building. A water-jacketed tile oven often requires an electric pump to move the hot water to the accumulation tank, but there are attempts to use gravity to move the water in water-jacketed stoves.

Wood-burning stoves are small stoves that are so light that the floor structure does not need to be reinforced. Wood-burning stoves

An air-jacketed pellet stove. There are pellet stoves on the market with automatic fuel feeds and efficient combustion obtained using a fan plus primary and secondary air intakes. They are an attractive alternative for home heating.

A sand oven from Stockholm's Kakelugnsmakeri. The aluminium or copper shell is filled with 750kg sand, which makes up the heavy mass around the combustion chamber and piping. Installation by a layperson takes about two hours.

are often used for peak heat in combination with another heat source.

Wood-burning stoves can be made of metal or of mineral materials such as soapstone, or of metal with an insert or a heat storage layer of mineral material (metal + mineral). A stove can have an air jacket and a fan to blow out the heat (called a hot air stove). Wood-burning stoves are most often made of cast iron or sheet metal. A metal stove heats up and cools down quickly and requires regular attending and refilling with fuel. Good combustion can be achieved in several different ways. The design of the combustion chamber and how the combustion air is brought into the fire are significant aspects that affect combustion.

In a soapstone stove, some of the heat is stored in the soapstone mass which releases heat for a while after the fire has gone out. Thus, heat is released over a longer time period, and some of heat peaks, which can cause excess temperatures in an energy-efficient building, can be capped.

There are also water-jacketed wood-burning stoves that can be connected to accumulation tanks.

A **hot-air stove** is in principle an airtight stove with an air jacket. Hot-air stoves provide both radiant heat and warm air. The stove is therefore equipped with double walls. Cold air is taken in at floor level, heated up in the air jacket and blown out with a fan from openings in the upper part of the stove. There are also hot-air stoves that get their combustion air from outdoors in order to avoid heating with air that is already warm. Some hot air stoves are covered with tiles and are called tiled stoves.

Pellet stoves can be used in energy-efficient buildings. One advantage of a pellet stove is that a special boiler room is not required. The stove can be placed in the living room. If there isn't a waterborne heating system, it is possible to install an air-jacketed pellet stove. If there is an open floor plan, the stove can provide 70 per cent of heating requirements, which means that they can be a good investment in homes heated with electricity.

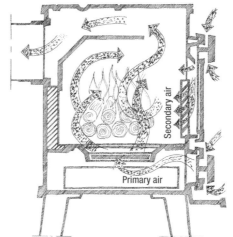

Environmentally certified stoves use downdraught and are designed to take in both primary and secondary air. The cast iron and ceramic combustion chamber provide even combustion. The combustion air is preheated in the double cast iron cover. The illustration shows a Morsö heater.

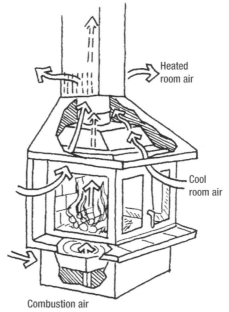

Hot air stoves are in principle air-jacketed fireplaces. Sometimes a fan is used to circulate the air in order to more quickly heat the house.

Water-jacketed pellet stove with 95 per cent efficiency.
Source: Wodtke.

A pellet stove is made up of a combustion chamber with a built-in fuel hopper that provides about 24 hours worth of fuel during the coldest part of the winter. There is a convection component (air jacket) that quickly leads heat out into the room air. Highest efficiency is achieved if a stove is centrally located. Modern pellet stoves are equipped with a speed-controlled fan and automatic fuel feed and can run unattended for several days. They come ready to be connected to a room thermostat.

If there is a waterborne heating system, a water-jacketed pellet stove can be connected to it. An advantage of these is that a special boiler room is not required. The stove heats a building both directly via radiant heat and convection and indirectly as it heats water that can be stored in an accumulation tank.

Stove inserts are built into open fireplaces since these always have a low level of efficiency and some open fireplaces may even cool more than they heat if the dampers aren't adjusted properly. Therefore stove inserts (cassettes) are available that are installed in open fireplaces so that the hearth can be closed off using doors. Stove inserts are usually made of metal and have an air jacket in order to heat room air more efficiently, or a water jacket to heat hot water. Some stove inserts have both air and water jackets. Glass doors are common so that the fire can be seen. If there is no existing fireplace, it is possible to build a new one with a stove insert. These are called masonry fireplaces/stoves and are available as do-it-yourself kits.

A water-jacketed stove insert installed in a beautiful fireplace.

A water-jacketed pellet stove is all you need in an energy-efficient house to get hot water and heating. The pellet storage is big enough for approximately one week's operation.

Source: Palazzetti

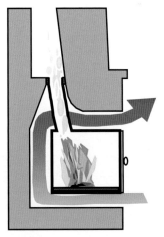

A stove insert.

Chimneys

In the past, chimneys were built of fireproof brick, which is of course also possible today. Chimneys built using ready-made modules are also currently available. These can be made of metal or mineral material. Metal chimneys are usually double-walled and insulated and are available in a variety of sizes. Chimneys made of mineral material often consist of modules with an inner liner and outer casing that are separated by an insulating air gap. The casing can be made of expanded clay or pumice and the liner can be made of ceramic material or pumice. The modules are notched and are either glued or held together with mortar. For resource reasons, chimneys made of mineral material should be prioritized; however, sometimes due to a shortage of space it is necessary to use metal chimneys. There are also chimneys with two pipes, one for air supply to the fireplace and one for the smoke. In this kind of chimney the combustion air is preheated by the flue gases.

In old buildings, chimneys may be too large or cracked. There are methods to repair them so that they work again. Either an insert pipe can be put inside the old chimney or the chimney can be sealed internally with ceramic mortar.

Chimney cowls may be required to protect chimneys from rain. There are chimney cowls that self-adjust according to wind direction and improve the draught in addition to protecting the chimney from rain.

A **draught limiter** can be installed in a chimney since a chimney also acts as a ventilation channel. This means that if left open, the chimney can cool down the boiler, the boiler room or the whole building. A draught limiter closes the chimney when there is no fire.

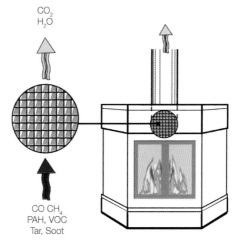

Wood-burning stove with catalytic flue-gas cleaning.

Spark-arrester. Sparks from the chimney flue and embers from wood ashes can cause fires. The sparks are made to rotate in a container where they are extinguished against the walls.

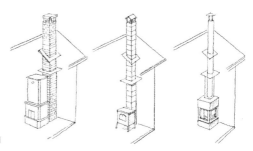

Chimneys can either be made of masonry, built of special chimney components, or made of double-walled metal pipes with insulation between the pipes. Masonry chimneys are heavier and are mostly used when several chimney pipes are needed.

The KMP Neptuni stove is intended to be the primary heating source. It can be combined with a normal chimney or KMP Drag, a fan-controlled waste gas flue that can go out through a wall.

Source: Ariterm.

3.1.2 Solar Heating

Solar heat can be used to heat tap water or for both space heating and hot water. The scale of solar heating applications varies. Most common are systems for single-family homes, where 5–10m² of solar collectors are put on the roof. Roof-integrated solar collectors are used on block of flats. Large solar collector systems connected to a district heating network and/or seasonal storage often consist of a solar collector field where solar collectors are mounted on the ground. Another way to use solar collectors is for heating swimming pools. Since the temperature required for pools isn't so high, these solar collectors are often simpler and unglazed. There are also solar collectors that heat air.

Solar Heating Systems

Advertising for solar collectors often mentions their energy efficiency, but it is not enough for the solar collector to be efficient. The whole system has to be efficient. Therefore the term solar fraction is used, i.e. the portion of energy for hot water and space heat that can be replaced with solar energy as compared to a conventional system. Thus the energy that is used to cover heat losses in the system is included in the calculation.

One of the world's largest single energy storage units is the aquifer reservoir at Arlanda airport near Stockholm. The water temperature in the aquifer is used to cool the buildings in the summer and to melt snow in the winter in the parking areas for the aeroplanes.

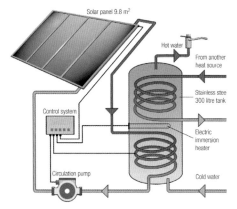

Solar collector systems consist of solar collectors, an accumulation tank and a system package that includes a pump, check valve, filter, expansion tank, filling device, safety valve, thermometer, pressure gauge and regulating equipment. The parts are connected together with insulated pipes.

Source: illustration Leif Kindgren

The Kungälv solar collector field is one of the biggest in Europe. There are 10,000m² of ground-mounted collectors that take up an area of 35,000m², more than five soccer fields. The solar collectors are made using technology including, among other things, anti-reflection treated glass that results in greater efficiency than other solar energy systems. The system has an output of about 4GWh and contributes about 4 per cent of the municipality's heat production per year.

Solar collectors on semi-detached houses in Smeden eco-village, Jönköping, Sweden.

Source: Architects: Jan Moeschlin and Peo Oskarsson. Photo: Lars Andrén

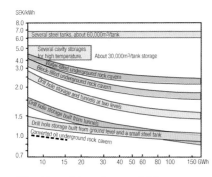

Different types of storage for seasonal storing.

Clay storage

Drill-hole storage

Steel tank

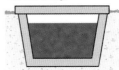
Cavity storage

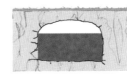
Underground cavern

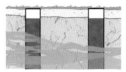
Aquifer storage

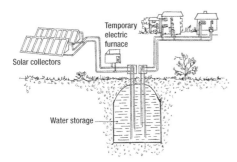

Solar heat with seasonal storage has been tried in Sweden. The bigger the project the more economical it is. The goal is 75 per cent solar energy and 25 per cent other energy. The illustration shows Lyckebo in Uppsala, built in 1983, which consists of 550 flats. A total of 4300m² of solar collectors heat up 100,000 litres water, which are stored in an uninsulated cave 30m underground. Solar heat provides only a small portion of the heat for the homes in Lyckebo.

Orientation and Angle

Naturally, the ideal solar collector orientation is facing due south, but a deviation of +/-30° reduces the collection of energy only marginally (maximum 10 per cent). If the deviation is greater it is possible to compensate somewhat by increasing the collector surface area. It is of course important that the solar collector area is not shaded, e.g. by deciduous trees.

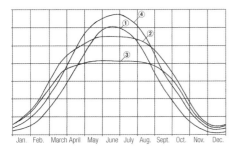

1) 0° angle to the horizontal plane
2) 70° angle to the horizontal plane
3) 90° angle to the horizontal plane
4) 45° angle to the horizontal plane

Estimated average insolation per 24 hours for south-facing surfaces at different angles. If solar energy is used during a large part of the year, 60°–70° is a good angle. Less comes in during the summer but not as much heat is needed then anyway.

Source: Gunnar Lennermo, Energianalys AB, Alingsås

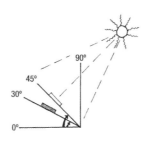

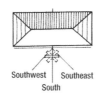

Solar collectors should be oriented between southeast and southwest.

In the Stockholm area solar collectors should be at a 45° +-15° angle, i.e. between 30° and 60°. A lower angle is better if the solar collector is only used for hot water during the summer, but if it is also going to be used for heating during as much of the year as possible a higher angle is preferable.

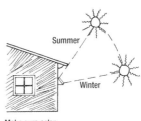

Make sure solar collectors are not shaded.

Flat-plate solar collector

Absorbers that capture the heat lie in an insulated box

Evacuated collectors

Absorbers are placed in vacuum tubes

Concentrating collectors

Solar radiation is focused on the absorbers with the help of mirrors.

There are three categories of solar collectors: flat-plate, vacuum and concentrating collectors.

How the Energy Efficiency of a Solar Collector is Affected by Its Orientation and Angle.

It is recommended that solar collectors be placed at an angle of between 15 and 65 degrees.

Angle above horizontal plane	15°	30°	45°	60°	90°
Orientation in relation to sun					
0 degree (= due south)	0.93	1.00	1.04	1.04	0.90
30 degrees	0.90	0.98	1.01	1.01	0.86
45 degrees (south-west/south-east)	0.90	0.95	0.97	0.96	0.82
60 degrees	0.88	0.91	0.92	0.90	0.76

Types of Solar Collectors

It is common for solar collectors to be confused with solar cells. Solar collectors are used to provide heat while solar cells produce electricity. The expression 'solar panel' is sometimes used in both cases.

There are three categories of solar collectors: flat-plate solar collectors, vacuum collectors and concentrating solar collectors.

In the early 2000s concentrating solar collectors began to be developed that are efficient enough for use when the sun is as low as it is as far north as Sweden.

Flat-Plate Solar Collectors

Flat-plate solar collectors that heat water are the most common type of solar collector. These are insulated boxes with absorbers inside that are covered with sheets of glass or plastic. There are also flat-plate solar collectors that can be integrated into roof construction. The cheapest way to get solar collectors is to build them yourself, for which there are special do-it-yourself kits.

A flat-plate solar collector consists of absorbers in which water or oil is heated. The absorbers are covered with glass or plastic and insulated on the back. Between the

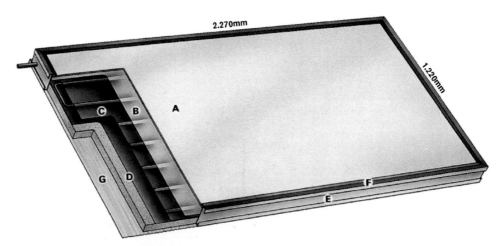

In Sweden, flat-plate solar collectors are most common. The illustration shows the construction of a TeknoTerm collector.
(A) toughened glass with low iron oxide content. (B) Teflon layer with good translucency. (C) absorber with world patented technology. (D) high temperature insulation with fibreglass surface. (E) aluminium frame. (F) EPDM rubber weather stripping. (G) back section of corrugated aluminium.

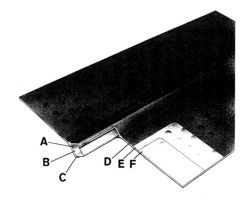

The development of the 'Sunstrip' absorber has resulted in Swedish solar collectors that are durable and of high quality. Aluminium and copper are compressed under such high pressure that there is metallurgic contact between the metals and thus corrosion between them cannot occur. (A) copper pipe, (B) small amount of water, (C) metallurgical contact, (D) selective surface, (E) stabilized surface, (F) aluminium flanges.

absorbers and the glass there can be a convection barrier and between the insulation and the absorbers there is a layer that prevents emissions from the insulation dirtying the glass.

Single-family home solar collectors for hot tap water are usually 5m^2 in size and provide half of annual hot water requirements. If solar heat is to be used for heating and hot water between 10 and 15m^2 are needed, which usually provides about one-third of the total requirements. Purchase of complete solar heating packages is usual. Prefabricated solar collectors are units that can be placed on roofs, walls or on the ground. Roof-integrated solar collectors are part of the roof construction itself, and are often both more attractive and cheaper than prefabricated ones. If roof-integrated solar collectors cover the entire surface of the roof, the way in which they are connected to the roof ridge, base and gable is important. How pipes are connected to the solar collector also affects the appearance. Solar collectors for summer cottages are simple and inexpensive systems that are not equipped with antifreeze and a pump. The water in such solar collectors is filled using the pressure in a conventional water system, is heated in the solar collector and flows by gravity to the hot water tank. Experience has shown that a reflector in front of a flat plate collector can increase the efficiency by approximately 50 per cent.

Vacuum Solar Collectors

Vacuum solar collectors are designed as oblong glass tubes with an absorber in the middle enclosed in a vacuum to reduce heat loss. The tubes are mounted beside each other in complete units. Vacuum solar collectors are in principle used in the same way as flat-plate collectors. Although they are more expensive per m^2 than flat-plate collectors, they provide more energy per unit, so the total economics are similar. Vacuum solar collectors are used where there is a shortage of space. They look different architecturally as well.

Vacuum solar collector.

There are several different types of vacuum solar collectors. The absorber can be made of metal or black painted glass. The water in some models is heated by circulating inside the absorber, and in other models the heat is taken from the absorber to the heating circuit in a heat pipe.

Swimming Pool Solar Collectors

One of the most economical ways of using solar heat is to heat swimming pools. Pool solar collectors often have a simple construction. They don't require antifreeze and they don't need to reach high temperatures. They are made of plastic or rubber material and are often used without insulation or a glass cover. Outdoor pools are only heated during the summer when there is lots of sun, and an accumulation tank is not needed since the pool water is heated directly in the solar collector.

A good supplement to a pool solar collector is a radiation loss cover to use at night to reduce pool cooling in the cold night air.

Solar Collector Implementation

In 2006 28,500m^2 of solar collectors were installed in Sweden, which is 24 per cent more than the year before. During the same period in Germany 1,500,000m^2 were installed, which was 58 per cent more than during 2005. Austria installed 300,000m^2 (a growth of 25 per cent). In Spain 175,000m^2 (a growth of 81 per cent) were installed and in Italy 186,000m^2 were installed (a growth of 46 per cent). But the biggest installer in the world is China with 14,500,000m^2 (during 2005). In Sweden vacuum solar collectors are more popular than flat-plate solar collectors, taking 30 per cent of the market in 2006.

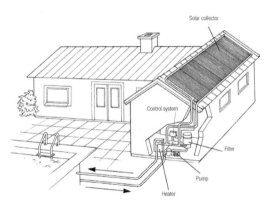

A solar collector area equivalent to slightly more than half a pool's surface area can replace all other additional heat, and at the same time, the swimming season can be extended a few weeks in spring and autumn. A large part of a pool's existing heating system equipment can be used. The pool water acts as heat carrier and heat storage.

Source: *Värmeboken – 20° till lägsta kostnad*, Lars Andrén och Anders Axelsson, 2000

Roofing tiles made of glass by SolTech Energy. They act as a roof, a watertight layer and as solar air collectors to heat water in the accumulator tank, via a heat exchanger (air to water).

Solar Air Collectors

In solar air collectors, the sun heats air and not water. Solar air collectors can be integrated into roof and wall constructions or purchased prefabricated.

Closed air solar heating systems are an application where air is heated in an air solar collector built into a south-facing roof or wall. Fans move the warm air down a gap in the outer wall or floor structure, which it heats and then returns to the solar collector. In the summer, with the use of a heat exchanger, the hot air can be used to heat water.

In open air solar heating systems, the intake air is preheated in an air solar collector on a south façade before being carried into the building, which reduces heating requirements. To avoid excess temperatures in the summer, it has to be possible to turn off the preheating.

A problem with summer cottages that are left unheated is that damp problems can occur. Solar air collectors for summer cottages are installed on a south-facing outside wall. Warm air is led directly from the solar collector into the cottage by a solar-cell-run fan that starts up when the sun shines. The warm air helps to dry and heat the cottage. Solar dryers are used in agriculture to dry grain and hay. They usually consist of a roof or wall designed as an air solar collector and fans that blow the heated air.

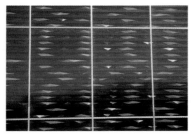

In Umeå, northern Sweden, there is a car dealership located in an environmentally designed facility called Green Zone. Intake air to the car dealership rooms is brought in through an air solar collector which preheats the air. In northern Sweden it can be warm in the sun despite below-zero temperatures.

Source: Architect: Anders Nyquist

Solar air collectors are also used in agricultural solar dryers to dry hay or grain.

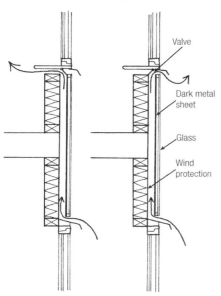

Intake air can be preheated in a window back designed as a solar wall. Fresh air is taken into a gap between the wall insulation and a black-painted sheet of metal that is glazed on the outside. It must be possible to short-circuit the system in the summer so that it doesn't overheat the room.

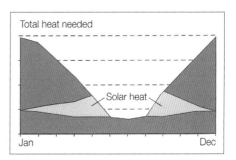

Solar air collectors can be used to preheat fresh air, and 10m² per apartment can cover 10–20 per cent of the heating demand.

3.1.3 Heat Pumps

The development of heat pumps has been very rapid. The more energy efficient a house is and the smaller the heat requirement, the less likely it is for a heat pump system to be profitable. Heat pumps can be a good investment for houses with an energy consumption over about 15,000kWh/year (for space heat and hot water). On the other hand, for houses with a smaller annual energy consumption, a heat pump may be an unwise investment. Large heat pumps are used to heat blocks of flats, neighbourhoods, or to contribute heat to the district heating network.

The Heat Pump Principle

Heat pumps employ technology that make use of the heat at low temperatures and by using a small amount of high-quality energy (e.g. electricity) increase this heat to a medium high temperature (e.g. 60°C), which is enough to heat a building. It is based on the principle that a gas/liquid (a refrigerant) has different temperatures and states at different pressures. Most heat pump compressors are electric but there are also heat pumps running on liquid fuel motors. The technology is used for both heating and cooling on both a small and large scale.

Size

It is wasteful to dimension a heat pump for a building's maximum heating requirement. The maximum need applies only a

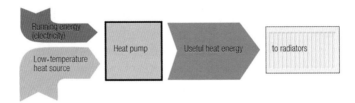

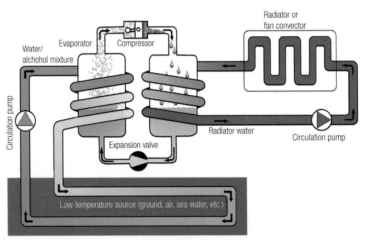

How a heat pump works. By changing the pressure, the temperature and boiling point of a refrigerant (gas/liquid) can be changed. A low-temperature source is used to make liquid in an evaporator change to a gas, the gas is compressed and the temperature rises. The warm gas heats radiator water and condenses to a liquid. The pressure and temperature of the liquid is reduced in an expansion valve and the cycle can begin again.

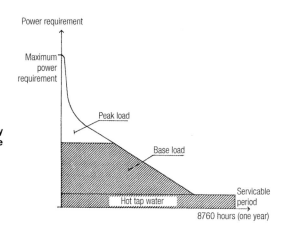

Design of a heating system should be based on a duration curve for the project in question. The annual heat and power requirement can be obtained from the curve, and this information can be used to determine the best way to meet hot water, base load and peak load requirements.

few days per year. Therefore heating is usually divided up into a base load and a peak load. The base load can, for example, be supplied by a heat pump and the peak load by another energy source, e.g. a combustion device or electric heater. A heat pump is specified according to energy consumption and power requirement. A general rule of thumb is to choose a heat pump with an output equivalent to about 50 per cent of a building's maximum power requirement. Then the heat pump provides 80–90 per cent of the annual energy requirement for space heating and hot water. Therefore, when it is extremely cold, heat pumps must be complemented with peak heat from a hot air stove or electric immersion heater.

The Heat Pump Debate

Heat pumps have been controversial for several reasons. The main reason they have not been considered desirable from an environmental perspective is their use of freon, which has a negative impact on the ozone layer. Another reason is their electricity consumption, and a third is reliability and cost.

Development of environmentally friendly refrigerants that can replace freon is under way. The goal is to find refrigerants that don't affect the ozone layer or contribute to climate change. Such heat pumps are already available for small-scale use (e.g. refrigerators) and for large-scale use in district heating plants. Work is also under way to develop heat pumps for homes that use environmentally friendly refrigerants. Currently, 'soft freons' are used, which don't affect the ozone layer but do affect global climate.

Regarding dependence on electricity, the reasoning is that electricity is three times more valuable than liquid and solid fuels, since using fuels to produce electricity only has an efficiency of about 30 per cent. This means that a heat pump should have a heating factor greater than three in order for its use to be justified. The heating factor describes the relationship between the energy obtained (for heating a home) and the high-quality energy supplied (to run the compressor), and is usually about 3:1. As heat pumps with a heat factor greater than three are beginning to emerge, heat pumps can also be justified from an energy viewpoint. However, if the extra electricity required is produced with nuclear power or coal, the energy isn't renewable. A better environmental alternative to avoid dependence on electricity is to use fuel-powered heat pumps, especially if they are run with biofuel.

Heat pumps have been expensive and breakdowns have been a problem. Insurance company statistics indicate that the transition to environmentally friendly refrigerants caused lubrication problems for heat pumps that resulted in a lot of breakdowns. Problems with breakdowns occurred more often for some models than others.

Refrigerants' ODP and GWP Properties.

Refrigerant	ODP = Ozone depletion potential	GWP = Global warming potential	Comments
R12	1	7100	Prohibited* 1998
R22	0.05	1600	Prohibited* 2002
R502	0.23	4300	Prohibited* 1998
HFC			
R134a	0	1200	Not for freezers
R404a	0	3520	
R407c	0	1600	
Natural refrigerants			
R290 propane	0	0	Inflammable
R600a isobutane	0	0	Inflammable
R717 ammonia	0	0	Poisonous

Note: * refers to refilling

Sizes and Types of Heat Pumps

Heat pumps are commonly available in two sizes, large heat pumps and home heat pumps. Electricity or motors can be used to run large heat pumps, which are placed in a heat pump station connected to a local or district heating plant. There are three types of home heat pumps: exhaust air, outside air and geothermal heat pumps. The latter can obtain heat from both the ground and watercourses.

Home Heat Pumps

Small home heat pumps using low-temperature heat from the environment are quite common. These bedrock, lake and ground-source heat pumps have a

Low-temperature heat for a heat pump can be extracted from outside air, the ground, sea water, or from a hole drilled in the ground. The higher the temperature of the low-temperature heat is when heat is needed, the more economical the heat pump system will be.

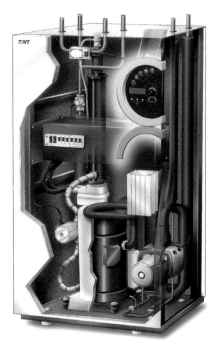

A home ground-source heat pump consists of a compressor, a pump, a heat exchanger and control devices, among other things.

collector pipe that is placed in a hole drilled into bedrock, or laid underground or on the bed of a lake. The collector pipe carries the low-temperature heat to the heat pump, which is often the same for all three types. There are two ways to capture the energy. Indirect systems are most common, where a water/alcohol mixture circulates and then exchanges heat with a refrigerant in the heat pump. The other method is called direct vaporization, where the refrigerant itself is pumped through the collector pipe. There are also exhaust air heat pumps that take low-temperature heat from ventilation exhaust air, as well as outside air heat pumps that take it from the outside air.

Geothermal, seawater and ground-source heat pumps use the heat from the sources their names refer to. A geothermal heat pump uses heat from 60–170m holes drilled into bedrock. Horizontal ground-source heat pumps extract heat from the ground via 200–400m-long collector pipes buried at a depth of about 1m. The pipe is laid to and fro across a plot, spaced about 1m between rows. This of course requires a certain amount of land. For lake heat, the collector pipe is weighed down on the bed of a lake or watercourse. Whether or not heat is extracted from bedrock, the ground, or water depends on the location and type of land.

Exhaust air heat pumps make use of heat in ventilation air for space heating and heating hot water. Exhaust air heat pumps are installed in homes with mechanical ventilation, but there have also been attempts to use them in naturally ventilated homes. Exhaust air heat pumps are approximately the size of an ordinary free-standing cupboard and noise levels vary greatly. Exhaust air heat pumps that only heat hot water are also available; they are cheaper but save less energy.

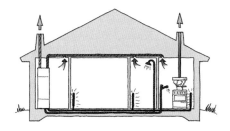

Exhaust air heat pump in a single-family home.

Air-source heat pumps take low-temperature heat from outdoor air. The heat is either released indoors using a fan and air-to-air heat pump, or the heat is transferred to the building's waterborne heating system by an air–water heat pump. An air–water heat pump can heat tap water as well as the building. For houses that would otherwise use direct electric heat, air-source heat pumps usually provide a 30–40 per cent saving on the annual heating requirement. The problem with air-source heat pumps is that the colder it is, the more electricity is needed for heating. When the outdoor air temperature drops to –10°C, an air–air heat pump doesn't provide a house with any appreciable heat, and direct electricity is used for heating. Air-source heat pumps therefore contribute to an increase in the peak load when it is coldest and the electricity grid is under greatest stress.

Operating experience with home heat pumps is not so good. Too many breakdowns occur early in the life cycle of the product (after less than five years) and affect heat pumps' reputation. Some models and makes have clear manufacturing faults. Better operating performance should be required of an expensive product, with an estimated operational life of 20 years. However, good heat pumps are available.

Heat pumps combined with solar collectors and boreholes. Since the late 1990s ground-source heat pumps combined with solar collectors have been used. The sun provides the energy during the summer and the ground-source heat pump provides energy during the winter. When there is a surplus of solar energy the ground is heated and when there isn't enough solar energy to heat the building the solar collector increases the temperature of the liquid pumped through the borehole and thus increases the efficiency of the heat pump.

Development of heat pump systems. A relatively new development is combined solar collector – ground-source heat pump systems. Energy is provided by the sun during the summer and the ground-source heat pump provides energy during winter. When there is a surplus of solar energy the ground is heated and when there isn't enough solar energy to heat the building the solar collector increases the temperature of the liquid pumped through the borehole, which increases the efficiency of the heat pump. There are also heat pump systems that combine solar collectors and heat pumps with heat recovery from intake and exhaust air.

Large Heat Pumps

Large heat pumps take low-temperature heat from lakes and rivers, sewage pipes, exhaust air in large ventilation systems, mines or industrial waste heat. Large heat pump facilities are usually custom-made for a specific situation. Heat pumps are especially of interest where both cooling and heating are required, e.g. large supermarkets that need to be heated but also have large cold-storage areas, or a skating rink where the ice needs

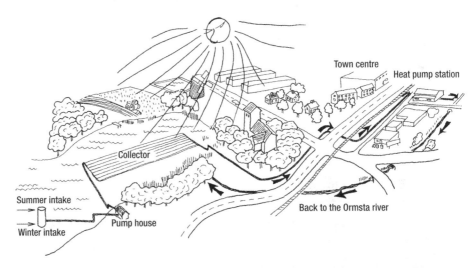

In Sweden, most heat pumps (about 90 per cent) are used in single-family homes. However, there are quite a few large heat pump systems connected to district heating systems (about 7TWh). The illustration shows the successful lake-water system in Vallentuna. Unfortunately, many large heat pumps still use environmentally hazardous refrigerants and leaks do occur.

to be cold but the stands and other spaces need to be heated. In Vattenfall's 'Project 2000' programme it was found to be worthwhile to invest in recovery of condensation heat from refrigeration equipment in grocery stores to use for space heating.

Another application for large heat pumps is heating indoor swimming pools. In this situation, the exhaust air is so moist that an exhaust air heat pump can remove heat from the exhaust air and condensation heat, i.e. the energy released when the moist air condenses to water. An important task is to develop heat pumps that use refrigerants without a negative environmental impact.

Heat Pumps, Thermal Mass and Boreholes

Office buildings often need both heating and cooling. One way of avoiding the use of external energy is to activate the building's thermal mass to even out heating and cooling needs by exposing the thermal mass to the interior. This effect can be increased by integrating radiant piping systems in the slabs to move heat from one part of the building to another. The latest development is to combine the thermal mass of the building with the thermal mass under the building by drilling boreholes into the ground and connecting the two masses with a heat pump. One example of this technology is a project by Steven Holl Architects in Beijing, where a new town block of 160,000m^2 is combined with 660 borehole heat exchangers to provide a total of 5000kW cooling and heating.

Geothermal Heat

There is a lot of free energy within the Earth. In different geological formations water of different temperatures can be found. In certain places, in volcanic terrain, steam emerges from the surface and this can be used to produce electricity, as in Iceland, Italy, Japan and the US. Geothermal heat sources may also be developed using water from other geological strata to heat power stations. Heat pumps are used to transfer the heat to the district heating network. The heat produced may be used for district heating.

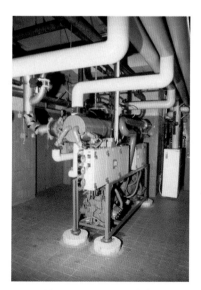

A large heat pump in Sundsvall, Sweden that extracts heat from the municipal sewage system.

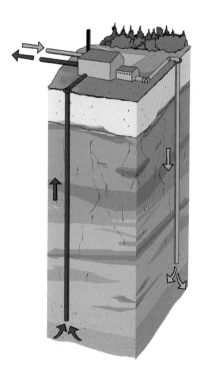

Geothermal heat is obtained from very deep holes. It is important for the production hole to be situated a considerable distance from the hole used to return the used water to the ground.

Source: Leif Kindgren

HEAT PUMPS OF THE FUTURE

A problem with conventional heat pumps is their dependence on electricity. There are gas-powered heat pumps, which are powered by what is in principle an ordinary car motor. There has been experimentation in Japan and Denmark with heat pumps powered by Stirling engines. The technology is of interest as Stirling motors and heat pumps complement each other. Both have a warm and a cold side, but the temperature levels are different. Stirling motors are heated with fuel on the warm side and cooled by a radiator circuit on the cold side. The warm side of a heat pump is heated by return water from the radiator circuit, and the cold side is cooled with a ground pipe. Compressor-driven heat pumps require a ground pipe that is half as long as that required by ordinary ground-source heat pumps. Test results show that the price of heat pumps driven by fuel (gas, liquid or solid) is somewhat higher than electric heat pumps. The energy efficiency of Stirling heat pumps is almost as good as for electric compressor heat pumps, but no electricity is used. What is more, Stirling engine-driven heat pumps are quiet and their exhaust gases are considerably cleaner than those of gas-powered internal combustion engines. Compared to heating directly with gas, there can be gas savings of 40 per cent.

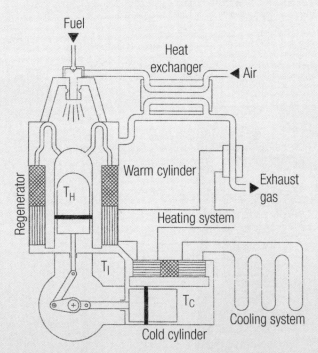

Diagram of a gas-powered Stirling engine that runs a ground-cooled heat pump.
Source: *Individuelle gasdrevne varmepumper*, Henrik Carlsen, H. C. Aagaard, Prøvestationen for varmepumpeanlæg, Denmark, 1998

3.1.4 Cooling Buildings

Cooling of buildings is common in warm countries, and in many countries air-conditioning is beginning to consume huge amounts of electricity. Even in northern Europe, where cooling is not a big problem, some buildings need to be cooled in summer, e.g. offices. This cooling requirement can be reduced using passive methods such as night cooling (ventilation during the night) and building with heavy construction materials that store the cold. The rest has to be done using active methods.

Methods of Cooling

The most common way to meet cooling needs is to install air-conditioning, but in recent years the use of district cooling is increasing. Cooling technology has also recently become available that doesn't require great amounts of electricity, such as evaporative cooling and solar cooling.

Free Cooling

Free cooling means that cooling is obtained from cold water on the bed of a lake, cold air in pipes lying deep in the ground, or from nearby boreholes.

Free cold can be stored from winter to summer as melting ice and snow. This is done using an old method where blocks of ice are cut out, stored and covered with sawdust to insulate them. Old icehouses can still be found. Snow may also be saved from the winter and stored in a pile. The pile is insulated with wood chips and the melt water produces cooling for a building during the summer.

One of the world's largest single energy storage unit is a groundwater aquifer reservoir at the Arlanda airport near Stockholm. Cold water is pumped up from a depth of from 15–25 metres and warmed-up water down a system of 11 groundwater wells. By use of a large heat exchanger, the groundwater provides cooling in the summer and heat in the winter.

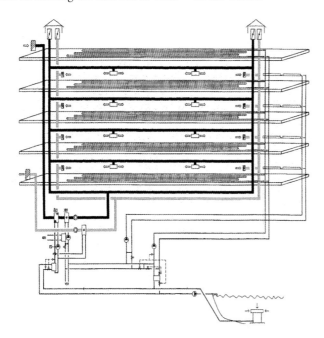

Principle for cooling/heating the floor structure in the new office building of White Architects in Stockholm (finished in 2003). This building uses free cooling via cold water from a nearby lake.

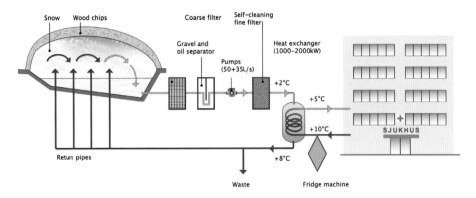

Basic diagram of the snow-storage system at the hospital in Sundsvall.

- 77–93 per cent of cooling needs in the summer are met, which is equivalent to 655–1345MWh
- the output is designed for about 2.5MW
- electricity savings are about 90 per cent compared to earlier cooling machines
- the snow storage area is the size of 1.5 soccer fields, about 9000m^2
- the storage area is filled during the winter with about 40,000m^3 of snow
- the snow is insulated with about 20cm of wood chips.

Source: Illustration Leif Kindgrent

District Cooling

District cooling is supplied almost exclusively to companies, but blocks of flats may also use district cooling. The biggest district cooling system in Europe is in Stockholm. Sea water is the source of 80 per cent of the cooling. The temperature of the cold water on the bed of the sea is between 1 and 5°C. It is taken into district heating plants (also district cooling plants) to a vaporizer in a heat pump where district heat is generated. There, the water is cooled a few more degrees. It then goes to a heat exchanger where it is cooled down further. District cooling in Stockholm is delivered in special pipelines at a temperature of about +6°C.

Air Conditioning

The most common air conditioners used around the world are based on heat pump technology. There are various types: the simplest are built into a wall or window with the warm part outside and the cooling part inside. The problem with these is that the heat released from the warm part can counteract the cooling of the room. Split type air conditioners solve this problem since they consist of two or more parts – one or more cold parts and one warm part – with pipes or hoses between them. There are also

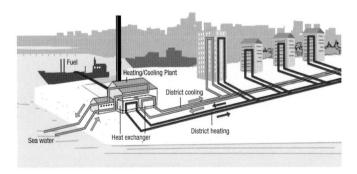

Many towns have an elaborate district heating system where the heat is produced in a district heating plant using biomass and/or heat pumps. More and more towns supplement district heating with district cooling. If heat is produced with a heat pump, heating and cooling systems can be combined and thus be more efficient.

Source: Illustration: Leif Kindgren

Technical Possibilities for Increasing the Efficiency of Air Conditioners

Technical improvement	Increased efficiency
Increase the size of the cooling flanges by 45%	11%
Two extra pipes with cooling flanges	16%
20% less room between the flanges	16%
Better flange design	11%
More efficient heat transfer (refrigerant to pipe)	8%
15% more efficient processor	8%
Speed adjustable compressor	10–40%
Electronic expansion valve	5%

Source: European SAVE study presented in *Appliance Efficiency* 3/99

air conditioners with systems of air intake ducts to different rooms. Cooling can take place with either cooled air or cooled water in cooling coils. The heat pump's warm side is cooled with water or outside air. The level of efficiency is measured in cooling (W) per added electricity (W). Water-cooled combined units are most efficient, while separate and air-cooled models, as well as water-cooled units connected to air-intake ducts, come next in terms of efficiency.

Evaporative Cooling

Evaporation cools. The Persians realized this thousands of years ago. Parks with fountains around a palace acted as refreshing cooling plants. At the World Exhibition in Seville in 1992 refreshing fountains and sprinklers between the buildings improved the microclimate in the heat. There are new buildings that are cooled via roof sprinkler systems.

Evaporative cooling can also be used indoors. As moist air is not desirable inside buildings, the cooling is transferred to the intake air with a heat exchanger. The combination of evaporative cooling and a heat exchanger results in energy-efficient air conditioning. This technology is cheaper than cooling with heat pumps.

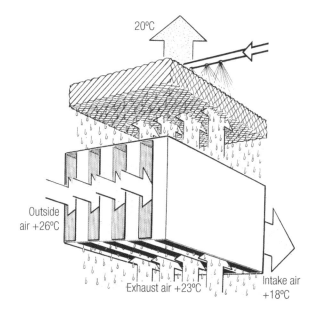

The principle of evaporative cooling for energy-efficient air conditioning.

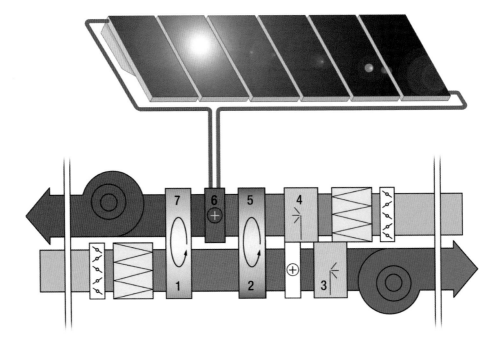

The cooling system works as follows:

1. The intake air is dried (with a heat exchanger with the exhaust air, which increases the heat of the intake air).
2. The intake air is cooled indirectly (with another heat exchanger with the evaporated, cooled exhaust air).
3. The intake air is further cooled and moisturized (using evaporative cooling). At this point the intake air has reached a comfortable temperature.
4. The exhaust air is cooled with evaporative cooling.
5. The cooled exhaust air provides cold to the heat exchanger used to cool the intake air.
6. Exhaust air is heated, e.g. with solar heat.
7. The now very warm exhaust air provides heat to the heat exchanger used to dry the intake air.

Source: Munters ingenjörsbyrå

Solar Cooling

Solar cooling sounds like a paradox, but the fact is that solar energy can be used for cooling. One method is to use electricity produced by solar cells to run a compressor. Another method is to use absorption heat pumps. As has already been mentioned, heat pumps have a cold and a warm side. The sun warms the refrigerant on the warm side which provides cooling on the cold side. Development of solar absorption heat pumps is under way.

The latest development in the area of solar cooling is to combine sorption (drying) and evaporative cooling (evaporation). The cooling system was developed by the engineering company Munters using the Swedish inventor Carl Munters' basic concepts. There are no heat pumps or freons used. Rather, drying, heating and humidifying take place, and a heat exchanger is used between intake and exhaust air.

Thermoaccumulator

Climate Well is a company that has put a new product on the market, a thermoaccumulator that absorbs solar energy and stores it to use for heating and/or cooling. It is an excellent method of cooling for warm countries. Heat is stored in a solution of metallic salts. Heat is generated when the metallic salt absorbs water and cold is produced when it releases water. The equipment is made up of two parallel systems: one absorbs and stores energy and the other releases heat or cold. Each system is made up of two parts: a condenser/evaporator and a reactor. The four units are connected with pipes and a little electricity is needed to run the pumps and valves. The reactor gets heat from the solar collector and the condenser/evaporator gets cold from a pool or outside air. The slave reactor releases heat and the slave evaporator releases cold.

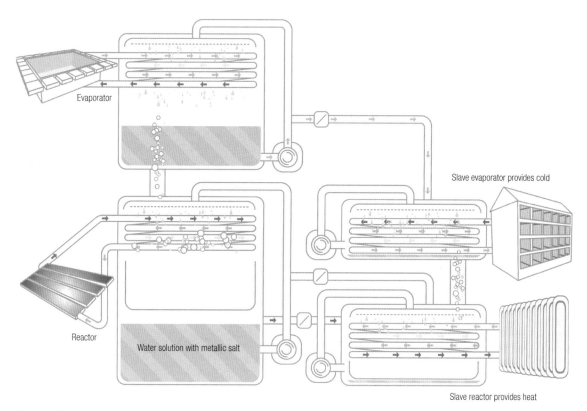

Diagram of how a thermoaccumulator works.
Source: Illustration: Tomas Hamilton

Wind turbines in Öresund.

3.2 Renewable Electricity

Production of electricity with nuclear power and fossil fuels should be avoided. So production of electricity from renewable sources needs to be expanded, for example district heating and power plants fuelled with biomass, wind turbines and hydroelectric power. With regards to hydroelectric power, it is mostly a question of increasing the efficiency of existing hydroelectric plants as well as developing small-scale plants. In order for this development to take place, both nuclear power stations and fossil-fuelled power stations must be forced to pay for the environmental damage they cause. Then renewable electricity production will not only be technically viable but economically competitive as well.

3.2.0 Production of Electricity in a Sustainable Society

Production of electricity with nuclear power and fossil fuels should be avoided. So production of electricity from renewable sources needs to be expanded, for example district heating and power plants fuelled with biomass, wind turbines and hydropower. With regards to hydroelectric power, it is mostly a question of increasing the efficiency of existing hydroelectric plants as well as developing small-scale plants. In order for this development to take place, both nuclear power stations and fossil-fuelled power stations must be forced to pay for the environmental damage they cause. Then renewable electricity production will not only be technically viable but economically competitive as well.

Electricity Production

We can't replace electricity production by fossil fuels and nuclear power with one single renewable energy source. The solution is to create systems that combine electricity production from all available renewable sources. Existing hydroelectric power can be made more efficient by new technology and can be complemented with wind and wave energy. The hydroelectric power can be used to regulate changes in wind energy production. If there are towns with heating needs, these can be heated with central heating systems and those should use co-generation plants that produce electricity as well. The fuel for these plants should be biomass fuel or organic waste from households, agriculture and industries. The biomass fuel for the cogeneration plants can be

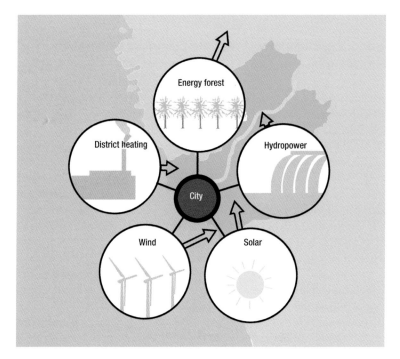

Falkenberg is a leading municipality when it comes to using renewable energy sources. There are two hydropower stations at Ätran. There is a district heating plant fuelled by wood chips, of which a large part come from local 'energy forest' cultivation. Adjacent to the main road is one of the world's largest solar collector facilities, producing heat for the district heating grid. In addition, a number of wind power generators are located along the coast.

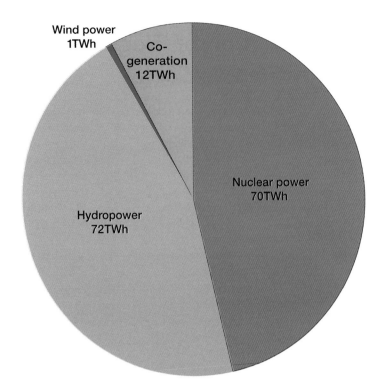

Electricity production in Sweden in 2007. Total net supply is about 145TWh. More than half the electricity in Sweden came from renewable energy sources. The cogeneration is mainly produced by biomass fuels.

Source: Swedish Energy Agency

waste from the wood industry or produced in fast-growing energy plantations. If this is not enough, solar energy can do the rest.

New Ways of Producing Electricity

Wave power is under development in several countries.

Electricity is produced using geothermal energy in countries with volcanic geological formations, such as Iceland, Italy and the US. Tidal power is an old technology that is being taken up again, as in France. There are plans to place wind power generators on the ocean floor to take advantage of tidal flows. Osmotic energy is a technology that produces electricity from the difference between fresh and salt water. It is possible to use the difference in osmotic pressure between saline ocean water and fresh river water to generate electricity in power plants placed at river mouths. Ocean thermal energy conversion (OTEC) makes use of the temperature difference between surface and water on the seabed. A water-ammonia mixture is vaporized by the warm surface water and runs a turbine.

Private Electricity Generation

It is possible to have a private electricity supply, e.g. for a summer cottage that is not connected to the electricity grid or for security in the case of power cuts. However, private electricity production is only profitable where there is no access to the electricity grid. The situation can change if, as in other countries in the EU, a feed guarantee is implemented where it is certain payment will be made for electricity fed into the grid over many years in the future.

The easiest way to obtain your own electricity is with solar cells, wind power or a combination of solar cells and wind power connected to a battery bank. Such systems are used to produce electricity for only the most basic purposes due to their high cost. So electric space heating is not an option, and kitchen stoves should be fuelled with gas or wood. It is important to be aware that solar cells provide very little electricity during the darkest winter months.

Very small-scale hydropower generators made for charging batteries are available, but they require access to running watercourses. Biomass-fuelled mini power stations, run with Stirling engines, are also beginning to appear on the market. It is also possible to get backup power systems made up of a motor connected to a generator. These can be run with renewable fuels such as rapeseed oil or biogas.

Security of supply can be dealt with by having a battery charger and battery bank that can run important services in a power cut. Examples of such services are a furnace and heating system circulation pump, a few light bulbs, a radio and TV, as well as a water pump. This method necessitates either a 12V system or a converter that converts direct current from the battery to 220V alternating current.

Summer cottage in the Swedish archipelago with wind turbine and photovoltaic panel.

A problem with private electricity production is storage. Of course development of batteries is ongoing, but they are still expensive and relatively short-lived. Therefore, it may be better to be connected to the electricity grid, buy electricity when private production isn't adequate and sell electricity to the grid when more is produced than is needed for private use.

There are electricity meters that measure both the electricity consumed and fed in.

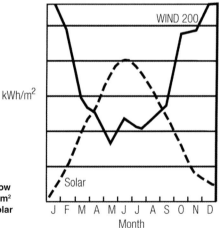

Local electricity production. The table and diagram show the average output kWh/day with 1m² solar cells and 1m² wind rotor, central Sweden. The data show how well solar cells and wind turbines complement each other.

	Jan	Feb	March	April	May	June	July	Aug	Sept	Oct	Nov	Dec
Solar	0.05	0.14	0.27	0.36	0.47	0.53	0.47	0.40	0.29	0.17	0.07	0.04
Wind	0.84	0.79	0.45	0.40	0.26	0.40	0.32	0.38	0.46	0.71	0.73	0.84
Total	0.9	0.9	0.7	0.8	0.7	0.9	0.8	0.8	0.8	0.9	0.8	0.9

Electricity Storage

There is a lot of research going on aimed at developing better batteries, but progress is slow. Storage facilities in the form of pumped storage power stations, flywheels and hydrogen gas are available. In pumped storage power stations water is pumped up into a water reservoir when there is excess electricity production capacity so that the stored energy can be accessed when needed. Well-balanced, low-friction flywheels are set in motion when there is excess electricity production and the energy from the rotating flywheel is used to run a generator. More and more people believe in the potential of hydrogen gas since it is extremely energy dense, and if the electricity is produced in fuel cells there are no environmentally hazardous emissions.

Batteries

Batteries are the weakest link in a private electricity system. They are expensive, often heavy, have a short life and poor capacity, require regular maintenance, are sensitive to methods for charging and discharging, require cool but not cold conditions and require ventilation.

The most common types of batteries in the past were lead batteries and nickel-cadmium batteries. Their heavy metal content is problematic. Today, nickel metal hydride batteries and lithium ion batteries are common. They have a higher energy density than lead and cadmium batteries, but they contain the toxic metals cobalt and manganese, though in small amounts. Metal hydride batteries have no 'memory effect', which is good as charging capacity does not deteriorate over time. However, they are more expensive, have a relatively short life and are also sensitive to low temperatures. Lithium ion batteries are even more expensive and are easily damaged by overcharging. Zinc-air batteries have the greatest energy density of today's batteries. A zinc-air battery

More and more self-installed small-scale wind turbines and solar cell panel systems are being produced. They can be connected to the electricity grid via an inverter than converts direct current to alternating current. There should be an open area of 75 metres around small wind turbines for them to function optimally. The photo shows an Energy Ball wind turbine, which does not require a building permit since it has a diameter of less than two metres. Building permit requirements should, however, always be checked.

Source: www.homeenergy.se.

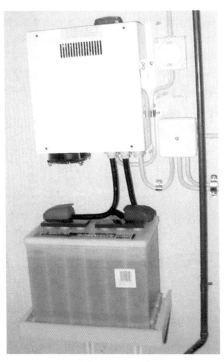

Emergency electricity system.

Comparison of Various Types of Battery

Battery type	Energy content WH/K	Energy content WH/l	Life (number of cycles)	Relative energy price
Ventilated lead batteries	20–45	40–100	200–2000	1
Sealed lead batteries	10–30	80	500–1000	1–2
Nickel-cadmium	15–45	40–90	>5000	3–5
Nickel-hydrogen	40–60	60–90	3000–6000	5–10
Adv nickel-iron	22–60	60–150	1000–2000	1–1.5
Nickel-zinc	60–90	120	250–350	2
Sodium-sulphur	100–200	150	900–2000	0.5–1
Lithium-sulphur	200–600			
Zinc-boron	55–75	60–70	600–1800	0.5–1
Iron-chromium (Fe/Cr redox)	–	–	20,000	1
Zinc-manganese (ZN/MnO_2)	70	160	200	1–2
Hydrogen gas fuel cell	–	–	–	40

Source: *Nye fornybare energikilder*, Norges forskningsråd i samarbete med Norges vassdrags- og energidirektorat (NVE), 2001

with a capacity of 1kWh weighs only 4.5kg. This is about eight times less than a normal lead-acid battery. There is a lot of development work being done on batteries. In hybrid cars nickel metal hydride batteries (NiMH) dominate, but for plug-in hybrid cars lithium batteries are considered to be the main alternative (there are several different types of lithium ion batteries). Sodium nickel chloride and bipolar lead-acid batteries are other alternatives under development.

Hydrogen Gas

The idea is to produce hydrogen gas with the help of electricity from renewable energy sources (wind power and/or solar cells). Using electricity, hydrogen gas can be produced from water via electrolysis. Hydrogen gas can also be produced by gasification of biomass, or in a reformer from natural gas, methanol or biogas. Research is under way on photo-chemical and thermo-chemical methods of production.

Hydrogen gas can be used in fuel cells where it is combined with oxygen from the air and electricity is created. The only by-product is warm water. Hydrogen gas can also be used in motors, e.g. to run a car, and to fuel boilers. The primary combustion product from combustion of hydrogen gas is water vapour. Hydrogen gas can also be used for catalytic combustion, with the advantage that no nitrogen oxides are created.

One idea is to produce hydrogen gas with solar cells in sunny, desert regions. The electricity is conducted to locations where water is found and there electrolysis is used to divide the water into oxygen and hydrogen gas. The hydrogen gas is then transported through pipelines or by sea to population centres.

(Fuel cells are described in Section 4.2 Social Structure under 4.2.1 Traffic.)

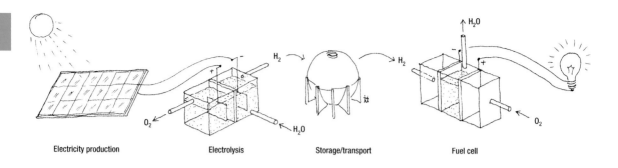

| Electricity production | Electrolysis | Storage/transport | Fuel cell |

An environmentally friendly hydrogen gas system can look like this.

Source: *Energi & Framtid*, Vattenfall, 1990

The Swedish company Morphic produces an energy system where wind energy is transformed into hydrogen gas that can be saved and transformed again to electricity via fuel cells when there is no wind.

3.2.1 Combined Heat and Power with Biomass

When electricity (power) and heat are produced at the same time from a fuel for municipal facilities, the technology is referred to as combined heat and power. The same technology used in industry is called back-pressure power. The technology uses fuel efficiently. The energy is used in two steps and an 85 per cent level of efficiency can be reached. Where district heating is used, conditions exist for production of combined heat and power. Many towns in Sweden and Finland are successfully running biomass-fired combined heat and power plants.

Combined Heat and Power Technology

A little more than 30 per cent efficiency is achieved when only electricity is produced from a fuel. If electricity and heat are produced at the same time, the level of efficiency is increased to about 75 per cent. Better planning and insulation can result in smaller transmission losses and an 85 per cent level of efficiency. With the latest gasification technology, a level of efficiency over 90 per cent can be reached. Furthermore, the percentage of electricity produced is greater. Combined heat and power is produced in relatively large facilities, which is good from an environmental perspective. Good combustion and flue-gas cleaning are achieved.

The technology for generation of electricity using the steam turbine cycle (the Rankine cycle) is well-established, and consists of four parts: a steam boiler, a steam turbine, a condenser and a feed pump. The fuel produces steam in the steam boiler. The steam runs the steam turbine, which in turn runs the generator. The steam is cooled back to water in the condenser with the help of the district heating water, which is heated at the same time. The feed pump pumps the water back to the steam boiler. The technology works well at a scale of 7–25MW electricity.

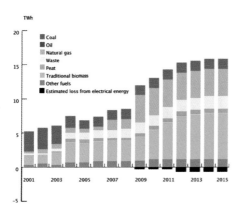

Planned electricity production in cogeneration plants for district heating and different fuels. In 2007 5.6TWh electricity was produced from industrial cogeneration, and 8.3TWh in the sector of district heating.

Source: Swedish Bioenergy Association, SVEBIO, 2008

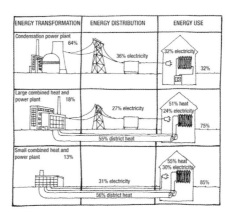

A combined heat and power plant (production of electricity and heat) has a much higher level of efficiency than condensation power plants (electricity production only), as heat isn't cooled down but instead used for district heating. Local heating (a small district heating grid) has smaller transmission losses and the level of efficiency is increased further if it is supplied by combined heat and power.

Electricity can also be generated using gas turbines (the Brayton cycle). This technology may be used in peak-load power plants (reserve power plants), which are run when the regular electricity supply system isn't adequate. They have a capacity of 14–25MW. For combined heat and power production the process has the following main components: compressor, combustion chamber, gas turbine and waste heat boiler. The turbine's waste gases are cooled with district heating water in the waste heat boiler.

Combined Heat and Power with Gasification

It is possible to extract combustible gases from biomass by gasifying the fuel using a pyrolysis process, i.e. combustion in an almost oxygen-free atmosphere. The gas is first used to run gas turbines which produce steam that runs steam turbines. Thus, electricity is generated at two points, first in the gas turbine and then in the steam turbine. Furthermore, waste steam from the steam turbine can release its waste heat to a district heating grid before being returned to the steam boiler to be reheated. With gasification of biomass together with a combination of gas turbine and steam turbine, an electricity efficiency as high as 53 per cent can be achieved. This is a new technology that makes it possible to produce electricity and heat from renewable energy sources.

Gasification of Caustic Liquor

Black liquor is a by-product of the pulp industry. Currently, black liquor is combusted in recovery boilers to produce steam, and some electricity production takes place using back-pressure turbines. By gasifying black liquor instead of combusting it in recovery boilers, the pulp industry's sulphate plants can more than double their electricity production. With gasification, process chemicals are recovered.

Biomass gasification combined heat and power plant in Värnamo. Energy is extracted in three stages: first as electricity in a gas turbine and then in a steam turbine, followed by district heating to heat the town.

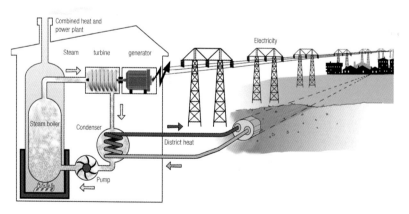

In a combined heat and power plant, the heat in the cooling water is used in a district heating system.

Source:: Leif Kindgren.

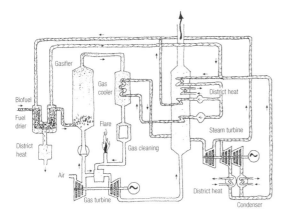

Gasification of biomass, electricity production with gas turbine and steam turbine as well as extraction of district heat.

Source: *Göteborgs Posten*, 23 April 1995

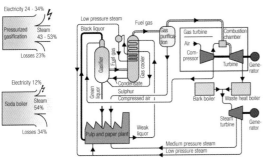

With gasification of black liquor, both a gas turbine and steam turbine can be used, which increases the opportunities for producing electricity.

Source: adapted from *Ny Teknik*, 25 January 1996

Small-Scale Combined Heat and Power

The Totem unit is a small, ready to use combined heat and power plant, consisting of a car motor (Fiat) that can be run using different types of fuel, e.g. biogas. They can be used in sewage treatment facilities where the sludge is stabilized in a biogas plant. The energy obtained is used to run the treatment plant. Biofuel-fired combined heat and power can provide both heat and electricity to an agricultural operation, e.g. using rapeseed oil-run motors. In Germany, Totem units are common in blocks of flats and ecocommunities, but usually use natural gas.

Being able to run a motor that produces both electricity and heat and is fuelled with biogas is a dream for many people building ecologically. The existing choices are steam engines, producer-gas generators and thermal engines. Producer-gas technology requires a lot of service and maintenance. Steam engines have low efficiency and require an operator. One type of thermal engine is the Stirling engine, which seems most promising. Small Stirling motor-run combined heat and power plants have started to appear on the market.

A small-scale combined heat and power plant run using natural gas provides electricity and heat to a neighbourhood in Berlin, Germany.

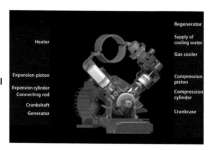

The German company SOLO Kleinmotor Gmbh develops small combined heat and power plants with Stirling motors. One of the variations is fuelled with pellets.

Stirling Engines

Sterling engines operate using a closed system in which a working gas is alternately heated and cooled. When heated, the gas expands and mechanically moves a crankshaft connected to a piston system. Combustion of the fuel takes place separately. The heat from combustion is transferred via a heat exchanger to the motor's closed gas system. Heat can be recovered from the motor's cooling water system and waste gases as in conventional combustion engines.

The Stirling engine is a heat motor that has many advantages for small-scale operation. It is quieter than a combustion motor and it has cleaner waste gases. A Stirling engine and a pellet burner connected to an accumulation tank can heat a single-family home. The burner is on one side of the tank and the Stirling engine on the other, so the combustion chamber is inside the tank. The system provides both heat and electricity for the house. Research is also being carried out on connecting a Stirling engine to a heat pump, which is an ingenious combination from an energy viewpoint.

Thermoelectric Generators

A thermocouple (Peltier element) is a technology that directly transforms heat to electricity or electricity to heat. A thermocouple consists of a circuit of two metals or semiconductor materials, joined at two junctions. If one junction is heated and the other cooled, electric current is created in the circuit. A thermoelectric generator uses the Peltier effect to create electricity from heat. However, the level of efficiency is poor, only about 2 per cent, so the technology is only used where there is excess heat and electricity requirements are minimal. One example of the technology is running a TV from the heat of a woodburning stove or a radio from the heat of a paraffin lamp. Thermocouples also act in the reverse direction. They cool if electricity is added to the circuit, and are used in small refrigerators on pleasure boats, for example.

The Swedish Defence Research Agency (FOI) has developed an electricity generating pan for a spirit stove (of the Trangia model). Electricity is produced at the same time as food is heated in the pot. A voltage-producing 'Peltier element' is inside the double-bottomed pan. With the right mixture of semiconductor materials, in this case bismuth telluride, it is possible to produce enough watts to run a laptop computer.

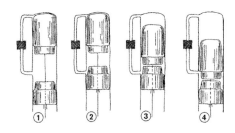

The principle of the Stirling engine. The most important parts are a cylinder, a piston, a displacer, a regenerator, a heating tube and a cooling tube. The illustration shows the four different phases in a cycle:

1. The piston at the bottom and displacer at the top. All the gas is in the cylinder's cold section.
2. The piston compresses the cold gas. The displacer remains in the top position.
3. The piston in the top position. The displacer pushes the gas via the regenerator and heating tube to the cylinder's warm section.
4. The warm gas expands. The displacer pushes the gas through the regenerator and cooling tube to the cylinder's cold section, and the cycle starts over again.

3.2.2 Hydropower

Despite environmental objections to developing further hydropower plants, existing large-scale hydropower stations can be modernized and made more efficient, and some possibilities exist for developing a number of small-scale hydropower stations.

Development of Hydropower

Large-scale hydropower together with small-scale has made it possible to generate large quantities of electricity, providing a good base for electricity systems, and these can be complemented by wind power.

Large-Scale Hydropower

Many large hydropower stations are old and in need of renovation. The technology has developed and it is now possible to rebuild old hydropower stations so that they produce more electricity. Surfaces in shafts and tunnels can be made smoother (i.e. improved/faster water circulation), new, more efficient generators can be installed, and controls can be improved. New generation and transmission technology has been developed, based on high-voltage direct current. Use of this technology would further increase efficiency.

In 2005 in Sweden there was a total of 2082 hydroelectric power stations of which 511 are larger than 1.5MW (1500kW). The total output was 16,200MW. The rivers that produce the most energy are the Lule River (15.9TWh) and the Ångerman River (13.6TWh).

Small-Scale hydropower

There are three things that are important to research when constructing a hydropower station, namely vertical drop, volume of water and environmental impact. The vertical drop multiplied by the volume of water determines how much electricity can be generated. The greater the vertical drop, the cheaper the production costs. It may be unrealistic to install hydropower where the vertical drop is less than 1–1.5m. Theoretically, 9.8kW can be produced for every m³/s that falls 1m. Since the efficiency of a turbine is less than one (e.g. 60–90 per cent), the energy obtained is somewhat less.

In Sweden in the beginning of the 2000s there were 1571 small-scale hydropower generators in operation that produced 1.7TWh

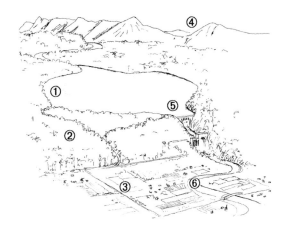

Large-scale hydropower often results in serious environmental disturbances:

1 Productive forestry and agricultural land is immersed under water.
2 Roads are built in unexploited natural areas.
3 There is a cessation of the annual floods that fertilize land with silt.
4 The flooded vegetation rots releasing greenhouse gases.
5 Waterborne sicknesses increase when water becomes stagnant.
6 If a watercourse dries up, migratory fish are killed.

Fish ladders make it possible for fish to bypass small-scale hydropower stations and locks, here at Sickla lock in Hammarby Sjöstad, Stockholm.

A small-scale hydropower station connected to the electricity grid.

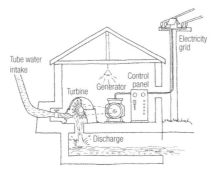

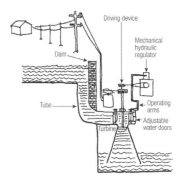

Small-scale hydropower stations can be used to produce more renewable electricity, but they must be designed so that they do not disturb the environment. The photograph shows the water tube at the far right where the water enters. The water then goes through the turbine, with the generator above it.

Regulation is important when a separate electricity grid is supplied by a small-scale hydropower station. For example, lamps may break if too much voltage is put into the grid.

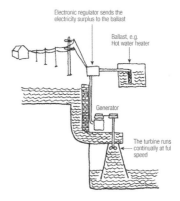

With the development of small and inexpensive electronic regulators, small-scale hydropower has become cheaper.

per year. There were, however, many more earlier. In the mid-1950s, before the shutdown period began, there were about 4000.

Taking the environment into consideration is of course important when building hydropower stations. Fish should be able to live in a watercourse, which can be ensured by building fish ladders. How much is dammed and whether land is flooded or waterlogged is important. Both these problems are easier to solve for small-scale hydropower stations than for large ones.

Damming and diverting watercourses and constructing hydropower stations is regulated by law. This means that water rights are owned by someone and a permit is required from a water rights court before a water project may be started.

Environmental objections can thwart development of hydropower to its full potential.

Mini-hydropower stations are hydropower stations less than 300kW in size. There are many watercourses with dams adjacent to old mills, etc. These places could be used for electricity production. In recent years, mini-hydropower stations have become significantly cheaper to construct, with the development of electronic regulators and the use of simpler and cheaper turbines.

Regulators

Small-scale hydropower stations have been regarded as uneconomical to build. With mechanical-hydraulic regulation, the flow of water through a hydropower station is controlled using gates, which open and close using expensive and complicated hydraulics. The smaller the hydropower station, the greater the proportionate cost of the hydraulic regulator.

With electronic regulation, a constant flow of water passes through the hydropower station and runs the turbine, the electricity being regulated by a small and inexpensive electronic regulator. It detects how much electricity is needed on the grid, transmits this amount and dumps the excess into a hot water heater or the watercourse.

Power stations connected to the grid have the advantage of operating concurrently

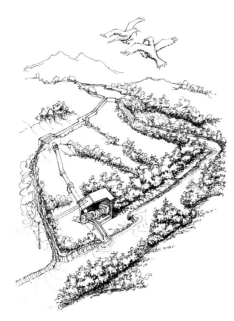

Small-scale hydropower stations can be built so as to minimize environmental problems.

Source: Illustration: Peter Bonde

Micro-hydropower

Hydropower stations with a few hundred watt capacity are considered micro-hydro. In the 1980s, especially in the US, extremely small-scale hydropower stations designed to charge batteries in the same manner as small-scale wind turbines and solar cell panels became available. This type of small-scale hydropower station provides 50–500W depending on the water flow and vertical drop. They usually consist of small Pelton turbines or Turgo turbines connected directly to the axle of a small generator. They can supply a summer cottage with enough electricity to run a radio, a TV and a few lights.

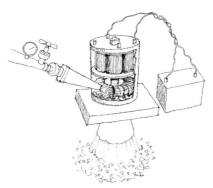

Micro-hydropower units are very small-scale hydropower stations used to charge batteries. A hose can be run from a creek to a micro-hydropower unit. 'Just add water' is the slogan for an American model.

with the grid and not having to store or regulate electricity at the station. They operate with asynchronous generators, that are magnetized with electricity from the grid, so that the electricity delivered to the grid has the same phase and frequency as the grid.

Turbines

The primary factor in determining which turbine is suitable is vertical drop. Advanced propeller and Francis turbines are three-dimensional, curved devices and are therefore expensive to make. The cross-flow turbines (Michell-Banki turbines) on the other hand are made of two-dimensional curved components and can be inexpensively produced in a workshop. This type of turbine does not have the same high level of efficiency as the others, but it doesn't make such a big difference if a simple and inexpensive method is required and extracting as much energy as possible out of the watercourse is not a priority.

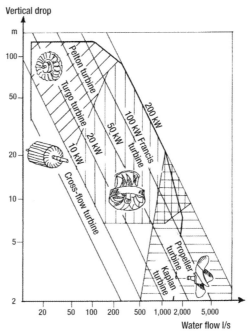

Choice of turbine depends on vertical drop and water flow.

Water enters a Francis turbine via a spiral pipe wrapped around the turbine. On the inside of the spiral pipe there is a column with adjustable guide vanes which can be used to adjust the water flow. Water flows into the spiral pipe, through the column and is directed towards the turbine by the guide vanes, so that power is transferred to the turbine wheel and water flows down through the intake pipe.

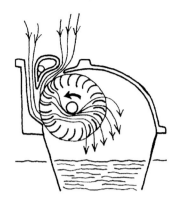

The efficiency of cross-flow turbines can be increased if they are equipped with a guide vane and an intake pipe. There has been a revival in the popularity of this type of turbine as they are simple and inexpensive to make.

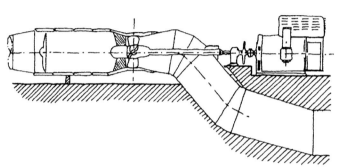

A propeller turbine consists of a propeller inside a pipe. When water flows through the pipe, the propeller is set in motion. Power is transferred via an axle to a generator outside the curved pipe. There are also propeller turbines with straight pipes where the generator is placed inside the pipe.

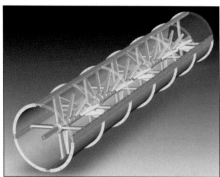

A promising turbine is the German–Norwegian Hamann which consists of a pipe, with a length of 6m and a diameter of 1m, containing a spiral-formed turbine, At its best it can produce an efficiency of 50 per cent, but 30 per cent is more realistic.

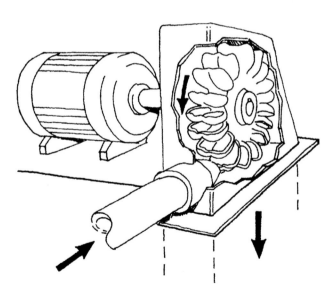

The Pelton turbine. Water is injected through a nozzle towards the edge formed where the two cups meet. The jet of water is divided into two equal parts and power is transferred from the water to the blade wheel.

Efficiency of Different Hydropower Stations

Type of turbine	Efficiency
Undershot waterwheel	0.25–0.40
Overshot waterwheel	0.50–0.70
Breast waterwheel	0.50–0.60
Poncelet waterwheel	0.40–0.60
Horizontal waterwheel	0.20–0.35
Impulse turbine (Frances)	0.70–0.87
Cross-flow turbine	0.60–0.80
Reaction turbine	0.65–0.90

3.2.3 Wind and Wave Power

There is a great potential for development of wind power in countries with plenty of wind. The best places for wind energy are often coasts, shores of large inland lakes and mountain areas. The electricity grid needs to be developed and have the capacity to deal with variations in power generation. There exists technical capacity to build and maintain wind turbines. Much of this development will take place offshore.

Wind Power Development

Most modern development in wind power has taken place in Denmark. It started in the 1980s after the energy crises of the 1970s. Development of wind power is difficult as the wind is erratic and difficult to forecast, so there has been much trial and error. Initially, small wind turbines of 55kW were used and the size has gradually increased. In the 1990s, wind turbines with a capacity of 200–500kW were built. Today the most common size is 1MW or greater. Denmark is still one of the world's largest producers of wind turbines. The industry tries to continually improve their design, for example, by working on the blades, gearing, generators and adjustment of blade position during operation.

It is expensive to service wind turbines once they are erected, not least when they are out in the ocean. Much development effort is going towards vertical wind turbines with low sensitivity to wind direction and low risk of sound disturbance. Further, the generator is placed on the ground which makes installation and maintenance much easier. But the design of the blades must be very precise to get good results.

Wind Power in Sweden

Development of wind power in Sweden has been slow due to the low cost of electricity and the difficulty in obtaining permission. However the situation has improved, and since the early 1990s there has been an increase in the

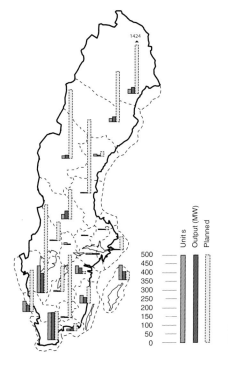

Built and planned wind turbines in Sweden.

One of the biggest wind turbines in Sweden at 3MW (Vestas) situated in Holmsund, Umeå. It produces 6–7 million kWh per year.

Source: Photograph: Lars Bäckström

number of installations. Half the wind turbines have been built by energy companies (about 35 per cent) and private individuals (15 per cent). New innovative ownership forms, such as joint stock companies, cooperative organizations and community associations are responsible for the rest.

In 2007, about 1000 wind turbines produced 1.43TWh. A national planning goal for wind power in Sweden is set at 10TWh by 2015. The goal is not for a particular number of installations, but rather an aspiration to create conditions favourable to future wind power development. The Swedish Energy Agency has proposed that the goal be increased to 30TWh by the year 2020, and the trade association *Svensk Vindkraft* (Swedish Windpower) wants to aim for 25TWh.

Siting Wind turbines

The appearance of the landscape is affected by wind turbines since they are so prominent. More and more wind turbines are located offshore. By avoiding important bird flight paths, important resting places for migratory birds, known fish spawning and fry locations, and areas where seals usually seek land to rest, much conflict regarding placement of wind turbines could be avoided.

The power output of a wind turbine increases rapidly with increase in wind speed (wind speed cubed). Power output increases eight times with a doubling of wind speed. So it is important to place wind turbines where there is the most wind. Factors to consider are problems with noise and safety distance, e.g. so that loose ice doesn't present a risk. New blade designs have reduced the aerodynamic sounds. The generators meet legally set sound-level requirements if they are placed within 250–400m of inhabited areas. The military may also have views on placement. There is concern that wind turbines may disrupt reconnaissance systems and satellite communication.

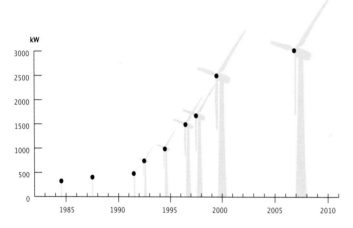

The size of wind turbines has been growing over time.

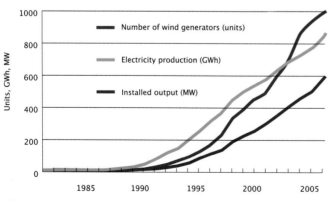

Wind power development in Sweden.

Source: Swedish Energy Agency

Connection to the Grid

Wind power does not need to be stored but is deposited directly into the electricity grid. The fluctuations in the amount of electricity generated from wind power are evened out by saving water in existing hydropower stations when it is windy and using more hydropower when there is no wind.

Small battery-charging wind power plants. There are many small wind power plants (50–500W) designed to charge batteries, which are durable and reliable. They are used on boats, for summer cottages, etc., and provide an excellent alternative or complement to a solar-cell installation.

International Comparison

Germany has the world's largest wind power production, nearly 19,000 turbines with an installed capacity of 20,500MW (2006). Annual wind-generated electricity exceeds 30TWh, which is 5.65 per cent of German energy demand. Denmark generates nearly 17 per cent of its electricity with wind power, 5275 wind turbines with an installed capacity of 3137MW generate more than 6TWh (2006). Spain installed in the end of the 1990s as much wind power as Sweden did in the year 2007, today they have as much wind power as Sweden needs in 2020. The US is the country with the fastest annual net increase in capacity with 2454MW.

Self-Installed Wind Power

According to the newspaper *Ny Teknik* (New Technology) (no 39, 2008) the following should be done when installing a wind turbine in Sweden:

1) Check local wind conditions with the Swedish Energy Agency.
2) Calculate the annual energy production by using the generator's output curve and wind speed throughout the year.
3) Make an economic estimate to determine if it pays off.
4) Apply for a municipal building permit. There are a variety of regulations among municipalities.
5) Apply for connection authorization from the utility company.
6) Once the building permit and connection authorizations are received, order the wind turbine.
7) Install the wind turbine and notify the utility company that it is ready to be connected.
8) Have an authorized technician make the connection and then begin producing electricity.

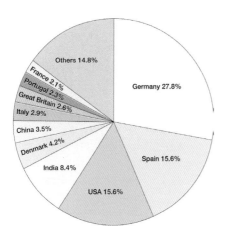

The diagram shows countries with most wind power in the world, 2007.

Source: Global Wind Energy Council, GWEC

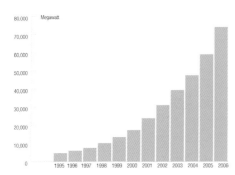

The development of wind energy worldwide by megawatts installed. The biggest producers of wind turbines are Vestas (Denmark), General Electric (US), Enercon, Siemens, Nordex and Repower (Germany), Gamesa (Spain) and Suzlon (India).

Source: Global Wind Energy Council, GWEC

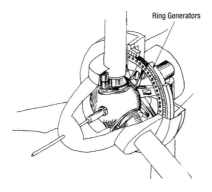

Wind turbine with direct-driven ring generators, an attempt to make wind turbines even more efficient.

Vestas wind turbine with a 27m-diameter turbine and 225kW generator.

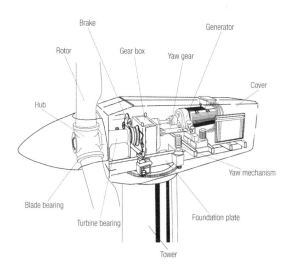

The English company Marlec is a world leader in the production of small wind turbines, which have been on the market since 1979.

This helix-formed wind turbine has height 5m, diameter 3m and is adapted for mounting on building roofs.

Source: Quiet Revolution

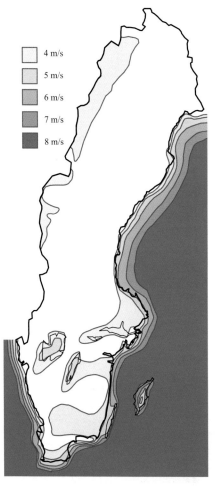

The map shows where there is the most wind in Sweden and naturally wind turbines should be sited there.

Vertical wind turbine from Vertical Wind.

The Canadian company Magenn has developed a system where a large helium balloon twists around in the air which generates electricity via the cable it is anchored to.

Wave Power

Energy prices and technical development have resulted in interest in wave power as an energy resource. 70 per cent of the Earth's surface is covered by water. The global potential for wave power is 10,000–15,000 TWh per year. Wave power has about the same economic potential as hydropower. The most suitable areas in Europe are off the coasts of the UK, Norway and Portugal. The potential in the Baltic Sea has been estimated at 24TWh.

Different kinds of wave power plants are being tested in several places in Europe. The Scottish company Ocean Power Delivery is testing three 130m-long 'seaworms' off the coast of Portugal. The Danish company Wave Dragon has technology resembling a floating hydropower plant, which is being tested off the coast of Wales. In Norway, the Fred Olsen Energy Company is testing a wave energy platform equipped with floating buoys in the sea.

In 2008, the world's first commercial wave power park was inaugurated in Aguçadoura, northern Portugal. The three wave power plants from the British company Pelamis Wave Power are about 150m long and produce a total of 2.25MW. The next phase is 25 more wave power stations.

On the west coast of Sweden the companies Seabased, ABB and Vattenfall are testing a wave energy concept developed by Professor Mats Leijon. This consists of buoys connected to linear generators on the seabed. The first wave energy park in Sweden will consist of 1000 wave power units covering an area of 600m². The output will be 10–12MW and production 50GWh per year.

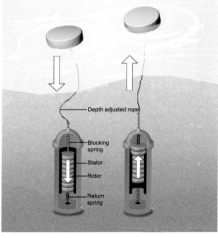

The Swedish wave generator consists of a buoy that moves up and down with the waves. About 20 per cent of the energy can be absorbed and converted to electricity. This movement is transmitted by wire to a linear generator on the seabed. The wire moves a strong magnet (rotor) up and down through the stator producing electricity. One buoy 3m diameter can produce about 10kW. The whole mounting weighs about 10 tonnes. This wave-power concept was developed by Professor Mats Leijon at the University of Uppsala and commercialized by the company Seabased AB.

Source: Illustration Leif Kindgren

3.2.4 Solar Cells

Solar cells (photovoltaics) provide electricity when exposed to direct sunlight and should not be mistaken for solar collectors, which only provide hot water or hot air, not electricity. Solar cells are used to charge batteries, mostly in places not connected to the grid, such as lighthouses, mountain settlements and summer cottages, and on mobile units such as boats, caravans and lawnmowers. Solar cells can be used for outdoor lighting at sites such as bus stops, to avoid having to install electricity lines. As solar cells become cheaper there will be more grid-connected applications, e.g. where solar cells are sited on south-facing roofs and façades.

How Solar Cells Work

Solar cells can be classed into three generations. The most common are based on silicon technology. Typically, they have an efficiency of about 13 per cent. The second generation is thin-film solar cells. They are technically just as good as silicon cells but cost about five times less to produce. The third generation is still in the research stage and is the nanostructured solar cells.

Solar cells transform light directly into electrical energy. Monocrystal solar cells are made up of two layers of extremely thin silicon wafer. Inside the silicon wafer is a strong inner electric field, which separates electrons and 'available spaces'. The top layer is treated with phosphorus and is the negative pole. The bottom layer is treated with boron and is the positive pole. The layers are separated by a blocking layer. When sunlight hits the solar cell and the amount of energy exceeds a certain threshold, electrons are freed from the silicon bonds, which creates moving electrons and holes allowing electrons to move through the blocking layer. The holes and electrons move in different directions, electric tension is created and current can be extracted.

Solar cells have a lifetime of 40 years. They generate direct current and thus require a battery or inverter that is plugged into the electricity net for further use.

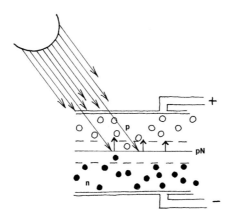

Photons in solar light release electrons from their silicon binding and create moving electrons and holes. The electrons flow through the electric wires.

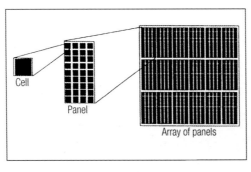

Solar cells are combined into panels, and the panels make up banks.

Different Types of Solar Cells

There are two types of solar cell on the market: monocrystalline silicon cells (accounting for more than 90 per cent of delivered solar cell modules) which have a high level of efficiency but are expensive; and polycrystalline cells, which have a poorer level of efficiency but are cheaper. The total cost for both of these solar cell technologies is about the same.

Monocrystalline silicon is historically the most frequently used material for solar cells. The disadvantage is that production is expensive and energy-intensive as the silicon must be absolutely clean and the crystals must be 'cultivated'.

Polycrystalline silicon is cheaper to produce but has a lower level of efficiency. The difference between monocrystalline and polycrystalline silicon is that the polycrystalline consists of many small crystals packed densely together instead of one individual crystal.

Different types of solar cells vary in colour, level of efficiency and other characteristics. Polycrystalline cells' colour varies a lot, while CIGS cells are naturally completely black.

Amorphous silicon has, despite the initial atoms being the same, completely different characteristics from crystalline silicon, since the atoms are not ordered in a crystalline structure but are instead randomly arranged. They are made as thin film cells, thus not as much material is needed, which keeps the cost down. The material is unstable and therefore it has a shorter life. They are mostly used in consumer products.

Solar cells made of **gallium arsenide**, a semiconductor material, have a high level of efficiency. They are often used in space applications. Production of the material and manufacturing is expensive and in addition arsenic is a poisonous substance.

Solar Cell Efficiency Records

	Cell	Module
Silicon		
Monocrystalline	24.7%	22.7%
Polycrystalline	19.8%	15.3%
Thin film		
Nano/amorphous silicon	10.1%	8.2%
CdTe	16.5%	10.7%
CIGS	19.2%	16.6%
Extreme cells for space applications		
GaInp/GaAs/Ge	40.8%	
New types of cells		
Grätz cells/DSC	8.2%	4.7%

Placement of Solar Cells

The angle of solar cells, their orientation and any shading is very important. There are movable solar cell panels which follow the path of the sun. Research is under way on optically concentrating sunlight on the solar cells. The higher the temperature of the solar panel, the lower the level of efficiency, and vice versa. On façades, a ventilated air gap can keep down the temperature of solar

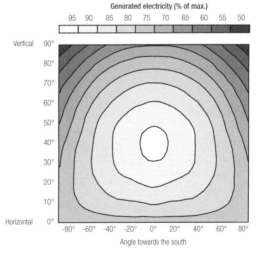

Optimal placement in Sweden is facing towards the south at a 45° angle. The diagram shows the reduction on deviation from this.

Source: adapted from 'Solel i bostadshus – vägen till ett ekologiskt hållbart boende?', Maria Brogren och Anna Green, 2001

panels. Such an air gap can lower the temperature by up to 20°C. The solar radiation towards a south-facing, sloping surface in northern latitudes is about 1000kWh per year. With an average level of efficiency of 10 per cent, solar cells can generate about 100kWh/m² annually.

Prices

A common solar cell system usually used for holiday houses consists of one crystalline silicon module (50 × 100cm) weighing about 5kg. The cost of solar cell electricity includes the costs of solar cells, the stand holding them, the battery bank, regulator, electricity line connection and electrical equipment. For a holiday-house in Sweden an 80W-system costs about €1350 (2009). Grid-connected solar cell systems are installed on building roofs and façades at a cost of about €5500/kW. Installations attract subsidies amounting to 60 per cent of the total cost.

The cost of solar cells themselves has gradually gone down as production has increased. The price development has, however, levelled out and is currently about €2.50 per Wt (peak watt, which is the number of watts the solar cell provides at maximum sunlight in the middle of the day). In order for there to be a big breakthrough for solar cells, the price has to be a lot less. For this to happen, new solar cell technology is needed and large-scale production.

Developed Countries

There have been special campaigns in several countries to stimulate development of solar cell facilities. The first was a programme in Japan in 1994 to install photovoltaics on 10,000 roofs, where the government paid one-third of the cost. That is why Japan has the highest installed solar cell output. Germany started a similar programme in 1999 where the goal was 100,000 rooftop PV systems. Germany has an attractive subsidy system: for every kWh produced by solar cells there is a guaranteed payment of €0.5 for a period of 20 years. Germany has become the biggest market in the world for photovoltaics with an annual growth of 53 per cent. Today Germany is second after Japan in installed PV capacity. In Europe, the greatest photovoltaic capacity is installed in Germany, Greece and Austria. In 2007 the government of California decided to invest US$3.4 billion over 10 years to install photovoltaics on a million roofs. The US is in third place. The amount of solar energy that can be made use of in Stockholm is about the same that can be used in Paris. The big difference is that in Sweden there is very little solar energy available for solar cells during November, December and

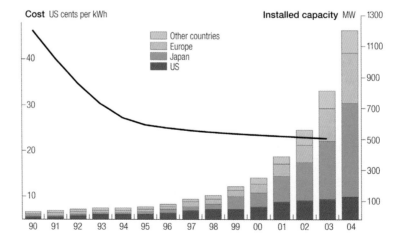

The production of solar cells is increasing rapidly year from year and the price is slowly decreasing. The diagram shows the price in US cents per kWh.

Source: *Svenska Dagbladet*, 2 February 2007

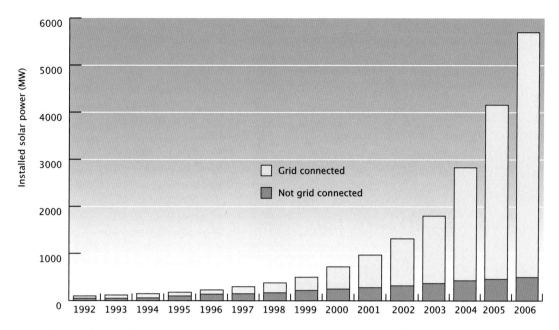

Installed solar-cell output in IEA countries (26 countries within the OECD) 1992–2006. The top part of the columns shows installed output for solar cells connected to the grid, and the bottom part shows installed output for solar cells not connected to the grid.

(Source: Report IEA-PVPS T1-16:2007)

January, while in southern countries solar energy for solar cells is available all year. Despite this climatic difficulty photovoltaics are installed in Sweden as well.

In Sweden in 2007 there were about 4MW of photovoltaics installed, mostly in small installations like cottages, boats and lighthouses. IKEA's head office in Älmhult was one of the first bigger installations in Sweden. It has a peak output of 60kW. The sports stadium Ullevi in Gothenburg has installed 600m² of PV. The tennis stadium in Båstad also installed a PV system. It is also intended to integrate solar cells into architecture. This has been done in several places, for example on dwellings in Hammarby Sjöstad (Stockholm), on the technical university in Malmö, and on the office building of Gothenburg Energy. The largest and most spectacular solar cell plant in Scandinavia is located in Sege Park in Malmö, installed in 2008. The plant has 1250m² of solar cells with a top output of 166kW.

Developing Countries

In many developing countries, solar cell systems cost less than extending the electricity grid, e.g. for cooling vaccines and lighting health stations. There is a great need for lighting, especially in rural schools. Telecommunications can then be maintained in areas without an electricity network. An interesting application is pumping water when the sun shines and storing the water in tanks or water reservoirs. The advantage of this application is that no batteries are needed.

Large-Scale Solar Energy

Solar energy can provide all the energy we need. If we covered 7 per cent of the Sahara desert with solar cells that would be enough to provide the whole world with energy. One problem is to make solar cells cheaper. Solar cells can be placed on our buildings so that a lot of solar electricity will be produced locally. If we produce solar electricity

in deserts the electricity can be used to produce hydrogen from water by electrolysis.

One of the world's largest solar cell parks (1GW) is in Qaidam, China.

Cleanergy in Åmål produces a solar power station that uses parabolic solar collectors of their own design to concentrate solar heat to a Stirling engine, which in turn runs an electric generator. The solar power stations are used in countries with an abundance of sunlight such as Dubai, Kuwait, Egypt, Spain, Italy and Greece.

The Stirling engine developed by Kockums in Malmö is of central importance for two large solar power stations in California. One is 500MW in size with 20,000 solar mirrors, and engines and generators in the Mojave desert; and one 300MW in size in Imperial Valley with 12,000 solar power units.

NAPS Magic Lantern. A lamp in the evening represents a significant improvement in the quality of life where there is no electricity grid.

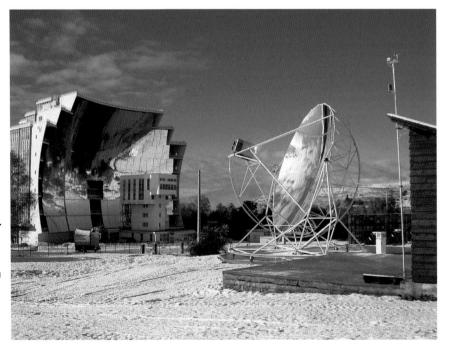

Photo from Odeillo, Font-Romeu in the French Pyrenees. On the right is a solar power station from Cleanergy with mirrors that focus the solar heat to a Stirling motor. This in turn produces electricity from the heat.

On the left is a building from 1970 that concentrates solar rays via its façade.

Source: www.cleanergy-industries.com.

440   3 ECOCYCLES | 3.2 RENEWABLE HEAT | 3.2.4 SOLAR CELLS

Different Systems

The simplest system consists exclusively of solar cells, batteries and a regulator to protect the batteries. Such regulators extend the life of the batteries. The electrical equipment is usually adapted to 12V, but a transformer can be built into the system so that 220V equipment can be used.

In a system connected to the grid, solar cell electricity is converted to 220V. Private electricity is used when the sun shines, and any excess electricity is sold to the grid. When it is dark or rainy, electricity is purchased from the grid. Such systems thus do not need batteries. This is a great advantage as batteries have the shortest life of any component in the system.

There are systems that have both batteries and a grid connection that function with both direct and alternating current. Such combined systems are good in locations that are subject to frequent power failures.

In Åmål the company Cleanergy builds solar power plants that are based on combining small Stirling motors with a parabolic solar collector (for use around the Mediterranean Sea latitudes).

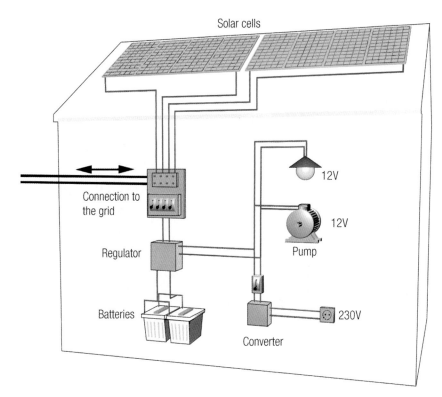

Grid-connected solar cells with battery bank.

Source: Illustration: Leif Kindgren

DEVELOPMENT OF NEW SOLAR CELL TECHNOLOGY

CIGS Cells

CIGS cells (copper-indium-gallium-selenide) are a relatively new type of solar cell. Sweden is in the forefront in developing this type of thin-film solar cell. Research is being carried out at the Ångström Laboratory in Uppsala, among other places. A problem with CIGS cells is that defects often occur during production.

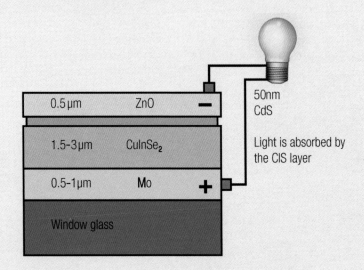

The Construction of CIGS cells

Plastic Solar Cells

Plastic solar cells are cheaper and require less energy to produce than other solar cells, but have a low level of efficiency (so far only a maximum of 6.5 per cent). They can be connected to a variety of surfaces and large areas. Many research teams are working on development in this area.

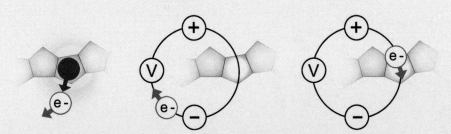

How a plastic cell functions: an energy-rich photon breaks loose an electron from a polymer chain leaving a hole behind, an electron shortage. The solar cell leads away the electron, which through an electric circuit is returned to the hole, which is then refilled.

Cheap and rapid production

1. Solar cells are pressed on thin plastic film with metal contacts using a rotation press.

2. Two different-coloured solar-catching plastics, which take better advantage of the energy from light, are put on the plastic film.

3. An electrically conducting and protective layer is put on top of the plastic layer.

4. The cell is folded so that light can bounce between the folds.

V-formed solar cells catch light several times

(i) Light with different-coloured photons falls towards the solar cell.

(ii) The green fold absorbs the energy-poor red light.

(iii) The red fold absorbs the energy-rich green light.

(iv) The remaining light reflects to the other fold. The larger the number of reflections the greater the chance that the light will become electricity.

Source: University of Lindköping, *Applied Physics Letters*.

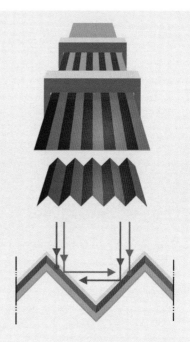

Hydrogen Gas-Producing Solar Cells

A solar cell that can split water into oxygen and hydrogen using sunlight is being developed. The goal is to store solar energy as hydrogen gas and then use it to produce electricity in a fuel cell. Such research is taking place at the National Institute of Advanced Science in Japan, at the National Renewable Energy Laboratory in Golden, Colorado, US, and in Sweden in a cooperation between the Universities of Lund, Stockholm and Uppsala. Some of the research involves letting sunlight strike complicated molecules that free electrons and result in the creation of oxygen. In Japan, work is being done using metal oxide particles and in Sweden using ruthenium compounds and manganese complexes.

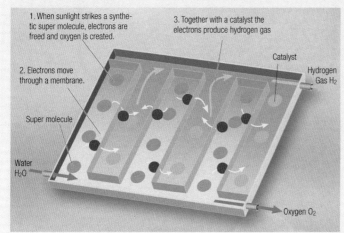

Hydrogen gas-producing solar cell.

Grätzel Cells

Grätzel cells are dye-sensitized cells that imitate the photosynthesis of green plants. The cells use the semiconductor material, titanium oxide. They differ from normal solar cells as the semiconductor surface is covered with a dye, and a liquid that conducts electricity (an electrolyte) is used as the electron bearer instead of a metal. Light is absorbed by the dye and its electrons move to the porous titanium oxide. They are inexpensive to produce and can be recycled. There is a problem with long-term stability. Research on Grätzel cells is being carried out at the Ångström Laboratory in Uppsala.

Artificial Photosynthesis

At the Arrenius Laboratory at the University of Stockholm researchers are trying to imitate photosynthesis. Ruthenium is used instead of chlorophyll. The metal acts just like chlorophyll, it absorbs light and releases energy. Solar energy splits water into oxygen plus electrons and protons. A manganese complex and tyrosine are then connected to the ruthenium. Tyrosine is an amino acid, that together with the manganese complex transforms sunlight into chemical energy. The result is hydrogen gas. The hydrogen gas can be used in fuel cells to produce electricity.

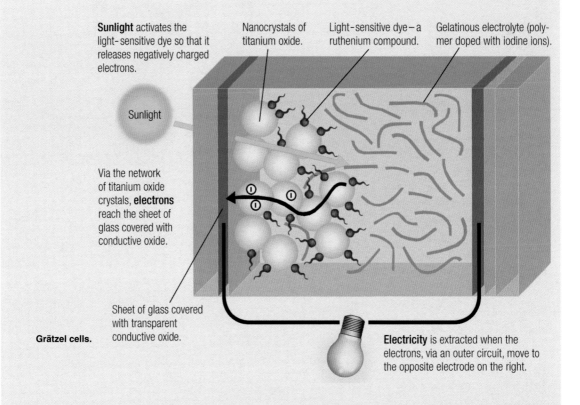

Grätzel cells.

EXAMPLES OF SOLAR CELL APPLICATIONS

Integration with Architecture

When solar cells become more common they will be a natural part of architectural design. Façade modules, roofing and roof tiles have been developed where the outer surface is made up of solar cells. There are also transparent solar cell panels that are placed with a little distance between them so that daylight filters in. There are products for screening out the sun that can be placed in front of windows that are also electricity-generating solar cells. They can be mobile or fixed, and provide protection from sun and rain.

Nieuwland at Amersfoort in The Netherlands, an urban district where all the buildings are equipped with solar cells.

Solar cells as roof covering.

Solar cells that screen the sun.

Solar cells as façade module.

Transparent solar panel bank on the environmental training centre 'Die Kliene Erde' near Boxtel in The Netherlands.

Movable solar panel bank that follows the path of the sun at a theatre studio in Copenhagen.

Solar cell sail sculpture beside a motorway in The Netherlands.

Solar electricity can be made in steam turbines by concentrating sun rays. The sunlight bounces off the mirrors and is directed to a central tube filled with synthetic oil, which heats to over 400°C (750°F). The reflected light focused at the central tube is 71 to 80 times more intense than the ordinary sunlight. The synthetic oil transfers its heat to water, which boils and drives the steam turbine, thereby generating electricity. Synthetic oil is used to carry the heat (instead of water) to keep the pressure within manageable parameters. The photograph shows Solar Energy Generating Systems solar power plants III–VII at Kramer Junction, California.

Source: Wikipedia

The fine settling pond, the last stage in the sewage purification process, at Boknäset, Bohuslän, Volvo's holiday village.

3.3 Sewage

Conventional sewage treatment plants focus on hygiene and the environment, i.e. rendering infectious matter harmless and ensuring that treated water does not pollute lakes and watercourses. In addition to these aspects, ecological sewage systems are designed so that the nutrients found in sewage, especially phosphorus, nitrogen and potassium are returned to the environment.

3.3.0 Sewage in Ecological Cycles

Conventional sewage systems have the following five goals:
1. Sanitation: rendering infectious matter, such as bacteria, viruses and parasites harmless.
2. Environment: preventing watercourses from becoming overgrown or odorous. Ensuring the continued existence of fish.
3. Energy: running an energy-efficient operation.
4. Reliability: the facility should be robust and be able to withstand operating disturbances.
5. Economical: the cost of operation and management of the facility should be reasonable.

In an ecological sewage system, the five goals should be supplemented by a sixth:

6. Nutrient recovery: returning nitrogen, phosphorus and potassium (NPK), etc., to agricultural land without polluting it.

Today's Sewage Systems

Sewage systems use mechanical, biological and chemical methods of purification. They take care of bacteria and viruses and are very good at removing phosphorus (through chemical precipitation), but they are not good at removing nitrogen. In order to prevent lakes and watercourses from becoming overgrown, the law may require that sewage facilities in coastal municipalities have an extra purification stage to reduce the quantity of nitrogen. In this stage a nitrification process is followed by denitrification which results in removal of 50–70 per cent of the nitrogen (a large portion of which is released to the air as nitrogen gas). The remaining nitrogen is still released into the sea.

Currently sewage from homes, industry, traffic surfaces and runoff are all mixed together in sewage systems. This results in pollution of the sludge with heavy metals and chemicals. In addition, most large sewage systems in Sweden leak to some degree and unpurified sewage leaks into the ground.

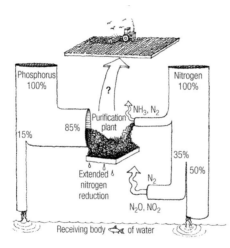

Conventional purification plants are good at removing phosphorus from sewage, but poor at removing nitrogen. Extended nitrogen reduction removes 50–70 per cent of the nitrogen, and the rest runs into the receiving body of water. The use of sludge from mixed sewage as fertilizer is questioned.

Source: Peter Ridderstolpe, WRS Uppsala; illustration: Peter Bonde

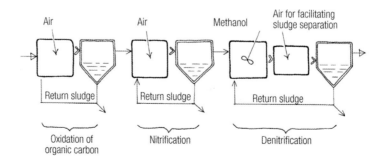

Three-step process for biological nitrogen reduction.

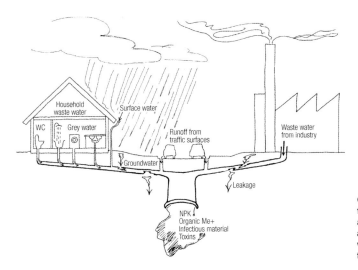

Currently sewage from homes, industry and traffic surfaces are mixed together, which results in contaminated sludge.

The Sludge Debate

There are attempts to make use of the nutrients from conventional sewage treatment facilities by returning sludge to agriculture. One problem is that the sludge contains heavy metals and organic environmental toxins. Farmers are unwilling to spread the sludge on their fields, and periodically there have been fierce debates between municipalities (that want to get rid of the sludge), farmers' organizations and nature conservation organizations, which work to protect arable land.

Another problem is chemical sludge. If iron or aluminium is used for chemical precipitation and to dewater the sludge, the phosphorus becomes so fixed in the sludge that recycling is impossible. In order to recycle the phosphorus either wet sludge or sludge where calcium has been used for chemical precipitation has to be used. These methods are more complicated. However, in a sustainable society, phosphorus must be recycled as it is a non-renewable resource. Another possibility is to avoid phosphorus in the sludge by prohibiting the use of phosphorus in detergents.

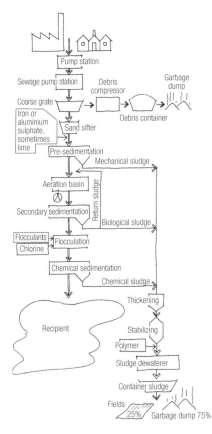

Sewage treatment in a conventional plant can go through the stages illustrated in the above diagram. Precipitation chemicals, flocculent chemicals and chlorine are added. Mechanical, biological and chemical sludge are produced. About 25 per cent of the sludge is spread on arable land and the rest ends up in rubbish dumps.

Use of Chemicals

The amount of chemicals in sewage water from homes depends on people's choice of household chemicals and what they dump into the sewage system. People need to learn to take leftover chemical products to collection points for environmentally hazardous waste instead of throwing them down the drain. One way of minimizing the use of household chemicals is to use dry cleaning methods or steam cleaning. A further factor to keep in mind is the choice of surface materials and the treatment they require, since these materials and cleaning products eventually end up in the sewage system.

FUTURE CRITERIA

Sewage expert Peter Ridderstolpe (WRS Uppsala) has listed the following criteria for ecological sewage systems:

Protection against infection

- Outgoing water should be bathing water quality.
- Hygienically acceptable management of waste products.
- Rules for sanitizing and spreading.

Protection for the recipient

- A reduction in phosphorus of at least 70 per cent, preferably 90 per cent, and emissions of <0.4kg/person/year.
- A 50 per cent reduction in nitrogen, and emissions of <2.5kg/person/year.
- Reduction of oxygen-consuming substances by at least 70 per cent, preferably 90 per cent.

Recirculation of nutrients

- Return of at least 50 per cent of the phosphorus to arable land, preferably 70 per cent.
- If technically and economically possible, other nutrient salts (e.g. nitrogen and potassium) should also be returned to arable land.

Other criteria

- The system should conserve energy and other natural resources.
- The system should be economical.
- The system should be user-friendly.

3.3.1 Sewage Separation at the Source

Solutions to current sewage problems include preventing heavy metals and hazardous chemicals from entering the sewage water and separating sewage into different fractions. The fractions from separated sewage water can then be purified and dealt with individually, which simplifies and improves purification. Runoff from traffic surfaces, surface water and sewage water are separated from each other, and a separation between residential and industrial sewage is also made. Systems are also being developed that separate sewage so that urine or black water is separated out as well as grey water (i.e. bath, dishwashing and laundry water).

In newly developed areas where conditions are created for sustainable development, runoff water is managed locally, runoff from roadways is purified on its own and buildings are equipped with double sewage systems, e.g. one for black water or urine and one for grey water or urine-free sewage water.

Sewage Contents

Sewage contains both hidden resources and pollutants. Types of pollutants include infectious matter (bacteria that cause sickness, viruses and parasites), heavy metals and organic environmental contaminants. The resources present in sewage are: nutrients, organic substances, heat energy and water. The resources can also be an encumbrance, e.g. when large amounts of organic substances end up in the recipient instead of on fields.

The nutrients can be used as fertilizer, as can the biological substances that are usually collected in a sludge separator. The mineral nutrients of greatest interest for agriculture (NPK) are found primarily in urine: about 80 per cent of the nitrogen, 50 per cent of the phosphorus and 60 per cent of the potassium. Thus, urine separation is a good method for utilizing a large part of the nutrients in sewage, especially if the urine is for the most part sterile and the heavy metal content is low. If urine, which accounts for only about 1 per cent of the total sewage volume, is separated and used as fertilizer, the greater part of the mineral nutrients in sewage can be utilized. If faecal matter is also separated, only a small amount of the mineral nutrients remain in sewage water. Faeces account for about 15 per cent of the total nitrogen, 30 per cent of the phosphorus and 25 per cent of the

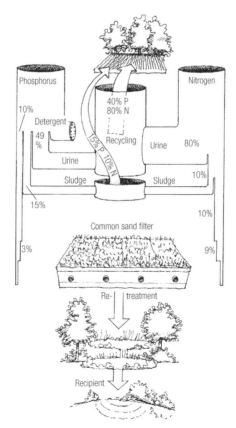

If urine is separated from sewage, 80 per cent of the nitrogen and 40 per cent of the phosphorus is removed. Use of phosphorus-free detergents results in 35 per cent less phosphorus. A sludge separator removes a further 10 per cent of the nitrogen and 15 per cent of the phosphorus. This results in better purification and possibilities for reuse of nutrients than when using conventional methods. The remaining sewage is so clean that it can be dealt with locally.

Source: Peter Ridderstolpe, WRS Uppsala AB; illustration: Peter Bonde

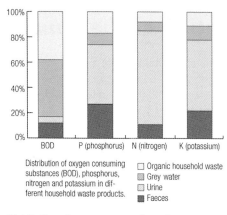

Distribution of oxygen-consuming substances (BOD), phosphorus, nitrogen and potassium in different household waste products.

Source: Jordbrukstekniska Institutet, 1999

Systems for Sewage Purification

Generally, sewage systems can be divided as follows: separation, preprocessing, purification (with technical and/or natural methods), sanitation and nutrient recycling. Purification methods can be combined in many different ways, and the different steps often overlap. Which approach is chosen depends on the number of people, the location (urban or rural), local soil conditions, proximity to agricultural land, as well as whether or not there are extra requirements to protect nearby lakes and watercourses. For industrial sewage, purification methods must be custom-designed, depending on the sewage contents.

potassium. Consequently, urine separation or black water separation (urine + faeces + a little flushing water) are two suitable methods to use in ecological sewage systems. The quantity of organic substances in sewage water is usually measured as biological oxygen demand (BOD), which is a measure of the oxygen consumption of organic substances during the breakdown by micro-organisms. A large load of organic material in waste water with a subsequent high BOD constitutes a risk of oxygen deficiency in the recipient. The amount is usually measured over a number of days (BOD7 means oxygen consumption during a seven-day period).

The amount of dissolved oxygen in a water sample is measured before and after incubation. The units used are mg/L.

Sorting

The very best alternative is if sewage is sorted in the home. It is then easier to provide suitable treatment for the different fractions. Sewage is usually separated in the toilet (dry toilets, urine-separating dry toilets, urine-separating flush toilets, and vacuum toilets). There is also the bath, dishwashing and laundry water (grey water) to deal with.

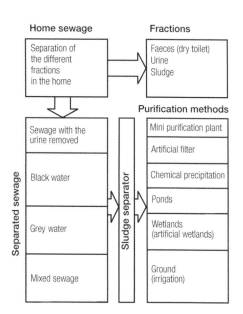

Different systems for sewage purification. Separated sewage can be purified using different methods, as well as a combination of methods.

Source: WRS, Water Revival Systems, Uppsala

LARGE-SCALE SEWAGE SEPARATION

Hammarby Sjöstad was the first urban district in Sweden to separate sewage into different fractions. The decision to use environmentally friendly façades and roofing helps to keep the runoff water clean, which is managed locally. Runoff from roadways is collected and purified using special filters. Sewage water from homes is not mixed with other sewage but is purified in a special purification plant, which produces cleaner sludge. The sludge is fermented in a biogas facility together with organic (digestible) waste and produces an energy-rich gas that is used in kitchen stoves in the dwellings. The sludge from the purification plant is an organic soil that is used as compost and high-quality fertilizer for energy crops. The purified sewage goes to the Hammarby district heating plant where a large heat pump removes the energy from the sewage water and supplies it to the district heating that heats Hammarby Sjöstad. Lake water from the district is circulated and oxygenated in water ladders, which keeps the water in the lake cleaner and healthier, and the water features between the buildings enhance the town's appearance.

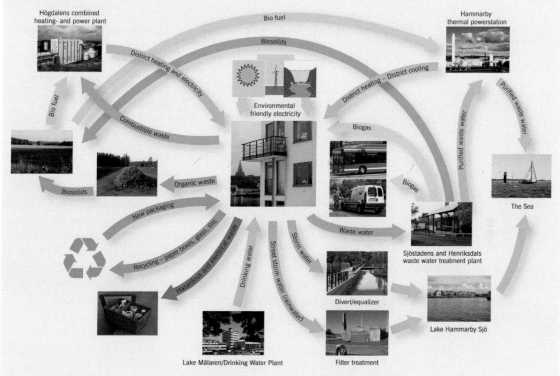

The various flows of water, energy and biological material in Hammarby Sjöstad, Stockholm, where the first phase was opened in 2002.

Source: www.hammarbysjostad.se

Ecological Sewage Purification can be Divided into Six Stages.

There are a variety of purification methods that can be combined in several ways.

Household measures	Preprocessing	Purification, technical	Purification, natural	Post-treatment, sanitation	Nutrient recycling
Separate in the toilet	Sludge separator	Compact filter	Ponds with plants	Storage	Watering
Save water	Separator	Mini purification plant	Artificial wetlands	Composting	Fertilizing
Choose good chemicals	Wet-dry	Chemical precipitation	Broad irrigation	Wet reactor	Resorption
	Sand trap			Aerobic	
	Sand filter			Biogas	
				Anaerobic	

Small-Scale Sewage Separation

Dry toilets can be practical in some situations. Modern dry toilets have a ventilation duct from the composting compartment in order to avoid odours and are designed to make emptying simple. It is important for the base to be sealed to avoid any leakage. The night soil from a dry toilet must be allowed to mature separately for a year in a night-soil compost. There are dry toilets with two compartments, where one is used and the other is closed off to mature for a period of one year.

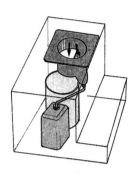

There are inserts for urine separation that can be installed in existing dry toilets, here the '*Dass-Isak*' model.

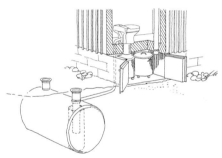

Mats Wolgast, MD (Wost-Man Ecology AB) developed urine-separating dry toilets.

Urine-separating dry toilets are in principle ordinary dry toilets equipped with an insert for urine separation. Separation of the urine results in a carbon/nitrogen balance that is more favourable for the decomposition of the faeces portion. The fraction is drier, takes up less space, has less of an odour and is easier to deal with. It can be advantageous to dig a container into the ground to collect the urine. This keeps the urine cool and problems with nitrogen loss and smell can be avoided.

Large *multrums* (composting toilets) treat faeces and urine together with organic kitchen waste. Good decomposition requires a good carbon/nitrogen balance. Therefore, dry toilets that include nitrogen-rich urine require the addition of carbon, e.g. in the form of kitchen waste or sawdust. Such *multrums* must be located in a warm place (indoors), preferably directly under the toilet seat, and be ventilated. They are expensive and require a lot of room. If they are not taken care of meticulously, problems with odour, flies and overflow can occur. Therefore they are rarely used today. The Clivus Multrum, one of the more effective models, has a tank for leachate under the decomposition chamber to take care of any surplus liquids.

Urine-separating *multrum* toilets have several advantages over large *multrum* toilets. They work better and there are no problems with overflow, odours or flies. They are small and inexpensive, and there is enough room for the container underneath a suspended foundation. There are two containers; one is in use while the other is used for maturing the compost. The urine tank is set into the ground outside the house. Ventilation with a fan is required in most cases. Smaller models where the container is placed under the toilet seat are available, primarily for use in summer cottages.

Urine-separating flush toilets are the most commonly installed ecological sewage system in Sweden. Many people do not want dry toilets, so urine-separating flush toilets are a suitable alternative. Ventilation of the toilet is not required, which eliminates noise from a fan. The urine flows into a tank. It is important to keep the quantity of water low, especially

Three types of urine-separating toilets. At the far left is one from Beijing (China). In the middle is a Swedish, wall-hung urine-separating flush toilet. At the far right is a Mexican model.

Source: 'World Toilet Summit 2003', Taiwan, Proceedings

when flushing the urine, in order to avoid the need for large urine tanks. At the same time, the design needs to prevent clogging of the urine pipe by urine crystals (which can, however, be dissolved using soapy water). Such flush toilets use 3.5–5 litres of water for a large flush and 0.2–0.7 litres for small flush. There are toilets that are even more water efficient (0.6–2 litres for a large and 0.2–0.4 litres for a small flush), but they require a steeper sewage pipe pitch and a maximum of 10m distance to the collection tank.

Urinals are one way to reduce the addition of water to urine when separating urine. There are a number of urinals that do not use water. They are equipped with odour barriers in the form of floats or small water locks with special blocking fluid (that is lighter than urine).

The urine-separating flush toilet *Dubbletten*. It has two bowls and two flush buttons, one for 'the large' and one for 'the small'. It also has an extra cover that can be lowered so that it can be used by children.

Source: *Invention and Development*, Bibbi Söderberg

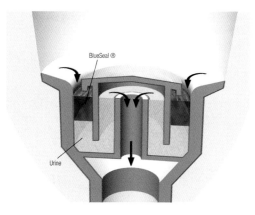

Cross-section of a urinal that does not use any water.

Source: Waterless Co. Illustration: Leif Kindgren

Vacuum toilet

Vacuum toilets are used in sewage systems where the black water (urine + faeces + a little flushing water) is separated out. They use about 1–1.2 litres per flush. There are more water-efficient vacuum toilets (about 0.5 litres per flush), but then the distance between the toilet and the vacuum generator cannot be more than 6m. Smaller dimensioned pipes can be used in vacuum systems than in ordinary flush toilets.

Vacuum toilets with urine separation are another way to reduce the amount of water in sewage. Vacuum toilets have earlier been associated with an unpleasant flushing noise. There are now quiet soft vacuum systems. There are different kinds of urine-separating vacuum toilets, some with both large and small flushes (0.5–1.9 litres for large flush and about 0.4 litres for small flush), and some where the urine is dealt with entirely without water (where the large flush requires 6L).

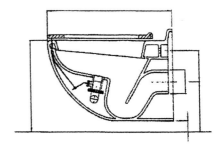

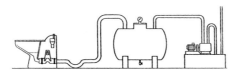

Vacuum toilets must always be equipped with one or more vacuum generators.

Source: BFR, R27: 1979

Quiet low-vacuum toilets, with or without urine separation, are the latest development in separating sewage systems. They can be used both for black water systems or urine-separating systems with faeces separation.

Source: Wost-Man Ecology AB

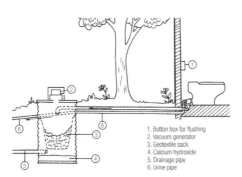

1. Button box for flushing
2. Vacuum generator
3. Geotextile sack
4. Calcium hydroxide
5. Drainage pipe
6. Urine pipe

The German urine-separating vacuum toilet Roevac No Mix Toilet from Roedinger. The toilet bowl has two outlets, urine runs down the front outlet and is transported away undiluted, faecal matter falls down the back outlet. As soon as the toilet is flushed, the cover over the urine outlet closes.

FOUR SMALL, LOCAL SEWAGE PURIFICATION SYSTEMS

1 Dry system based on urine-separating dry toilets

The first modern urine-separating toilets developed were dry toilets. Many people still believe that urine-separating dry toilets are best from a hygienic and environmental viewpoint. Although they are mostly used in summer homes, they are equally suited for use in year-round dwellings. They are presumably a good type of toilet for use in many developing countries.

The urine is collected in a tank, best dug into the ground. This keeps the temperature down, which helps to minimize the release of ammonia (and thus nitrogen loss). The urine can be used diluted in a local garden or collected to be used on agricultural land. The grey water is purified separately after sludge separation, e.g. in a compact filter and/or a wetland.

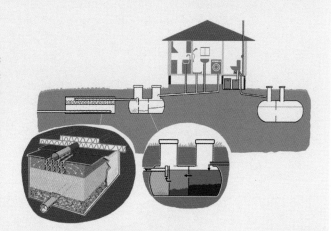

This system is based on a urine-separating dry toilet. The faeces falls down into a small, ventilated composting chamber under the toilet. The urine is collected in a tank. The grey water goes to a sludge separator and then to a closed compact filter where it is purified before being released.

Source: Illustration: Ingela Jondell

2 Urine-separating flush toilets and mini purification plants

Many people prefer flush toilets, and urine-separating flush toilets are used in many eco-buildings and eco-communities. In single-family houses, purification in a compact filter is often sufficient after the sludge has been separated from sewage water that has had the urine removed. In eco-communities, where several homes are connected to the same system, they usually use mini purification facilities. In order to reduce phosphorus leakage from mini purification facilities, chemical precipitation or an aquatron (a Swedish invention designed to separate solids from a flow of liquid) may be installed behind the toilets so that the faeces are separated from the urine-free sewage water at an early stage.

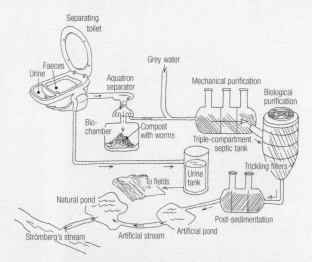

The sewage system in the Smeden eco-community, Jönköping, Sweden. There are urine-separating flush toilets and the urine is collected in a tank sunk into the ground. The urine is utilized by farmers in the Jönköping area. The urine-free black water is taken to a separator, an Aquatron, where faeces and paper are separated out to a small composting chamber. Then the flush water is mixed with the grey water and taken via a sludge separator to a mini purification facility. After purification, the waste water is taken to a second sludge separator for post-sedimentation. Finally, it is run through an artificial wetland before being released.

3 Black-water Systems

In larger and denser population centres, urine and night-soil management can be difficult. In these places, double sewage systems are used instead, one for black water and one for grey water. Vacuum toilets are used to reduce the water content in the black water. Organic waste can be added to the system from rubbish disposal grinders. The black water is treated and sanitized aerobically in wet reactors or anaerobically in biogas digesters. The grey water is often purified locally in sand filters and artificial wetlands.

Black-water systems with a vacuum vary: either the black water is carried to a tank or directly for treatment. The system can also be divided up into several stages, e.g. the sewage can be transported with the help of a vacuum from the toilets to a sump. From the sump, the sewage can be pumped through pressurized pipes using grinding pumps to a collection tank or to treatment. The black water is a nutrient-rich resource that after sanitization can be used in agriculture.

Rubbish disposal grinder in a dwelling

Rubbish disposal grinder in a large-scale kitchen

The black-water system in the Lübeck-Flinten-Breite residential area in Germany, built in 2003 and designed for 350 people. The sewage is divided into three fractions: one for black water and organic kitchen waste, one for grey water, and one for runoff water. The black water and organic waste are removed in a vacuum system for treatment in a biogas facility. The grey water is purified in an artificial wetland and the runoff is dealt with locally.

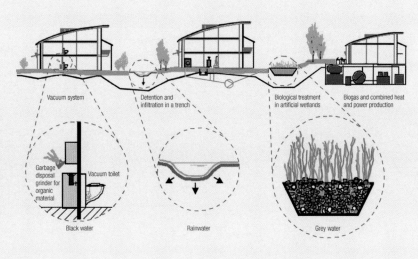

4 Chemical Precipitation of Unseparated Sewage

Many private sewage systems in rural areas use sludge separators and infiltration into the ground. If a more effective purification is desired, these can be complemented with a dosing feeder for chemical precipitation. This increases the amount of sludge, but reduces the phosphorus leakage. The next step is to improve the infiltration system, e.g. by complementing it with compact filter units. Other possibilities are building a mini purification facility and/or installing urine separation.

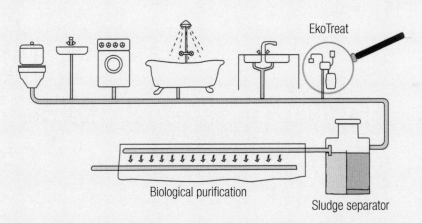

Chemical precipitation of unseparated sewage, followed by sludge separation and a compact filter.
Source: EkoTreat

Comparison of Different Sewage Alternatives

	Municipal water and sewage	Black-water mini purification facility	Black water transported to the municipality	Dry urine-separating system	Wet urine-separating system
Protection against infectious disease	Very good	Good. Some uncertainty when grey water infiltrates	Good. Compact filter results in good grey water purification	Good. Faeces management may occur	Good. Expanded clay aggregate bed results in good grey water purification
Protection of the recipient	Good for phosphorus, acceptable for nitrogen	Probably good (untested)	Very good (especially regarding nitrogen)	Very good (especially regarding nitrogen)	Very good (especially regarding nitrogen)
Recirculation	Acceptable. Recirculation of phosphorus can take place	Acceptable. Recirculation of phosphorus may take place	Very great potential for returning all nutrients	Very great potential for returning all nutrients	Good potential for returning all nutrients
Economics	Investment per property: 154,000 SEK. Operation: 4100 SEK/year. Annuity costs: about 15,300 SEK/year	Investment per property: 127,000 SEK. Operation: 5000 SEK/year. Annuity costs: about 14,200 SEK/year	Investment per property: 124,000 SEK. Operation: 1700 SEK/year. Annuity costs: about 10,700 SEK/year	Investment per property: 93,500 SEK. Operation: 2400 SEK/year. Annuity costs: about 9200 SEK/year	Investment per property: 135,000 SEK. Operation: 800 SEK/year. Annuity costs: about 10,700 SEK/year
Resource consumption	Relatively high. Chemicals and energy for operation	Relatively high. Chemicals and energy for operation and transportation	Relatively low. Energy for operation and transportation	Relatively low – transport the urine	Relatively low – transport the urine
Reliability/robustness	Very good	Unclear (not tested)	Good. Some risk of clogged pipes	Relatively well-tried technology. Robust	Relatively well-tried technology. Robust
Flexibility/adaptability to the site	Poor flexibility. LPS system required in rocky terrain	Relatively high flexibility for the future	High flexibility for the future	High flexibility for the future	High flexibility for the future
User aspects	Very good. No dealing with waste products	Good. No handling of waste products	Good. No handling of waste products	Relatively good. Some work is required by the user	Good. No handling of waste products
Organization and responsibility	Municipality	Home owner association or entrepreneur	Home owner association, municipality or entrepreneur	Home owner. Urine and faeces collected by municipality	Urine and sludge can be collected by municipality

Source: Älgörapport, Nacka kommun, 2003, VERNA

3.3.2 Technological Methods of Purification

Sewage can be purified in a purification plant using mechanical, biological and chemical technologies. Most purification facilities use a combination of mechanical and biological methods, with the addition of chemical purification in the form of a dosing feeder for precipitation chemicals. This section primarily describes small-scale purification.

Small-Scale Purification

Currently, the most common methods of small-scale purification are infiltration or sand-filter trenches. The problem with these methods is that it is difficult to check the purification efficiency, and reuse of nutrients is difficult to achieve. A wide range of new purification methods have been developed including mini purification plants and artificial filters. At the same time, new products to supplement purification with small-scale chemical precipitation have been developed.

Pretreatment

Almost all purification methods begin with separating out solid particles, usually with a sludge separator. Sand traps are one way to remove large particles in large purification plants. Filters used to catch solid particles must be replaced at intervals. There are also filters that can be emptied.

Sludge separators, three-compartment septic tanks, are included in most sewage systems to remove particles and small objects from sewage water. This is usually done through sedimentation, i.e. the particles sink to the bottom. The sludge must regularly be emptied from sludge separators. It can undergo post-treatment locally or in municipal purification plants. There are several types of ready-made plastic sludge separators with a varying number of chambers, which are just as efficient as traditional three-compartment septic tanks made of concrete.

Aquatrons are separators that separate solid from liquid fractions. In an aquatron, the sewage water moves in a spiral in a snail-shaped pipe that widens towards the bottom. The liquid is separated by surface tension (i.e. the liquid follows the walls of the container), while the faeces and toilet paper fall straight down into a decomposition chamber (biochamber). Mulch and compost worms can be added to the biochamber to speed up the decomposition process. Urine and flush water that are separated out are infected with bacteria and viruses from the faeces and are not as easy to reuse in a hygienically acceptable manner as the urine from a separating toilet. Aquatrons need to be supplemented with a sludge separator to remove sludge particles from the waste water.

Sludge separation can take place efficiently in a three-compartment septic tank made of concrete. Most modern sludge separators are made of plastic.

Source: Illustration: Ingela Jondell

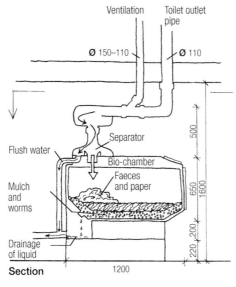
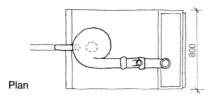

An aquatron is a separator that separates solids from liquids. Faeces and paper fall down into a small *multrum* (biochamber) for composting. Urine and flush water is removed for purification.

Mini Purification Plants

The treatment methods are based on the same processes found in municipal purification plants, i.e. mechanical, biological and chemical treatment. Sedimentation is a mechanical purification method used to separate solid material, e.g. in a sludge separator. Biological treatment of organic material and nitrogen is carried out using micro-organisms, primarily bacteria in the form of active sludge or as biofilm on a carrier material (packing material). Chemical precipitation is used to separate phosphorus and suspended material (broken-down solid material suspended in the sewage water). Most mini purification plants use all three of these stages, though some only use mechanical/biological or mechanical/chemical treatment.

There are a number of different mini purification plants for sale. They come ready to use and complete with a reactor tank, control equipment, etc. They are often made to suit a normal household (five people), but many manufacturers produce larger sizes for use by a number of households. Mini purification plants are often dug into the ground, but there are some that are intended to be sited indoors, e.g. in a cellar. The latter are easier to inspect and maintain. Treatment can either be done batchwise with sequencing batch reactor (SBR) technology (which means that a certain volume of sewage water is treated each time), or as a continuous flow in submerged biobeds.

This is what an above-ground mini purification facility can look like. This one is in the Smeden eco-community, Jönköping, Sweden.

In a **mini purification plant with batch treatment**, the sewage water is first collected in a tank. A certain amount of sewage water is then pumped into the reactor tank. The reactor is ventilated, and micro-organisms break down the organic material. The air flow is then closed off, the sludge sinks to the bottom and the purified water remains in the upper portion of the reactor tank. The purified water is drawn off, excess sludge is pumped out of the reactor and the process begins again. The advantage of this technique is that it is less sensitive to variations in flow and load and all the water gets the same treatment.

Mini purification plants with continuous flow have a sludge separation phase before the water goes to the reactor tank. The biological processes take place on an artificial carrier material in the reactor tank, e.g. perforated pieces of plastic. The carrier material is designed so that it creates as large a surface area as possible for micro-organisms to feed on. The reactor tank is ventilated and the sewage water circulates over the carrier material. The micro-organisms break down the organic material and some nitrification takes place at the same time. To make denitrification possible, the water is circulated frequently to an unventilated chamber (where the nitrogen is released into the air). Sludge forms in both the sludge separator and the sedimentation tank. Therefore, an additional sludge separator is often placed after the reactor tank.

Biovac's mini purification plant, with batch treatment, placed accessibly in a cellar.

Source: Photo: Maria Block

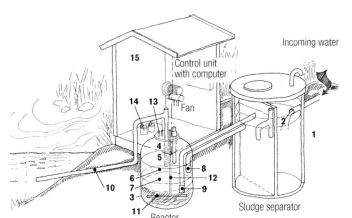

Mini purification plant with batchwise treatment.
1 sludge separator with two chambers
2 buffer volume
3 reactor tank
4 air volume
5 clean water zone
6 safety zone
7 active sludge zone
8 intake from sludge separator
9 recirculation of excess sludge
10 extraction pipe for purified water
11 bottom ventilator for biological decomposition
12 air line for the bottom ventilator
13 intake air for, among other things, extraction of clean water
14 collection container for water samples
15 structure for fan, dosage and automatic control equipment

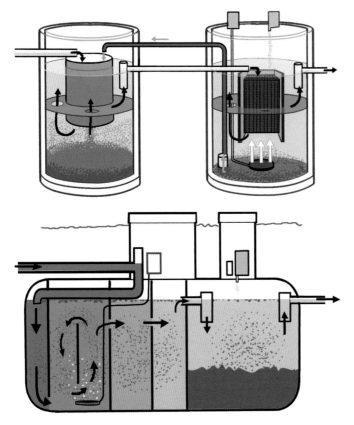

Two types of continuous mini purification plants. The upper one (Alfa) consists of two units; the left-hand one is a sludge separator. A ventilated bioreactor is in the middle of the other container, into which precipitation chemicals can also be added. The excess sludge is pumped back to the sludge separator. The plant on the bottom (Biotrap) is preceded by a separate sludge separator, but sludge is also formed in the reactor tank, which must be emptied at regular intervals.

Source: Illustration: Ingela Jondell

Purification in Filter Beds

Many infiltration and sand filter trenches work poorly and nutrients cannot be recovered. There are several ways to improve such facilities. The sewage water can be spread out in a better way, e.g. by sprinkling or in diffusion layers. The trenches can be enclosed so that rainwater does not get in, resulting in a more stable treatment process. It is also possible to use different types of filter materials. When the above three methods are combined, they form what is known as a compact filter.

Enhanced infiltration can be achieved by using artificial diffusion layers, which makes it possible to reduce the size of the facility. The method is suitable for grey water or sewage water with the urine removed.

Filter beds that contain phosphorus binding material instead of sand are a way of improving sand filter trenches. Crushed expanded clay or sandy marl are especially good at absorbing phosphorus, but the material eventually becomes saturated and then has to be replaced. The filter material can be spread on arable land as phosphorus fertilizer. It is also possible to combine enhanced infiltration with phosphorus binding material.

Compact filters (artificial filters) are an improvement over sand filter trenches; they both work better and are smaller. The filter beds are enclosed in boxes or geotextiles to prevent ingress of unwanted water. There is a diffusion layer on the top, a filter bed underneath and a draining layer at the bottom. The distribution layer is made of an artificial material designed to spread water over the surface and to provide a large surface area for

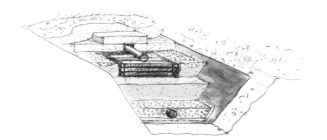

There are different makes of compact filters, but features they all have in common are that they can be checked to see how they are working and discharge takes place at one spot. The illustration shows the Infiltra model compact filter.

Source: Swedenviro, Peter Ridderstolpe, WRS Uppsala AB

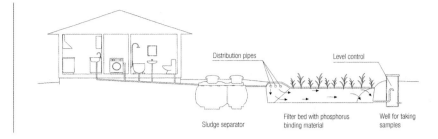

It is possible to have plants in filter beds with phosphorus binding material.

Source: *Småskalig avloppsrening – en exempelsamling*, edited by Birgitta Johansson, Formas, 2001

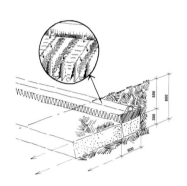

Enhanced infiltration with a diffusion layer on top.

well as adsorption of dissolved substances. The dosing of a precipitating agent can be time-controlled or associated with toilet flushes or filling of a pressurized tank. As the amount of sludge increases, either a larger sludge separator or more frequent emptying of the sludge is required. Chemical precipitation must be supplemented by other treatments.

bacteria to feed on (bioskin). The filter beds are made up of different sand materials. Compact filters result primarily in a reduction of oxygen-consuming substances and infectious matter. The sewage water must go through a sludge separator before it is treated in the filter.

Chemical Precipitation

Chemical precipitation is found in mini purification facilities, but can also be used as a supplement to sludge separation. The process of chemical precipitation involves adding a precipitating agent to sewage water, resulting in the precipitation of phosphorus and the formation of gelatinous tufts or lumps. The precipitated phosphorus is bound to these lumps, which settle to the bottom and form a chemical sludge. The precipitating agent also contributes to precipitation of suspended substances as

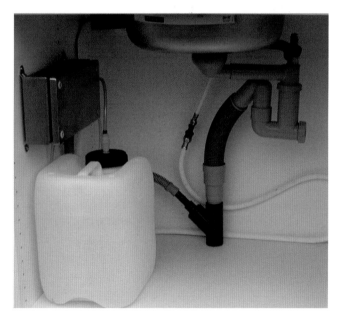

One problem with local sewage purification has been phosphorus leakage. To reduce this, small sewage treatment facilities can be supplemented with chemical precipitation. The method improves older infiltration facilities, new filter beds and mini purification plants.

Source: Eko Treat

3.3.3 Natural Methods of Purification

The natural cycle (earth, vegetation and micro-organisms) has a good capacity for purifying sewage water, so a small creek can act as a purification system if the amount of sewage isn't too great. Natural biological purification takes place through the breaking down of pollutants by a great variety of micro-organisms. Mechanical purification also takes place naturally since the ground acts as a filter. Vegetation also plays an important role in nature's purification process. Vegetation absorbs water and nutrients, increases evaporation and pumps oxygen down to micro-organisms in the upper layers of the soil.

Purification in Constructed Natural Systems

Sewage purification facilities have been developed where constructed natural systems are used for purification, e.g. as the last stage in a purification process (effluent polishing). Nature-imitating purification systems can be divided into the following categories: ponds, wetlands, artificial wetlands and the ground (that is irrigated).

Purification Ponds

Ponds can be used to purify sewage water. Purification takes place through the settlement of sludge on the pond bed and the breaking down of pollutants by micro-organisms in the pond. Aeration of the water increases the amount of oxygen and then nitrification takes place. In areas of the pond where there is less oxygen, denitrification takes place whereby nitrogen is released to the air as nitrogen gas. Water plants contribute to purification by absorbing nutrients and stimulating the vitality of micro-organisms. Nitrogen in the form of ammonia is transformed by the plants to nitrates. Several ponds can be connected to each other, and the water becomes successively cleaner in each pond. The plants and micro-organisms needed for each particular stage of the purification process are naturally established in each pond. By making use of the sludge that settles on the pond beds and the plants that grow there, nutrients are ecologically recycled.

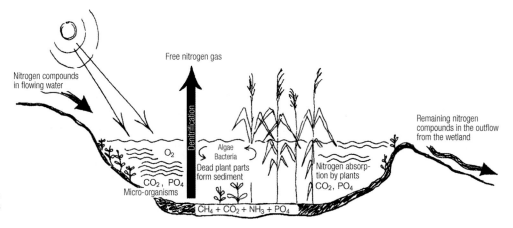

This is how a pond purifies sewage.

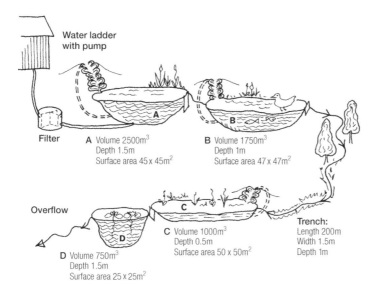

Sewage plant in Järna with a series of ponds, wetlands and water ladders.

Wetlands

Wetlands act as nutrient traps. Nitrogen is transformed to nitrogen gas or stored in the sediment. The phosphorus that follows along with particles in running water stops short, settles to the bed and is bound. As agriculture has become more and more efficient with the draining of wetlands, straightening of streams, and filling in of trenches, there has been a decrease in natural purification and leaking nutrients flow down to the ocean. By re-establishing wetlands, and allowing water meadows and bogs around lakes and streams to re-establish, it is possible to reduce the nutrient leakage to the ocean.

Today, artificial wetlands are used in a number of different ways: as a last step in municipal purification facilities, to purify runoff from motorways, to buffer runoff after heavy rainfall and as purification systems for small residential areas.

Artificial Wetlands

Artificial wetlands are concentrated, constructed wetlands without open stretches of water. They consist of water plants in sand beds. The biochemical processes are the same as in sand filter trenches and wetlands. Water plants carry oxygen down to their roots where a good environment is created for sewage-purifying micro-organisms. The water level is adjusted with an outlet well, which should be located under the earth surface in order to avoid odours. The sewage water can be introduced at one end of the artificial wetland and taken out the other (horizontal artificial wetland), or introduced on the surface and taken out at the bottom of the artificial wetland (vertical artificial wetland).

Nutrients bound in the plants and bottom bed can be returned to agricultural land by removing the bottom material and spreading it on fields (and replacing it with new material), and by harvesting and composting the plants from artificial wetlands.

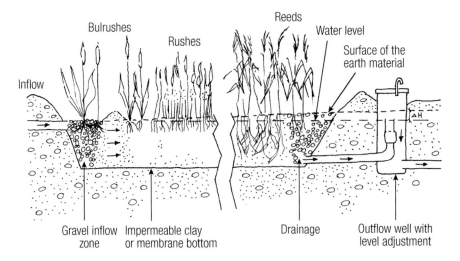

Artificial wetland with horizontal water flow.

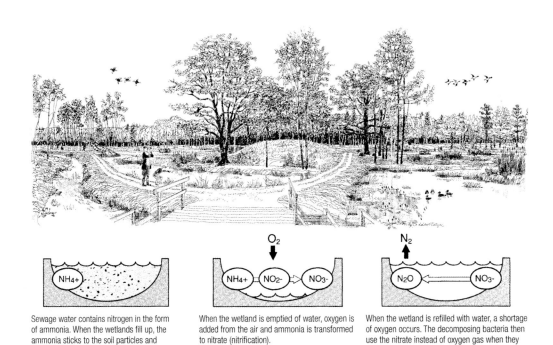

Sewage water contains nitrogen in the form of ammonia. When the wetlands fill up, the ammonia sticks to the soil particles and plant parts.

When the wetland is emptied of water, oxygen is added from the air and ammonia is transformed to nitrate (nitrification).

When the wetland is refilled with water, a shortage of oxygen occurs. The decomposing bacteria then use the nitrate instead of oxygen gas when they breath. Thus, plant nutrient nitrogen is transformed to common nitrogen (denitrification).

Artificial wetland in Oxelösund, Sweden, the last stage at the Oxelösund sewage purification plant. The amount of nitrogen is reduced and at the same time, a beautiful wetland park with rich birdlife has been created.

Source: Illustration: Fritz Ridderstolpe

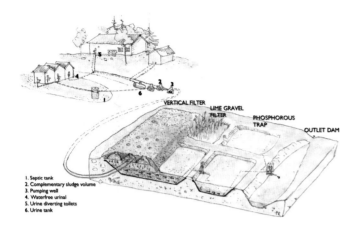

A sewage system with urine-sorting water toilets and purification in a constructed natural system. Urine is collected in a tank and then used as a fertilizer by local farmers. Waste water is pretreated in a septic tank and pumped to a constructed soil filter where water is distributed through spray nozzles. Phosphorus is then absorbed in horizontal filters with calcium-rich reactive media. Overall the treatment capacity is 97 per cent BOD-removal, 90 per cent P reduction, and 65 per cent N reduction. Bacteria are reduced by 99.9 per cent. The urine diversion contributes to 40 per cent of P and N removal. Operation is simple and maintenance costs are low.

Source: WRS AB, Uppsala/Peter Ridderstolpe

Flood Irrigation, Resorption

There are purification facilities with water meadows where sewage water is allowed to flood meadows. Nutrients are not reused in such systems since the plants in the system are not harvested.

Aquaculture in Greenhouses

Aquaculture in greenhouses is used to purify sewage and produce heat, biomass, fish and crustaceans in closed systems. The technique was developed in the US at the New Alchemy Institute. Such facilities exist in Denmark (Thy, Kolding, Grynebækken) and have been tried in Sweden (Stensund's Folkhögskola in Trosa and Överjärva farm in Solna). All the Scandinavian facilities were expensive to build, are difficult to run and there have been problems with purification.

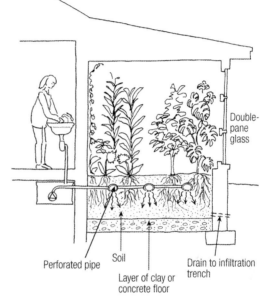

Grey water can be purified, after sludge separation, in a greenhouse attached to a building. Purification can take place most of the year since the growing season is extended by the greenhouse and waste heat from the building and sewage.

A mull filter for treatment of water from bathing, dishwashing and clothes washing (greywater), before the water infiltrates into the ground.

Source: Peter Ridderstolpe, Water Revival Systems.

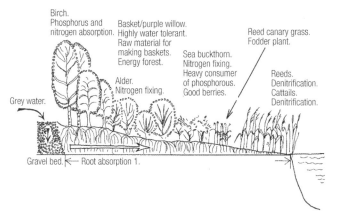

Sewage purification using flood irrigation in Kungslena, Västergötland.

Details of an artificial wetland system (or plant-covered sand trench) for purifying grey water into good quality surface water. Various plant varieties are chosen to satisfy different needs in the constructed ecosystem.

Source: Folke Günther

Irrigation

Irrigation with sewage water provides plants with the water and nutrients they need while simultaneously purifying the sewage water. One problem in northern climates is that the growing season is limited while sewage is created all year round. A solution is to store the sewage water in a pond during the winter and use it for irrigation during the summer. As there are concerns about infectious material, such effluent can be used to irrigate fenced-in energy forests. On the island of Gotland, for example, using effluent to irrigate crops is relatively common, and there is a lot of knowledge about which crops to irrigate and when irrigation should take place in order to avoid the spread of contagion. Separated sludge or filtered grey water can be used for irrigation in private greenhouses, which as a rule takes place below the soil surface (resorption) to avoid odours. Since sanitary waste-water is not included in grey water, spread of contagion is not a concern.

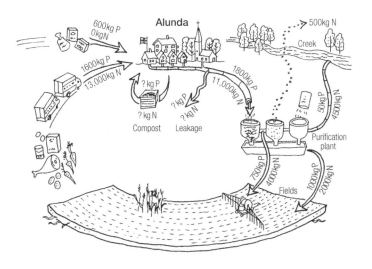

The nutrient flow in a proposal for sewage management in Alunda, Uppland, Sweden, with a population of 2000. By spreading wet sludge and sewage water on fields near the sewage treatment facility, a 50 per cent reduction in nitrogen and good integration with natural cycles can be achieved.

Source: Peter Ridderstolpe, WRS Uppsala AB

3.3.4 Nutrient Recycling

In order to use waste water or its fractions for crop cultivation, the waste water must be treated to some degree in order to be regarded as hygienically acceptable. It is often a matter of storing the sewage for a certain time (considered a type of treatment) prior to spreading it on gardens and farm land. If sewage is stored at a temperature of +8°C, a storage period of 6–12 months is recommended. At 40°C, storage for one month is considered adequate, and at +65°C one week is sufficient. There is a big debate about how to spread sewage water without contaminating agricultural land and food.

Cooperation with Farmers

Creation of ecological sewage systems requires cooperation with farmers and their involvement in development. Above all, the sewage sludge must be sufficiently clean so that it will not spoil agricultural land; dispersion methods that reduce nitrogen loss and odours must be developed; and fertilization methods (on which crops, when and how) need to minimize nitrogen leakage and the spread of contagion. Sewage should be applied to fields at the beginning of a crop's life cycle, which is when most nutrients are absorbed. Sewage water is spread via irrigation and resorption (irrigation under the surface using pipes) or on the soil, using hoses connected to a tank vehicle or by ploughing it under. Sewage can be used to fertilize fodder grain but not grazing land. Biodynamic farmers believe that food grown for people should not be fertilized using sewage from people, but that the cycles should be separated. EU regulations forbid the use of urine or black water for organic cultivation.

Sludge Management

Sludge from private households is often much cleaner than sludge from large purification facilities. The sludge from three-compartment septic tanks does not contain as many pollutants as sludge from a purification facility, and is a familiar product in the farming industry. It requires sanitation before use and conventional agricultural technology can be used for dispersion. Grey-water sludge is a relatively nutrient-poor product. After dewatering, it has a high dry content and is easy to deal with. Chemically precipitated sludge has a

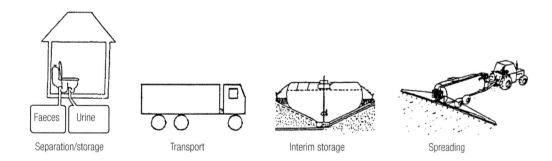

Dealing with separated sewage has several phases: transport, storage, sanitization and spreading. Suitable methods are required for each phase.

Source: 'Det Källseparerande Avloppssystemet – ett steg mot bättre resurshushållning', unpublished thesis 1995 Stockholm University, Mats Johansson and Jan Wijkmark

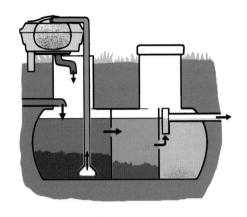

A sludge pump sucks up the sludge to a filter unit placed above the well. The sludge is dewatered by gravitation and then put in a composter. In this way, the sludge does not end up being mixed with other waste in a central purification facility.

Source: Illustration: Ingela Jondell

The last stage of Biovac's mini purification plant consists of barrels for dry sludge where the moisture in the sludge evaporates. Wet sludge removal is not required.

Source: Photo: Maria Block

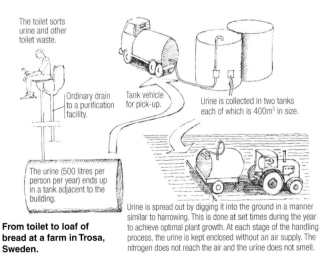

From toilet to loaf of bread at a farm in Trosa, Sweden.

Source: DN, 950511

high phosphorus content. Since the phosphorus is chemically bound, its availability to plants is uncertain and the nutrient content is unbalanced due to low levels of nitrogen and potassium. In the course of the sludge debate, chemical sludge has received a relatively poor reputation.

Sludge removal is usually done using a sludge separator. Then the sludge is transported to a municipal treatment facility. However there are also methods for managing sludge locally. The sludge is pumped from the sludge separator to a container.

There is a filter in the container that acts like a large coffee filter. The sludge is dewatered and then remains for maturing. There are mini purification systems where the dewatered sludge is collected and can be easily removed and composted. Sludge can also be dewatered, mineralized (composted) and dealt with by spreading it on the surface of an artificial wetland facility. The leachate from the artificial wetland is returned for sewage purification.

Fertilizing with Sludge

Returning sludge to ecological cycles is not entirely simple. A lot of research is taking place at agricultural universities on which crops may be fertilized and when fertilization should take place in the cultivation cycle. Research is also being carried out about fertilization of energy crops and forest land. One way to reduce the risk of transferring hazardous substances is to fertilize crops intended as green manure crops, e.g. stinging nettles or leguminous plants that can be composted. Green manure crops are either ploughed under or composted to add plant nutrients to the soil.

Combustion of Sludge

In some places, the sludge is so polluted that it is an environmental problem, e.g. in the Falun area where leakage from mines and slagheaps releases heavy metals into the water. In such places, the sludge can be safely deposited in rock caverns. There are methods for combusting sludge, extracting the nutrients and depositing the toxic ash. Innovators are also working on alternative purification processes for polluted sludge, e.g. using calcium hydroxide to remove heavy metals. The methods are expensive, but perhaps necessary to protect the environment.

Management of Faeces

Night soil from dry toilets must be matured in special night-soil composters which are equipped with rainproof covers and two compartments with sides that let through air. The bottom consists of a waterproof concrete or plastic bin so that contaminated leachate does not get into the groundwater. Such night-soil composts can be used for private night-soil management. One compartment is filled while the other sits and matures. The process takes about six months.

Management of Urine

Urine is very nutrient rich and has a nutrient content balance suitable for plants. Urine can be safely returned to arable land since it contains few pollutants and the hygiene risk is considered minimal. Urine management requires a new system for management and use, but existing agricultural equipment can be used for spreading it. Urine can be stored and sanitized in buried tanks so that it is kept cool by the ground. The tank should not be ventilated. Since urine is antiseptic, sanitization takes place during storage; nutrients are retained and almost no nitrogen is lost. The tank should be big enough so that it only has to be emptied a few times per year. The volume of the tank for urine/flush water depends on the size of the household, how much water is used per flush and how often the tank is emptied.

Urine often requires storage in interim storage facilities as it needs to be accessible to farmers in sufficient amounts well in advance of when it is going to be spread. For private facilities, the nutrient solution can be used in private gardens or for forest fertilization. The solution needs to be heavily diluted. Urine should not be spread on anything destined for consumption in the near future, both because the nitrogen is of no use in that phase and because some bacteria and viruses may still be present.

A night-soil composter made of plastic, with ventilation.

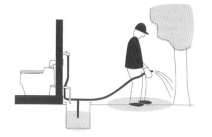

An ejector tank for collection of urine can easily be connected to an irrigation hose for crop fertilization.

Source: Separett AB.

Storage tanks for human urine at the Björnsjön research facility, Sweden, which has been managed by Stockholm Water since 1996. The tanks are filled by gravity from a tanker. Each of the three tanks holds 150m³. The tanks are placed near fields where the urine is spread using feeder hoses on spring grain used as fodder.

Source: Photo: Maria Block

The black-water system at Tegelvik school in Kvicksund, near Eskilstuna, Sweden.

Source: Illustration: Leif Kindgren

Black Water

Black water is nutrient-rich with a nutrient content balance that is suitable for plants. Black water can be dealt with together with organic household waste and agricultural manure. It requires sanitization prior to use and conventional agricultural technology for spreading. Direct black-water treatment followed by recovery of nutrients would increase the level of recovery to over 90 per cent.

Wet Reactors

One way of dealing with black water is to use a wet reactor. In a wet reactor, composting takes place, i.e. the breakdown and sanitization of the black water. Retention for four to six days is sufficient for sanitization and the system is inexpensive in relation to biogas facilities.

The difference between a biogas facility and a wet reactor is that the biogas process takes place anaerobically (with access to little or no oxygen), while a wet reactor has a fast aerobic process (with access to oxygen).

A farm near Tegelvik school in Kvicksund. The black tank is a wet compost reactor.

At the Agricultural University of Norway in Ås, outside Oslo, a student residence has been built with a separating sewage system. Quiet, low-vacuum toilets result in concentrated black water. The black water is carried to a wet compost reactor and the wet compost material is then used in agriculture.

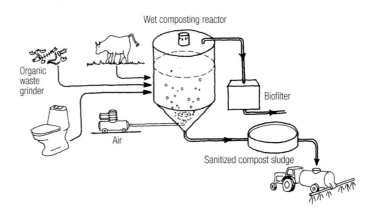

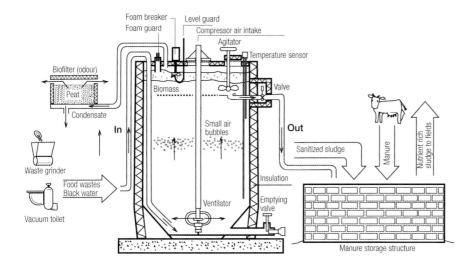

How a wet reactor works.

Source: Illustration: Leif Kindgren

Biogas Facilities

Biogas is created when organic material is broken down by micro-organisms in an anaerobic environment. The process is useful in three ways: it produces good fertilizer plus an energy-rich gas, and sanitizes the decomposed material.

The biogas process can be used to treat organic waste such as black water from sewage facilities, manure from agriculture, compostable household waste and waste products from the food industry, as well as sewage sludge from sewage purification plants. Organic household waste can be added using waste grinders in order to increase the amount of dry material in the sewage system. In agricultural biogas facilities, energy crops can be added for this purpose. The sanitization that occurs during the process is very effective. In general, the treatment destroys all pathogenic micro-organisms.

Decomposition takes place in closed containers at 35–40°C (mesophilic treatment), or at 50–60°C (thermophilic treatment). The temperature in a biogas chamber should be kept constant and stable, the pH value should be neutral and there should be a satisfactory carbon/nitrogen (C/N) balance. The initial mixture should contain at least 65 per cent water. The retention time in the digester is about 15–30 days. Biogas facilities are usually large scale as investment costs are high, and experienced operators are needed to run them.

Biogas usually contains 55–65 per cent methane by volume, 30–40 per cent CO_2 by volume, as well as water and small amounts of hydrogen sulphide and ammonia. The gas is energy rich since methane has a fuel value of 50MJ/kg or 18–22MJ/m^3. Another way of expressing this is that about 1m^3 biogas with this composition is equivalent to 0.5 litre oil. Biogas is used to produce heat, for combined heat and power production and as a vehicle fuel. For use in cars, biogas must be purified and compressed.

The digested sludge is dewatered to a level of about 35 per cent dry material. As all the nutritive salts and trace elements remain, the residual product is an excellent fertilizer, reinforced by the presence of humus substances and other biological material that is recovered in the residual product (which is an excellent soil improvement material). It is also odour free. In order to use the sludge for soil improvement, it is very important that heavy metals and other toxins are not added with the initial material.

The waste water is reused to soak the initial material. The water that is not used goes to a purification facility. The waste water can also be used as liquid fertilizer and the digested sludge can be composted together with other organic material.

All types of organic material can be added to a biogas digester. The output is energy-rich biogas (methane gas) and nutrient-rich fertilizer.

Artistically designed biogas plant, Ryaverken (GRYAAB), in Gothenburg.

Source: Svenskt Gastekniskt Center

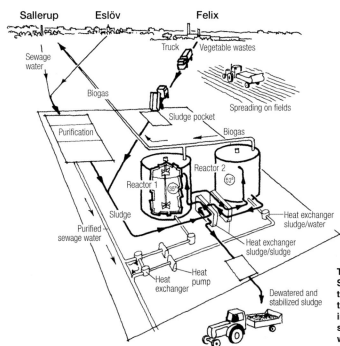

The biogas facility in Skåne, Sweden. It handles the organic waste from the Felix food manufacturing industry and sewage sludge from the sewage water from Salerup and Eslöv.

BIOGAS USE

The potential of biogas is significant, using available waste products from sewage, households, industry and agriculture. The available waste from agriculture is much bigger than other sources. If special energy crops were utilized, the energy potential could be expanded considerably. An equivalent of 10–30TWh/year could be produced from biogas crops grown on 500,000ha of land. Biogas technology reduces CO_2 leakage from biodegradation of organic waste.

Sludge Stabilization

The most common use of biogas is for sludge stabilization in municipal purification facilities. The gas can be used to power motors for small-scale combined heat and power production. A third of the energy is delivered as electricity, and the rest as heat. So a purification facility can be totally self-sufficient in fuel.

Industrial Waste Water

Biogas can be produced from industrial waste water, such as from sugar refineries, distilleries and paper mills. The primary goal is to achieve a highly purified concentrated waste water, though extraction of energy is also important.

Organic Waste

Separated organic waste from households, restaurants and industries can be treated in biogas facilities. The waste can be combined with sewage sludge and treated in existing digestion chambers to increase gas production. Household, restaurant and industrial waste can be digested together to make liquid fertilizer for agriculture. Separated household waste can be digested to make a soil-improvement material with a wider variety of uses.

Landfill Gas

Spontaneously produced gas is collected for environmental reasons, to increase safety and to make use of the energy. Biogas can also be extracted from specially designed digestion cells where conditions for anaerobic decay are controlled by regulating temperature and moisture content.

Agriculture

Use of biogas facilities in agriculture has tended in recent years to move in the direction of shared facilities, i.e. facilities where maybe 15–20 farmers deliver animal manure to a large biogas facility. Such facilities produce and deliver electricity and heat to the surrounding population using the local grid and district heating network. It is more difficult to produce biogas from energy crops, but the technology is under development, especially for fuel production.

Sewage Management

Biogas facilities that deal with compostable waste and sewage from toilets are a good way to return urban biological waste to the ecocycle. Large-scale biogas facilities are more profitable, and a facility for less than a few hundred people is hardly worth considering.

Combined Facilities

Biogas facilities can provide parts of the municipalities with heat, and produce fuel for buses and electricity. The farmers are encouraged to grow grass for the production of biogas and separated household waste and sludge from the sewage network will also be used by the biogas facility.

Biogas Facilities in Sweden in 2005 and Annual Gas Production

Type of facility	Number	TWh/yr
Sewage purification plant	139	0.56
Landfills/biocells	70	0.46
Industrial sewage	4	0.09
Combined waste digestion	13	0.16
Agriculture	7	0.01
Total	233	1.28

Source: Svenska Biogasföreningen

One of Europe's largest biogas facility is located in Helsingör, Denmark. It treats 20,000 tonnes of waste per year from 70,000 households. The 3 million m³ per year of gas produced can heat 700–800 single-family homes.

The Kennedy family's patio, an extension to their single-family home in the Lebensgarten eco-community, Germany.

3.4 Vegetation and Cultivation

People like to have flowers, trees and green plants around them for aesthetic reasons, but green plants are also the basis for all life on Earth. The only part of the natural world that uses sunlight, CO_2 and nutrients to build organic material, all other living things live directly or indirectly on green plants. Both waste and sewage contain organic fractions. These should be returned to arable land to close the ecological cycle.

3.4.0 Permaculture

In summary, permaculture involves consciously designing the environment, natural and built, in a way that succeeds in the long term and resembles nature's own ecosystem. The permaculture concept was formulated by the Australian, Bill Mollison, for which he received the Right Livelihood Award in 1981. The term means 'permanent agriculture'. Building up a permaculture system is labour-intensive in the construction phase but in return should require little time to manage.

Function and Design

The core of permaculture is design. Design addresses the relationship between different parts, and having the right thing in the right place. An example is water reservoirs that are placed higher than buildings and gardens so that the water runs down by gravity. Vegetation that serves as a windbreak is placed so that it blocks winds but does not shade buildings. A garden is placed between the home and henhouse so that garden wastes can be taken to the henhouse and chicken manure to the garden on the way back.

Every element should serve many functions. Chickens are one example. People need food and chickens provide eggs and meat. The garden needs fertilizer but can provide green fodder; the chickens eat green fodder and provide fertilizer. The greenhouse needs CO_2 and provides warmth; chickens release CO_2 and periodically need heat. Fruit trees need weeding and removal of insects and snails; the chickens can feed on weeds, insects and snails. Trees are exposed to insects and larvae, and beneath them chickens get protection and food. A piece of land needs to be prepared for cultivation, and chickens walk around and simultaneously peck, scratch and fertilize.

Every function should be able to be met by several elements. Every important basic need such as water, food or energy should be able to be satisfied in several ways. A good design should include perennials, annuals and tree crops, so that if one crop fails there is no shortage of food or fodder. If a solar heating system is used, it should also be possible to heat water with biomass if the sun does not provide enough heat. Electric pumps can break down and then it is good to have a hand pump and a pond as backups.

- Instead of trying to maximize one element, total yield is optimized, e.g. a forest where there are not only trees but mushrooms, berries and animals as well.

- Site drainage is the starting point and elements are organized so that water is used as efficiently as possible.

- Perennial or self-seeding plants are used to a great extent.

- The plant kingdom's natural succession is used, i.e. some plants and herbs establish themselves quickly on newly cultivated land to eventually be replaced by other types of plants.

- Diversity is increased by having trees, bushes, animals, different kinds of cultivation, ponds and pasture.

- Companion planting is carried out instead of establishing monocultures, and crops, trees, bushes and small animals live together in the same place.

- Edge zones are increased, such as forest edges, lake shores and beaches, since nature's productivity and diversity is greatest in these zones.

- Work goes on at different levels in terms of height, for example by growing plants

Water that zigzags through sloping terrain can be used for irrigation on its way through the landscape.

Source: *Villrosene – Økologi i hagen*, Marianne Leisner, 1996

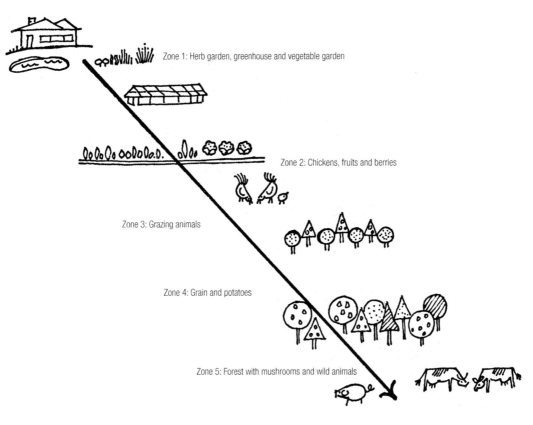

Zone 1: Herb garden, greenhouse and vegetable garden
Zone 2: Chickens, fruits and berries
Zone 3: Grazing animals
Zone 4: Grain and potatoes
Zone 5: Forest with mushrooms and wild animals

The foundation for energy-efficient planning is zoning. In zone one, which is closest to the house, things requiring access several times a day are located, such as the herb garden and vegetable garden. In zone two there are things that require attention sometime during the day, e.g. chickens, fruit and berries. Constituents of zone three need to be checked every few days, e.g. sheep and beef cattle. In zone four are things that require attention several times a year, such as grain and potatoes. Zone five can be left wild, for hunting, picking mushrooms and berries, and tree-felling.

Source: *Permaculture One. A Perennial Agriculture for Human Settlements*, Bill Mollison & David Holmgren, 1981

Permaculture cultivation at the home of Marianne Leisner and Rolf Jacobsen on Tjømø, Norway.

at the ground level, in raised beds, on walls and by planting bushes and trees.

- Intensive land usage. Every possible piece of land is use for cultivation or raising animals. So a lot can be produced locally in a very small area.

- Small-scale rather than large-scale approaches are chosen, where the land is used efficiently and a certain amount of manual work is needed rather than large inputs of fossil energy and machines.

- Animal power is used as much as possible, e.g. chickens or pigs work the land ('chicken and pig tractors') while at the same time their manure is used.

- The system should be long-term and durable.

- It should be possible to adapt the system to totally different climates and types of terrain and to urban, rural and suburban locations.

Description of an existing farm by the sea in Lista, Norway:

1 dwelling
2 barn
3 workshop
4 garage
5 greenhouse
6 henhouse
7 studio
8 driveway
9 lawn
10 rose hedge
11 willow hedge that creates a sheltered area
12 artificial wetland for purification of grey water
13 composting and outlet for solid waste from separating toilets in the home
14 paved inner courtyard
15 chicken run
16 fruit trees
17 viewpoint in a sheltered area
18 windbreak of berry bushes and a mixture of local wild and hardy species
19 vegetable garden
20 herbs
21 playhouse
22 barbecue area
23 sandbox
24 well
25 grass paths
26 'wild zone' – uncultivated area

Source: *Villrosene – Økologi i hagen*, Marianne Leisner, 1996

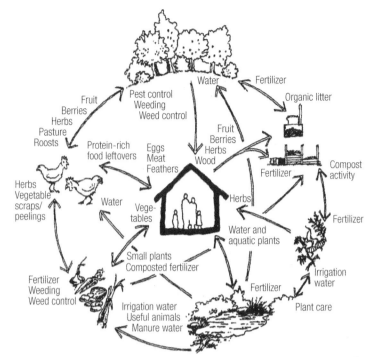

A fruitful dialogue occurs when both parties have something to give each other. In a garden, contact can be established between various elements in order to benefit from give-and-take relationships.

Source: *Villrosene – Økologi i hagen,* Marianne Leisner, 1996

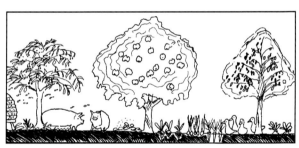

Within permaculture and agroforestry, cultivation systems are created where trees are mixed with crops and animals so that the whole provides more than the sum of the individual parts.

In permaculture, cultivation is planned three-dimensionally. Plants that grow to different heights and that have different root depths are combined in a way that makes efficient use of surface area. It is a question of making use of light from the sun and the volume of soil as well as putting plants together that do well rather than compete with each other.

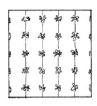

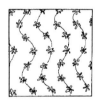

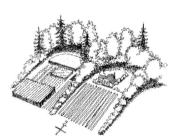

A permaculture designer makes use of the microclimate to protect against winds and to open up towards the sun. The illustration shows a solar pocket, or sun trap, that is used for both cultivation and dwellings.

Edge zones are especially productive. To maximize their length, permaculture design uses wavy 'natural' shapes.

3 ECOCYCLES | 3.4 VEGETATION AND CULTIVATION | 3.4.0 PERMACULTURE

3.4.1 Vegetation Structures

Vegetation is a natural part of ecological building. It should be taken into consideration at all levels of planning from where the town meets the countryside, where green areas wind through development, all the way to the planning of a private home with its garden, outdoor sitting area, balcony, climbing plants and green roof.

Green Cities

Biological diversity is also important in cities, so different biotopes in the urban environment should be protected. By using natural features on site as a starting point for planning, it is possible to protect and integrate valuable environments in new areas while at the same time preserving natural diversity using corridors and passageways for plants and animals linking the various natural areas. New biotopes, e.g. ponds and wetlands, can enrich the outdoor environment. There are green areas that require a lot of maintenance, e.g. lawns for football and gardens with sculpted bushes and trees. Flower beds also require regular tending, care and watering. Other types of vegetation can get by with less attention, e.g. natural areas, certain berry bushes and 'wild' flower meadows.

One strategy for bringing vegetation into the town is to use green wedges. These are green strips that are preserved and not developed. They can begin as large wedges of natural wilderness that extend in towards towns and suburbs, where they connect with smaller wedges extending from these into urban areas. The idea is to maximize the edge zones where built-up areas and green areas meet.

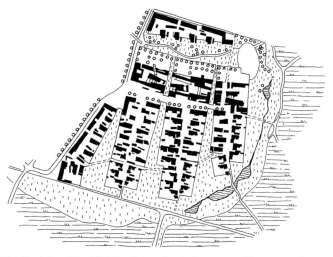

The Finnish architect Petri Laaksonen's winning proposal in a competition to design an ecological housing area in Vik, Helsinki, Finland, 1995. He worked with three levels of green wedges: large wedges that bring the agricultural landscape into the town, medium-sized wedges that define neighbourhoods and small wedges that introduce vegetation between buildings. Garden plots are situated where neighbourhood courtyards open towards the green areas.

Stockholm's green and blue wedges bring connected natural areas in towards the city centre. They are a prerequisite for a good urban environment and should be kept free from development.

An 'ecoduct', a bridge where nature and the park cross a highway, has been built in the German city of Mainz. Ecoducts are found in both Austria and Germany. Similar bridges can be built over motorways so that wild animals can move more freely. It is also possible to build tunnels, e.g. for frogs.

Source: Illustration: Eva Külper

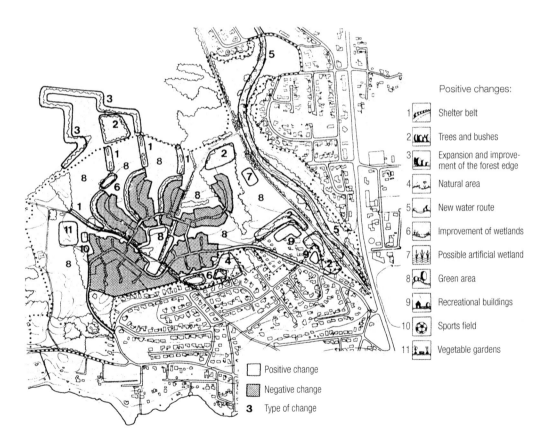

Positive changes:
1. Shelter belt
2. Trees and bushes
3. Expansion and improvement of the forest edge
4. Natural area
5. New water route
6. Improvement of wetlands
7. Possible artificial wetland
8. Green area
9. Recreational buildings
10. Sports field
11. Vegetable gardens

Positive change
Negative change
3 Type of change

Planning of an environmentally adapted residential area, where the guiding principle has been balance between ecological areas, i.e. a district that is species rich and diverse with regard to plants and animals, and where valuable biotopes are undisturbed.

Source: *Dyrking og grønt i boligområder*, Frederica Miller, NBBL, 1993

3.4

Trees and bushes along streets and roads greatly influence the level of particles in the air. There can be a reduction of 50 per cent or more.

Vegetation on Different Scales

Trees and other vegetation can even be brought into the 'concrete desert'. Trees can be planted along streets on pavements and if there is enough room, continuous green spaces between streets and pavements can be established, so that streets become boulevards. Forecourt areas can be turned into green areas between pavements and buildings. The character of parks can be highlighted by making ecological cycles visible. For example, ditches, creeks and ponds can deal with rain, and there can be compost heaps, fruit trees and berry bushes, and gravel pathways instead of asphalt. Taking it a step further, there can even be vegetable plots or urban farms.

Vegetation in Residential Areas

In ecological town planning, vegetation in residential areas concerns integration of vegetation and buildings. Between buildings and natural areas there is a planned green zone that needs to fulfil many functions. There should be room for play and leisure, gardening and recreation. Architects can work with the microclimate and protective vegetation, as well as with views of open landscape and other valuable environments. The greatest difference between an ecological green area and other green areas is that there is a focus on increasing its biological diversity and trying to make the area 'useful', i.e. food-producing. People even talk about the 'edible' landscape, which can involve planting fruit trees and berry bushes instead of bushes with poisonous berries or prickles.

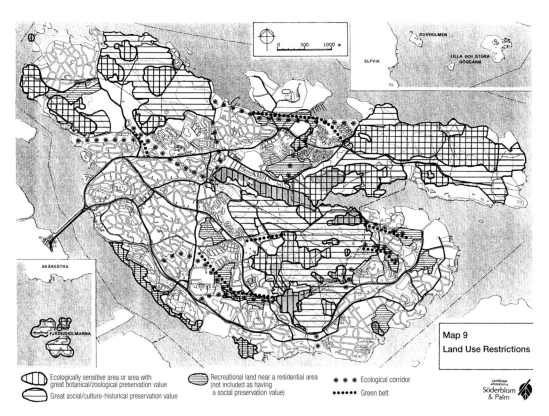

The municipality of Lidingö incorporated green areas into its general plan in an exemplary manner. The general plan demarcates ecologically sensitive areas of great social or cultural and historical value as well as recreational and green zones close to residential areas.

Ecological Green Areas

Here are guidelines for ecological green areas, as outlined by the landscape architect Bengt Persson:

- Design of house plinths and other parts of the building that touch the ground.
- Making a gradual transition between the building, the immediate private area and the immediate shared neighbourhood.
- Making the whole plot cultivable.
- Natural (i.e. 'ecological') materials for ground cover, walls, edges and equipment.
- Establishment of fruit trees and berry bushes in areas for cultivation and enjoyment.
- Water that is utilized and visible.
- Terracing is used for shape, character and practicality, e.g. in cultivation.
- Concentration on a framework with plenty of space for residents to enjoy the immediate environment and to design the 'interior decor' themselves.
- Garden character in parts of the site dominated by cultivation together with a broad range of biotopes and plant communities arranged on a natural scale.
- Use of plants and choice of species that provide a feeling of ecological growing power in the whole environment.
- Protection from the wind from all round the property border towards the buildings, with particular stress on shelter from the north.
- Use of trellises to create space-saving wind protection towards the north and to make use of protected growing beds on the west, south and east.
- Integration of functions for playing, living, etc. in a holistic environment with very little division between functions.
- Making use of existing vegetation, rocks, cultural elements, etc., in the design (and protecting them carefully during the building phase).
- Investment in many trees.
- Strict functionality in surface area division and especially in use of vegetation.
- Vegetation designed for rich fertility and a garden-like layout.
- Broad spectrum of species, variation in and adaptation to different biotopes on the microscale.
- Gradients in biotopes between sparse–abundant and dry–damp, a wealth of damp and water biotopes.
- Space for composting.

Natural Landscape

There are many different types of natural landscape, e.g. forest edges, bedrock, wetlands and pasture lands. Natural landscape provides an experientially rich environment, but is inexpensive to care for. However, it cannot be abandoned, but requires thinning, clearing and rejuvenating. Some ground cover is very sensitive and can be destroyed by wear and tear. In such cases it may be suitable to increase the hardiness of the vegetation prior to building. This may be the case for both large areas such as copses or forest edges as well as individual objects, such as beautiful trees.

Thinning natural land cover in residential areas can improve growing conditions for specific protected trees. Rocks and wetland areas, as well as all surfaces near entrances and walkways, are the only land that requires attention prior to building. The rest of the land can wait until it is clear where wear is occurring. Wetlands located where they will be trampled should be drained. Of the individual tree species, beech is the most sensitive to trampling. Building in natural areas subjects the land to great stress.

Felling trees that provide protection from the wind leaves the remaining trees exposed to harder winds. Spruce are especially sensitive, particularly on wet land. Covering root systems with earth greatly stresses trees primarily because the roots do not get enough air and can suffocate. Vegetation can be damaged if surface groundwater flows are cut off and the water is diverted, e.g. because of roads or service trenches.

Natural landscape that is going to be preserved must be actively protected during the building phase, which is most often done using fencing. Trees that are going to be preserved are numbered and if they are damaged, expensive penalties are incurred.

Choosing Trees for Towns

Trees that grow to be very large, such as poplar, willow, ash and elm should be planted on the largest, widest streets. Smaller trees, such as whitebeam, mountain ash and fruit trees, can be planted on smaller streets and locations. Linden and maple trees are examples of park trees that can be trimmed and shaped as desired. Willows are fast-growing and can be used for bowers, trellises and hedges (woven together) as wind breaks. A rule of thumb is that a free-growing tree needs an area for its roots the size of its crown. Trees such as poplar and aspen have far-spreading roots and therefore should not be planted near building walls. Near car parks, willow, ash, hornbeam or plane trees should be chosen. Other kinds of trees can cause problems for vehicle finishes since they drop sticky seeds or secretions.

It is a question of introducing vegetation at all levels, not just as pieces of natural vegetation towards the centre of towns and populated areas, but also as green elements in most street and garden spaces.

Source: *Town Planning in Practice*, Raymond Unwin, 1909

Flower Meadows

A flower meadow can take 10 years to establish, but once it is established, almost no maintenance is required. It needs to be cleared a little in the spring, dead grass and twigs need to be raked up, and once in the late summer the meadow needs to be cut with a scythe. Meadows require poor soil. The levels of nitrogen and phosphorus in an ordinary lawn are too high for the growth of meadow flora. A meadow should not be fertilized or limed. Meadow seeds can either be manually collected from a suitable place and then sown, or ready-mixed meadow seed can be purchased. It is also possible to plant 'seed plants' to guarantee establishment of e.g. bluebells, daisies.

Weeping willow in Monet's garden, Giverny, France.

Allergies

Unfortunately, for people with allergies, access to the outdoors may be limited by the presence of plants and animals. This is the case when certain types of trees spread pollen. Even strongly scented bushes can cause problems. Immediately next to dwellings, wind-pollinated plants and deciduous trees should be avoided. Although coniferous trees are wind-pollinated, their pollen is not allergenic.

Examples of trees and plants suitable for people with allergies: spruce, raspberry bushes, chestnut, clematis, bell flowers, kitchen plants (dill, chives, carrots, lettuce, etc.), cherry, dog rose, whitebeam, pear, most crowfoot plants, roses, mountain ash, rockery plants, pine, currant bushes and apple trees. Late-blooming types of grass should be chosen.

Examples of plants and trees to be avoided by people with allergies: alder, elm, aspen, birch, broom, cowslip, hazel, bird cherry, hyacinth, jasmine, composite plants such as chrysanthemum, ox-eye daisy and marigold, lily of the valley, linden (has a strong odour), mock orange, spiraea, large lawns (especially with mugwort, dandelions and timothy grass), lilac and willow.

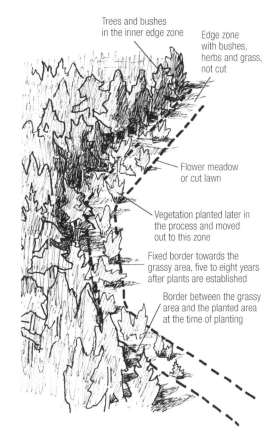

Example of how a forest edge can be laid out. From an ecological perspective, the transition zones between different biotopes are important.

Source: adapted from 'Naturlika grönområden', Bengt Persson, BFR T22:1981

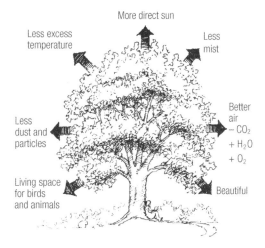

Trees have many good characteristics. They provide shade, their roots prevent erosion and draw up water and nutrients from the soil. Trees can provide timber, firewood, fodder, nuts, berries and fruit. Trees affect the climate, clean the air and produce oxygen. Trees provide a home for birds, insects and animals. Trees are beautiful, they can blossom, smell good and provide a place to play. They also provide a time perspective on life.

3.4.2 Vegetation on and in Buildings

A building's architecture does not end at the outer walls. It includes and creates a border zone with the outdoors. Verandas, porches and landings extend a building. Plant window sills, winter gardens and greenhouse extensions are elements that bring plants and growth into a building. Pergolas, canopies and outdoor sitting areas draw living activities out of the building. Green roofs, façade plants and hedges enhance the architecture with biological elements.

Plants on Walls and Roofs

By bringing vegetation into towns and cities, planting trees along streets and in gardens, and planting vegetation on walls and roofs, the urban climate can be improved significantly, especially during the leafy, verdant time of the year. Plants clean the air, produce oxygen and help to reduce the problems of high temperatures and dry air. Covering about 5 per cent of all horizontal and vertical surfaces in a town with vegetation is assumed to be enough to create a healthy urban climate in spring, summer and autumn.

Climbing Plants

Some appropriate climbing plants are Chinese wisteria, Russian vine, firethorn, hops, honeysuckle, clematis, climbing hydrangea, climbing rose, ivy, Dutchman's pipe, Boston ivy and virginia creeper. Climbing plants usually flourish in ordinary, well-fertilized garden soil. Most climbing plants require something to climb on: only Boston ivy, virginia creeper, ivy and climbing hydrangea climb by themselves. Both wires and trellises should be fixed about 10cm out from the façade. Some climbing plants can also be used as quick-growing groundcover in half-shaded areas, e.g. honeysuckle, climbing hydrangea and ivy.

Vertical Gardens

An interesting development is vertical or hanging gardens, introduced by French botanist Patrick Blanc. He has developed a system where vegetation is planted in pockets of 3mm thick acrylic felt fixed on boards to cover a façade. The wall is irrigated automatically several times a day, sometimes with nutrients added to the water. He chooses vegetation that doesn't need much soil and he avoids climbing plants because they climb and cover other plants. Such a vertical garden weighs 15kg per m². The result is spectacular, like the vertical gardens in the Pershing Hall Hotel and the Musée du Quai Branly in Paris. In Singapore vertical gardens are used as a cooling device. Experience shows that a façade covered with plants in a temperate climate is about 4°C cooler than a façade without vegetation. In a hot, humid climate the temperate difference can be even larger, up to 14°C.

In Sweden, the company Green Fortune has been inspired by Patrich Blanc. Vertical plant walls are also used successfully

In ecological architecture there is vegetation in, on and beside buildings. This is especially important in urban environments, where there is often a scarcity of green areas.

Source: adapted from *Byggeri og Økologi – begreber og forslag, Eksempelsamling,* BUR, København, 1988

With knowledge and a little imagination, fruit trees can be grown on façades.

Vertical garden on the Musée du Quai Branly in Paris by Patrick Blanc.

Source: Photo: Emma Hedberg

Vegetation has its place even in narrow inner courtyards.

indoors. The plants improve the air quality and capture dust. Air humidity is increased, which reduces static electricity and makes it easier to breath. Oxygen is released and plants remove hazardous substances from the air. The plants need to be dusted off and tended to every once in a while.

Pergolas and Trellises

Buildings can be equipped with pergolas and trellises to create green spaces in densely built-up areas, and to create a gentler transition between indoors and outdoors. They make it easy to pick fruit. Trellised trees start to bear fruit early, often by the second or third year. If trellised trees are planted against a warm south-facing wall and protected from the wind, it is possible to grow species that

Vegetation wall in the musician's Green Room in the conference center Uppsala Konsert- och Kongress.

Architect: Henning Larsen Architects.

probably wouldn't survive if they were free-standing. Such a location can represent a microclimate improvement equivalent to one cultivation zone (i.e. the minimum temperature is 6°C warmer). Apple and pear trees are the easiest to trellis.

Green Roofs

Roofs with vegetation (green roofs) have a long history of use in Scandinavia. A green roof has a marked water-regulating effect. As much as 60 per cent of total annual precipitation can be handled by a 2cm sedum roof. Transpiration and evaporation increase humidity in the air, which is beneficial in otherwise dry urban air. About 80 per cent of total precipitation falling on green roofs evaporates. For other types of roofs, 20 per cent evaporates and 80 per cent runs off. The temperature difference in a green roof between winter and summer is about 30°C. For other roof surfaces such as metal, the seasonal difference in temperature can be 100°C, which puts great stress on the roofing material. Vegetation surfaces act as filters. A vegetation-covered roof muffles background noise and the mass in the substrate layers has a noise-dampening effect for the building itself (e.g. muffles noise from aeroplanes). An important function of green roofs is that they act as 'stepping stones' in spreading of animal and plant species. (See also Section 1.1.3 Assessment of Materials.)

Construction

In principle, green roofs are made up of four layers: uppermost is a vegetation layer that grows on or in a plant substratum, underneath this is a drainage layer and below this is root penetration protection and a weatherproofing layer.

There are two different construction principles for grass roofs. The most common is that the weatherproofing layer is placed directly under the soil layer and that there is an air gap between the weatherproofing layer and the underlying insulated construction. It is also possible to build a 'reverse roof' which has no air gap and the weatherproofing layer is placed under the insulation. In such cases it is important to use a moisture-tolerant insulation material. Reverse roofs are used primarily on terrace roofs. Weatherproofing may be air-gap-forming plastic sheeting, polyolefin sheeting or rubber mats.

Roof Slope

In order to avoid damage to the vegetation, all grass roofs should have a slope of at least a few degrees. For slopes greater than three degrees, a thicker substratum is required to compensate for the runoff. For slopes greater than 20 degrees, as a rule, some type of anchoring of the substrate is required. For slopes greater than 30 degrees, this problem increases and battens or netting are usually added to keep the soil in place. Slopes greater than 45 degrees are not recommended.

Weight and Thickness

Traditional grass roofs weigh 300–400kg/m^2, and it is assumed that the roof construction can bear the weight. A way to reduce the weight and at the same time increase the insulating capacity is to use a mixture of half earth and half loose expanded clay.

It is possible to build thinner and lighter green roofs. The weight is related to the thickness of the roof and different plants

Grass roof on the national park building in Tyresta, Stockholm.

Source: Architect: Per Liedner, Formverkstan Arkitekter AB

require a growth substrate of different thicknesses.

- Moss-sedum roof (2–6cm): 25–75kg/m^2
- Sedum-moss-plant roof (6–10cm): 75–100kg/m^2
- Sedum-grass-plant roof (6–15cm): 75–150kg/m^2
- Grass-plant roof (>15cm): >150kg/m^2

A critical load limit for existing roofs is usually about 50kg/m^2. When constructing a green roof with a weight under 100kg/m^2, the loft's bearing capacity must be taken into consideration. Such a construction can be made with ready-to-use plant mats that are fastened to the underlay.

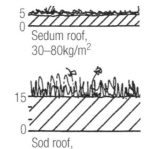

Sedum roof, 30–80kg/m^2

Sod roof, 300kg/m^2

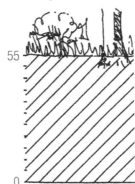

Trees and bushes on roof, 1000kg/m^2

An ordinary grass roof should have an earth layer of at least 15cm. If trees are going to be planted on a roof, a thickness of 60cm is needed (40cm for bushes). Sedum roofs require only a few centimetres.

Source: *Sedumtak*, Pär Söderblom, 1992

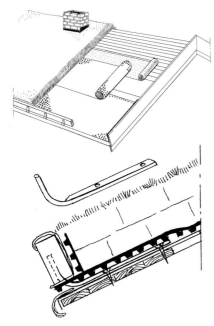

A common method of building grass roofs is to use roofing felt on top of tongue-and-groove board, with air-gap-forming overlapping plastic sheeting over it. A 15cm thick layer of earth, in which the grass is sown, is put on top of these. The earth should not be so nutrient rich that other plants can establish themselves. Expanded clay balls can be mixed into the earth to make the roof lighter and increase its draining capacity. The board holding the earth in place should be fixed so the screws or nails do not make holes in the weatherproofing layer.

Source: a brochure by Platon

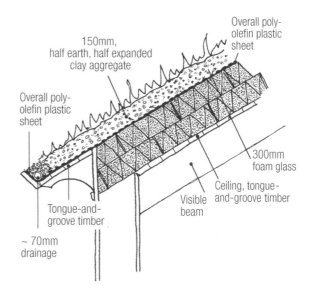

To make full use of the insulation capacity of the earth layer, the roof can be constructed without an air gap between the insulation and the earth layer. In this case, the insulation should be moisture-tolerant and able to bear the weight of the earth without compressing it and losing its insulating capacity.

Sedum Roof

If sedum plants are used instead of grass, roofs can be made thinner and lighter as sedum plants require much less water. Sedum plants vary in colour and it is possible, if desired, to have a roof that looks different during different seasons of the year. A sedum roof only requires a 2.5–6cm-thick earth layer. Sedum roofs should have a slope of at least a few per cent. Thin sedum roofs do not require special roof constructions.

Sedum roof at the 2001 housing exhibition in Malmö (BO-01).

Roof Gardens

In order to have trees on a roof, about a 60cm-thick earth layer is needed. For bushes, about 40cm is sufficient. For grass only, a thickness of 20cm is adequate. To have a garden on a roof, the roof must be able to support a heavy load. A 45–55cm thick roof-garden bed weighs between 800 and 1000kg per m^2. In addition, the weight of snow and machines must be included.

Greenhouses

The most important factors to consider with greenhouses is how they fit in and how easy it is to access them. The direction a greenhouse faces is of less importance – they don't have to face south. Greenhouses do not require sun all day in order to grow plants in them. A few hours of sunshine are sufficient. However, continual access to water is a necessity. Although there can never be too much light for plants in a greenhouse, there can be too much warmth. There are several ways to limit the heat in a greenhouse. Suitable shading can be achieved with white curtains. The most important way to lower the temperature is to ventilate. Appropriate ventilation consists of windows that open at the highest point of the roof and provide a total openable surface area equivalent to about 25 per cent of the greenhouse's floor area, along with ventilation windows situated down low that let in fresh air. Operating a greenhouse without automatic ventilation is almost impossible. Consideration should also be given to the appearance of a greenhouse during the winter when nothing is growing inside. There is a difference between moist, dirty greenhouses and dry, clean, glassed-in verandas with plants in pots.

Unheated greenhouses are very useful in the northern climates for starting plants that need to reach a certain size before being planted outdoors (e.g. celery, sweetcorn and cabbage). During the summer a greenhouse is ideal for growing tomatoes, as well as aubergine, melons and sweet

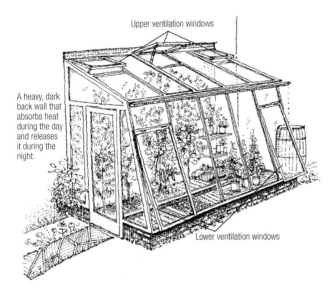

A greenhouse lean-to is a common feature in eco-architecture. It is a good way to make use of passive solar heat and at the same time produce some food.

Source: Illustration: Peter Bonde

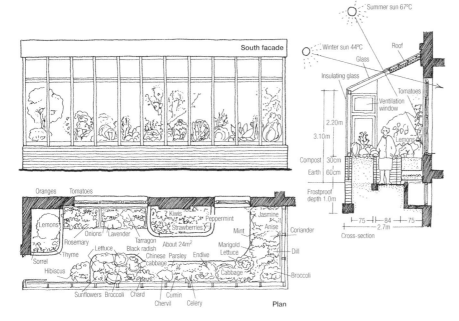

In New Hampshire, US, Dr Sonia Wallman developed a cultivation system for a greenhouse lean-to. It is 20m² in area and can supply a household of four people with 70 per cent of their vegetables and 30 per cent of their fruit with a work input of only 15 minutes a day. The greenhouse also contributes heat to the house.

peppers. Lettuce can be grown almost all year round.

Heated greenhouses maintain a temperature of +4°C at night during the winter, and thus present great opportunities. If a greenhouse is big enough, it is possible to grow peaches, pears, nectarines, grapes or any type of exotic fruit, depending on the microclimate. The best way to heat a greenhouse is with waterborne heat. If an inside wall in the greenhouse is painted black, it will absorb heat during the day and release it later.

Vegetation Inside Buildings

Besides being beautiful, green plants inside buildings provide a better indoor climate. Many people find it pleasant to have plants around. Plants affect the indoor climate by helping to keep the humidity at a higher level and by cleaning the air. Pollutants are absorbed by a leaf's stomata, and root microfibres can break them down. Plants also have a positive effect on odours and airborne microbes. Research by NASA on the capacity of plants to improve spacecrafts' indoor climate yields information about which plants clean air best.

Different plants have different purification characteristics. For reducing formaldehyde levels, the most efficient plants are sword fern (*Nephrolepis exaltata*), chrysanthemum (*Dendranthema x grandiflorum*), gerbera daisy (*Gerbera x cantabrigiensis*), pygmy date palm (*Phoenix roebelenii*), striped dracaena (*Dracaena deremensis 'Warneckei'*), bamboo palm (*Chamaedorea erumpens*) and fern (*Nephrolepis obliterata*). The best plants for reducing levels of xylene and toluene are golden cane palm (*Chrysalidocarpus lutescens*), pygmy date palm and *Phalaenopsis*. For reducing ammonia levels, it is best to use lady palm (*Rhapis excelsa*), queen of hearts (*Homalomena Wallisii*), creeping liriope (*Liriope spicata*) and linden (*Anthurium andreanum*). The peace lily (*Spathiphyllum wallisii*) is good at purifying the air from several unhealthy substances.

Other plants that are beneficial for indoor air are the spider plant (*Chlorophytum comosum*), table palm (*Chamaedorea elegans*), rubber fig (*Ficus elastica*), devil's ivy (*Epipremnum pinnatum*), Madagascar dragon tree (*Dracaena marginata*), common ivy (*Hedera helix*), umbrella plant (*Schefflera actinophylla*), sword fern (*Nephrolepis exaltata*) and mother-in-law's tongue (*Sansevieria trifasciata*).

3.4.3 Gardens

Gardens for household needs are one of the most productive places in the world. In a living garden, nature itself does most of the work. All the organisms that flourish there play a role in the ecological cycle. Working in a garden affects this delicate balance and the environment in general. Gardens should be a source of pleasure and relaxation.

Intensive Gardening

Even with a small area it is possible to achieve both beauty and large yields. It is important to take advantage of the available space, if possible on different levels, and to plant a garden that it is easily maintained. One method is to use raised beds, which can also be cared for by those in wheelchairs or with other disabilities. Another method is to surround gardens with walls to improve the microclimate. Keyhole gardening minimizes the area required for paths. Growing herbs is easy and inexpensive and they are a welcome addition to food. Most herbs prefer well-drained earth and lots of sun, and most flourish if their leaves are regularly harvested.

Plant Protection

Instead of using classified biocides it is possible to choose plants that are resistant to different kinds of attack, practise crop rotation, grow plants next to others that protect each other from attack, grow strong-smelling herbs that keep insects away, and use water with a temperature of 54°C.

Terraced house with rainwater barrel.
Source: Illustration: Heinrich Tesssenow, 1908

The private outdoors: behind the fence, residents are completely free to garden and potter about.

Architect Kjell Forshed's ecological housing area Kappsta, Lidingö, Sweden. There, not only the buildings but also the plots were planned in an architectural way. In a small area, there is room for a kitchen garden, fruit trees, composting, bicycles, cars and an outdoor seating area. The inside and outside spaces are attractively integrated with each other.

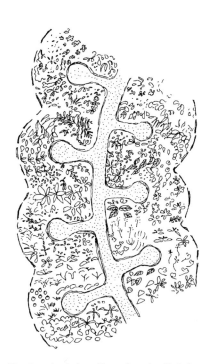

Plan for a keyhole patterned garden that also works well with raised beds.

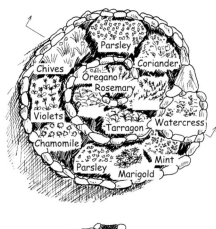

An herb spiral can be a beautiful garden patch that saves both space and water. It has a diameter of about 2m and is about 1m high. There is room for most important kitchen herbs. The cultivation bed ends up being about 9m long.

Source: adapted from *Introduction to Permaculture*, Bill Mollison, 1991

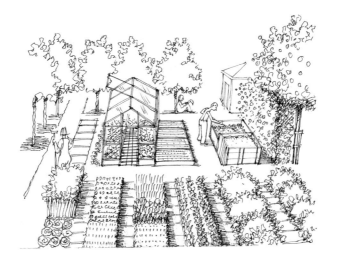

Space for composting is needed adjacent to gardens where there are compost containers.

Source: illustration Peter Bonde

Gardens for Blocks of Flats

A garden for a block of flats should have the following characteristics: a measure of privacy, a relationship with the indoors, some greenery, and some form of demarcation. Different studies have indicated that at least 20 per cent of the residents in multi-family dwellings would like to have access to land to plant their own gardens. An important experience from projects is that interest in gardening usually increases once a garden is established, even if only a few residents expressed an interest in the beginning. Garden plots on land associated with blocks of flats reduce the surface area that needs to be maintained by the property management. Establishing good conditions for apartment block gardens requires an initial investment in soil, water, planting of hedges and trees, and a clear division of responsibilities between the gardeners and property management. Access to good storage facilities for vegetables, e.g. root cellars or food cellars, increases the possibility of gardening for household needs. A large part of the undeveloped land in a residential area is needed for roads, parking, storage, etc. A rough estimate is that 0.6m² of land is needed per m² of floor area for these purposes. In order to have space for gardens, there needs to be a greater availability of land.

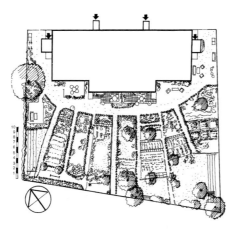

A block of flats in Enskede, Stockholm, where gardens for the flats are arranged on the plot. There are eight flats in the building and eight gardens.

Source: illustration adapted from *Lägenhetsträdgårdar*, Charlotte Horgby, Lena Jarlöv, BFR T1:1991

Gardens associated with a block of flats, Seseke Aue, Emscher Park, Germany.

Source: Joachim Eble Architektur

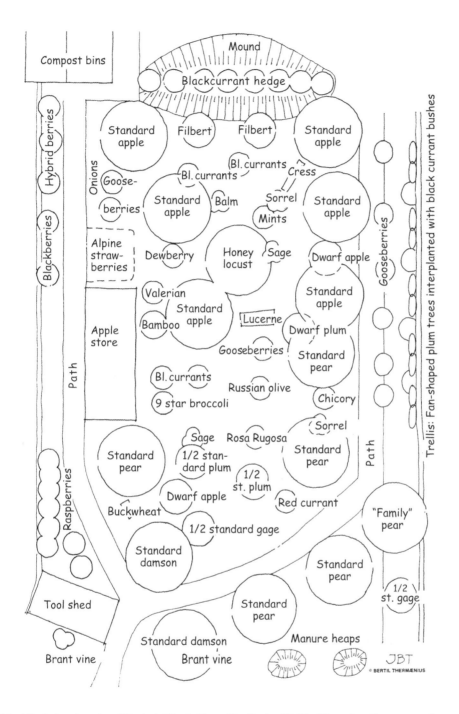

Robert Hart's forest garden at Wenlock Edge in Shropshire, England. Tall fruit trees form the uppermost layer, nut trees and grafted fruit trees make up a lower level, berry bushes and vegetables form the shrubbery layer with raspberries and crawling herbs underneath forming the ground layer. At the bottom, in the earth, are the root vegetables. Climbing plants in the garden form a vertical layer. The system is self-preserving, self-propagating, self-fertilizing and self-watering.

Source: illustration by architect Bertil Thermænius, in *Forest Gardening*, Robert Hart, 1988

Allotment with an ecological allotment cottage in Esbo, outside Helsinki.

Source: Architect: Bruno Erat

A kitchen garden in 18th-century style at Gunnebo castle in Mölndal, Sweden. The kitchen garden is in a protected location and has a good microclimate.

Source: Tema Architecture. Photo: Maria Block

Much of the knowledge about gardening comes from the monastic system. In a walled-in monastery garden, a good microclimate is created and the wall keeps out hares and deer. Plants for both basic needs and pleasure are grown in monastery gardens. Flowers, herbs, medicinal plants, fruit, berries and vegetables are grown. At the same time a restful and beautiful place is created with murmuring water, fragrances and colours.

Source: illustration Peter Bonde

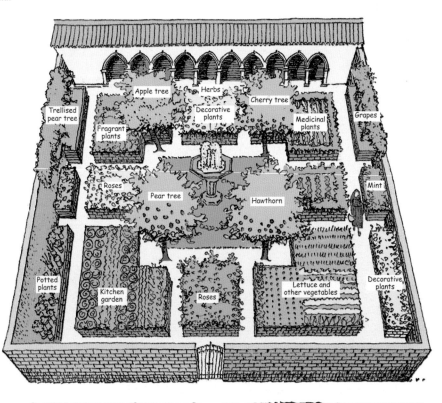

Ponds

Ponds are important as water reservoirs and are often a requirement for irrigation. In addition, ponds have a charm of their own. Ponds can be used to raise fish such as carp, pike-perch, tench and crucian carp, as well as crayfish. They also serve as a habitat for birds such as ducks and geese. Green algae, sea lettuce, chlorophyta, waterweed and duckweed are aquatic plants that can be harvested and used as fodder or fertilizer. Ponds that can be emptied can be easier to use and are sometimes necessary, for raising carp and pike-perch, for example. A pond needs an inflow and outflow in the bottom, because if the same water remains in a pond over the winter, a serious oxygen shortage may arise.

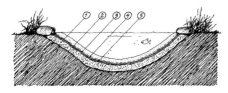

1. About 30cm loose grass
2. A thin layer of animal manure (e.g. from cows)
3. About 2mm of scattered bentonite clay
4. A thin layer of paper bags and newspaper
5. 20cm of sand or earth

Dig the pond at least 30cm deeper than the finished depth. Pack down the bottom and remove stones and roots. Build up the bottom layer as in the illustration. Pack it down again above the sand and earth layer. Fill with water.

Source: *Villrosene – Økologi i hagen*, Marianne Leisner, 1996

Small Animals

In ecological planning, there is an attempt to integrate food production with settlements. Gardens adjacent to blocks of flats and small animal husbandry are components that are included. Modern, efficient forestry and agriculture have made it difficult for many small animals to find a habitat. To help small animals, appropriate nests/dens can be arranged for them.

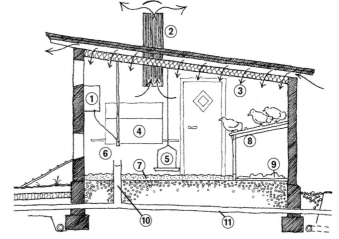

A small-scale poultry house where chickens can live and exhibit their natural needs and behaviour.

1 water, 2 exhaust air, 3 intake air, 4 common nest, 5 automatic feeder, 6 water nipples, 7 floor litter, 8 chicken roost, 9 manure bin, 10 floor drain, 11 drainpipe (150mm in diameter).

Source: *Naturligare hönsskötsel*, Detlef Fölsch, Kristina Odén, LT:s förlag 1989

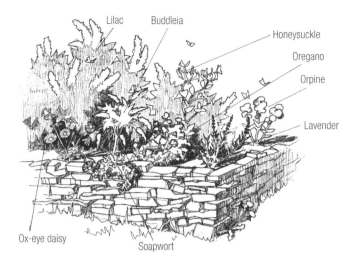

A 'butterfly restaurant' can be made to attract butterflies. Buddleia with red-violet flowers are butterflies' number one favourite. Other plants that butterflies are attracted to are blackberry, cornflower, gypsophila (baby's breath), forget-me-not, borage, cowslip, hawthorn, honeysuckle, oregano, Sedum telephium (orpine), lavender, marjoram, ox-eye daisy, lilac, thyme, thistle, violet, field scabious.

There are several positive aspects to beekeeping including efficient pollination which contributes to diversity, and honey and wax production.

Bats are becoming rare in densely populated areas. The picture shows how a bat house can be built. Another approach is roof tiles that allow access to bats.

Source: Vilda grannar, Per Bengtson and Maria Lewander, SNF, 1995

3.4.4 Ecological Agriculture and Forestry

Large-scale industrial agriculture causes problems in many different ways, affecting the environment, resource consumption, and economic and social spheres. Organic agriculture avoids artificial fertilizer and chemical pest and weed controls. Although there is about a 20 per cent reduction in harvest, there is an increase in quality. Several studies show that organically grown food contains from 10 to 30 per cent more vitamins and minerals than conventionally grown food.

Agriculture

Modern industrial agriculture is not compatible with ecological principles. If fodder is grown in one area and cattle are in another, recycling of dung is almost impossible. This means that you get a leakage of nutrients both from the cattle dung and from the artificial fertilizer used for the fodder. In sustainable agriculture you aim to produce most of the fodder for your cattle on your own farm or in the neighbourhood.

Industrial Agriculture

The problem exists on all three levels, local, regional and global.

On the local level, monoculture as well as chemical pesticides are employed, both of which undermine diversity. It is difficult for an individual farmer to achieve any economic gain. Operations are dependent on EU subsidies. Rural areas are being depopulated, resulting in a deterioration of services and a lonely, hard and monotonous life for farmers.

On the **regional** level, leaking of the nutrients nitrogen and phosphorus, primarily from livestock, results in eutrophication of inland waterways, and groundwater is polluted by leakage of nitrates. About 40 per cent of nitrogen leakage comes from agriculture. Transport distances are long, expensive and energy-intensive because

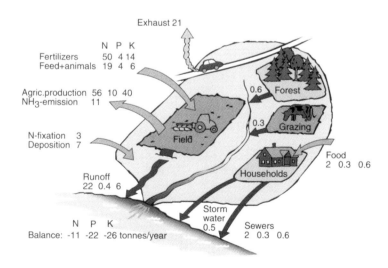

The figure shows the complexity in circulation of substances on a farm. Todays farming leaks ten times more nitrogen than buildings.

Source: *Water Use and Management*, edited by Lars-Christer Lundin, Uppsala University

Feed is grown in the Mälar Valley and hogs are raised in the Laholm area. This makes recycling of nutrients from animal farms to fodder farms impossible.

Source: Steineck et al, 1997

specialization is regional as well. Socially, direct contact between consumers and farm food production is lost.

Globally, agriculture contributes to climate change. About 20 per cent of the energy consumed is used for food production. Production of artificial fertilizer is energy-intensive and phosphorus will end up being in short supply if artificial fertilizer continues to be wasted as it is today.

Global industrial agriculture is extremely dependent on access to cheap energy. To produce one unit of energy in the form of food, 10–15 units of energy are required.

More than 50 per cent of all fertilizer spread on the ground ends up in the atmosphere or oceans. Fertilizers contain nitrous oxide, which has 300 times more impact on climate than carbon dioxide.

In the last 50 years, use of artificial fertilizers has increased from 10 to 80 tonnes/hectare. The nutrient content in grains is, however, only 50 per cent of what it was 30 years ago. Artificial fertilizer mainly adds nitrogen, phosphorus and potassium, while levels of vitamins, minerals and essential amino acids are inadequate.

Cattle and sheep, due to their digestive system, release an enormous amount of methane gas. Every kilo of beef produced results in 13kg of greenhouse gases, and every kilo of lamb meat results in 17kg of greenhouse gases. The production of pork and chicken results in releases of about half the size. Natural grazing can, however, reduce releases since pasture land holds a lot of carbon.

Food patents are becoming more and more common, which threaten both food security and biodiversity.

Ecological footprint refers to the total area required to support a population (i.e. the total amount of land per capita used in a specific country to sustain the population). National lifestyles result in ecological footprints of varying sizes. With the help of such a concept it is possible to see which changes are necessary in order to achieve sustainable development.

Ecological footprints for regions and income groups in 1999

Source: *Den gode skole – Økologien med i skole* ('The good school – ecology included in schools' – in Danish only), Dansk center for byøkologi (Danish center for Building Ecology), 2002

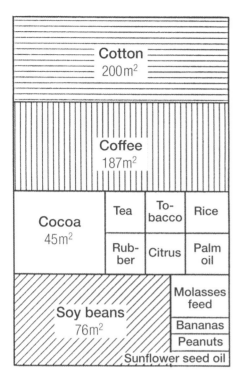

The land area required in other countries to supply the annual consumption for each Swede.

Eutrophication caused by nutrient leakage from agriculture is a major problem for the Baltic Sea.

Source: The Baltic Ecological Recycling Agriculture and Society (BERAS) project

Over the last 50 years, use of artificial fertilizers has increased from 10 to 80 tonnes per hectare. Economically, the EU agricultural policy is a catastrophe. Large agricultural subsidies are paid to produce surpluses that are dumped on the world market. The result is that farmers in developing countries cannot compete and are kept in poverty.

BERAS

Baltic Ecological Recycling Agriculture and Society (BERAS) is a research project investigating why the Baltic Sea is dying and what can save it. There are serious eutrophication problems that cause the blooming of poisonous algae and therefore dying sea bottoms because of the lack of oxygen. The problems are caused mainly by conventional agriculture, but also by inadequate sewage cleaning processes and the use of detergents that contain phosphorus. The main reason for the increased nitrogen

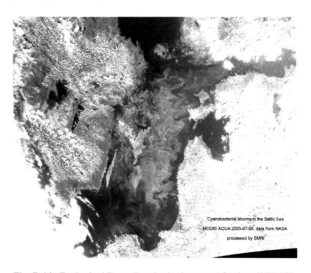

The Baltic Ecological Recycling Agriculture and Society (BERAS) research project has shown that the main cause of algae blooming is the modern monocultural agricultural system. Instead of recycling manure to crops, livestock farming leaks nutrients from manure and arable farms leak nutrients from artificial fertilizers.

Source: 'Cyanobacterial blooms', *MODIS AQUA*, 6 July 2005, data from NASA processed by SMHI

and phosphorus load from agriculture to the Baltic Sea is the specialization of agriculture with its separation of crop and animal production. Large animal farms leak nutrients from the dung, and big fodder farms leak nutrients from artificial fertilizers. The recommendation from BERAS is to transform to ecological agriculture with a local balance between animals and farming. Leys with both clover and grass would have to be produced on all farms. Agriculture based on the principles of ecological recycling would lead to a decrease in the leaching of nutrients by half. Local production, processing and distribution of food products from ecological recycling agriculture can diminish primary energy consumption per capita by 40 per cent and greenhouse gas emissions by 20 per cent. As well, the energy required for food production can be reduced by half and greenhouse gas emissions by almost as much if consumption of vegetarian food is increased so that meat from grazing animals is eaten only once a week at the most. An ecological and locally oriented food chain leads to freedom from chemical pesticides, greater diversity in production and more grazing-based animal husbandry.

Organic Farming

Artificial fertilizers and chemical pesticides are not used in organic farming. It is estimated that about 60 per cent less energy is consumed than in conventional cultivation. However, it is not enough to convert to organic cultivation. Specialization must also be reduced, and mixed farming (raising of crops and animals) is needed so that ecological cycles can be completed in an economically viable manner. Food processing should be decentralized. This would reduce transportation needs and create the conditions for returning wastes from the food industry to ecological cycles. Food production should be primarily oriented towards satisfying local needs. Production would decrease somewhat but quality would increase. The surplus would disappear, the land could be kept open and conditions for biological diversity would improve.

KRAV is a certification association for organic production. The association works for sustainable development by developing standards for organic production, verifying that they are being implemented and promoting the KRAV label. To fulfil the KRAV criteria, production must take place without chemical biocides and without artificial fertilizers, domestic animals must be treated well and be free to move freely in an outdoor environment, and no known genetically modified organisms (GMOs) are allowed in the production.

The KRAV certification association does not accept GMOs because the health risks are not clear. The effects of spreading GMOs

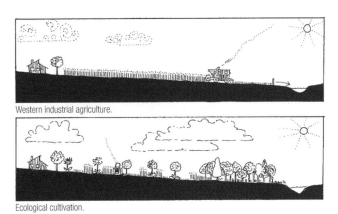

Conversion from monoculture to ecological cultivation.

Western industrial agriculture.

Ecological cultivation.

The land area required for self-sufficient small farms as compared to agribusiness. With small farms, an area of 12,000m² can support nine families (36 people). The same amount of land used in agribusiness can support one family.

are unknown. The risk of undermining biological diversity both within species and with regard to the number of species is obvious. Currently implemented uses of GMOs in agriculture only benefit a one-sided, chemical-dependent use system and multinational companies that own and sell both the seeds and biocides.

KRAV certification also takes into consideration carbon dioxide releases.

Biodynamic Agriculture

Put simply, in organic farming people talk about what should not be done, while in biodynamic agriculture people talk about what should be done. Those who practise biodynamic agriculture do not regard agriculture as a purely mechanical system, where the most important concerns are that nutrients are returned to the ecological cycle, that the earth is not depleted and

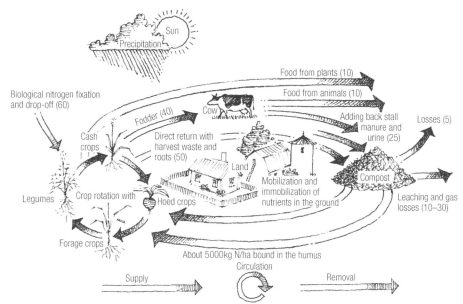

In organic farming, an attempt is made to close ecological cycles.

that infestation and damage are reduced through crop rotation. In addition, they feel that agriculture is also influenced by cosmic forces. Therefore, biodynamic agriculturalists follow a sow and harvest calendar since they believe that celestial bodies affect the results, and they use biodynamic preparations to strengthen the cosmic forces. For composting, compounds of medicinal plants such as yarrow, chamomile, nettle, oak bark, dandelion and valerian are used. A diluted compound of composted cow dung and quartz is sprayed on the crops. The compound helps the plants to develop and creates life in the earth.

Above all, agriculture is regarded as a 'living organism', where the four different elements, i.e. animals, plants in crop rotation, compost and the soil, must be in balance with each other.

Local Food Production

Local food production is an important part of closing ecological cycles and having high-quality, fresh food on the table. Agriculture in the area surrounding population centres should include both raising livestock and crops, fodder and green manure. Vegetables and fruit should also be grown in the area. Processing should take place locally, thus a local dairy, slaughterhouse, mill and bakery are needed. Products can be sold directly by farmers, via the internet or in local shops. Ecological cycles can be closed by taking waste from the food industry and agriculture, as well as the compostable waste from cafeterias, restaurants and households, to biogas facilities for example. Biogas facilities can produce solid fertilizer that can be composted and put on fields, liquid fertilizer that can be used in the same way as animal urine on cornfields and meadows, as well as biogas that can be used to run farm tractors and vehicles. Human waste, such as urine and faecal matter, should not be used as fertilizer for food production but rather as fertilizer for energy forests, for example.

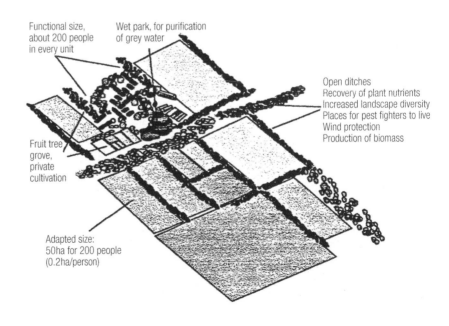

The energy and plant nutrient problem can be solved by reducing the distance between farms and residential areas. This means, according to systems ecologist Folke Günther, that towns need to be rebuilt so that they are more sparsely populated. Residential areas should have their own water and sewage systems and be 80 per cent self-sufficient in food.

Source: Folke Günther

A private biodynamic vegetable garden can provide a large part of the annual needs for one family.

Source: The Wärn family's garden in Kungshall, Småland, Sweden

Ecological Forestry

The WWF and the Swedish Society for Nature Conservation have developed criteria for environmentally certified forestry. They are in agreement with the forest industry about the goals. The following is a summary. A number of key biotopes (e.g. mountain forests, old-growth forest and some wetland forests) should be excluded from forestry. The deciduous tree stock should account for at least 5 per cent of the total stock. The felling method should be adapted to forest conditions. In an environmentally labelled forest, all deadwood is left to provide habitats for a variety of insects and animals.

The criteria do not allow the draining of forest land or the use of chemical biocides. In Scandinavia, special consideration needs to be given to the reindeer industry, and forestry roads are prohibited in mountain forests and in key biotopes.

The Forest Stewardship Council (FSC) was founded in 1993 and is made up of about 300 organizations around the world. The board consists of people from environmental and humanitarian organizations as well as companies with an economic interest. The FSC has established criteria for 'good' forestry practice. The criteria deal with forest management that promotes environmental sustainability and social progress, while maintaining economical profitability. The main tasks of the FSC today are to develop criteria as well as to certify companies that in turn can certify forest management.

The Programme for the Endorsement of Forest Certification schemes (PEFC) organization was established in 1999 and is supported by family forest managers throughout Europe as well as by many European forest industries and trade organizations. The purpose of PEFC is to develop sustainable forest management that maintains a good balance between production, environmental and social interests. 35 countries are members. Products marked with the PEFC logo contain at least 70 per cent certified wood fibres.

There are about 30,000 species in forests in Sweden, which make up 60 per cent of the area of the country. Many plants and animals are threatened. Large-scale forestry reduces biological diversity. Perpetual use of forests is preferred, without clear-cutting, with felling intervals of 10 to 15 years. Perpetual use of forests can absorb 30 per cent more carbon from the atmosphere than clear-cutting with ground processing.

Source: *Vägskäl för miljön*, Naturvårdsverket, 1995

Endangered species in the forest landscape.

Fungi: 195 species
Lower animals: 435 species
Higher animals: 20 species
Moss and lichen: 110 species
Vascular plants: 25 species

4 Place

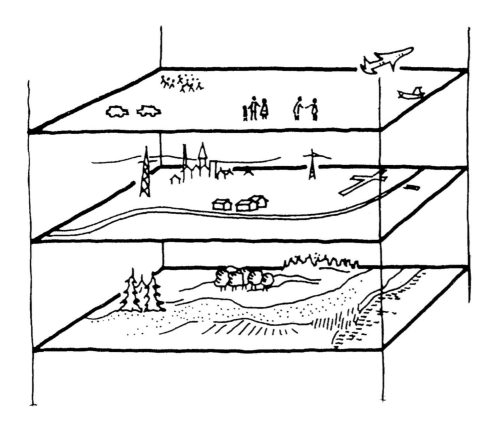

During physical planning, the built environment and infrastructure are related to human and economic activity on the one hand, and natural features, land and vegetation on the other.

Source: Lars Orrskog, KTH

Adapting to the Location

To make buildings suit a location, studies must be made to obtain an understanding of local conditions. Important aspects to consider are the natural surroundings, climate, social structure and human activities.

Adaptation to Nature –
How to Adapt to Natural Surroundings and Climate

The local geography, geology, hydrology, flora and fauna, as well as microclimate should be studied. Let the results from these studies be the basis for building and layout. Adapting buildings to the local climate saves energy. Adapting to natural surroundings is positive for biological diversity and land-based industries.

Social Structure –
Development of a Sustainable Society

Infrastructure and buildings should be planned so that transport needs are minimized. To achieve this, a multifunctional community connected by a well-developed public transit system is required. Pedestrian and bicycle traffic should serve as a starting point for planning. Towns, communities and rural settlements must be integrated into a network structure.

Existing Buildings –
Environmental Adaptation of Existing Buildings

New buildings make up a small part of the total building stock. To achieve sustainable development, existing buildings must also be more healthy and resource efficient, better adapted to ecological cycles, to local conditions and the rest of the community. A huge effort is needed for this type of conversion.

People –
Building with People as the Starting Point

The goal should be to build a humane society with room for everyone. Safety, comfort and beauty make us feel good. Sometimes we need to be alone and sometimes with others. There should be opportunities at different levels to participate in community development. How can segregation and fear be avoided, and how do we make towns and communities flourish?

Making an Inventory

Site adaptation begins by studying different aspects of the area in question. An area has to be understood in order to provide the conditions for planning. What resources are there at the site? What uses are these resources most suited to? What are the limitations they place on exploitation?

Maps

Specialized maps, e.g. for vegetation, hydrology, geology or climate, are used as a basis for analysis. Any missing information needs to be obtained.

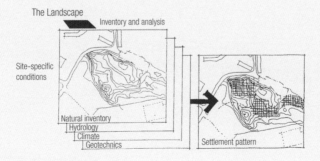

Superimposing maps is a suitable method for obtaining a good basis for development.

GIS (geographic information system)

GIS is a computer-based information system with functions for input, processing, storage, analysis and presentation of geographic data. The information is arranged in layers with different themes, e.g. roads, watercourses, land use, property boundaries, settlements, etc.

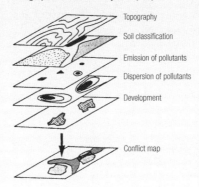

Many municipalities use GIS in order to work more efficiently, especially in the areas of physical planning and environmental protection.

Analysis

When the basic data is assembled in map form, the maps can be laid one on top of the other. This combined data provides the basis for an analysis.

Landscape Analysis

Analysis starts with a planning area. Most landscape analyses show the character of the vegetation, soil and topography. Landscape analysis is supplemented by studies of the geology, hydrology, climate and cultural history. Together these provide an overview of the combined characteristics of the

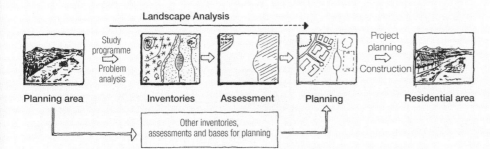

Stages of landscape analysis.

Source: *Tänk efter före – landskapsanalys för bostadsområden*, Bengt Isling, Tomas Saxgård, BFR, T22:1982

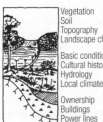

Landscape analysis and landscape information.

A landscape analysis is done for a geographical area and is supplemented by other studies. The results of the inventory are assessed and provide the basis for planning.

Source: *Tänk efter före – landskapsanalys för bostadsområden*, Bengt Isling, Tomas Saxgård, BFR, T22:1982

4 PLACE 513

landscape, which are integrated with information about infrastructure, activities and ownership to provide an overview of the regional characteristics.

Infrastructure Analysis

The flows into and out of a town or community are examined. At which level are the flows taken care of?

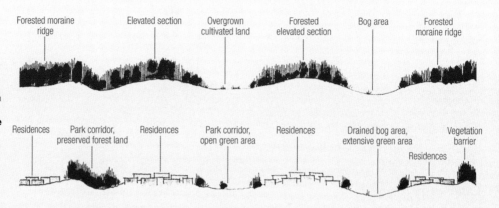

Section through the landscape at Ersboda, Umeå. By analysing the landscape it was possible to situate buildings in the landscape in a sensitive manner and even preserve some natural features, such as a park corridor, open green areas and vegetation barriers.

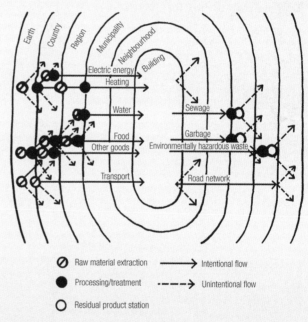

Every household has an effect on both near and distant natural systems. In sustainable development, flows are dealt with as close to the source as possible for practical, economic and environmental reasons.

Source: *Planering för uthållighet*, Lars Orrskog, 1993

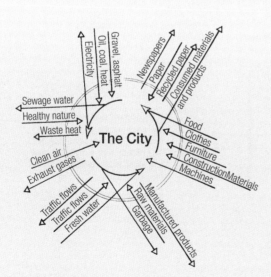

There is a continual exchange of materials and energy between towns and the area that surrounds them. Raw materials and basic necessities flow in to urban residents. Manufactured goods as well as residual products and flows of used water and heat flow out.

4 PLACE

Fallingwater is a famous example of how a building can be adapted to the natural surroundings in a way that accentuates the distinctive characteristics of a site.

Source: Frank Lloyd Wright, architect, US

4.1 Adaptation to Natural Surroundings

For ecological construction it goes without saying that settlements and infrastructure are adapted to the natural surroundings of a location. It is the natural characteristics, i.e. the geography and environment, that guide construction and planning and not the opposite.

4.1.0 Local Conditions

Every place has a distinctive character – people even talk about the spirit or soul of a place. In the Roman Empire, the term *Genius Loci* was used, referring to a local divinity said to inhabit every place and which must be taken into consideration and cared for. In environmentally adapted architecture, people talk about studying and understanding a place, and use this understanding as a starting point for planning.

Natural Conditions

It is important to examine the geology and topography of a place in order to determine suitable locations for laying a foundation or siting new buildings at a future date with a minimum of interference. A location's hydrology is investigated, water flows are integrated into the planning and the water table level is maintained using runoff management. Flora and fauna are inventoried. Beautiful natural areas and important areas for animals and plants are preserved. The microclimate is studied and buildings are adapted to the climate by siting them in sunny, sheltered areas, avoiding shade, windy places and frost pockets.

Environmental Impact Assessment

When an activity that has a 'significant impact' on the surrounding area is going to be established in a municipality, an environmental impact assessment (EIA) is carried out to determine how the activity will affect the environment. Currently, there are not only requirements for detailed development plans, but legal requirements for an impact assessment of the consequences of general plans. Many municipalities today do environmental impact assessments even if a development plan does not meet the 'significant impact' criteria. They do this because an EIA is a good way to integrate environmental issues into planning. An EIA should be started in the initial stages of a development plan. Three aspects of an activity should be described: location, design and scale. An action programme for avoiding, reducing and dealing with damaging effects should also be included, as well as alternative locations for the activity. By identifying and describing effects upon people, animals, plants, climate and the landscape, etc., an overall assessment is provided and a better basis for decisions that affect the environment and health. The new Environmental Act also includes increased public participation as it provides an opportunity for the public to express their views on the environmental impact assessment before a decision is made. The following are some questions that an environmental impact assessment should answer:

- Does the project reduce energy consumption and provide greater opportunities to use renewable energy sources?

ENVIRONMENTAL IMPACT ASSESSMENTS

An environmental impact assessment is required for all activities that have a significant environmental impact. It should include:

- a description of the project in question;
- the local environmental conditions to possibly carry out the project;
- probable environmental consequences of the project;
- possibilities for reducing negative environmental impacts;
- unavoidable negative environmental impacts;
- alternatives to the project (including not carrying it out) as well as the impacts of the alternatives;
- analysis of how the local and long-term use of the environment relates to the ambition of long-term improvement of the environment; and
- the occurrence of irreversible impacts.

- Does the project increase biological diversity and allow the opportunity to generate more natural resources?
- Does the project contribute to closing ecological cycles?
- Does the project stay within the limits tolerated by the environment and people?
- Does the project solve more problems than it creates?
- Does the project take into consideration the precautionary principle?

Use of environmental impact assessments has been criticized for being too defensive, i.e. that environmental thinking isn't used as part of planning but that it is used only in the evaluation phase of a project. Therefore strategic environmental assessment (SEA) is discussed as a further development of environmental impact assessment (EIA). Strategic environmental assessment is already used in conjunction with pre-planning, planning and as a basis for political decisions. A strategic environmental assessment differs from an environmental impact assessment in the following ways: the context studied is often greater than in an EIA, work begins earlier in the process compared to an EIA, a longer period of time is used for gathering and analysing information, and the project contents are not so fixed that they can't be changed according to the results of the study.

Land Restoration

Contamination of soil is a growing problem. In many places land has been contaminated by industrial facilities such as petrol stations and timber creosote-treatment sites, to name two examples. As previously developed industrial land, called brown land, is being redeveloped, more and more contaminated land areas are being discovered. Polluted masses can even be found on lake beds. In such cases, the lake floor sediment can be removed and decontaminated, but the toxins are often spread during the digging. Another method is to cover the lake bed with a material that confines the contaminants.

Land restoration is a complicated and expensive process. It can be done by transporting contaminated earth to a landfill (moving the problem to another location), by digging it up and decontaminating it, or by decontaminating it in place. Four different methods are used to restore polluted land: biological, physical, chemical and thermal.

Biological decontamination is used if the earth is polluted with organic pollutants, e.g. oil. Bacteria and fungi are used to break down the contaminants. Vegetation that absorbs contaminants from the ground can be planted and then destroyed when the task has been accomplished.

Physical purification involves washing the earth and is used to remove metals and organic pollutants. The principle is the same as in an ordinary washing machine. It is carried out using water or steam.

Chemical methods involve washing or leaching out organic or inorganic pollutants with detergents. Another method is to oxidize or reduce pollutants by exposing the ground to an electric current or by putting chemical additives directly into the ground.

Thermal methods are suitable for organic pollutants and mercury. The method requires advanced flue-gas cleaning and involves warming up the masses or combusting the organic contents of the material.

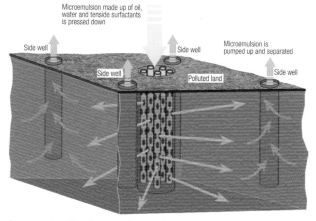

An example of land restoration on site using chemical methods. On its way through the earth, the microemulsion pushes through sand and soil pores, taking the toxic organic pollutants with it.

EXAMPLE OF ADAPTATION TO NATURE

Development of the Marby Area, Norrköping

Marby is located near Norrköping, south of Bråviken. HSB (The Swedish National Association of Tenants, Savings and Building Societies) bought the area to build housing on it. One of the fundamental ideas was to adapt the new buildings to the location by investigating the most suitable building sites and avoiding unsuitable ones. Before the project was carried out, a careful inventory was made proceeding from ecological principles.

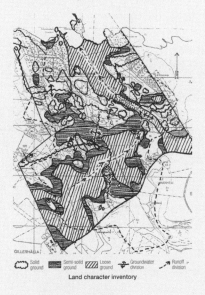

Land character inventory.

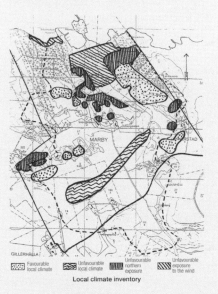

Local climate inventory.

Inventory of important natural areas.

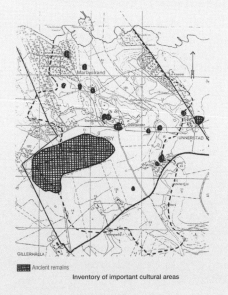

Inventory of important cultural areas.

Source: *En naturresursanpassad utbyggnad av Marby*, Norrköpings kommun, 1991

4.1.1 Geology and Topography

It is important to have an idea of what is hidden beneath the ground surface. Geological maps are a good beginning. They show where clay, sand, moraine and rock are located. If any uncertainty exists, the remaining option is a geotechnical investigation using probes or drilling for samples. For planning purposes it is important to know where buildings should not be placed (e.g. on land suitable for cultivation or where condition of the ground is poor for building, etc.). The geology also influences the choice of local sewage methods. When an area is developed, the activities that have the greatest impact on the land are road construction, laying water, sewage and power lines, as well as the foundation method used. Disturbances to the ground should be minimized, blasting should be avoided, and careful mass balances should be made.

Basic Investigation

All physical planning is preceded by a basic investigation. The investigation should provide information about geological, geotechnical and hydrological conditions, as well as show suitable areas to build according to conditions for laying foundations, permitted soil pressure and settling conditions.

The investigation is often divided up into three parts: gathering existing maps and drilling documentation, carrying out the necessary field studies, and putting together all the material. The information makes it possible to avoid problems such as landslides, high levels of radon, high water table, risk of flooding, settling problems, thick peat layers and very hilly terrain. The investigation should clarify the possibilities for using different parts of the area for various purposes, such as construction, roads, power lines, sewage treatment facilities, runoff infiltration, cultivation, etc.

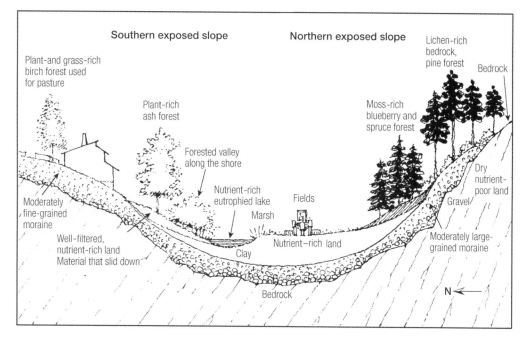

Illustration of a typical valley in central Sweden showing the location of rock and different types of land. The types of land determine the plant communities that can be established in this kind of valley.

Land Types and Foundations

There are four main geological classifications of land types: rock, moraine, sand and clay. There are great differences between them in terms of a surface to build on.

Construction and foundation laying can be carried out on rock at a moderate cost. However, costs for laying power lines and sewage pipes and streets are high in rocky areas. It is better to plan using contour lines instead of blasting away rock that is millions of years old.

All types of buildings can be built on moraine at a moderate cost. However, if the boulders are large and the density high, the value of a moraine area for building on is reduced. The costs for power and sewage installation and road building are high in large-boulder moraines. Moraine areas with few boulders are regarded as one of the most valuable land types for development. Gravel and moraine allow excellent runoff and waste-water infiltration. The existence of radon emissions should be checked.

Sand and fine sand lands that are firm throughout and stratified to a great depth, with moraine on bedrock underneath, can be used to build up to four- or five-storey buildings at a normal or slightly above normal foundation-laying cost. Street and service costs are comparatively low. In terms of development, this type of land is just as valuable as moraine land with few boulders.

Clays can be expensive to build on since it is more difficult to lay a foundation on clay than on other surfaces. The alternatives are to drive piles or build a floating foundation.

When the ground consists of thick enough dry-crust clay or other solid outer layer resting on looser clay, the foundation laying costs for two- and three-storey buildings is moderate. However, a strong enough shear strength is not the only prerequisite for laying a foundation on clay. There shouldn't be too much settling. Areas where the dry crust is so thin that only single-family dwellings can be built directly on the surface of the ground is considered poor land for development. In such a case, a method often used to lay a foundation is the compensation method, i.e. that as much clay as the building weighs is dug out and taken away. Excavation difficulties occur if the solid outer layer is less than 2m thick. In such cases houses should not have basements. On this type of land, heavier buildings require a foundation of piles or plinths. Piles are used at depths greater than 5m, otherwise plinths are used. Exceptional measures often have to be taken for service pipes, and costs can be substantially greater than normal. Infiltration into clays is difficult, but by using certain methods (e.g. compact filter) even some clays can be used for infiltration.

Mass Balancing – Blasting

The smaller the excavation volume the better. In particular, blasting should be avoided as it is risky, expensive, often unnecessary and results in irreparable damage to the natural environment. Mass balance calculations can be used to conserve the masses found on site so that large amounts of earth do not have to be brought in or taken away.

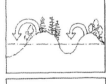

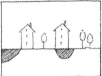

During the Million Programme (1965–1975) in Sweden, little consideration was given to the environment. Rather, expedient building was the primary objective. At that time it was common to defy natural forces.

Source: *Ett ekologiskt synsätt i översiktlig planering – en kunskapsöversikt*, Birgitta och Conny Jerkbrant, Björn Malbert, BFR, R98:1979

A building dug into the earth is one way to adapt to a site. Examples include hobbit homes in *The Lord of the Rings* and this daycare centre near Hanover, Germany.

Source: Architect Gernot Minke

Foundation-Laying

There are four ways to lay a foundation: with a basement, suspended with a crawl space, with a slab on the ground, and using plinths. Buildings should be situated so that the foundation conditions are as homogeneous as possible under the whole building. In hilly and sloping terrain, buildings should be placed so that they follow the contour lines lengthwise as depth contours often follow the land's contour lines. A building that has direct contact with the ground usually loses less heat than buildings on plinths or with suspended foundations. Along with better insulation for foundations, types with better moisture resistance have been developed.

Suspended foundations can be insulated and designed as warm suspended foundations. Such a foundation cannot be raised more than a few centimetres from the ground. Slab foundations are insulated underneath and certain kinds have heating channels within the slab.

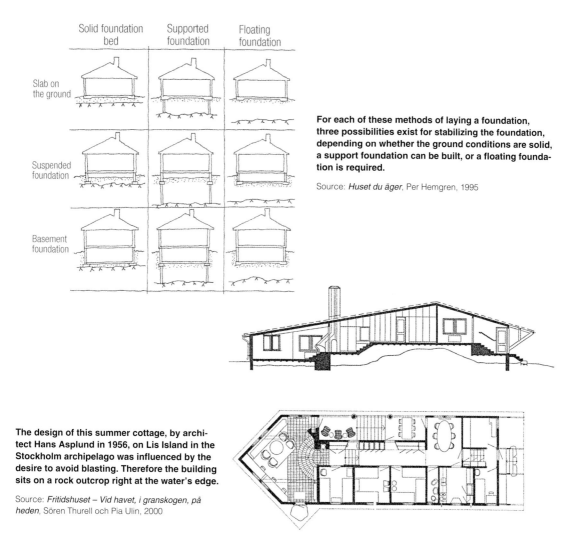

For each of these methods of laying a foundation, three possibilities exist for stabilizing the foundation, depending on whether the ground conditions are solid, a support foundation can be built, or a floating foundation is required.

Source: *Huset du äger*, Per Hemgren, 1995

The design of this summer cottage, by architect Hans Asplund in 1956, on Lis Island in the Stockholm archipelago was influenced by the desire to avoid blasting. Therefore the building sits on a rock outcrop right at the water's edge.

Source: *Fritidshuset – Vid havet, i granskogen, på heden*, Sören Thurell och Pia Ulin, 2000

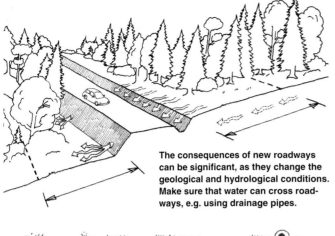

The consequences of new roadways can be significant, as they change the geological and hydrological conditions. Make sure that water can cross roadways, e.g. using drainage pipes.

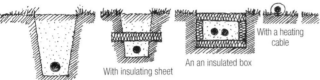

Comparison between traditional and surface systems for laying water and sewage pipes.

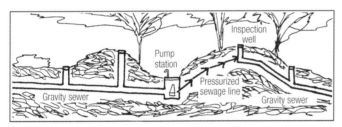

A pressurized line can be laid so that it follows contour variations. However, a gravity sewage pipe must be laid with a certain minimum slope so that it doesn't clog up.

Earthquakes

For construction in potential earthquake zones, there are ways to build that minimize damage from an earthquake. Buildings can be constructed with a flexible frame that can absorb movement without collapsing. Systems of reinforced beams and pillars can make a construction so strong that it can withstand stress even from large movements.

Erosion and Buildings

Entire foundations are sometimes washed away by rain. Thus, it is important to plan for efficient diversion of rainwater and to protect land from erosion before, during and after construction. The most common way to protect against wind erosion is to plant vegetation. Pavements and roads should have a hard surface so that they don't turn into rivulets. A filter provides good protection against all forms of erosion. They can consist of gravel, pebbles or macadam, and are built up by covering the ground with several layers that have a gradually increasing particle size, so that the top layer has the largest particle size. Such filters or ground covers have a greater capacity to absorb and infiltrate a downpour.

Roads

Roads should not have too much of a gradient, otherwise they affect land alteration work. Roads also affect natural drainage, and it is therefore important to plan what will happen to roadways and roadside land and vegetation when there is heavy rainfall or flooding. Will the vegetation below the road get damaged when groundwater conditions change? Will part of the road wash away? Will road construction cause erosion?

Sewage Pipes

To reduce impact on the land it may be suitable to gather together and lay all service pipes under roads. Placing sewage pipes underground at a shallow depth results in little environmental impact and may also keep the cost of laying sewage pipes down, especially when there are difficult conditions. The depth of a trench can be reduced by insulating the upper side which reduces the depth of frost penetration. Trenches for pipes can be even shallower or pipes can be laid above the ground if they are equipped with electric heating or placed in an insulated housing. If heating pipes are also included inside the housing, electrical heating is not necessary. In cases where gravity sewers require deep trenches, sump pumps and pressurized pipes can be used to reduce the impact.

CASE STUDY: UNDERSTENSHÖJDEN, STOCKHOLM

When the Understenshöjden eco-community in Stockholm was going to be built (in the 1990s) in forested and relatively hilly terrain, the following criteria for adapting to the natural surroundings were drawn up:

- a landscape inventory and analysis and tree measurements would be carried out early in the project and used to guide the location of roads and buildings in the terrain;
- blasting and trenching would be minimized;
- as little landfill as possible;
- service pipes would be laid with the least possible intrusion on the natural surroundings;
- to protect the vegetation during construction a high priority was given to information, obligatory penalties for damage incurred, fencing and protecting of certain trees (nonetheless, five or six trees had to be taken down because of damage from blasting); and
- follow-up treatment of damaged vegetation was important.

Natural areas that were fenced off during construction. Marked trees were protected both with fences and with an obligatory economic penalty if damaged.

Plan showing where it was necessary to excavate and blast.

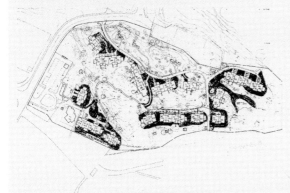

Plan showing where it was necessary to fill up the land.

Carefully laid service pipes to minimize damage to nature.

4.1.2 Hydrology

Water and its movement is often a starting point for construction. Homes are not built where there isn't any drinking water or where there is a risk that buildings will be washed away. In ecological planning, it goes without saying that site hydrology is a primary consideration and is more carefully examined than usual. The discussion includes different types of water such as the following categories of clean water: (i) surface water, (ii) groundwater, (iii) rainwater (runoff); and the following categories of polluted water: (A) traffic water, (B) greywater, and (C) black water. Planning needs to consider all of these types of water.

Drainage Basins

Water and air are the principal media for transport of environmental pollutants. It is therefore common in ecological planning to proceed from the drainage basin and try to gain an overview of what is taking place.

Water Table

The water-table depth influences both the choice of foundation method and of any local sewage treatment system. Slab-on-grade or plinth foundations are usually used in areas where the groundwater levels are high. Infiltration systems and sand filters can be difficult to construct in such places. Construction should be avoided in areas with large amounts of surface and/or underground moving water, especially if there are loose earth layers (fine sand or silt) or steep slopes. The groundwater level and soil conditions, as well as the capacity and load of the recipients (such as lakes, watercourses and wetlands) determine how and where waste water can infiltrate, be purified and returned to the ecocycle.

Surface Water

Surface water includes rivers, streams and lakes. They are of course affected by environmental disturbances, e.g. acidification which can lead to fish mortality and leaching of metals. In the long term, groundwater can become unfit to drink because of acidification. All that can be done in the short term is to lime lakes, but in the long term, air pollutants that cause acidification must be reduced.

Over-fertilization of watercourses is most often due to the leaching of nitrogen from agriculture or from sewage treatment facilities. This can be counteracted by implementation of better fertilization methods, urine-separating sewage or using nitrogen traps. Nitrogen traps can take the form of constructed wetlands or overland flow systems. The problem of over-fertilization has been accentuated by the draining and reclamation of natural wetlands.

Drainage basin – the natural unit for studying the interaction between land and water. 1 Precipitation, 2 evaporation, 3 airborne salt, 4 biological effects, 5 weathering, 6 use of artificial fertilizer, 7 irrigation, 8 emissions from populated areas and industries, 9 water supply, 10 power production, 11 surface water, 12 soil water and 13 groundwater.

Source: *Forskning och Framsteg*, 1974:5

Runoff

Redirecting runoff has several disadvantages. It lowers the groundwater level which can lead to damaged vegetation and deteriorated conditions for vegetation on site as well as settling, which can cause crack formation and other damage to buildings. If runoff is carried away in drainage pipes, which is often the case, treatment facilities can overflow during heavy rainfall and unpurified sewage water can reach the recipient bodies of water. Knowledge is currently available about how to deal with runoff locally in most situations, which has many advantages.

Local management of runoff is environmentally appropriate. Instead of carrying runoff together with waste water via sewage pipes to purification facilities, waste water is transported by the sewage pipes and the runoff is dealt with locally and released onto the land, or into ditches and creeks. Sometimes it is necessary to build detention (percolation) reservoirs for runoff in order to avoid flooding.

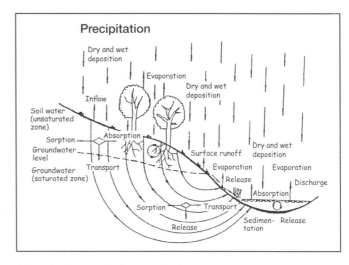

Schematic diagram of an inflow and outflow area along with the underground movement of the water.

Source: *MARK OCH VATTEN i 1990-talets kommunala naturresursplanering – Det långsiktiga forskningsbehovet*, Castensson, Falkenmark, Gustafsson, BFR, G23:1989

The most important concern for local management of runoff is to direct water away from buildings and foundations. This can be accomplished using drainage channels that lead water away from drainpipes.

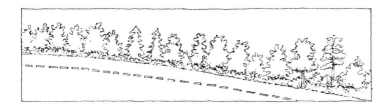

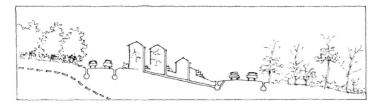

Interference in a sloping area to divert runoff disrupts groundwater transport and may result in a lowering of the water table, which results in damage to buildings and vegetation below the exploited area. By collecting the runoff in the exploitation zone and infiltrating it in the sloping area towards clumps of trees downhill, interference in the hydrological balance can be reduced.

Source: *Daggmaskens dilemma*, Arkitektur, CTH, 1991

Surfaces covered with asphalt prevent infiltration of runoff, although water-permeable asphalt is also available. The diagram on the left shows a traditional gravel-bitumen pavement, and the diagram on the right shows an open pavement (modular pavement).

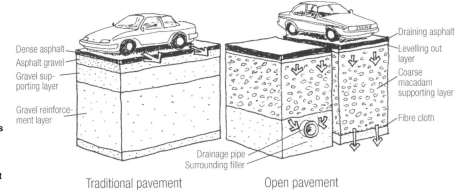

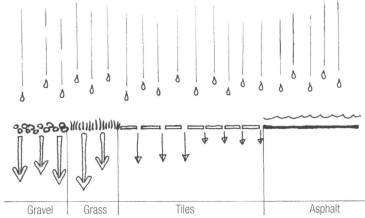

Ground cover or pavements with large spaces between slabs and high permeability allow infiltration of precipitation.

Drainage channels with enough gradient can lead water away from buildings.

When artificial water surfaces are included in developments, the water must be able to flow and be oxygenated so that it doesn't start to smell. The photo shows a stairway for water to flow over in Hammarby Sjöstad, Stockholm, in order to oxygenate surface water.

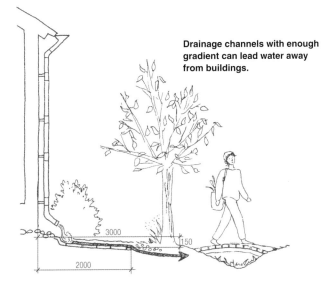

Areas in immediate proximity to buildings must slope away from the foundation (a height difference of at least 15cm over a distance of 3m). Rainwater can be utilized for watering plants or, in areas where there are water shortages, for washing and toilet flushing. It is important to avoid laying hard surfaces unnecessarily and to ensure that there are green areas to absorb rain. Suitable hard surfaces include gravel or tiles with large spaces between them. Grassy spaces can be reinforced with concrete tiles that have holes in them. Car parks can be paved using modular pavement and permeable asphalt. Water from surfaces covered with asphalt can contain 30–70mg oil per litre of water. If permeable surfaces are used the oil is trapped by the upper part of the modular pavement. There the oil is intensively biodegraded by bacteria and fungi. This degradation can be increased by putting a geotextile in the modular pavement. This kind of arrangement can reduce the oil content in the runoff by up to 97 per cent. Ditches, ponds and wetlands can act as water distributors and delaying reservoirs. Grass roofs help to reduce the amount of water that has to be dealt with. It can even be advantageous to use local management of runoff in densely populated urban areas. Where there is clay soil or fine silt, or at construction sites with compacted land, coarse macadam needs to be mixed into the uppermost layer of soil (permeable covering).

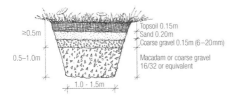

A ditch can act as both a percolation and delaying reservoir for large amounts of water that arise during a heavy downpour.

Local use of runoff involves making use of rainwater, which can be an option in places where water is scarce. Local use of runoff is common in Germany and Denmark.

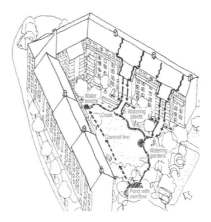

There is local use of runoff system in the Tusenskönan block of flats area in Västerås, owned by HSB (The Swedish National Association of Tenants, Savings and Building Societies). Runoff is collected in a stream that runs into a pond. Water is pumped from the pond up to a high point where the stream rises. The stream babbles and contributes to the beauty of the area, while the water slowly infiltrates into the ground during the course of its journey. The pond also acts as a delaying reservoir.

Source: HSB, Västerås

The Tusenskönan block of flats area in Västerås, Sweden.

Water stairway for water purification in a residential area in Schafbrühl, Germany.

Source: Architect: Joachim Eble Architektur, Tübingen

Traffic water is often so polluted that it causes problems. Particles containing environmentally hazardous oils are released from car tyres. Brake linings release metals, e.g. copper. Exhaust gases contain many environmentally hazardous substances. Traffic water is not wanted in sewage purification facilities, but it is too dirty to be dealt with as runoff. Filters are now available that can purify traffic water. Such filters can either be placed in gutters or in special treatment facilities that deal with traffic water from a bridge or stretch of highway. Currently, artificial wetlands that catch and purify traffic water are being constructed adjacent to new highway constructions.

Insert for existing runoff manholes with filters for oil and heavy metals.

Source: Consept, Stadspartner AB, Linköping

Wetlands

Wetlands are important living environments for many water-dependent organisms, both plants and animals. Wetlands act as water reservoirs, which balance flows and reduce the risk of drying out and flooding. They help to maintain the hydrological balance and water table in an area. They can also act as nutrient filters that reduce the impact that land use has on lakes and watercourses.

Moist land is difficult to develop and wetlands are very difficult to develop. Some of the moisture problems that have occurred in sick buildings come from building on moist land and failing to drain off the water.

Wetlands have disappeared and the water purification capability of those remaining has been reduced because of human intervention in the form of drainage, covered dykes and straightening of ditches. As a result of these interventions, surface water

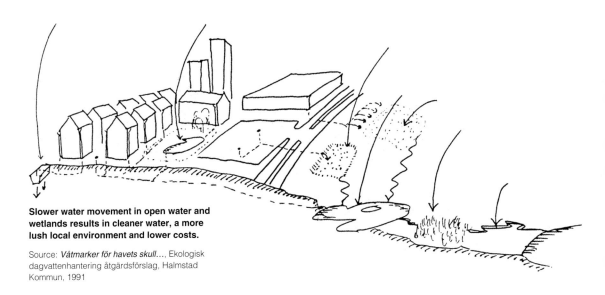

Slower water movement in open water and wetlands results in cleaner water, a more lush local environment and lower costs.

Source: *Våtmarker för havets skull...*, Ekologisk dagvattenhantering åtgärdsförslag, Halmstad Kommun, 1991

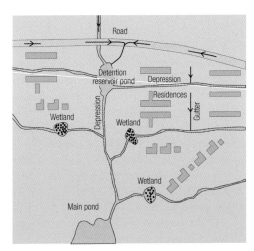

Ecological planning includes planning on-site water management. The illustration shows the plan for water flows in the area around Fornebu, the old airport in Oslo, Norway.

runs too quickly into recipient bodies of water. In order to reduce nutrient leakage, wetlands are being recreated and bends are being added to ditches. New artificial wetlands are created for different purposes. Small ponds or wetlands are created to increase biological diversity. Artificial wetlands are built at sewage plants to remove nitrogen from the cleaned sewage water. Wetlands in agricultural and forestry areas can reduce the leakage of nutrients. Wetlands are used in built environments to catch rainwater, delay run off and to clean rain from streets and car parks (traffic water).

Flooding

Flooding occurs at sporadic intervals and brings great economic consequences in its wake. Many of these consequences could be prevented with more rigorous requirements in town plans. A lot is known about the river valleys and lakes that are subject to floods. Extreme weather situations should be planned for. There should be certain restrictions for building on areas liable to flooding, and methods to ensure that water can drain off or flow into lakes, wetlands or percolation reservoirs. If climate change continues, extreme weather conditions will become more common.

Impoverished people in many developing countries build their homes on land subject to flooding as they cannot find other places to build. Flooding is made worse if trees are cut down in a drainage area, if vegetation is removed from riverbanks, and if wetlands are drained by ditches. Planning based on knowledge about water movements can help to lessen the problem.

Snow Management

There is every reason to plan for local management of snow, so that drifting snow doesn't block entrances and roads. Entryways should be placed on the windward side. Snow collects in leeward zones. Snow removal requirements can be reduced by using knowledge about how snow collects. The goal should be to minimize the surface area that requires snow removal and to deal with all snow locally. To reduce the need for large machines and transportation, plans should include surfaces for piling snow. These should be located in sunny areas on well-drained land.

On a low building with a gently sloping roof, less snow collects on the leeward side. Placing a snowbreak a short distance away in the predominant wind direction cooperates with a building. Entrances and garage doors should be placed so that the predominant winds blow snow away from them. Windbreaks should be placed about 10 times the obstacle height from entrances and sidewalks. Patios on the leeward side should be protected by roof projections, extended walls or wind barriers. Where possible, buildings should be planned so that they create a 'room' around the patio/outdoor area.

A properly designed roadway can remain almost snow-free. Closely spaced snow breaks are placed 10–15 times the obstacle height from the road. More sparsely spaced breaks are placed 15–20 times the obstacle height from the road.

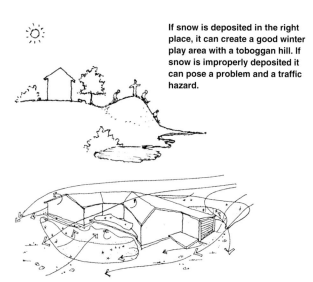

If snow is deposited in the right place, it can create a good winter play area with a toboggan hill. If snow is improperly deposited it can pose a problem and a traffic hazard.

Wind patterns and snow accumulation around a climatically adapted home. In the mountainous regions of Norway, they consciously plan the shape and location of buildings so that snow doesn't collect at entrances and on patios or result in more snow removal than necessary.

Source: *Hus og husgrupper i klimautsatte, kalde strøk, Utformning og virkemåte*, Anne Brit Børve

4.1.3 Flora and Fauna

Striving for diversity is a key ecological concept. It involves trying to create the conditions for as great a biological variation as possible. An attempt is made to leave some areas untouched, so-called biological core areas, and to have green areas, green zones and green wedges that reach as far as possible into developed areas. An effort is made to link the various green areas with ecological dispersion corridors, so as to facilitate the movement of plants and animals between them. Trees and bushes greatly influence air quality and wind conditions. In addition to their ecological functions, green areas also have an important social function.

Biotopes

Get to know the different biotopes. There are forest biotopes, such as bedrock pine forest; oak forest; mixed deciduous forest; old virgin forest and mixed coniferous forest. There are water and wetlands, such as marsh forest, open swamp, inland lakes or sea bays; open landscapes such as fields and fodder-growing, meadows and pasture lands

Types of Trees	Use
1 brush	Good in the long term
2 spruce	Barrier towards the west
3 pine, sparse	Barrier towards the west
4 clearing, seed pine	Green area
5 spruce, dense + deciduous	Barrier along the boundary
6 pine forest on thin soils on bedrock	Green area
7 pine, young	Difficult to maintain, a lot of landscape work for a road
8 spruce, pine, dense	Green area
9 spruce, pine, dense	Green area
10 spruce, pine, dense	Green area
11 spruce, pine, dense	Green area
12 spruce	Conserved
13 alder, spruce, birch	After transformation as park zone
14 spruce, pine	Green zone
17 birch, young	Hardy in green zone
0 open land	

The mapping was done with the help of air photo interpretation and field controls. The same people carried out the analysis and town plan. Optimum use could therefore be made of the natural conditions.

Vegetation mapping in the Northern Kråkegården area in Vetlanda municipality

or small-scale mosaic landscapes. There are urban biotopes, such as parks, mixed forest, or barren land in developed areas. Sensitive biotopes include bedrock pine forest and pasture lands threatened with becoming overgrown, as well as wetlands.

In the inventory stage, the flora and fauna are mapped. Biotopes, possible preservation areas and environmental loads are analysed to provide a basis for planning. Certain areas deserve special attention. With regards to biology, there are ecologically sensitive areas, or botanical or zoological resources that merit protection. Important social resources to pay attention to include areas that are experientially attractive, cultural historic areas worthy of preservation, or areas that are good for outdoor recreation. Great importance should also be attached to zones that protect against air pollution and noise, areas for infiltration into the ground, areas for runoff, and areas that protect against strong winds.

Finally, with the help of green zones and ecological corridors, the designated areas are fused into a functioning whole. Ecological corridors and passageways form dissemination routes for plants and animals, while simultaneously providing good access to the outdoors, lake shores, beaches and coastal areas.

Natural Parts of Developed and Cultivated Landscapes

Developed and cultivated landscapes contain many interesting biotopes created by human intervention that currently require protection to keep them from disappearing. Examples include avenues, willow pasture areas, deciduous copses, forest edges, pasture islands, juniper hills, grazing lands, ponds, springs, meadows, natural pastures and summer upland pastures. Some of these biotopes are disappearing due to highly efficient, monoculture-type agriculture and forestry.

Nature Conservation

Nature conservation is maintained by use of many different instruments, including national parks. The purpose of a national park is to preserve large continuous areas of a certain landscape type. The state owns all national park land. Nature reserves are created for the purpose of preserving biological diversity, protecting and preserving valuable natural environments or satisfying a need for outdoor recreation. Nature reserves can be established on land owned by the state, municipalities or private interests. Biotope protection is designed to preserve small areas for endangered species in forested or agricultural areas. It is possible to protect nesting birds, for example, or seal colonies by prohibiting access to an area during certain months of the year (wildlife sanctuaries). Economic support can be given to farmers managing valuable parts of rural heritage

Examples of environmental resources meriting protection: beaches and wetlands act as biological filters and provide important habitats for many species. Wetlands have a significant hydrological function. Areas with stands of old hardwood trees and large coniferous trees provide habitats for hundreds of plant and animal species, and are resources meriting protection.

Source: *Stadens utveckling – Samråd om förslag till Översiktsplan 96 för Stockholm*, Stadsbyggnadskontoret och Stockholm inför 2000-talet, 1995:7

areas. Some types of landscape have almost disappeared with the common use of modern farming methods.

Diversity

Large areas of land used for agriculture and forestry as well as the outdoor environment in population centres are often characterized by monocultures and a lack of diversity in both plant and animal species. In order to support diversity, a variety of biotopes must be maintained and developed. This is equally important for both the rural and urban landscape.

Regarding biological diversity at ecosystem level, from an international perspective, each country has several resources that merit protection.

Protection in law means that a plant or animal species is protected – picking, catching, killing or in any other way gathering or damaging individuals of that species is forbidden. In many cases it is also forbidden to remove or damage the seeds, eggs, roe or nests of the species. A species can be protected by law in the whole country or parts of the country, depending on the organization which made the law.

The publication 'The 2005 Redlist of Swedish Species' reports on the animals, plants and fungi that do not have long-term, vigorous populations in Sweden. The assessments are based on IUCN's international criteria. The redlist is the result of a two-year project by the Swedish Species Information Centre and its 15 expert committees that analysed the conditions of 20,000 species. It covers 3653 species, of which 1664 were classified as threatened. All the protected species listed on the global redlist, and in international conventions or EU regulations, are listed in a table of their own. The lists are revised every five years.

The Number of Redlisted (Endangered) Species in Sweden 2005

Species group	Unknown	Extinct	Critically endangered	Strongly endangered	Vulnerable	Threatened	Total number
Vascular plants	3	23	41	124	96	95	382
Green algae (*Charophyta*)	3	3	5	5	10	8	34
Mosses	38	17	11	24	57	69	216
Fungi	100	5	23	98	180	226	632
Lichens	36	18	34	39	63	64	254
Mammals	1	2	3	3	5	3	17
Birds	0	8	5	7	31	37	88
Reptiles and amphibians	0	0	1	1	4	2	8
Fish	6	1	5	6	7	6	31
Total							3653

Source: ArtDatabanken, 2005

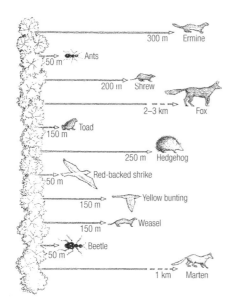

There is a big difference in the distance that different animals travel from their habitations, their biotopes. In some situations, viaducts are built over or under motorways so that animals can cross traffic routes.

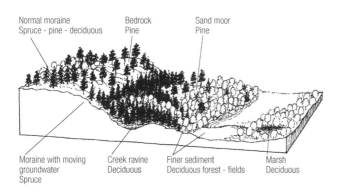

Where conditions for vegetation vary, it is easy to achieve variation. The differences between various plant locations can be preserved in natural areas.

Ecological area principles. Design and size of an area in relation to richness of species and possibilities for diffusion.

Source: adapted from Diamond

Worse		Better
	Distance	
	Size	
	Shape	
	Variation	
	Rest spot	
	Corridor	
	Diffusion network	
	Road	
	Diversity	
	Protected zone	
Young biotope	Age	Old biotope
Small	Age variation	Large
Planted biotope	History	Remains of original nature

4.1.4 Adaptation to Climate

Climate-adapted construction is when an effort is made to find or create a good microclimate when designing buildings and their surroundings. Phenomena that must be studied are: sun–shade, shelter–wind, hot–cold and dry–humid.

Climate Adaptation

Buildings have always been adapted to climate, but modern architecture and an international architectural style have resulted in buildings round the world looking more and more alike. Glassed-in buildings in northern countries require a lot of energy for heating, and glassed-in buildings near the equator require a lot of energy for cooling. In Scandinavia, when buildings are constructed with flat roofs, the rainy and cold climate often results in leakage problems. Construction should be specific to the climate zone where a building will be located.

Learn from Tradition

Traditionally buildings have always been adapted to the climate. A study of traditional construction clearly shows many examples where people have had a detailed understanding of the climate. For instance, in desert areas, people have built heavy buildings that help to even out the temperature between hot days and cool nights, and buildings are often dug into the earth. In hot, humid climates, the cooling properties of the wind have been taken advantage of by constructing buildings on piles, having verandas that provide shade and high ceilings that increase ventilation. In very cold climates such as Greenland's, the Inuit people built igloos, which are small, well-insulated homes with a small outer surface area and an airlock in the entrance.

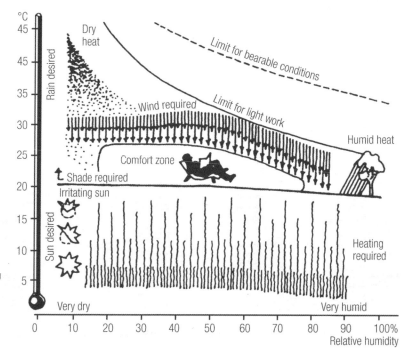

Olgyay's bioclimatic diagram can be used for climate analysis to provide information for climate-adapted construction. A location's climate during the different months of the year is plotted in the diagram, and the need for air movement, shade and extra heating or cooling can be read off.

Source: adapted from Olgyay, 1963

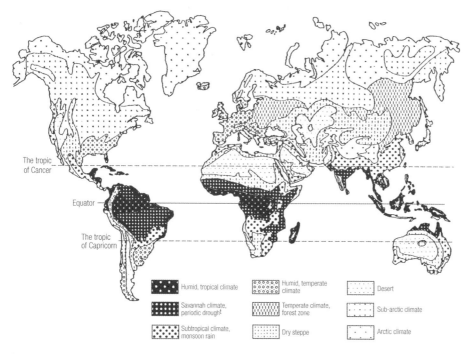

Climate zones of the Earth

Source: *Kompendium i stadsbyggnad, allmän kurs*, KTH, Stockholm, 1968

This Pueblo village at Mesa Verde in southern Colorado, US, is sited so that it is shaded in the summer and heated by the sun in the winter. The heavy material stores heat from the sun.

Climate Adaptation in Sweden

In Sweden, the seasons each have a very distinctive character, which should naturally be reflected in seasonally adapted buildings. Buildings must be able to withstand the worst weather conditions, and it is possible to build them so that they take advantage of passive solar heat in spring and autumn, using cheap and simple methods. A glassed-in veranda can increase the usable indoor area as soon as the sun is high enough in the spring and lengthen the season long into the autumn. In summer, as glassed-in verandas can get too hot, it should therefore be possible to shade the glass surfaces and to ventilate away any excess heat. In the winter, the well-insulated parts of the house can be used, where a fireplace can provide extra warmth and cosiness. (See also section 2.1 Heating and Cooling.)

In Sweden, climate adaptation involves utilizing southern slopes warmed by the sun and avoiding shaded northern slopes. In particular, windy locations should be avoided and outdoor sitting areas should be shielded from the wind. It is preferable to build homes/buildings that are light and sunny, and wind and air pollution sources should be taken into consideration when locating fresh air intakes. Good climate conditions should be arranged at entrances and approaches, as well as around pedestrian and bicycle routes. Snow removal and storage should be facilitated. Building in cold air sinks should be avoided.

In Sweden, the energy consumption of a building located inappropriately from a climate perspective can be 25 per cent higher as compared to the same building located in the best location. The goal is not only to save energy, but most importantly, to create pleasant outdoor areas and sunny homes.

The artist family Ekengren's home in Järvsö, Sweden, is a good example of a building adapted to the local climate. The house was build in the 1950s. The forest protects against the north wind and the meadow in the south lets the low, winter sun shine in through the large, south-facing, triple-glazed windows. The house is well insulated and has large roof projections towards the south. The floor is heavy and can store both solar warmth and heat from wood heating.

Source: Photo: Sture Ekengren

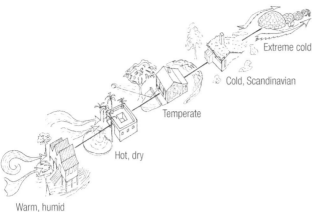

Climate-adapted buildings from south to north. From the left, buildings for a humid and warm climate, dry and hot, temperate, cold, and arctic climate. Buildings in a cold (Scandinavian) climate should be well-insulated, south-facing and protected from the wind, but glassing-in southern facing areas is not as important as it is for buildings in a temperate climate (e.g. US or Austria).

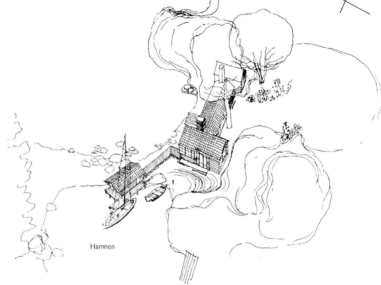

Architect Rolf 'Jefta' Engströmer's own summer cottage on Söderholmen in the Stockholm archipelago, built in the early 1950s. The bedrooms are on the east side towards the morning sun and a morning swim. Towards the south, there is a sunny outdoor area protected from the wind and a meadow between the buildings and the rocks. The living room faces west towards the evening sun and the harbour where guests arrive. The kitchen with its larder is on the north side, and the kitchen door leads to a jetty where there is a corf (i.e. a wire cage fixed to a jetty pier where fish is kept fresh and submerged).

Source: Byggmästaren 1957, A12

There can be a 25 per cent difference in energy consumption between a building located in a sunny, dry and leeward location and a building located in a shaded, damp and windy area.

Source: Mauritz Glaumann

Sun and Shade

A significant portion of a building should be in the sun during a large part of the heating season. It is also important for balconies, outdoor sitting areas and play areas to be located in sunny, wind-protected places. One possibility is to create sun traps by placing vegetation on the north, east and west to form an area that catches the sun's rays.

Times when Sunlight is Wanted

Aside from the character of light from different directions, it is important to think through whether or not the morning, afternoon or evening sun is wanted in different rooms and for how long.

Kitchens and living rooms should be able to have at least four hours of sunlight at autumnal and vernal equinoxes. Alternatively, a kitchen and living room facing different ways should together have five hours of sunlight. Private outdoor sitting areas and balconies should be able to receive five hours of sunlight after nine in the morning, or four hours after 12 noon at the autumnal and vernal equinoxes. The greater part of children's open play areas and adult recreation areas within 50m of entrances should be situated so as to receive five hours of sunlight between nine in the morning and five in the afternoon at the autumnal and vernal equinoxes.

The amount of development influences the access to sunlight. For example, there is a big difference between a floor-space ratio of 2.2 and 1.6. (The floor-space ratio is floor area divided by total land area.)

Shade from Buildings

A certain distance is needed between buildings so that enough sunlight can get in. The required distance between shading objects is determined by latitude, time of year, and time of day. For sunlight to get into a ground-floor apartment in Stockholm (60°N latitude) in December, the distance to a building in the south must be 8.4 times the building's height. In June, the distance only needs to be 0.7 times the building's height.

Studying the Sun

It is easiest to use computer programs to study the sun. There are several such programs, and they are often included in CAD architecture packages. The sun conditions can be studied in other ways as well, either on plans or blueprints using diagrams of shadow-length and sun maps, or in models using sundials. A good way to study sun conditions is by using a sequence of times in a 24-hour period at the autumnal and vernal equinoxes. It is useful to check what happens at certain times during the winter as well, since studies done at equinoxes don't give any indication of what happens during the winter.

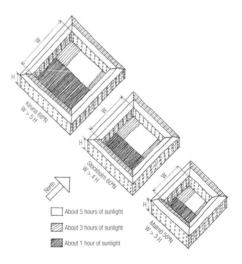

The size an enclosed courtyard has to be at different latitudes in Sweden in order to get five hours of total sunlight at the autumnal and vernal equinoxes.

Source: Mauritz Glaumann

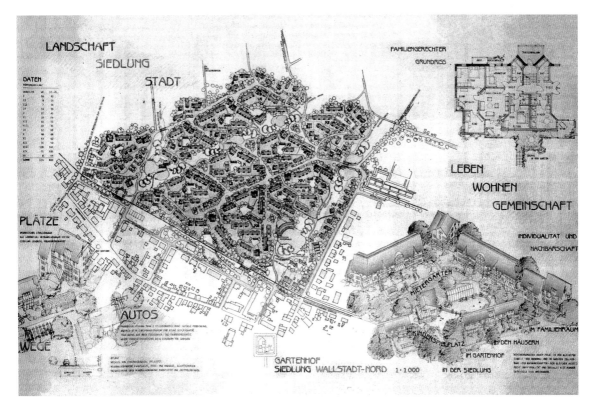

In Gartenhof Siedlung, Mannheim Wallstadt-North, Germany, a rhombic grid pattern has been used in the town plan in order to have as many sunny façades as possible.

Source: Joachim Eble Architektur, Germany

Wind and Lee

There are general wind maps, but they provide very little information about local wind conditions at a particular site. One method of documenting local wind information is to make a wind rose, which is a compact way of showing information about the wind climate. There are also computer programs that give an indication of the wind conditions around buildings planned at a particular site.

The strategy for ecological construction is to take local wind conditions into consideration and plan accordingly. Initially, the on-site wind conditions are assessed. Then the need for wind protection is estimated. General wind protection should be provided at a certain distance from the building, as well as by the location, shape and grouping of buildings. Building details should be designed so that wind problems at and close to buildings are avoided. In addition, outdoor sitting areas should be protected.

Almost all high-rise buildings give rise to wind problems since they pull the wind down to the ground. The goal should be a maximum of 1.5m/s at patios, balconies, outdoor restaurants, etc., and an average of less than 3m/s on pedestrian and bicycle routes, sports fields and similar places.

Wind Protection

Wind protection can be created in different ways: in the landscape by using windbreaks; in residential areas through building low, densely placed buildings; by grouping buildings suitably; as well as through design of entrances, fencing and placement of vegetation.

If the shape of the terrain or the unevenness of a surface blocks the wind, the strength or direction of the wind can change. In valleys, the prevailing winds are often those that blow in the direction of the valley. Wind speed increases significantly where there are clearings in the forest and terrain. In forests and behind rows of trees, there are less forceful winds.

Lee planting of vegetation is an excellent way to protect from a distance. The vegetation should be placed at right angles to the direction of the prevailing wind. The distance between windbreaks should be 100–150m. The more wind there is, the closer together they should be. Several thin breaks work better than one thick one. The design of a windbreak is important. They can be built up in several layers with both high trees and low bushes, e.g. tall poplar trees and medium-height hawthorn trees. In slopes below buildings, breaks should be thinner so that cold air is drained away. Near coasts, spruce, pine and larch can endure the strong winds that blow. Plant growth can be increased by 5–10 per cent behind windbreaks.

A windbreak can be made up of the following vegetation.

1. Nursery trees grow quickly and protect other, slower growing trees and bushes. They can be kept or thinned out after 5–10 years. A suitable proportion of nursery trees is 5–10 per cent. Examples: poplar, alder, willow, birch.
2. Examples of slower growing tall trees that may need the protection that nursery trees provide include maple, elm, birch, oak, lime, beech, wild cherry, hornbeam and bird cherry.
3. Undergrowth made up of layers of shrubs that are hard pruned every three to five years.

A building exposed to the wind needs more energy for heating. Therefore, buildings should be built in sheltered areas or shelter from the wind should be created around buildings. In general, the easiest way to reduce wind speed is to build enclosures, fences, walls, slopes, sheds and other outbuildings. Windbreaks can also be created by planting trees and bushes.

Development planning that takes the wind into account has to consider location, how the buildings are grouped and the height of the buildings. In an exemplary town plan, buildings should slow down the wind and establish areas that are sheltered from all wind directions. As this is almost impossible in practice, protection from the wind should be maximized for the directions of the harshest winds. In general, an irregular town plan results in a calmer wind environment than a town plan where all the buildings are oriented in the same wind direction.

Building design is important. High-rise buildings are exposed to the wind and

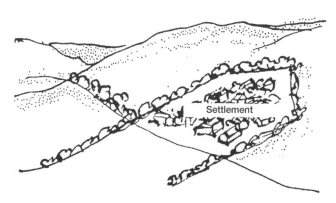

Windbreaks (or shelterbelts) provide general wind protection to an area.

Mollösund in Bohuslän, Sweden. In fishing villages in windy places along the coast, buildings were placed so that they slowed down the wind and protected each other.

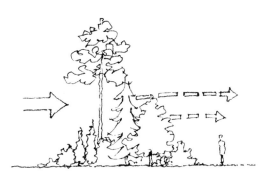

For a vegetation barrier to provide good wind protection it should have both high and low trees as well as bushes. The vegetation belt should also be adequately dense.

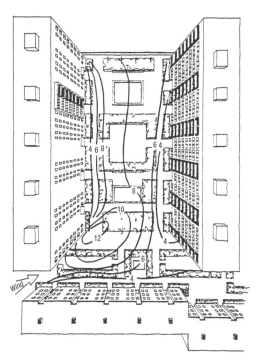

Certain development forms, such as high-rises with narrow gaps between the buildings, can result is such strong winds that it is unpleasant to be in the area. The illustration shows Krokbäck in Malmö and examples of relative wind speeds (m/s) within the area when there is a south-westerly wind.

Source: *Klimatstudier som underlag för bebyggelseplanering*, Mauritz Glaumann, 1993

direct the wind down to the ground. Elongated buildings or buildings that form a long row can create barrier effects and direct the wind to the leeward side, especially if the wind blows at an angle towards the row of buildings. Buildings that are parallel to each other, but are offset in relation to each other, can result in an increase in wind strength.

Buildings that form corridors, funnels or narrow openings create excessive wind speeds, as do buildings on piles or with large porticos. Therefore, it is wise to group buildings in an irregular manner.

Design details on or near buildings are important for reducing the wind. Extending partition walls and roofs over entrances can create sheltered areas. Irregular roof contours should be used, as well as roof extensions, railings and walls. Pergolas, grass roofs and vegetation on the walls of buildings are other possibilities. Protection nearby can reduce the wind by 70–80 per cent within a distance of 3–5 times the obstacle height. Especially bad are: balconies over corners, corners on piles and openings through buildings. On the upper parts of buildings, snow and rain can be carried from below up against the façade and eaves.

4 PLACE | 4.1 ADAPTATION TO NATURAL SURROUNDINGS | 4.1.4 ADAPTATION TO CLIMATE

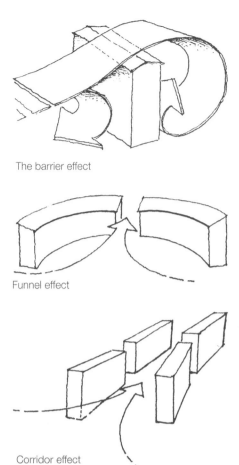

The barrier effect

Funnel effect

Corridor effect

The barrier effect draws strong winds down from higher air layers.

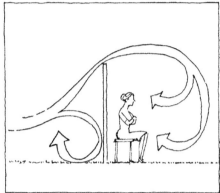

Windscreens provide better wind protection if they are perforated than if they are solid. A perforated screen dulls the wind, divides it and reduces turbulence, which creates a leeward area behind it. A solid screen can give rise to a barrier effect where wind is drawn down on the other side.

Patios can be protected from wind with windscreens. Dense screens result in a shallow leeward zone; a porous screen (30–50 per cent porous) creates a long leeward zone with less turbulence. It is possible to reduce excessive speeds around edges and corners with trellises, perforations and vegetation. Angular barriers are more efficient than straight sheet barriers, since wind direction always fluctuates (sailors are well aware of this).

Moist and Dry

Moisture evaporates and cools. Therefore, it is usually cooler where the land is damp.

Cold air is heavier than warm air, and therefore has a tendency to collect in low areas creating cold air sinks. Cold air sinks often occur where there is damp land, low-lying land, where there is a watercourse and where fog, hoarfrost and dew form easily. For energy and comfort reasons and to avoid damp problems in foundations, building on such areas should be avoided. It can be difficult to drain off moisture.

In the north it is best to build in dry places. It is the opposite in hot climates, where it is well known that evaporation has a cooling effect. Trees slow down the wind and provide shade, fountains humidify the air, and the evaporation cools parks and buildings.

In desert climates, gardens with fountains and ponds are part of climate adaptation. It is believed parks were originally built adjacent to palaces in desert climates as a type of air-conditioning.

Cold and Warm

Maps are available to show the variations in average temperatures. The outside temperature is affected by topography, proximity to the coast and building density. One way to classify local climate is to give the number of the degree-days per year. For every day in an average year, the difference between indoor temperature (e.g. +20°C) and outdoor temperature in an area is calculated. The total is added together to give a degree-day figure.

The microclimate varies from place to place even with in a small area and therefore it is interesting from an energy point of view to locate buildings where it is warmest. For example, note should be taken of where snow first melts on a plot in the spring.

Sea level also plays a significant role: temperature decreases with height above sea level at an average yearly rate of about 0.5°C per 100m. For these reasons, during the winter it is often colder down in valleys and on mountain tops than on slopes. Temperature differences of 0.1–0.3°C per m difference in height have been observed. Oceans and lakes affect temperature conditions by moderating the variations. Close to bodies of water, the daily and annual maximum and minimum temperatures are not as extreme as inland. In general the temperature moderation is a few degrees.

Surface Temperature

Differences in colour, humidity, conductivity and mass affect the amount surfaces are heated by solar radiation. The temperature of

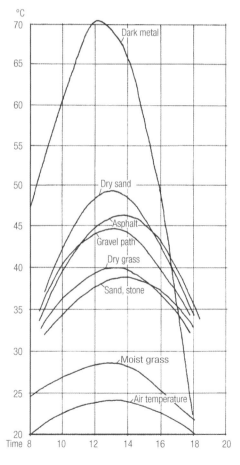

Ground surfaces get warm at different rates since they have varying characteristics with regards to absorption, reflection and evaporation. The illustration shows examples of the difference in temperature at ground level on a clear day due to the characteristics of the material on the ground.

Source: adapted from Gajzagó

surrounding surfaces affects a person's experience of warmth. The nature of the ground's surface affects temperature. Trees, above all, have an effect on radiation from the surface of the ground. Dry sand, asphalt and gravel paths get very warm, while moist grass is coolest. Dark metal gets hottest, e.g. on roofs.

Cold Air Lakes

Cold air forms on slopes at night, in clear and calm weather, when cold air sinks into depressions and valleys due to its relatively heavier weight. In a slightly hilly landscape it can be a question of a number of degrees. In larger valleys the temperature drop can be 10°C or more.

Where the cold air drains away on the upper part of a slope, there is a relatively warm zone called a thermal belt. The cold air flows sluggishly along slopes and therefore is easily blocked by obstacles. It can wrap large obstacles, such as trees and small buildings, in cold air. Condensation occurs earlier in cold air lakes than in other places because of the low temperatures there. Thus, there is an increased frequency of fog, hoarfrost and dew. However, a wind speed of as little as 3–4m/s provides enough turbulence so that the temperature differences mostly disappear.

Placing buildings where there is a risk of cold air lakes should be avoided. Some hollows, however, are located so that cold air can be drained off by removing vegetation or by excavating. Although cold air can be easily caught by buildings and clumps of trees, it can also be diverted around houses and buildings by strategically situated obstacles.

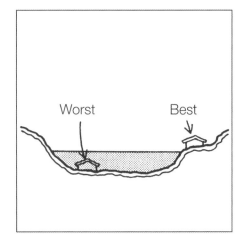

Avoid building in cold air lakes. Conditions are often worst in valley bottoms and best a little bit up the sides of valleys.

Source: *Bygg rätt för väder och vind*, P.-G. Andbert, 1979

Heat Islands

The average temperature in large towns is often a few degrees higher than in the surrounding countryside. In winter, when the weather is clear and calm, there can be a temperature difference of 5–10°C. The phenomenon, referred to as a heat island, is due to many factors, such as the storage of solar heat in the heavy material in the city, reduced surface evaporation due to drainage, and waste heat from buildings, enterprises and communications.

A phenomenon that occurs is that the temperature increase causes air to rise over the city. The temperature, cloudiness, precipitation and amount of dust in the city increases, while the wind together with radiation gain and emissions is reduced.

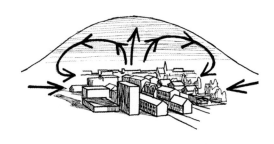

Heat islands are formed where there are many heated buildings. The buildings radiate heat towards each other and the temperature in the area is higher than elsewhere.

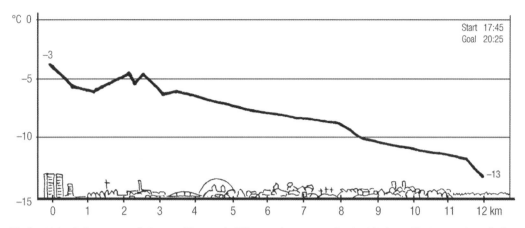

The heat island phenomenon in towns with many buildings and accompanying heat leakage. The temperatures in the curve were measured on a 'measuring journey' to and from Skärpnäck through the southern suburbs of Stockholm to the centre of the city on a clear and calm December night.

Source: Measurements by Mauritz Glaumann and Margitta Nord using a bus equipped with measuring equipment

Another phenomenon, an inversion, occurs when cold air forms over a city like a lid. Air movement slows down and vertical air exchange is impeded, which results in an accumulation of air pollutants. Suitable green spaces can improve city air and ventilation. Ground covered in vegetation acts as an evaporation surface to cool a city. Vegetation also changes air movement patterns, which can at least partly break an inversion.

4.2 The Social Fabric

The goal is ecologically, socially and economically sustainable societies. A social fabric that places a greater and greater reliance on transportation, continues the dependence on fossil fuels, and that is dominated by the use of private cars, is not possible. The solution is bringing about societies with a mixture of functions, where it is possible to walk and bicycle just about everywhere. The transportation that is nevertheless required can be done using environmental vehicles and public transportation systems. A city cannot be sustainable on its own. For ecological cycles to occur societies and the surrounding countryside must be planned as a single unit. Cultural values and locally distinctive features are important to take into consideration so that everything doesn't look the same everywhere.

4.2.0 The Sustainable Municipality

When trying to imagine a sustainable city we run into constant conflict. Should the city be concentrated or spread out? Should we think and plan on a local or a global scale? Should we rely on experts or on local democracy, etc.? Maybe it's not a question of conflict but of differing viewpoints in a complex world. Perhaps sustainable development is about looking at problems in a variety of ways simultaneously.

What is Sustainable Development?

When economists talk about sustainable development they usually refer exclusively to economic growth. Another interpretation may be that besides meeting the requirements of ecological sustainability we must ensure production demands and people's need to support themselves. When sociologists talk about socially sustainable development they are concerned with counteracting segregation and social maladjustment. Another approach is to work towards narrowing the gaps between social classes and increasing participation in community development. Ecologists talk mostly about biological diversity and how we can preserve various species and biotopes. People are also a part of this, and we could have a more human-ecological approach to the problem. Experts use their own jargon and may have trouble understanding one another. We must work to break down barriers between different specialities. Technical experts must supplement their knowledge with social awareness and reflective competence that enables them to note social tendencies towards unsustainability and the various threats arising around us.

We must begin to question the institutional circumstances that determine this relationship and within whose framework we are forced to work to remedy environmental problems. We must begin to discuss our lifestyle, as well as production and consumption as such.

A Global or Local Perspective

Much discussion about sustainable development has been about local communities and self-sufficiency in various goods. Politics usually deals with local or national issues. At the same time, the economy has become increasingly globalized. More and more of what happens locally has been decided elsewhere without those affected being able to influence the process. This gap between local residents and decision makers naturally affects the whole idea of sustainable societies.

Today many economic activities are less dependent on location than those based on raw materials. The globalization of production has in turn enabled companies to pit towns and regions against each other and to exploit the advantages of a particular location. Most towns have thus become globalized and compete with each other for investment and jobs.

The new information-driven global economy has entirely new consequences that are at the same time spatial, social and political. The result is a division into an elite and the masses. The elite have access to the network while the great majority are confined to one location and powerless. We must create bridges between these two groups, and continue to build new structures and create new forums for discussion, reflection and creation of identities.

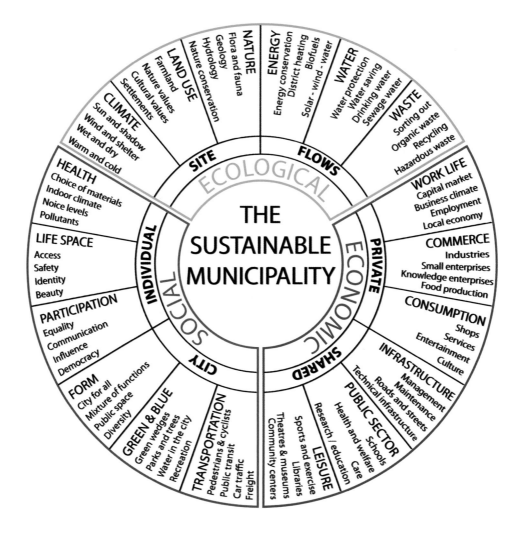

Aspects of a Sustainable Municipality

This weather rose is a holistic view and is about ecological, social and economic sustainability.

Ecological Sustainability

Achieving ecological sustainability means adapting buildings to the local site, and integrating newly created flows into local ecological cycles. Adapting utilization to the local site requires careful analysis of site conditions and their qualities and limitations. The goal is to:

- minimize encroachment into nature, and to strive for biological diversity;
- protect valuable nature, culture as well as productive forest and cultivatable land;
- adapt the buildings and living space between the buildings to the site microclimate;
- conserve energy and use renewable energy sources;
- preserve local water sources and use ecological sewage treatment; and
- organize minimization, categorization, reuse and recycling of waste.

A streetcar in the ecological district Vauban in Friburg, Germany. The streetcar is passing in front of the Solarschiff building, planned as a mixed-use building. The ground floor is intended for shops, the following floors for offices, and the top two floors are made up of terraced housing with solar cells on the roof.

Social Sustainability

Achieving social sustainability means creating good welfare for individuals, and to design cities so that they provide the conditions for a good life. Mixing functions, creating neighbourhoods as well as extension of sidewalks, bicycle routes and public transit are conditions that stimulate city life and reduce the dominance of private cars. The goal is to create:

- a healthy, waste- and toxic-free environment, clean air both indoors and outdoors;
- a city that is safe, pleasant and beautiful, and has its own identity;
- a society where it is possible to influence the local environment and development;
- a city with a diversity of functions and public spaces that stimulate city life;
- a city where green areas and water, parks, cultivated areas and wetlands are emphasized in the townscape; and
- a city for pedestrians and cyclists with a very well-developed public transit.

Economic Sustainability

Economics sustainability means creating a business-friendly climate and a city where it is easy for small-scale enterprises to become established. At the same time, infrastructure and the public sector is well developed and functions efficiently. Further, a life with appealing recreational activities, culture and entertainment that attracts people and enterprises should be strived after. The goal is to:

- create a good business climate that results in employment and a strong local economy;
- use regulations in the local development plan to make it easy for different types of enterprises establish themselves;
- integrate services into the city structure so as not to create unnecessary transport flows;
- see to it that an efficient infrastructure is developed based on a holistic perspective of sustainability;
- a public-sector (schools, social services and healthcare) that respects the individual and operates ecologically; and
- to provide easy access to exercise and recreational activities, culture and conference halls.

Compact Town or Green Town

In discussions about sustainable towns, two seemingly incompatible trends are apparent. One trend talks about a compact town with a high population density and a high degree of concentration, and the other about a green town where town and country are integrated.

Arguments in favour of the compact town are that it provides the foundation for: effective heating and sewage systems; good purification technology for treating sewage and flue gases; short distances between homes, workplaces and service centres which reduces travel; and close proximity between people and a variety of services, providing conditions for a stimulating urban life.

In a compact town, the town centre serves as a model. The town is built 'inwards' by increasing development density and by developing harbour areas and old industrial

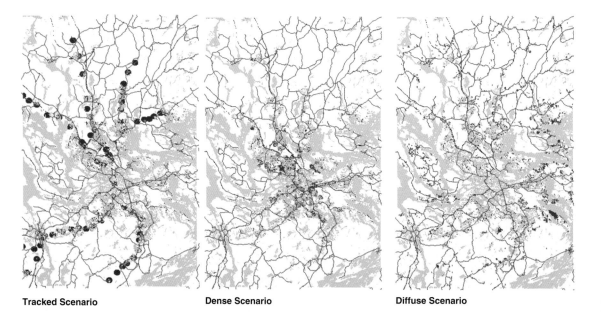

Tracked Scenario **Dense Scenario** **Diffuse Scenario**

Three scenarios for developing the Stockholm area: the compact city, development along the infrastructure and diffuse development. The red dots symbolize future population density and transportation routes.

Source: LEA-projektet (Landscape Ecological Analysis and Assessment), Berit Balfors och Ulla Mörtberg, KTH

sites. One problem with a compact town is that it often results in an increase in property prices and an increased demand on investment returns. This in turn leads to a segregation that forces emigration since it is often less expensive to live on the outskirts of the town. In the end, the compact town defeats its own purpose.

The green town provides closeness to nature in everyday life: fresh air that blows through the urban landscape, green areas of a size to supports plant and animal reproduction, local opportunities to meet basic needs and return nutrients to the soil, opportunities to take part in environmental protection and in cultivation, the conditions for local renewable energy production, primarily of biofuel, as well as a reduction in the need for two homes. System ecologists discuss a town adapted to ecological cycles. Country and town are integrated and the boundary zone is maximized. This is accomplished by building in strips along public transportation routes and allowing green areas, farms and forests to remain as undeveloped corridors within the town. Land directly adjacent to towns is set aside to meet the basic needs of the town dwellers. A new infrastructure is prepared that links town and country.

Decentralized Concentration

Is it necessary to choose between a compact and a green town or can the two be combined in a sort of concentrated decentralization, a network that links town and country, population centres and suburbs that aren't too heavily developed?

The network town involves concentrating towns along transportation routes with the junctions forming towncentres. If a network of population centres connected by fast transportation systems is developed, new opportunities for work and settlement are created. In addition to diversity in labour markets and higher education, the districts in the network would have direct contact with employment opportunities within about 40km. With new, fast transportation systems it would be possible to attend cultural activities in the regional centre while at the same time living close to nature and leisure activities. For example, people could live in garden towns, small-scale towns with traditional streets and varied housing styles (small blocks of flats, terraced houses and detached small houses), or in denser towns with close proximity to town life, culture and services.

The Planning Process

Planning is in a crisis. On the one hand, town planning has become too detailed. There is an attempt to solve too many

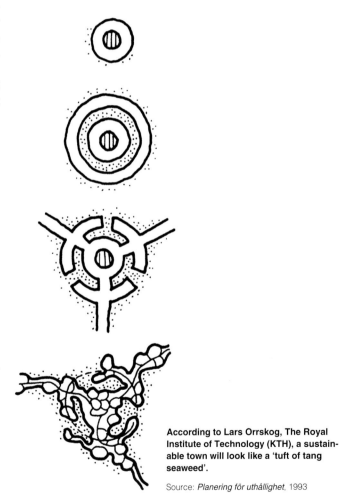

According to Lars Orrskog, The Royal Institute of Technology (KTH), a sustainable town will look like a 'tuft of tang seaweed'.

Source: *Planering för uthållighet*, 1993

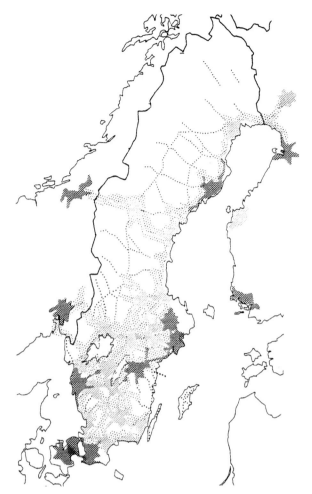

Instead of seeing a green map with red dots for cities we can more realistically see a network. This provides another perspective on planning. Instead of town planning, we need a holistic perspective in planning.

Source: *Sverige 2009 – förslag till vision*, Boverket Report 1994:14

In many municipalities there is a strong desire to invest in sustainable development, but within the administration there are contradictory signals coming from various management sectors, which creates confusion. It seems that while the entire administration is organized in a sectorial way, planning for sustainable development requires a holistic approach. A forum is missing where decision makers from different sectors can discuss such holistic approaches. It is easy for everyone to defend the interests of their own sector and for no one to defend the sometimes uncomfortable decisions that sustainable development requires. Discussions are under way about the possibility of creating forums for 'the friendly conversation' where different sector interests can come together to work on creating sustainable development.

The result of these conversations could also result in tangible planning measures. More work might perhaps be required to achieve overall planning that draws up general outlines for sustainable development. In this case, it is important that the wider picture is held in mind, rather than the interests of the various municipalities and companies. This comprehensive planning must go beyond municipal and county borders as well as national borders. The size of the planned area must be defined by the problem, not by administrative borders. We must look at the environment not only as an economic resource but also as a life-sustaining system that exists everywhere; and realize that it is up to all enterprises to take long-term responsibility for it.

Perhaps reaching consensus or creating conflict should not be a first priority, but rather let different competencies step forward and be examined at work.

There are no simple solutions. To create a sustainable community, the goals for town planning should be to ensure that all sectors of a modern regional town (inner town, suburban and fringe areas) offer good and attractive living environments for its citizens. The greatest threat to an urban environment is the emergence of poor quality, unattractive areas.

problems at the same time. On the other hand, the planning tends to be too narrow. There is a belief that all problems can be solved through centralization. The planners work with politicians, with representatives of various special interests, keeping the public in the background. Industry, the state, research and bureaucracy are seldom seen as problems when it comes to creating a sustainable community. There are large, inbuilt institutional inertias. New ways need to be found to deal with the problems.

4.2.1 Traffic

Community development is moving towards large towns becoming even larger. Traffic systems can't cope with such a heavy load. This results in enormous traffic congestion, long delays, environmental problems, poor air quality, and even complete traffic gridlock. At the same time, there is depopulation of the countryside, where it is becoming more and more difficult to provide basic services.

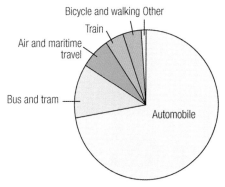

Distribution of domestic travel in Sweden by different modes of transportation measured in number of kilometres travelled.

Source: "*Färder i framtiden,*" (Trips in the Future) Peter Steen, Jonas Åkerman and others. The Centre for Environmental Strategies Research, KFB-Repport 1997:7.

Traffic Problems

There has been a type of fragmented community development that divides communities up in an unfortunate way, into, for example, housing–working areas, shopping centres–recreation areas, town–countryside, instead of making mixed-use communities.

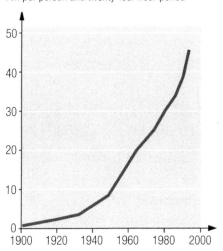

Km per person and twenty-four hour period

The average distance of travel has increased substantially during the 20th century. Today we travel four times further per person and year than we did in 1950. Car journeys account for the greatest part of the increase.

Source: *Att bygga ett hållbart samhälle* (Building a Sustainable Society), Birgitta Johansson, Lars Orrskog, 2002

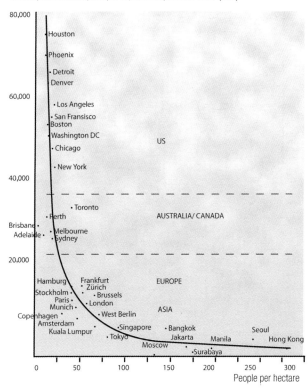

Annual petrol consumption per capita in major cities 1980 (**MJ**)

Relationship between town densities (people per hectare) and annual petrol consumption per capita (research done in 1980, the numbers indicate megajoules).

Source: *Solar Energy in Architecture and Urban Planning*, Thomas Herzog, 1996

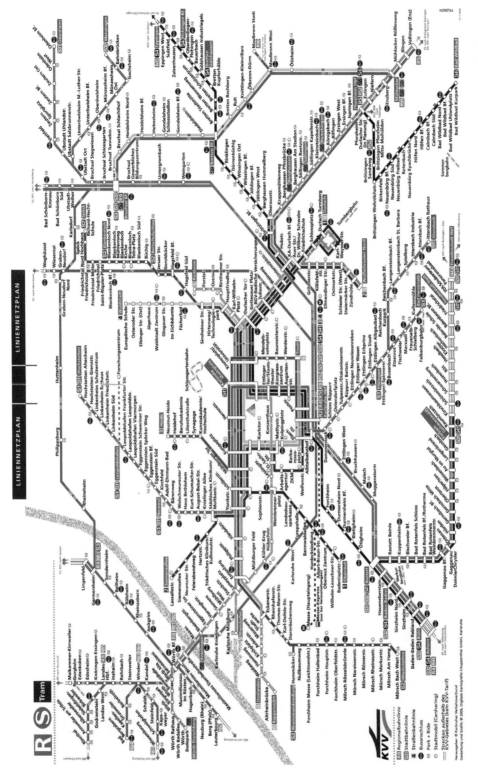

Karlsruhe, Germany, has one of Europe's most well-developed public transit systems. It is based on a train-tram system. The same train acts as both a tram inside the city and a commuter train outside of the city. KVV (Karlsruher Verkehrsverbund GmbH) has a very customer-friendly policy:

* The density of trips is high and there are electronic notice boards that show when the next train-tram will arrive. * School passes can be used on public transit year-round. * An individual with a monthly pass can take their whole family on a Sunday excursion. * The coaches have large windows that provide a good view. * There is no advertising inside the coaches, as this is believed to reduce graffiti and vandalism. * Bicycles may be brought along outside the rush hour. * There is a total of 650km of train-tram lines, which connect all the centres around Karlsruhe. The public transit system is in general quicker than taking a car, especially in rush-hour traffic.

The train-tram commuter train travels quickly on its own railway embankment between communities.

The train-tram runs slowly as a tram in the city between pedestrians and cars.

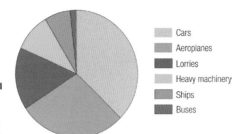

Emissions of greenhouse gases from the transport sector in Sweden, 2005. The transport sector contributes 45 per cent of the total load (approximately 35 million tonnes of CO_2-equivalent) when international Swedish transport is included.

Source: *Tvågradersmålet i sikte*, Jonas Åkerman, 2007

Much production, processing and warehousing is becoming increasingly large scale and centralized, which leads to a substantial increase in transportation needs. Even though the environmental impact is reduced in other areas, the impact of the traffic sector still increases. There is a tendency to mimic energy-intensive US models and to obliterate cultural and local distinctive features. Mass motoring doesn't just lead to large expansions of motorways and roads but also to larger land areas being claimed for parking and cars dominating streets and town squares at the expense of people. US cities consume the most resources and Hong Kong and Moscow are the most densely populated cities. It's interesting that European cities such as Amsterdam and Copenhagen seem to have struck a more happy medium in their city planning and have succeeded in combining reduced petrol consumption with enough space for city dwellers.

The Disadvantages of Traffic

The transport sector is one of the main contributors to CO_2 emissions. It uses 30 per cent of the total energy but causes 45 per cent (including international flights and shipping) of CO_2 emissions. Private cars cause the biggest emissions, followed by aeroplanes, with lorries in third place. Transport is the sector of society with the fastest increasing volume and CO_2 emissions. To achieve a reduction of 85 per cent in CO_2 emissions by 2050, which is the EU goal in order not to increase global temperature by more than 2°C, radical changes have to be made in the transport sector. These changes include more efficient technology, a change from fossil to renewable fuels, and a change in lifestyle.

Traffic not only causes energy problems, but also noise, bad smells and air pollution. Traffic involves, of course, safety risks, but motoring also takes up a lot of space, to the detriment of both the natural and cultural heritage. Obviously, densely built-up areas should be planned so as to minimize the drawbacks.

Solutions to Traffic Problems

Planning that maximizes integration of various community activities can solve traffic problems. One model is to first plan for pedestrian and bicycle traffic, then ensure that affordable and easily accessible public transportation is thoroughly developed. Making it more difficult to take the car than to take public transport helps the situation. This in turn leads to the need for a new form of government direction. There must be incentives to switch most transportation to the most environmentally friendly modes, to coordinate transportation on national, regional and local levels, considerably improving facilities for transfers and reloading between different modes of transportation, and to quickly develop more environmentally friendly engines and alternative fuels.

There are many examples worldwide which prove that it's not possible to eliminate traffic problems with new roads. The better the roads, the more motorists use them. Instead, methods for limiting journeys, such as tolls and congestion charges, are being used. Singapore was first to impose electronic toll collection, and in 2003 a congestion charge was introduced in central London. Experiences have generally been good. A toll system is also used in Stockholm.

Pedestrianized Zones

Pedestrianized zones should form a continuous network linking the whole town – a system with pedestrian streets in the town centre, pavements along busy streets with good pedestrian crossings, and pedestrian paths on the outskirts. An important aspect of pedestrian traffic is that various community amenities, such as services and recreational and cultural facilities, should be located within a reasonable distance. Pedestrianized areas should be pleasant: they should be easy to navigate with a pram, a rollator or a wheelchair. There should be benches to rest on, covered passage ways, arcades and bus shelters, and glazed-in pedestrian streets in shopping centres.

Copenhagen has one of the world's best pedestrian walkway systems. As early as 1962 Ströget was turned into a pedestrianized area. Since then, the network of such zones has been extended from 16,000m^2 in 1962 to nearly 100,000m^2 in 2000. In contrast to many other cities, there are no pedestrian subways or pedestrian bridges with steps that many people find troublesome. Included in the pedestrianized areas are town squares, public areas, places to sit and open-air cafés. In recent years, these areas have been supplemented with a system of so-called pedestrian-priority streets, a kind of combined pedestrian and cycle street, where traffic may enter only under pedestrian regulations and terms.

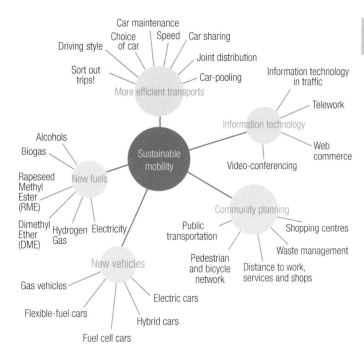

The transportation sector is one of the most difficult sectors to steer in a sustainable direction. But there are many things that can be done. In combination, they can make a positive difference.

Source: *Att bygga ett hållbart samhälle* (Building a Sustainable Society), Birgitta Johansson, Lars Orrskog, 2002

Copenhagen's network of car-free streets and town squares.

Source: *Byens rum byens liv* (Town Space, Town Life), Jan Gehl, Lars Gemzøe, 1996

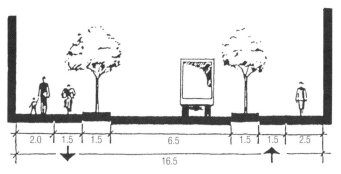

A section of a street in the Norwegian town of Tönsberg with room for pedestrians, cyclists and motorists. In Tönsberg, making the town more environmentally friendly is a priority. The goal is to make the entire town accessible by bicycle without causing any dangerous traffic situations.

Source: *Sykkelbyen Tönsberg-Netteröy*, Projektrapport *1991–94* (The Bicycle Towns Tönsberg-Nötteröy, Project Report 1991–94)

Cycle Paths

The bicycle is in a class by itself as the most energy-efficient means of conveyance ever invented. To make it easier for cyclists, planning should provide cycle paths that are well sited and safe. In Davis, California, the cycle paths are designed so that cyclists can ride on the lee side, there is shelter from the hot sun and there are drinking fountains at regular intervals. Why not prioritize bicycles in planning so that cars have to make detours and take second place? Why not produce more bicycles with luggage racks or cycle rickshaws

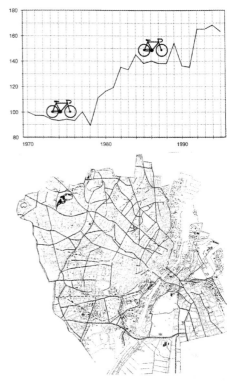

The graph shows how cycle traffic in central Copenhagen has increased by 65 per cent over the last few decades. The map shows the network of cycle paths along city streets in the municipalities of Copenhagen and Fredriksberg.

Source: *Byens rum byens liv* (Town Space, Town Life), Jan Gehl, Lars Gemzøe, 1996

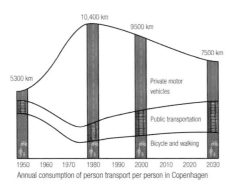

Invest in bicycle and pedestrian traffic and public transportation in order to reduce private motoring. Annual transportation distance and mode of travel per person.

Source: *Energihusholdning husholdning holdning* (Energy Management, Conservation, Economy), Jörgen Stig Nörgård, Bente Lis Christensen, 1982

Bicycle stand in an inner courtyard in a city block, Copenhagen. Note the grass roof.

so that a fellow traveller could be picked up? Bicycles can be equipped with electric auxiliary engines. There are even examples of bicycle lifts (similar to ski lifts) on steep slopes. What is important in this context is to plan areas with bicycle stands, locking possibilities and bicycle stores.

Denmark gives an inspirational lead when it comes to cycle paths. Work with bicycle traffic in Copenhagen has focused on building systems of cycle paths along all the city's main arteries so that, in principle, the network reaches all parts of the city. The cycle network was started in 1930 with 80km of cycle paths and by 2000 there were about 300km. Intersection problems, which are the greatest problem with cycle paths, have been solved in the same manner as pedestrian crossings for pedestrians. Blue asphalt is used at intersections with cycle paths. In 1995 a system of city cycles was introduced, where bicycles are placed throughout the city that can be borrowed in the same way as shopping trollies can be borrowed in large grocery stores,

Bicycle garage in the centre of Freiburg. Here you can park your bicycle safely, get it repaired or rent one. There is also a café at the top of the building and a photovoltaic panel on the roof.

i.e. by depositing a coin which is returned when the cycle is returned. In Copenhagen there are 125 bicycle stands where cycles can be picked up and returned.

Houten in The Netherlands with its 30,000 inhabitants is a town that was planned from the outset for pedestrians and cyclists. The railway station and shopping centres are located in the town centre. Footpaths and cycle paths start and fan out from the centre. Roads for vehicles were the last to be planned. They encircle the town with cul-de-sacs that extend partway into the residential areas. There are no big differences in level (no steps) on the cycle paths and there is weather protection at locations where waiting times may exceed one minute. From the outskirts to Houten's centre is no more than 1.5km.

Public Transportation

In Europe public transportation accounts for only 20 per cent of road travel. An increase in public transportation in built-up areas can be achieved if the will exists, but good public transport in sparsely populated areas requires coordination. There are many ways to do this and IT provides new opportunities. Currently, there are usually separate public transport services for the general public, for school children, for elderly and disabled persons, for postal services and for other transport tasks. Possibilities for coordinating services should be taken advantage of. Light rail and train-tram refer to trams that can run both on existing train rails and on new street rails. As public transport must be increased, the rail infrastructure should be improved. The energy use of express trains can be improved by 45 per cent and at the same time the speed can be raised from 200 to 250km/h (Åkerman, 2007). Light rail and train-tram systems combining the advantages of trains and trams have been built in many places in Europe. In Karlsruhe in Germany they have implemented a train-tram system that is 800km long and that covers the whole region with a rail network.

Comfortable public transport is a prerequisite if public transportation is to surpass or supplement motoring, in which case it must be developed according to motorists' wishes and needs. Public transport must be

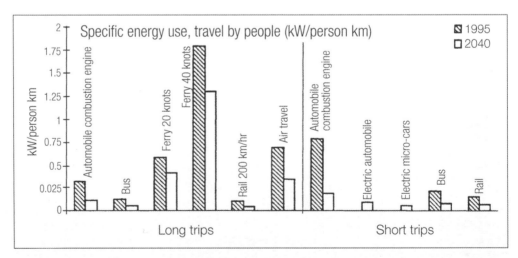

Presumed amount of energy consumption for the average passenger journey in about the year 2040, compared to the corresponding values in 1995. Unchanged occupation rate (persons per vehicle) and speed has been assumed.

Source: *Färder i framtiden* (Trips in the Future), Peter Steen, Jonas Åkerman and others, The Centre for Environmental Strategies Research, KFB-Repport 1997:7

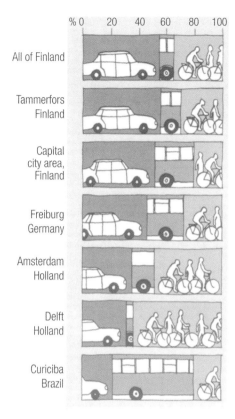

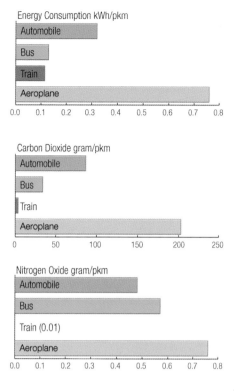

Division of trips by transportation mode. It is clear that investing in public transportation and bicycle traffic changes travel patterns.

Emissions and energy consumption for various modes of personal transport. The numbers apply per person and kilometre (pkm) for a trip of 250km.

Source: *Råd och Rön*, 5/98 (Advice and Findings)

The new tram system in Stockholm has been successful. It links the octopus-like arms of underground routes that extend out from the city centre.

Personal Rapid Transit. A driverless and fully automated vehicle on a network of tracks. Just enter your destination and enjoy your ride.

Source: Illustration: SkyTran

The bus stops in Curiciba are themselves reminiscent of a bus. Passengers enter and pay at one end and can then proceed to a comfortable seat. The advantage of this is that when the bus arrives, everyone has already paid and is ready to get on, which shortens the bus stopping time. The buses are easily accessible for people in wheelchairs.

Source: *Curitiba-A Revolucão Ecológica* (Curitiba – An Ecological Revolution), Prefeitura Municipal de Curitiba, 1992

available both when you need it and where you need it. No one should feel forced to have a car to cope with their everyday journeys; not on weekdays, the weekend or holidays. During passenger journeys, anyone so inclined can keep busy by working, doing homework, etc., while others would have the opportunity to rest. It should be possible to vary the lighting according to individual needs. Establishing special bus-only streets can increase the speed, comfort and appeal of public transportation. Most people will accept a total of one hour's travel time per day provided that it seems convenient. Distance to the stop must be moderate. The journey should have a high average speed and few stops along the way. A wait or transfer of four minutes or less is considered as no time at all. Transfers should be simple, preferably on the same level and made, for example, just by crossing a platform. Any waiting time could be positively utilized in a pleasant environment. Electric signs can display the number of minutes until the arrival of the next vehicle. It should be easy to bring along a bicycle, wheelchair, skis, etc., and to keep an eye on them for the duration of the trip.

Copenhagen has a public transportation system that works well. It is composed of an efficient system of express trains for long-distance trips, a bus system for the city centre and a bus network that connects the suburbs to the express train stations. Thus far the system has been supplemented with one underground line. Copenhagen is one of few cities where private car use in the city centre hasn't increased in many years. This is due to a conscious decision to plan the city centre so that travel by car is awkward (obstructed by pedestrians-only streets) and parking is scarce.

Curitiba, Brazil, is a town of over 1 million inhabitants that is a good example of deliberate investment in a smoothly functioning public transportation system. Fast shuttle buses and slower local buses on special bus-only streets have resulted in public transportation making up 70 per cent of all personal journeys. Shuttle buses have their own lanes and red lights turn to green when they arrive. This means that shuttle buses avoid traffic congestion and can keep to their timetables without much difficulty. The city also has first-rate cycle paths.

Car Traffic

Eliminating cars in towns altogether won't be possible other than in exceptional cases. However, work can continually be done to reduce traffic in towns and to ensure that cars are as environmentally friendly as possible. More efficient cars with an extra electric motor (plug-in hybrids) driven by electricity and biofuels seems to be the solution.

This is all what you see of an underground garage at the block of flats Silodam (Architects MVRDV) in Amsterdam. In this parking garage, the driver places the car at the entrance and an automatic mechanism parks the car.

Car share clubs are usually associations with a varying number of members. The club jointly owns a fleet of cars, and they can be booked in advance. Members have access to a car as often as they want, but pay only about half what it would cost to privately own a vehicle. Members share

Reserved parking spaces for car-share vehicles at the Understenshöjden eco-village in Stockholm.

responsibility for maintenance, taking the vehicles to a mechanic when necessary, MOT certificates and so on. Members are guaranteed a vehicle even on bank holidays and summer holidays, because if there aren't enough vehicles then the association rents more from an ordinary car rental firm. Sometimes car share associations collaborate with, for example, municipally owned companies in order to keep the cars fully booked. Municipally owned companies may need daytime, weekday access to cars while most car share club members need a car for weekday evenings, weekends and holidays.

Fuel-Efficient Cars

Fuel-efficient cars, i.e. vehicles that consume less fuel, are a partial solution to traffic problems. According to some studies, the specific energy use of cars can be reduced by 60–85 per cent. Fuel consumption varies greatly, weight being a determining factor. Use of more plastic, carbon-fibre-reinforced plastic, aluminium and magnesium keeps weight down. A fuel-efficient car can also manage with a smaller tank. An aerodynamically designed front with a small front area and smooth surfaces is another determining factor, as are narrow tyres that run easily with a minimum of friction and air resistance. Automatic transmissions, continuously variable transmissions and computer-controlled 'manual' transmissions can reduce energy losses. Motors can be made more efficient through friction reduction, accomplished by reducing the weight of moving parts, using direct injection, turbocharging, and having more valves with variable valve timing and cylinders that can be shut off. A dynator is a combined starter motor, auxiliary motor and generator which can deliver a 15 per cent fuel savings. A car with a dynator doesn't consume any fuel when stationary. The fuel consumption of a four-wheel drive SUV (sports utility vehicle) is 15L/100km and for an ordinary car (Volvo V70) it is about 9L/100km. However, fuel-efficient cars do exist. The most efficient cars today use 4–5L/100km. This can be reduced to 2–3L. Already in 2002 Volkswagen had designed a small car using 1L/100km but the concept proved too expensive for mass production.

Electric cars are quiet and don't emit exhaust. Electric vehicles have been in existence as long as petrol-powered cars, but they have three problems: a short range; large, bulky battery packs; and battery recharging time.

An electric car has more than twice the efficiency of a fuel car. The development of electric cars has been rapid in recent years. Many companies are examining new batteries. The lithium ion battery is one of the most promising concepts. Several companies are on the verge of launching new electric cars. The Norwegian model 'Think City' can run at 100km/h and travel up to 180km on one charge. The American model 'Tesla Roadster' can reach a speed of 210km/h and can drive 320km on one charge. The Californian model 'Aptera' can run at 150km/h and travel a distance of 200km on one charge. There is also a hybrid variant of the Aptera.

Small electric vehicles for city-centre use is one idea that's been tried. In Livorno, Italy, electric rental cars take precedence. They have access to areas where petrol-powered cars aren't allowed, including bus lanes, as well as to special car parks. Electric cars are inexpensive to rent. There are electric outlets where the cars can be recharged, which is paid for with a special credit card. The credit card registers information from the car's computer and evaluates how much recharging is required. There are high-speed rechargers that take 20 minutes, but they are more expensive than the normal rechargers which take 4–6 hours.

The future of the electric car depends on the development of batteries, which is proceeding apace. Lead batteries (Pb/A) have an energy content of 30–40Wh/kg, nickel-cadmium batteries (NiCd) 50–60Wh/kg, nickel metal hydride (NiMH) 50–80Wh/kg, sodium nickel chloride (Zebra) 95–115Wh/kg, lithium ion (Li-ion) 80–150Wh/kg and lithium metal polymer (LMP) 120–140Wh/kg.

Hybrid cars or hybrid electric vehicles (HEV) are one way to escape the drawbacks associated with electric cars. In the hybrid car the electric car and the energy-efficient car are united. The car has two engines, one burning fuel and one electric, and a battery

The Norwegian company's 'Think City' electric car.

The electric Vectrix Scooter with a top speed of 100km/h and a fully charged range of up to 110km. The onboard charger recharges the scooter in about 2 hours from a standard 110/220V (3-pin) power socket.

system. The hybrid car uses the fuel engine and braking to charge the batteries and drives at slow speeds with the electric engine. Both engines are used during acceleration. Both engines stop when the car stops and braking is done with the generator so that the battery is charged. The hybrid technology can save 20–30 per cent of the energy needed.

Plug-in hybrid cars or plug-in hybrid electric vehicles (PHEV) have a much better battery than HEV models that can be charged when the car is not used. This means that the car can be driven entirely on electricity for short distances. When the battery loses its charge the fuel engine will take over. By using plug-in hybrids with a reach of 50km on electricity, 75 per cent of all trips (in length) can be made with electricity. One very interesting solution is a plug-in hybrid car with a fuel engine burning biofuel.

Hydrogen gas cars have been developed by several car manufacturers. A hydrogen gas car that is run with a conventional motor releases water and nitrogen oxides via the exhaust pipe. They have the advantage of both cleaner exhaust emissions and the large amount of energy per volume contained in hydrogen gas. A difficulty with using hydrogen gas is the risk of explosion, but a method for storing hydrogen gas in metal hydrides instead of in pressure tanks has been developed. A main problem with hydrogen cars is their low energy efficiency.

Fuel cell-powered cars with electric motors may be the most environmentally friendly alternative, although not as efficient as the plug-in hybrid using electricity from the grid. The exhaust consists only of water vapour. Hydrogen gas and air are introduced into the fuel cell and electricity is produced. The environ-

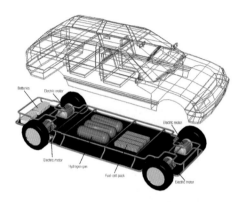

Fuel cell-powered car. One concept consists of building the entire power unit (pressure tanks, fuel cells, batteries and engines) into the chassis. Electricity drives small electric motors at each wheel and is also stored in batteries. Different types of auto bodies can then be attached to the chassis.

Source: Illustration: Leif Kindgren

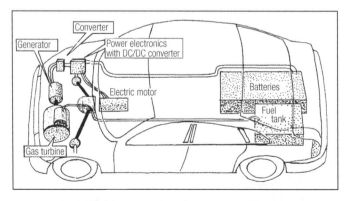

Volvo's hybrid concept car. The car runs on electricity all the time and has a small battery bank. But it is also equipped with a small fuel-burning gas turbine that runs a high-speed generator, which can generate electricity when the battery bank isn't sufficient.

mental impact depends on how the hydrogen gas is produced. For example, a sustainable method is production through electrolysis with electricity from solar cells. Fuel cell cars may be a solution for the future.

Car Engines

Diesel engines are more efficient than petrol engines, but there are ongoing discussions about the dangers of diesel exhaust. Carbon dioxide emissions are 20 per cent lower than for a petrol engine, but diesel emissions contain more nitrogen oxide and considerably more particles. Especially worrying from a health point of view are the smallest particles, which can cause cancer, heart disease and provoke asthma attacks. Using diesel engines requires better exhaust cleaning to lower the total amount of particles and CO_2. The total efficiency from fuel to wheel in a diesel car is not more than 22 per cent. For a petrol car it is even smaller (18 per cent). In hybrid cars the efficiency is about 37 per cent and for fuel cell cars it is about 55 per cent. The production of hydrogen for fuel cell cars is energy-demanding, so total efficiency for hybrid cars and fuel cell cars is about the same (32 per cent). In electric cars and plug-in hybrids driving on electricity the efficiency from fuel to wheel is about 73 per cent. The total efficiency for electric cars using electricity from wind power and hydropower is about 67 per cent. The movement towards electric motors means a much more efficient use of energy.

Alternative Fuels

Ethanol, biogas and rapeseed methyl ester (RME) belonged to the first generation of alternative fuels in Sweden. Ethanol is a common alternative fuel, produced from sugar cane, maize or wheat. This has been heavily criticized because fuel production thus competes with food production and because the whole process is not energy efficient. RME was used to replace diesel, but has received much criticism due to its polluting exhaust emissions. RME is even less energy efficient than ethanol. Biogas is made from organic waste and sludge from sewage plants. The second generation of alternative fuels includes ethanol and methanol made from forestry waste and DME (dimethyl ether) made from forestry or organic waste.

The small French car Airpod is extremely cheap to operate. It runs on pressurized air, is 2m long and has enough space for three people. Top speed is 70km an hour and 100km can be driven on a tankful of pressurized air. It takes a couple of minutes to fill up the tank.

Source: IMD Press.

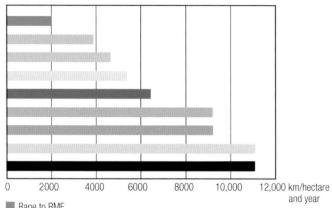

- Rape to RME
- Wheat to ethanol
- Energy forest (Salix) to ethanol
- Energy forest (Salix) to synthetic diesel
- Energy forest (Salix) to synthetic diesel via black liquor
- Energy forest (Salix) to methanol
- Energy forest (Salix) to DME
- Energy forest (Salix) to methanol via black liquor
- Energy forest (Salix) to DME via black liquor

One way to look at the energy efficiency of alternative fuels is to compare how far a vehicle can be driven with fuel made from the harvest of 1ha of farmland. This kind of comparison shows that one of the most promising technologies is production of biofuel via gasification. The fuels that are calculated to give the highest energy content by gasification of biomass are DME, methanol, methane and hydrogen. However, for practical and economic reasons, DME and methanol seem to be the most promising fuels.

Source: *Volvo lastvagnar och miljön*, 6 August 2007

Biogas has a future but it can't replace all the fuel needed. If all organic waste is used to produce biogas, 20 per cent of all cars can be driven by biogas. In the current production of biogas there is a leakage of methane in the cleaning process and the engines release nitrogen oxides. A new technology for biogas production uses a freezing technique where methane leakages are avoided. With new gasification technologies even more biogas can be produced. Better engine catalyst systems can solve the problem with nitrogen oxides.

Ethanol production from cellulose by using enzymes and a modified form of yeast is being developed. The energy efficiency of this method is not so high (25 per cent), but if the production is combined with a cogeneration plant (producing both electricity and heat) the energy efficiency is increased (65 per cent). The waste products from ethanol production can also be used to produce biogas. Commercial production in Sweden is expected by 2015.

Methanol production from cellulose is another technology under development. The process is done by gasification and the energy efficiency is higher (60 per cent) than for ethanol production.

DME (dimethyl ether) production from black liquor uses this waste product from paper production. Biofuel can be made from it by gasification under high pressure and at high temperature. In the gasification process a synthetic gas is produced that can be transformed into DME, methanol or synthetic diesel. DME requires special engines, while methanol can be used in modified petrol engines and synthetic diesel can be used in diesel engines. This has proven to be one of the most energy-efficient methods to produce biofuel.

Synthetic diesel (bio-diesel) production by gasification of cellulose uses the Fisher Tropsch method; a catalyst is used to convert the gas into synthetic diesel. The advantage of synthetic diesel is that it can be used in common diesel engines. Factories producing synthetic diesel have been built by Neste Oil in Finland.

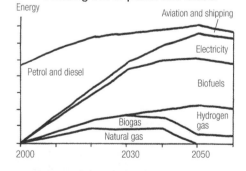

The Swedish Society for Nature Conservation's prognosis for vehicle fuel development.

Source: *Ny Teknik* (New Technology) 1996:46

Fuel Cells

Fuel cells can be compared to a battery that, when supplied with hydrogen gas at the anode and oxygen (air) at the cathode, generates electrical energy. At the same time 80–1000°C heat is generated, depending on the technology used. Fuel cells are light and compact, have no moving parts, are highly efficient (about 47 per cent efficiency for electricity generation) and produce minimal emissions. If natural gas-powered fuel cells are compared to fossil fuel-powered electricity/heating plants, CO_2 emissions are significantly lower and nitrogen oxide emissions are almost totally eliminated. The cost of fuel cell plants is, so far, relatively high.

A fuel cell's three essential parts are the anode, the cathode and the electrolyte. There are many different types of fuel cells. Fuel-cell technology can use a number of different electrolytes: phosphoric-acid (PAFC), molten-carbonate (MCFC), solid-oxide (SOFC), alkaline (AFC) and polymer (PEFC).

Environmental emissions from fuel cells are usually considered negligible since water is the residual product. However, certain smaller releases of nitrogen oxide have been recorded. Many consider fuel cells to be one of the energy technologies of the future.

Fuel cells can be used as a power source in electric hybrid vehicles. By using a fuel cell to charge a battery that drives an electric motor, reasonable performance is attained with low environmental emissions. With the help of the battery, the energy generated when braking and the surplus energy from the fuel cell can be utilized.

There are two great advantages to hydrogen gas propulsion of aircraft: first, there is a weight savings (the amount of energy in liquid hydrogen is three times greater than the energy in jet propulsion fuel), and second, emissions consist of water vapour which has no effect on climate change. In order for it to be practical to handle hydrogen gas as a fuel, it must be cooled to a liquid state at –253°C. The technology for this exists today and is used in the space industry.

Hydrogen gas can be used to generate electricity and heat in buildings. A hydrogen gas-powered engine runs a generator. Heat from the engine's cooling water and exhaust is used for heating and to produce hot water. In this way, a totally self-supporting, environmentally adapted system is obtained.

Extremely small fuel cells for cell phones and laptop computers are being developed. Such a fuel cell could run a cell phone for one month.

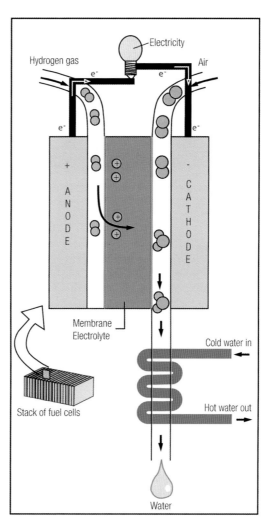

This is how a fuel cell works.

Hybrid car technology is introduced into lorries and buses.

Source: Volvo

The hydrogen gas city (Hydricity) is an interesting concept where vehicles supply buildings with energy. Fuel-cell cars contain a super-efficient power plant which could deliver electric current to the electric mains or to an individual's house during the 96 per cent of the year that the car isn't being used for transportation.

Hydrogen gas and electricity are two important interactive energy sources and the potential exists for integrating them into one power network. In a reversible fuel cell, energy conversion works in both directions. Electricity (from a renewable resource) splits water in the reversible fuel cell into hydrogen gas and oxygen. Electricity is generated from the air and hydrogen in the fuel cell. The car runs on electricity generated from hydrogen gas in the fuel cell. A fuel cell-powered car with hydrogen gas in the tank can be connected to a house and provide it with electricity and heat when the car is parked at home. Debits and credits are handled on the internet. With reversible fuel cells, a significant part of the issue of hydrogen gas distribution is taken care of at the same time as hydrogen gas cars (small, mobile power stations) contribute to the city's electricity and heating requirements.

Goods Traffic and Long-Distance Traffic

Goods traffic can be reduced by increasing local production. Necessary goods transport should use trains, ships or pipelines, which consume relatively little energy compared to long- and short-distance lorries. Logistics are important in order to avoid unnecessary driving and loss of time during transfers between transportation modes. Systems for delivering everyday commodities can be developed, but must service all households within a defined area in order to be economically viable.

The increased numbers of **lorries** on the roads is a development that makes it increasingly difficult to achieve environmental and traffic safety goals both in the short and the long term. More heavy vehicles also means a substantial increase in wear and tear on the roads. For lorries the potential efficiency is smaller than for private cars, because the weight of the cargo is big compared to the weight of the car. Modern lorries already use diesel engines that are more efficient. The regulatory restrictions on diesel engines are due to the emissions released. Better aerodynamics, lightweight materials, more efficient engines and hybrid solutions can make lorries more efficient; studies show that they can be 30 per cent more efficient for long journeys and 40 per cent for short deliveries. Lorry producers are developing a hybrid concept with a super-condenser charged by brake energy. The super-condenser is then used to get the lorry moving again.

Political measures are needed so that freight traffic is transferred from lorries to trains and boats. All modes of transport should bear their own costs, yet lorry traffic has benefited from significant tax advantages. Extensive lorry transportation has benefited at the expense of railways and shipping. A kilometre tax would make it possible to tax a vehicle according to the actual distance travelled in the country where the environmental impact occurred. A kilometre tax has already been introduced in Switzerland and lorries over 28 tonnes have been forbidden. Similar policies are being implemented in several other countries in the European Union. These countries have in common that the tax will be used to pay for roads, railways and canals.

Trains are energy efficient; only 0.06kWh/tonne-km is required for handling freight. With top speeds of 160km/h, trains can travel up to 1250km in one night. If rail transport is to be competitive, it will be important to organize transhipment stations where redistribution of freight between different modes of transport can be managed quickly and smoothly (through horizontal freight transfers).

Boats can transport goods much more efficiently than train and lorry transport. International boat transport makes inexpensive, long journeys possible. Coastal shipping may become even more important in the future. Harbours have a strategic value and should not be turned into something else. Canals will come back into favour. There are several canals being built in Europe (for example, between the Danube and the Rhine, and between Paris and Belgium). New types of boats mean new possibilities. Container ships have made shipping more efficient. However, shipping is by far the dirtiest type of transportation with regard to sulphur and nitrogen oxide emissions since low-quality fuel is often used. There is, however, no lack of technical methods to significantly reduce these. High-speed ferries reduce transportation times but consume more energy and have a greater environmental impact.

Transporting cargo by ship is already energy efficient, while passenger ferries use more energy. This has a lot to do with speed. Energy use increases with the square of the speed. Slimmer hull design and more efficient engines and transmissions can reduce energy needs by 35 per cent in ships. In ferries, this combined with a 10–15 per cent slower speed can reduce fuel needs by 50 per cent.

Aeroplanes have become more energy efficient during the last decades. Yet they can be made even 54 per cent more efficient, in part by using lighter material and redesigning the shape of the plane and wings. It also means better organization, fewer empty seats,

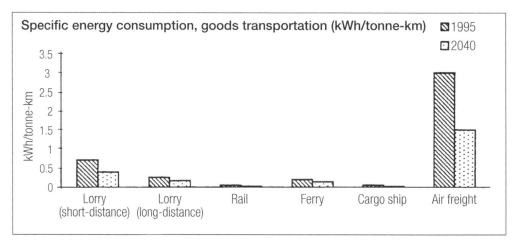

Assumed energy consumption figures for average goods journeys in about the year 2040, in relation to the comparable figures for 1995. Unchanged load factors and speed have been assumed.

Source: *Färder i framtiden* (Journeys in the Future), Peter Steen, Jonas Åkerman and others, Forskningsgruppen för miljöstrategiska studier, KFB-Rapport 1997:7

The picture shows Club Med's cruising ship, which combines engines with sails. There are several ongoing trials with new types of sailing ships that make it possible to reduce dependence of ships on fossil fuels.

Some of the aircraft of the future, which are still in the prototype stage.
Source: Illustration: Tomas Hamilton

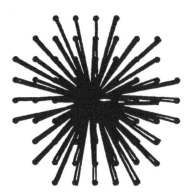
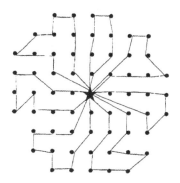

The picture on the left shows the pattern of travel around an average grocery store on a Saturday. A total of about 6000km is travelled. The picture on the right shows how two delivery services, using cars, can accomplish the same job spread evenly throughout the week. A total of about 300km is travelled. That amounts to about a 15t decrease in the amount of CO_2 released.

Source: Claës Breitholtz, 1997

A stairwell with boxes for delivery of goods ordered via internet.

shorter flight routes, optimal flight heights, and 'green' landing procedures. Propeller aeroplanes with slower speed (640–700km/h) instead of turbo jet aeroplanes (820–920km/h) can reduce fuel requirements by an extra 25 per cent. Aeroplanes using hydrogen were used in the Soviet Union and synthetic paraffin can be made from biomass. An aeroplane that runs on hydrogen gas could be a solution for the future.

For heavy and unwieldy loads, Zeppelins may experience a renaissance for transportation of both heavy loads and passengers, despite their slow speed. Another energy-saving technology is special low-altitude aircraft that fly about 6m above the surface of the ocean. A pressure wave occurs between the aircraft and the ocean surface that helps to hold up the aircraft.

Pipelines are an energy-efficient way of transporting liquid and gaseous materials in large quantities. There is a place for pipelines in a sustainable community.

Electronic Commerce

One way to reduce transport is to utilize information technology, for example, for shopping or holding meetings over the internet. Increasing numbers of people order food and other commodities over the internet. Since a person isn't always at home when a delivery is made, Electrolux, for example, is developing a box for home deliveries (UDT = unattended delivery unit). The box contains three separate compartments: a fridge and a freezer for food deliveries and a third compartment for other goods. The box holds 200L, which is the equivalent of five bags of provisions. A single-use code for the box is sent with every order placed. Once the delivery-person uses the code to open the box and put in the goods, the code can no longer be used.

4.2.2 The Holistic Town

Development trends have been towards an increasingly greater fragmentation: people live at one place, work at another, spend their spare time at a third and shop at a fourth. In other words, we've built up a society that's becoming increasingly dependent on transport, which represents a large proportion of energy consumption and environmental pollution. One way to reduce the transportation sector's negative environmental impact is to integrate the various functions and build 'the holistic city', not dormitory suburbs separate from industrial areas, shopping centres and recreational facilities.

The Town of the Future

From a historical perspective, society has developed as follows: in agrarian, pre-industrial society everything was located in or close to the home. During the rise of industrialism, people commuted to work and everything else was done at home. In industrial society, services were moved from the home to businesses. In the service society, people live in dormitory suburbs and all other functions have been moved out.

How should we shape the 'holistic town' of the future? How do we build up an urban structure that doesn't require movement of people from one place to another all the time? There are different ideas about this. The idea of nodes is about living close to recreational activities (which are either near or in the home). Everything else takes place at the network community's nodes. The other model is the IT community, where a large part of work, leisure and service takes place in the home: shopping is done over the internet and computers are connected to workplaces. However, work, recreation and services may also occur outside the home.

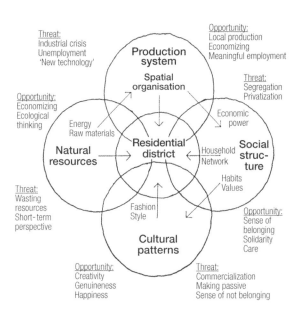

The 'holistic town' contains residential districts, production systems, social structures, cultural patterns and natural surroundings.

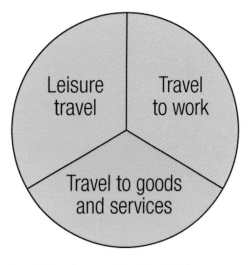

Travel in Sweden is roughly divided into three parts: 1/3 travel to work, 1/3 travel to obtain goods and services, and 1/3 leisure travel. There is a trend towards an increase in leisure travel.

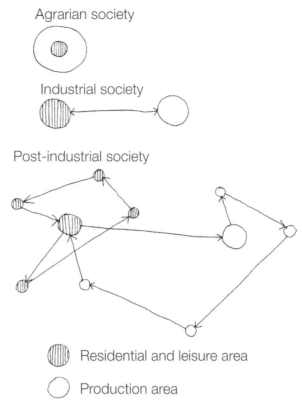

Travel has changed during the course of history: in agrarian society there was very little travel, in industrial society people commuted to work, while travel today follows more complicated patterns and also involves a lot of leisure travel.

Source: *Planering för uthållighet* (Planning for Sustainability), Lars Orrskog, 1993

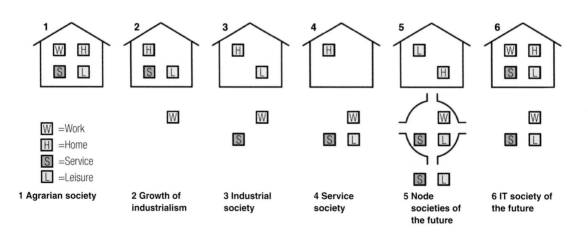

W = Work
H = Home
S = Service
L = Leisure

1 Agrarian society
2 Growth of industrialism
3 Industrial society
4 Service society
5 Node societies of the future
6 IT society of the future

Out-of-Town Shopping Centres

The establishment of superstores and outlets outside urban areas are typical examples of how towns are being fragmented. Consequently, city-centre and smaller shopping centres are driven out of business. The city becomes a somewhat shabby place with fewer people moving about, and therefore a slower pace. Additionally, problems arise for all those who don't have an car to access the services they want to use. In several countries, for example Denmark and Norway, the establishment of out-of-town shopping centres faces either prohibition or tough restrictions. Services should be integrated and accessible to all in a sustainable community. In 'The New Urbanism' new construction of out-of-town shopping centres is forbidden. These are placed instead on a town boulevard in the town itself, which offers a mixture of services. It is still easy for motorists to get to stores and superstores, while those without cars have access to the same selection of goods and services. This is called transit-oriented development (TOD).

Car Streets

Cars shouldn't be as dominant as they are today. Cars can either be excluded from residential areas and placed in car parks at the entrances, or be allowed access to roads designed to discourage high speeds with a risk of accidents. In The Netherlands, for example, in the eco-town district Ecolonia in Alpen-an-Rhein, cars are allowed to drive right up to the homes, but the landscaping is not a conventional street environment, but rather an area for pedestrians with lots of trees, benches, lighting and a lush ground cover. There are 'car-free,' 'car reduced' and 'traffic-calmed' areas. Here cars move totally on the pedestrians' terms. In garden cities in Germany and Austria the streets aren't broad, straight thoroughfares, but rather care has been taken with the width and profile of the streets so they meander, which makes travel through the area more interesting. Here and there are small squares where one must slow down to continue finding one's way.

The New Urbanism

The fragmented city leads to an increase in transportation work and time spent in cars. This has become especially clear in the US where the per capita petrol consumption is greater than in other parts of the world. A movement called 'The New Urbanism' has developed, which seeks to connect the broken city where you have to get on a motorway as soon as you need to get from one function to another. The New Urbanism has European cities as an ideal. An attempt

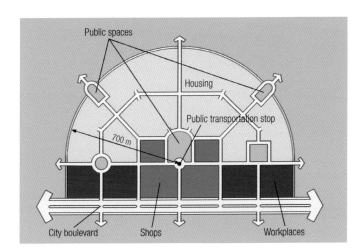

A TOD or transit-oriented development is a mixed town where the various services and the town centre are accessible on foot, by bicycle or by car. In the centre, shops and public transportation are located along a boulevard.

Source: *The Next American Metropolis – Ecology, Community, and the American Dream*, Peter Calthorpe, 1993

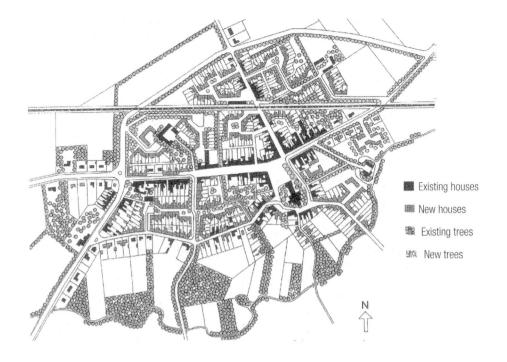

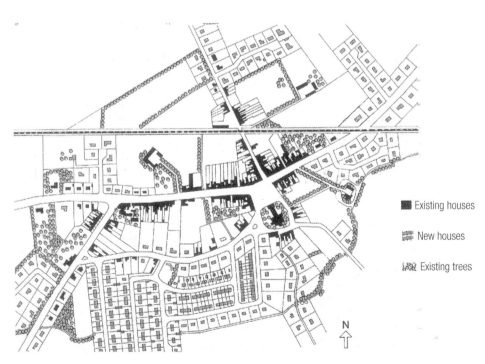

The top diagram shows examples of concentration in a small town, where the green structure has been developed. The bottom diagram shows what often happens in smaller communities, where there is an expansion of single-family homes on large plots (urban sprawl), which often destroys the small-town feeling of intimacy.

Source: *Duurzaam en Gezond Bousen en Wonen – volgens de bio – ecolgische principes*, Hugo Vanderstadt, 1996

is made to create comprehensive town districts, recreate public areas and remove the gaps between different functions.

Well-functioning town districts with a distinct town centre is a goal. It should be possible to walk to the outskirts of the area in 10 minutes. Town districts should be integrated and include housing for various income groups, workplaces, shops, schools and playgrounds within walking distance for children. Streets should form a grid, cul-de-sacs should be avoided and driving everywhere and in both directions should be possible. But the streets should be narrow to ensure slow driving and should provide space for pedestrians and cyclists. Street areas should be designed as boulevards with trees. Buildings in the town centre should be built close together to create a town feeling and the best sites should be reserved for cultural activities. Parts of this urban development ideology are applied in Sweden, for example in Lomma harbour, a new town district near the Öresund shore.

There are two structural dimensions in suburbs that can create problems, the differentiation of traffic and the separation of functions. Cul-de-sacs are socially isolating since they lack destination points for outsiders and complicate the siting of workplaces.

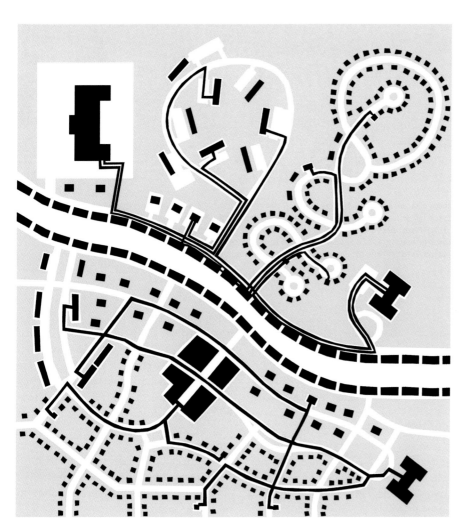

The top part of the diagram shows the traffic patterns in the fragmented city, where even short trips involve driving on the motorway with all its queues. The lower part of the diagram shows an amalgamated town where short trips are taken within the local neighbourhood.

Source: *The Charter of the New Urbanism*, essays by Randall Arendt and others, 2000

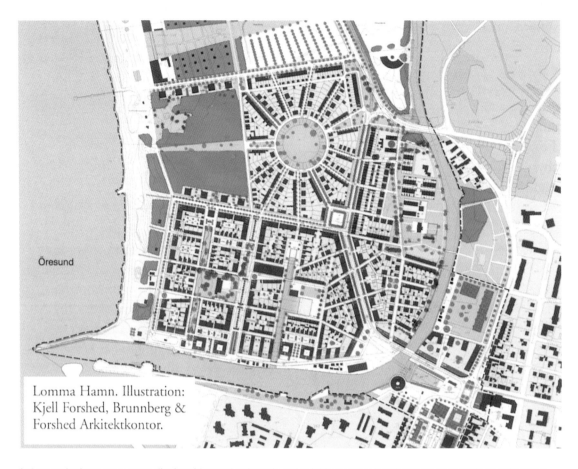

In Lomma harbour, a new town district with town boulevards and neighbourhoods is being planned that will be built over a 15-year period. The floor-space index is e = 0.5 (e = the total floor area divided by the land area) and most of the buildings will have two or three storeys. The keywords are low, compact, urban, rich and robust. The structure of the neighbourhoods creates the conditions needed for a mixture of housing types and forms of tenure, as well as making gradual expansion possible. The relationship between the public street environment and private property is clear.

Source: Brunnberg-Forshed Architect's Office

Cul-de-sacs also inhibit efficient public transport services. A mix of functions and the integration of traffic are prerequisites for social services and commerce, as well as for the sense of security created by busy streets.

Moderately Sized Towns

The tendency in development is towards enormous metropolises – so called mega-cities, which place huge demands on resources. Achieving a society that is balanced both within and in relation to its surroundings may be more easily attainable in appropriately sized ecological towns. Construction of energy-efficient housing and use of renewable energy sources have reached a high standard. A major problem is traffic, which is steadily increasing. There is less everyday travel in medium-sized towns than in metropolises and rural areas. In a resource-conserving society medium-sized towns should be aimed at, instead of expanding metropolises or building in unspoiled rural areas.

ECOLOGICAL CITY DISTRICTS IN EUROPE

In order to achieve sustainable development it is not enough to build sustainable buildings. These buildings must be incorporated into sustainable cities, and this is taking place in some parts of Europe.

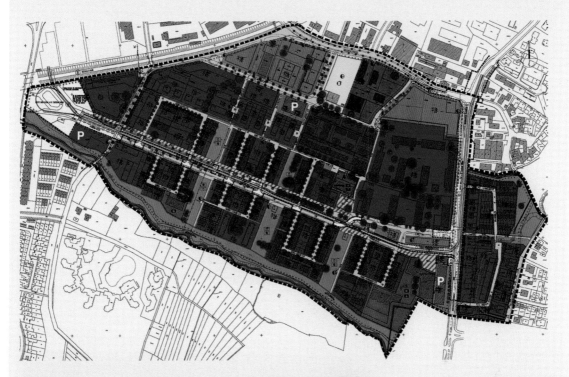

Vauban area in Freiburg is one of the most interesting new townships in Germany. A tram connects the area with the centre of the city. Cars are restricted in the housing areas: cars may be driven slowly up to a house, but must be parked in garages. The streets are used more as play areas by children than as car routes. The houses are energy efficient and of passive design. Many buildings have solar collectors and solar cells on the walls and roof. In spite of the high density in the area there are many beautiful green areas and gardens between the buildings.

Source: Stadt Freiburg im Breisgau

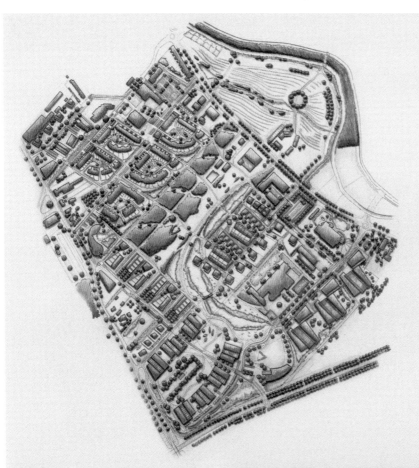

Culemburg in The Netherlands has been extended into a former water protection area. The houses are passive with solar collectors and solar cells. Car use in the area is restricted by manual gates. Water planning is divided into rainwater, traffic water, grey water and black water. There is an orchard and city farm in the area. The black water and organic waste from the farm are processed in a biogas plant; the grey water is processed in artificial wetlands. The area is divided into blocks where the inhabitants plan and tend a shared garden.

Source: Stichting E.V.A., Lanxmeer, Culemburg

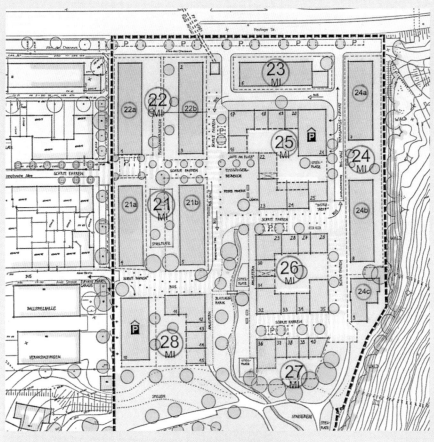

Französiches Viertel [the French quarter] in Tübingen in Germany is an extension of the town into an area that earlier was used by the French army. This concept is to make the town larger without creating suburbs. The principles behind this are mixed use and short distances. Mixed use means housing, workplaces and shops in every building. Short distances means pedestrian and bicycle connections to the centre, good public transport and restricted car use. An interesting aspect is that instead of selling to developers, land is sold to groups of people to use the buildings for purposes that fit the local city plan. The units in the plan are small and flexible so that they can be adjusted to the wishes of the building groups.

Source: Entwicklungsbereich Südstadt, Tübingen

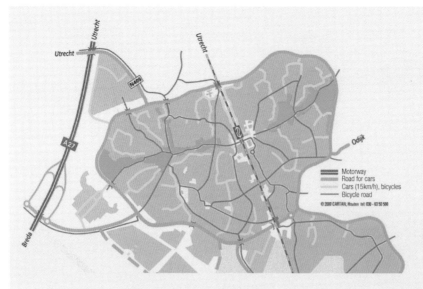

Houten near Utrecht in The Netherlands with 50,000 inhabitants is called the bicycle town. It has a unique traffic system where bicycles are the main mode of transportation. In the centre of the city there is a railway station connecting Houten with Utrecht. At the railway station there is a garage for 2000 bicycles. The traffic system is set up in such a way that cars and bicycles are separated from each other. The main roads in the city are for bicycles only, and pedestrians use the pavements. The city is surrounded by a ring road for vehicles, which can only access the outer parts of the city from the periphery.

Source: The Houten Council

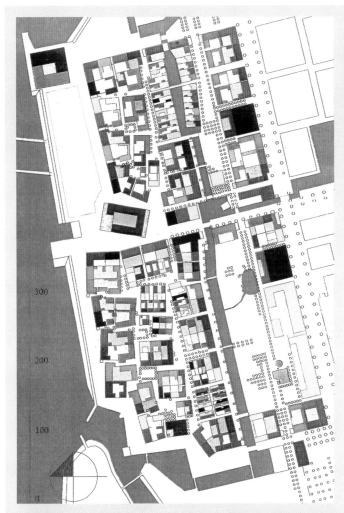

BO01 in Malmö, Sweden, is a new part of the city by the sea replacing an old shipyard. This project started as a building exhibition in 2001. Energy comes from renewable energy sources. The buildings, streets and water management in the area are artistically integrated. The most interesting aspect of BO01 is the plan, where diversity and human scale are the main themes. The architect Klas Tham got inspiration for the town plan from Roman and medieval cities.

Source: Plan Architect, Klas Tham; photo: Ronny Bergström, Malmö City Planning Administration

Hammarby Sjöstad in Stockholm, where building started in the beginning of 2000. This new development replaced an old industrial area by a lake. The aim was to half the need for water, electricity and heat compared to conventional buildings. One of the most interesting achievements is the waste-water planning. There are different systems for rainwater, traffic water and sewage water. Special care has been taken to keep the organic waste so clean that it can be made into soil for cultivation. Biogas for car fuel is produced in the cleaning process. Another area in Stockholm with an environmental image is the Norra Djurgårdsstaden district.

Source: Stockholm City Planning Administration

GARDEN CITIES

The garden city idea was first introduced in 1898 in the book *Garden Cities of Tomorrow* by Englishman Ebenezer Howard. It had to do with creating healthier living conditions for people living in crowded conditions in towns built of stone. Raymond Unwin was the English architect who put into practice the garden city idea. He wrote a book called *Town Planning in Practice*. Construction of the first garden city, Letchworth, 50km north of London, began in 1903. In Germany, people became interested in the garden city idea very early on, and the first garden cities built there are named Hellerau (1908) and Staaken (1914–1917), both in Berlin. The first garden city development in Sweden was Gamla Enskede in Stockholm. The Swedish garden city is characterized by the inclusion of several types of dwellings, such as detached single-family houses, semi-detached houses, terraced houses and small blocks of flats. The net floor space index e (= total floor area divided by the ground area) is between 0.15 and 0.4. When new city structures are built with low-rise, compact development, such as in Lomma harbour, there is an attempt to keep the floor space index under 0.5.

According to professor Johan Rådberg, Lund Institute of Technology (LTH), the garden city can serve as a model for ecological development. What characterizes the garden city are its structural attributes, which are predetermined in a city plan. The most important criteria are moderate density, gardens for all the buildings, a three-storey height restriction and a traditional street grid.

There are several reasons for the three-storey height restriction. One reason is to be able to establish gardens even for multi-family dwellings. In a three-storey block of flats, it's still possible for those living on the top floor to keep in touch with those living on the ground floor.

Garden cities are built up using a traditional street grid. All the buildings, in principle, are built along a street. In this way, houses have different sides, a public side facing the street and a private side facing the garden. Another component of the structure of garden cities is that there should also be gathering places or squares for shops and public meeting places. It's these small places that create much of the character and atmosphere found in garden cities.

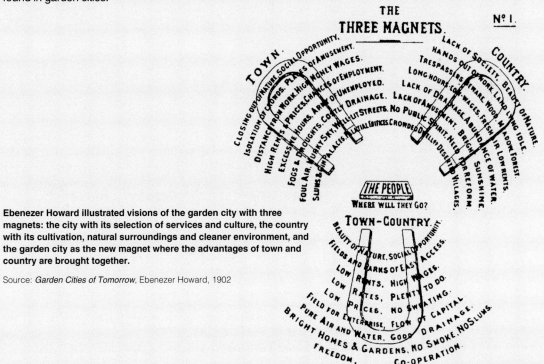

Ebenezer Howard illustrated visions of the garden city with three magnets: the city with its selection of services and culture, the country with its cultivation, natural surroundings and cleaner environment, and the garden city as the new magnet where the advantages of town and country are brought together.

Source: *Garden Cities of Tomorrow*, Ebenezer Howard, 1902

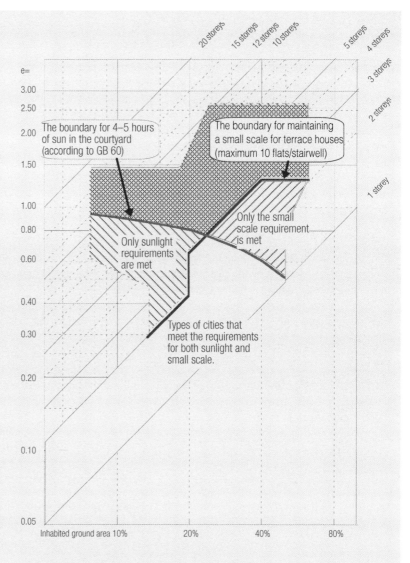

Diagram by Johan Rådberg, architect and professor, Lund Institute of Technology (LTH)

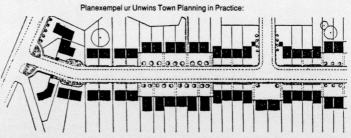

City planners were very inventive when it came to street layout in garden cities. It was a matter of placing the buildings in relation to the street and different types of hedges and fences to 'emphasize the effect' of the garden city, how a street perspective could be created at intersections, how to lay out open spaces and squares, and how these could be used to slow down cars.

Source: *Town Planning in Practice*, Raymond Unwin, 1909

4.2.3 Town–Country

Town planning has traditionally only been concerned with planning urban areas while rural areas have had to manage mostly on their own. In an ecologically adapted society, both urban and rural areas must be planned, as well as the connections between them.

Interaction between Town and Country

The town can't survive without the surrounding countryside, unless there are unreasonably long transportation distances. The old way of seeing cities as red dots on a green map doesn't apply anymore. In a sustainable society, rural areas are seen more as a structure of networks with more or less dense population centres and areas of countryside. The town needs food and biomass. Rural agricultural activities close by can supply the town with these necessities. The town produces biodegradable refuse and nutrient-rich waste water, nutrients required by rural agriculture. People in the town need to be close to green areas and natural surroundings. People in the country need to be close to social and cultural activities.

Proximity of Production and Consumption

International commerce which generates increasingly more transport is inconsistent with sustainable development. We have to realize that certain commodities should be produced locally and others in an international arena. Local production of food is the most obvious. A comparison of the life-cycle analyses of various meals shows that locally produced foodstuffs have a significantly reduced environmental impact. If nutrients are to be returned to agriculture, then proximity is an advantage. Additionally, as many foodstuffs are perishable, shorter transport often means higher quality. So locally grown apples should not be allowed to rot while apples are imported from New Zealand. Even production of biomass should take place locally. This is an indication that in the quest to achieve sustainable development, agricultural activities should be distinguished from other commercial pursuits, and general economic rules should be applied to them.

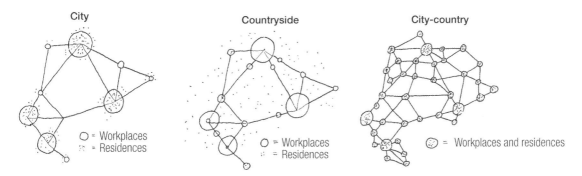

Decentralized concentration according to the urban–rural concept (top diagram)

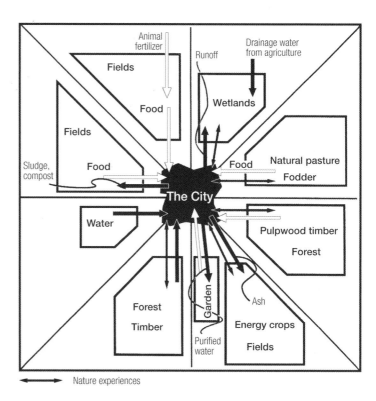

The new ecocycle arenas. Example of the kinds of spaces that surround an urban area and that will be needed in a sustainable society.

Source: *Sverige 2009 – förslag till vision* (Sweden 2009 – proposals for a vision), Boverket Rapport 1994:14

It is also preferable to produce heavy construction materials that are used in large volumes within the country in order to avoid unnecessary transport and environmental impacts. It's becoming increasingly common to import instead of buying locally produced building materials. Yet another commercial sector that benefits from local connections are recyclers, e.g. of building materials.

A Sustainable Municipality

A town can never be sustainable if it doesn't interact with the surrounding countryside. Planning for a sustainable future has to deal simultaneously with urban and rural contexts. The municipality might be the right planning area for this purpose. However, the boundary of a municipality is not always appropriate from the perspective of sustainable planning. In any case, planning within municipal boundaries requires cooperation with surrounding municipalities.

Land Use

Arable land should be protected and used mainly for food production. Food production must be ecological, which mainly means avoiding chemical pesticides and the use of artificial fertilizers. Rotation of crops, fodder, grazing land and fallow should be used to enhance soil fertility, soil structure and humus amount. Another important aspect is to maintain a balance between production of cattle and fodder to avoid leakage of nutrients into the water recipients. Parts of the arable land can also be used for growing energy crops. Orchards, commercial gardens, garden cities and edible parks connect the city to the countryside.

Forests are important resources that can provide timber, paper and bioenergy. Forestry practices should be carried out in an

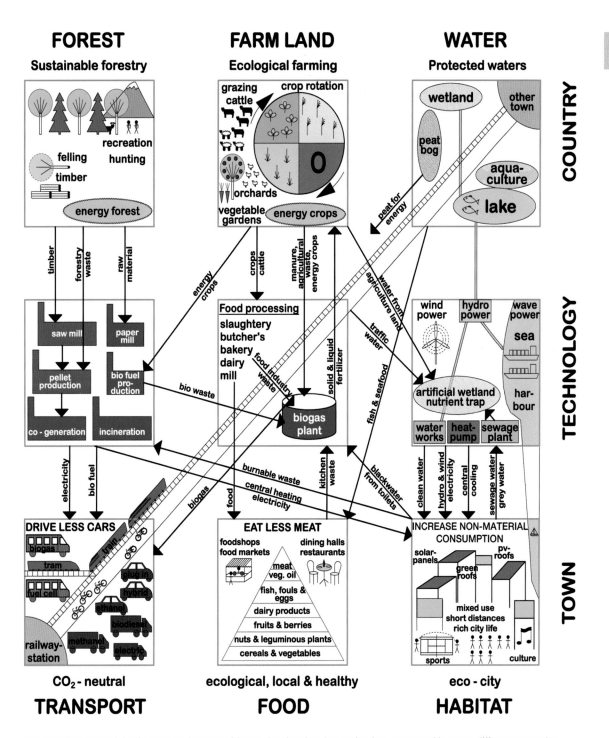

In a sustainable municipality town and countryside are closely related to each other, connected by many different ecocycles to provide society's basic needs. The aim is not to be self-sufficient. The aim is to use local resources more efficiently, to avoid leakage of nutrients, to minimize climate change, and to create a municipality where local culture, local production and local economy are stimulated.

ecological manner with mixed species and felling by thinning. Biodiversity and beautiful parts should be protected. Forests can also supply mushrooms, berries and the possibility to hunt wild animals. The importance of forest areas and other natural areas for recreation must not be neglected. Trees in steep mountain areas and along riverbanks are important to avoid soil erosion.

Water and wetlands are essential for human survival and must be protected so they can be used for producing drinking water. Wetlands act as natural nutrient traps and could improve the function of sewage works. Water planning has to be done in a much more diversified way than usual, which means that plans have to be made for the protection of groundwater, surface water and rainwater. Careful studies must be done on how to produce drinking water and where to situate water works. Aquaculture for producing food and energy crops could be part of the system. Possibilities to use hydropower and wave power should be considered. Wind power can be utilized in open land, on seashores and on shallow waters where strong winds are blowing.

Cities

Cities should be well integrated with the surrounding region and be in harmony with nature. Valuable natural and productive land should be protected. Development should be concentrated in suitable places and a balance between density and decentralization should be aimed at. The urban structure should be integrated with the traffic infrastructure where the goal is to minimize transport. This can be done by planning a mixed-use city with short distances to most functions. The city should also be planned in a bioclimatic way to improve the local microclimate. Green areas and water in the city are important for air quality and recreation, and as part of the bioclimatic design. Bioclimatic corridors, boulevards and bioclimatic parks should penetrate the city structure.

Basic Needs

Transport systems should be diversified into pedestrian and bicycle routes, public transport and trains as well as car and lorry systems. The emphasis has to be on pedestrians, bicycles and public transport systems so that cars don't dominate the city. The transport system should be planned for environmental vehicles (energy-efficient vehicles using hybrid technologies) and CO_2-neutral fuels.

Food should be produced locally using ecological farming methods. Food quality and taste should be the main target. One of the big problems in the Western world is that people are getting fatter. Much industrially produced food contains chemical pesticides and the content of nutrients and vitamins are lower than in ecologically produced food. One consequence of this is that meat consumption should be reduced, as meat production requires a large input of resources.

Habitat should be adapted to the local climate. Strong energy regulations should be applied. Houses can be built that don't need heating systems (so called passive houses) or houses where the heating needs are very small. The heating and hot water needs must be supplied with renewable energy resources. Electricity- and water-saving devices should be used as well as garbage sorting and recycling. A good indoor climate is aimed for, which means the use of healthy materials and well-functioning ventilation systems.

City life where human scale and urbanity is combined should be the goal for urban planning. A city should be a safe, healthy and beautiful place where people can live, work, receive services and take part in recreation. It should be a city for everyone, a city with many functions and diversity that stimulate city life. A city should be a place where cultural and social identity are preserved and developed. A democratic city has a planning system where its inhabitants influence planning and management. A sustainable lifestyle means more non-material consumption instead of exaggerated material consumption.

Technology

Fuels for the transport sector should be made from renewable energy sources. Technologies that can be used are biogas produced from organic waste, ethanol from energy crops, methanol made from raw material from forests and bio-diesel from organic materials. Future cars and lorries should be made energy efficient and use plug-in hybrid technologies. Electric vehicles can be used for public transport, and for short-distance transport. Fuel cell technologies and hydrogen are other possibilities.

Energy for the city should come from renewable energy sources and the houses in the city should be mainly be heated or cooled by central systems. Electricity and heat or cooling should be produced in co-generation plants. The cogeneration plants should be fuelled with burnable waste from forestry, agriculture, industry and households. Sawmills that produce building material and factories making pellets from wood waste are important local industries. Heat pump technologies could be utilized as parts of the energy system. Photovoltaic and solar collectors should cover south facing roofs and façades. With these kinds of technologies cities can produce their own electricity.

Food production should be ecological and local with the appropriate infrastructure. Among the local, ecologically run facilities required are slaughter houses, butchers, dairies, flour mills, bakeries and breweries. Also needed is a distribution system so that small, local producers can distribute their products, e.g. farmers' markets, health food shops, or direct distribution from local farmers to customers. Political strategies should be enhanced to stimulate purchase of local products by schools, hospitals and other public institutions. Cooperation between restaurants and local farmers is one way to popularize good and healthy food.

Waste is a resource. It can be divided in two main parts: organic waste and non-organic waste. Non-organic waste should be minimized by design, sorted in different

In the town of Enköping, Sweden, there is a good example of how different technologies cooperate, linking town and country. The cogeneration plant that supplies heat and electricity to the town is fuelled with chips from an energy plantation. The energy plantation is irrigated with sewage water. The sewage plant is heated with return water from the district heating system. In the future the system could be complemented with a biogas plant. The photograph is taken from the roof of the cogeneration plant and shows the sewage plant and energy plantation.

Source: Photo: ENA Energi, Enköping

fractions, reused and recycled. Toxic waste must be handled with care. Organic waste has an important role in recycling nutrients and in maintaining soil fertility. It should be kept clean from pollutants, treated in biogas plants or composters and distributed back to arable land. Maintaining high humus content in arable land is important for ecological farming. Burnable waste can be used in incineration plants for cogeneration. Paper can be recycled, i.e. as insulating material (cellulose fibre).

Sewage should be sorted into different fractions such as black water (toilet water), grey water (from showers, washing and dish washing) and traffic water (from parking lots and streets). Each fraction should be cleaned in an appropriate way, and the black water (rich in nutrients) can be treated in biogas plants.

Summary

In a sustainable region, towns and the countryside are woven together via ecological cycles. It is not a question of self-sufficiency but rather using local resources so that a large portion of basic needs are produced locally. Efforts are directed towards reducing transportation, using resources in an environmentally friendly manner and reducing loss of nutrients by using closed cycles. The goals are to minimize climate change and environmental impacts and to create a society where local culture, production and economic activity is stimulated. A sustainable society must be planned for. It will not occur on its own.

Flows in Sustainable Societies

When designing a sustainable society, plans are made not only for development and roads but also for climatic adaptation, green structures, water flows, waste-water flows, waste disposal, energy supply, transportation and for the city's social and cultural life. Each individual plan deals with its own flows.

Planning often begins with climatic adaptation and uses the local microclimate as a starting point, e.g. avoiding windy places, shady sites, damp areas and cold sinks. Outdoor areas, balconies and flats are placed to best capture the sun and be sheltered from the wind. Naturally, climatic adaptation is totally different in each climate zone.

A landscape plan includes plans for a variety of green structures. The work involves both private and public land. Planning includes consideration of vegetation in the city, e.g. in the parks, street environments, and squares, as well as the vegetation surrounding and on buildings, and the vegetation surrounding the city, i.e. agriculture and forestry, gardens and uncultivated areas. Biological diversity and local production of food and energy are the goals.

A water plan ensures that the various sources of water such as groundwater, surface water, rainwater and clean water are kept pure and used wisely. The waste-water plan provides methods for sorting and purifying the various waste-water flows: runoff, water from roads, domestic waste water and sewage. The plan also provides for the return of nutrients and purified water to the ecocycle.

A waste plan details which fractions will be sorted, how to take care of organic waste, what should be recycled, how to collect hazardous waste, and how to handle the remaining waste (which can be divided into combustible waste and landfill waste). The plan includes care of the organic waste (sludge, compost, leachate and biogas) and care of the remaining waste (where and how combustible waste will be incinerated and where unusable waste will be disposed of). Systems for handling hazardous wastes so that they pose no threat must be included in the plan.

An energy plan covers local production of renewable energy (e.g. energy plantations, solar collectors and solar cells, wind and water power) and distribution of electricity, district heating, district cooling and fuel.

It's difficult to fit biogas into the above-mentioned structure with its various flows, since biogas production, use and disposal are bound up together. Energy crops for anaerobic digestion are produced in the green structure, sewage sludge is recovered from the water flow, the organic waste is found in the waste flow. The final products end up in different flows, for example, gases in the energy flow and solid and liquid fertilizer in the green structure. Biogas is seen by many as having the greatest potential for managing and organizing different flows in a sustainable society.

A traffic plan deals with passenger traffic (pedestrian, bicycle, public transportation and car traffic) and cargo traffic (road freight, rail, boats, planes and pipelines). There are various hierarchies within the traffic structure, for example, high-speed trains, local trains, commuter trains and trams. So design of the junctions where the different levels in the hierarchy meet is also important.

A development plan is concerned with locating the different functions (housing, work, service and recreation) in close proximity to one another. Housing areas should be made so that segregation is discouraged. Workplaces should be located in residential areas and be easily accessible via the traffic structure. Shopping centres shouldn't be placed outside the town so that it is necessary to drive an car on the motorway in order to reach them. Schools, day-care centres, and recreational activities are placed within walking and cycling distance so that children don't always have to be driven. The city centre should be planned for vitality and activity, and include cultural institutions, cafés, squares and other beautiful public areas. Existing development should, if possible, be preserved.

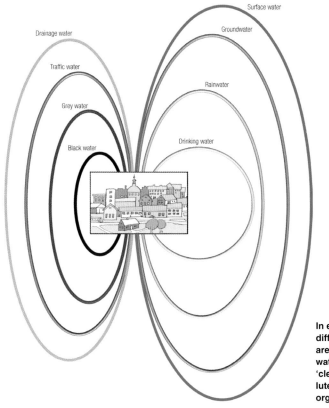

In each plan different flows are studied. In a water plan both 'clean' and 'polluted' water is organized.

Large synergy effects can be achieved by connecting together different ecological technologies where the waste from one progress is the raw material for the next progress. The illustration shows the possibilities by connecting together the ideas behind the energy systems at Händelsö in Norrköping and in Enköping, Sweden.

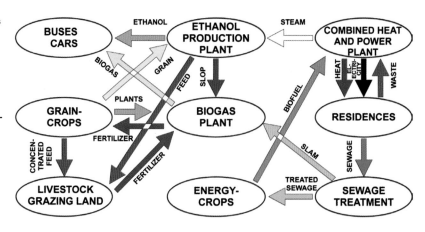

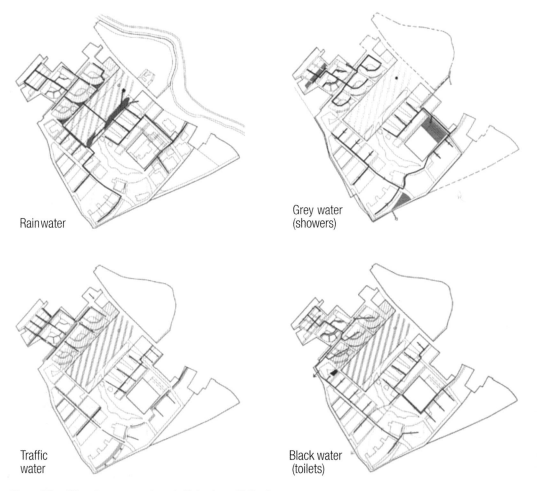

Rainwater

Grey water (showers)

Traffic water

Black water (toilets)

Plans of the different sewage systems in Culemburg, Holland.
Source: Joachim Eble Architecture

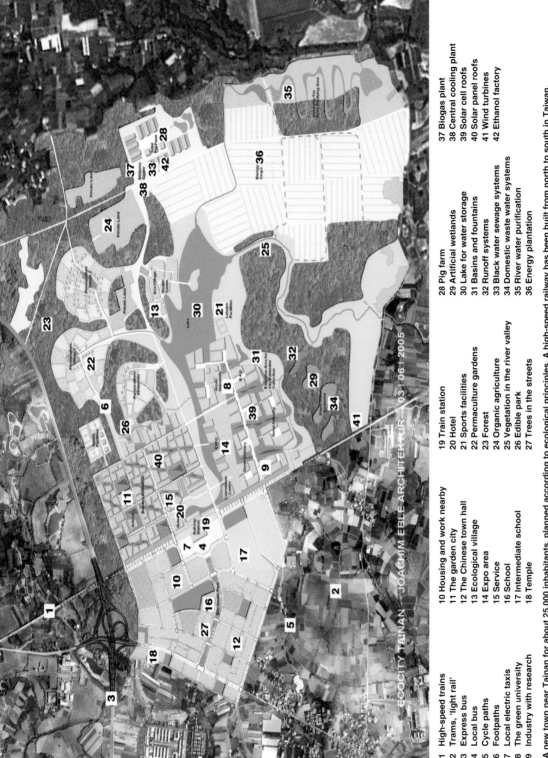

ECOCITY TAINAN – JOACHIM EBLE ARCHITEKTUR – 03.06.2005

1 High-speed trains
2 Trams, 'light rail'
3 Express bus
4 Local bus
5 Cycle paths
6 Footpaths
7 Local electric taxis
8 The green university
9 Industry with research

10 Housing and work nearby
11 The garden city
12 The Chinese town hall
13 Ecological village
14 Expo area
15 Service
16 School
17 Intermediate school
18 Temple

19 Train station
20 Hotel
21 Sports facilities
22 Permaculture gardens
23 Forest
24 Organic agriculture
25 Vegetation in the river valley
26 Edible park
27 Trees in the streets

28 Pig farm
29 Artificial wetlands
30 Lake for water storage
31 Basins and fountains
32 Runoff systems
33 Black water sewage systems
34 Domestic waste water systems
35 River water purification
36 Energy plantation

37 Biogas plant
38 Central cooling plant
39 Solar cell roofs
40 Solar panel roofs
41 Wind turbines
42 Ethanol factory

A new town near Tainan for about 25,000 inhabitants, planned according to ecological principles. A high-speed railway has been built from north to south in Taiwan. New towns will be built at the train stops, since the railway doesn't pass through the existing towns. The plan was made by GAIA International.

Source: Architects: Varis Bokalders (Sweden), Chris Butters (Norway), Joachim Eble (Germany) and Rolf Messersmith (Germany); together with EDS-Design, a Taiwanese group of planners; and Archilife Foundation, an interdisciplinary group of Taiwanese researchers.

PLANNING SUSTAINABLE TOWNS

The sustainable town is a town:
- in harmony with nature, people and the future

The town and the surrounding countryside
- in balance with nature
- where valuable nature and productive grounds are protected
- that's well integrated into the region
- where development is centralized to suitable places

The town's resources and flows
- that uses its resources economically
- adapted to its climate
- that produces renewable energy
- where the watercourses are protected and used
- that sorts and purifies its sewage
- where waste is sorted and recycled
- that is green and characterized by biological diversity
- with a strong local economy

The town and dependency on cars
- where everything is nearby
- in balance between dense settlement and decentralization
- for pedestrians, cyclists and users of public transport
- that's connected to international networks

Quality criteria for the town's buildings
- where the buildings are in the spirit of the town plan
- that's climate adapted
- with economical use of resources
- that's connected to the ecocycle
- with healthy houses

The town and people
- for everyone
- that's complete and characterized by diversity
- where town life is stimulated
- where human scale and urbanism are brought together
- that is safe, comfortable and promotes health
- for a sustainable lifestyle
- built and administered by the inhabitants
- with a cultural identity and social variety

The town and the future
- that contributes to sustainable development and gives hope for the future.

Illustration of a new town near Tainen in Taiwan (see previous page)

Ostia, near Rome

In the ecological town plan for Ostia, near Rome, by Massimo Bastiani, Vlerio Calderaro and Joakim Eble, the natural conditions at the site were used as a starting point. A hydrogeological analysis, as well as analyses of the vegetation and cultivation, of the microclimate and of the area's history and urbanism were used.

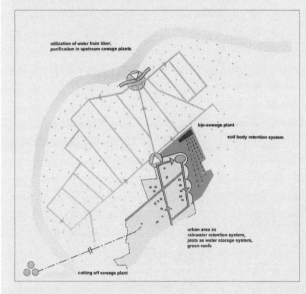

A settlement where there is rainwater management using ponds as reservoirs. Irrigation canals are laid in cultivation areas and the ground is used to retain precipitation. Pumping stations and biological purification plants are nearby.

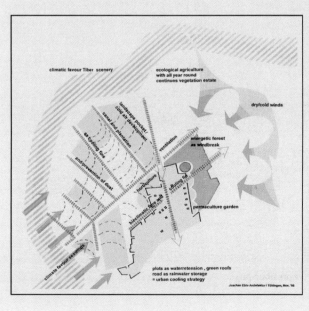

Climatic adaptation. Shelter belts protect against dry and cold winds. Open corridors for favourable sea winds. Vegetation and canals to capture dust and create coolness through evaporation. Courtyard formation creates good microclimates.

Ecocycle for organic materials. Agriculture, gardens and permaculture cultivation for production. A market for processing and selling food. Compost plant and distribution of organic material. Creation of an edge zone between the settlement and cultivated land for interaction between town and countryside.

 agriculture workshop, developing of existing form to ecological orientated plant

 biomarket as part of agricultural workshop to sell ecological and healthy products

 eco-station : energy production and consultation (workshops, school, exhibition, information)

agri-urban interaction :, sensual & social experience, exchange of demand & supply, communication

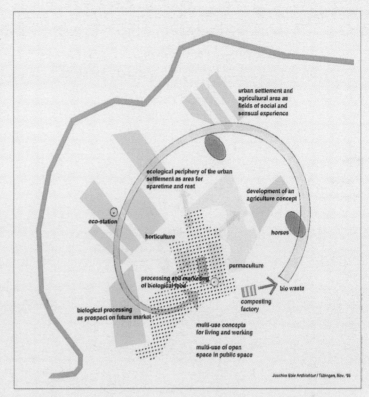

Energy plantations in the form of energy crops, energy forest and material from trees along the canals, that supply a central combined power and heating plant which supplies the neighbouring area with heat and electricity. The biogas plant is supplied with manure from horse farms and biological waste. Solar collectors and heat storage. The biogas and solar heat are connected to the district-heating system.

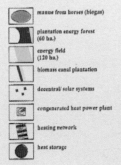

manure from horses (biogas)

plantation energy forest (60 ha.)

energy field (120 ha.)

biomass canal plantation

decentral/ solar systems

congenerated heat power plant

heating network

heat storage

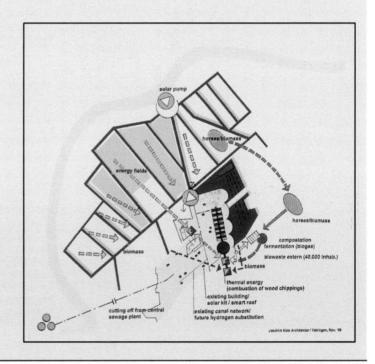

PROPOSAL FOR THE EXPANSION OF PALMA DE MALLORCA

4.2

The architect Richard Rogers was in charge of the 'Solar Village' proposal, an expansion of Palma de Mallorca that can serve as a model for planning an ecocommunity.

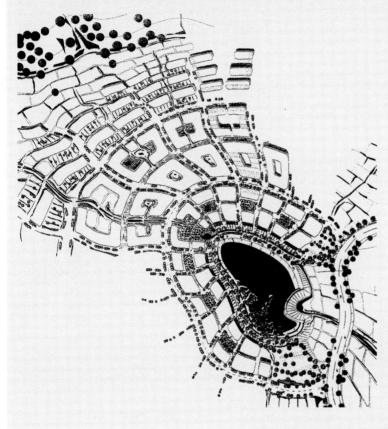

Topography and water routes have been allowed to influence the city plan. Vegetation and development are interwoven.

Source: *Sustainable Architecture, Principles, Paradigms and Case Studies*, James Steele, 1997

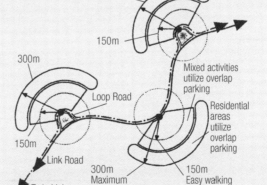

The diagram shows the transportation system and the comfortable walking distances.

Source: *Sustainable Architecture, Principles, Paradigms and Case Studies*, James Steele, 1997

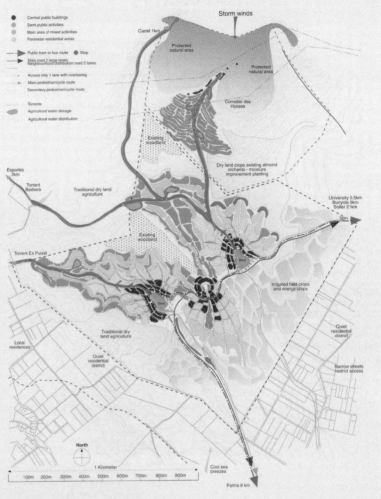

The urban pattern has been shaped by the characteristics of the location and by the objective of integrating the different elements with each other. The systems for water, cultivation, traffic movement, social life and energy work together. Energy plantations are included as a part of the plan proposal.
The energy supply in each of the three town districts is generated using solar energy, biomass and refuse incineration. The topography and the movement of water in the drainage basins have been a starting point for the planning work. All water is made use of and is used in cultivation and in settlements. The outflow is purified and is considered a resource.

Source: *Solar Energy in Architecture and Urban Planning*, editor Thomas Herzog, 1996

A photo of a model of the 'topographical' town plan. Three town districts were the result of the topographical studies of the area. Easy walking distance to the town centre was the criterion used to determine the size of each section. In the town centre, activity is the objective and housing is mixed with services and various enterprises. Development farther away from the centre is dominated by calmer residential areas that are integrated with vegetation and cultivation.

Source: *Solar Energy in Architecture and Urban Planning*, editor Thomas Herzog, 1996

4.2.4 Cultural Values

It's difficult to find a place where the landscape hasn't been affected by human beings. Areas with a rich cultural heritage influence us and our aesthetic values. Therefore it should be just as important to make use of, preserve and develop the aesthetic and cultural and historical values that human beings have created, as it is to utilize, preserve and develop natural values. A cultural adaptation involves trying to preserve existing cultural values, trying to develop and integrate them into the society of the future, as well as seeking inspiration and models in the local culture and specific types.

Building Traditions

In building ecologically with the intention of adapting the construction to the climate and utilizing local materials, local building traditions can serve as an inspiration.

Traditional Types of Buildings

In Sweden, building types have varied from region to region depending on, for example, local materials and building techniques. There are half-timbered houses in Skåne, log houses in Dalarna, houses with exterior boarding and decorated with gingerbread work in Medelpad, stone houses in Gotland and post and plank timber houses in Öland. There are typical structural components such as thatched roofs in Skåne, slate roofs in Dalsland and iron chimneys in Västmanland. There are different types of houses like the *Blekingestuga*, which is a high-loft house with a lower part ('*stuga*') in the middle, long-loft houses like the *skånelänga* with buildings surrounding a courtyard, the *hallandslänga*, farms with many log buildings in Dalarna, large Hälsinge farmhouses joined to equally large outhouses, and the Tornedal house, a *pörtebyggnad*, where everyone lives in the largest room (*pörtet*) during the winter.

The Ornäs house, a rich man's house from the 1500s situated between Borlänge and Falun.

4.2

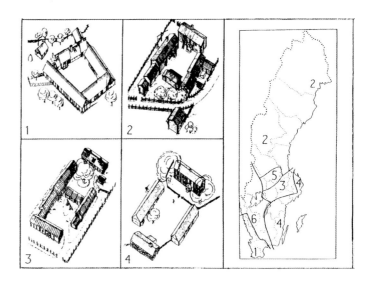

Division of Sweden by building tradition. Four types of farmyards can be discerned:

1. The southern Swedish farmyard in Skåne and Halland.
2. The northern Swedish farmyard that is found in Dalarna and farther north.
3. The central Swedish farmyard in Uppland and the Mälar and Hjälmar regions.
4. The Geatish farmyard in Blekinge and along the east coast.
5. Mixed types of farmyards.
6. Irregular and loose farmyard formation.

Source: *Geografiska notiser* (Geographical Notices), Erixon, 1949

New Old-Style Houses

In several regions, county museums have asked architects to design buildings that blend in well with the older buildings. In such cases, architects have used the local vernacular building types as a starting point and redesigned them into modern residences.

Traditional Layouts

Many of the old types of houses developed according to the length and characteristics of the timber and from a need to conserve resources. Traditional layouts include the single house (two rooms), three-part or pair house, long-loft houses, cross-formed houses with a four-room layout, and the traditional six-room manor house layout. All of these types can be found in one-storey, one-and-a-half-storey, and in two-storey variations. A chimney was expensive to build and was required to service as many rooms as possible. The two-room layout has a chimney in the middle of a long, narrow house. The four-room layout has a chimney that services four fireplaces. Larger houses for the aristocracy and clergymen have two chimneys in a six-part layout.

Village Configurations

Characteristic old village configurations are still found all over Sweden. The row-villages in Öland, with their breezeways toward the street and inner courtyards surrounded by buildings with the dwellings farthest back, are one example. Other examples include: church villages in Uppland where the dwellings lie along roads leading to the church (often from the Middle Ages); the coastal communities in Bohuslän where houses sit close together to protect against the strong winds; fishing villages on the east coast

The Swedish building tradition has been affected by the local climate, access to materials and by cultural characteristics. The picture shows a south Geatish building from the 1700s, Kyrkhult in Blekinge, which has been moved to Skansen, an open air museum in Stockholm. This type of building is characteristic of the south-eastern part of Sweden.

In Gotland, it was common to build stone houses, even in the countryside.

An old Närke house.

Source: A survey by Jerk Alton, architect

A wing of the Von der Eckerska farmhouse in Värmland, painted with the typical Swedish distemper paint (Falun Red paint), and with two iron chimneys.

Tornedal houses (here, Kommes, Niskanpää, Övertorneå), in the Swedish far north, have a porch for a weather lock and during the cold season, only the (*pörtet*) largest room is heated.

1 Single house
2 Three part or pair house
3 Long-loft house/southern Swedish long low house
4 Cross-formed house
5 Six-room manor house

Source: *Landskapshus-svensk byggtradition*, (Provincial Farmhouses – Swedish Building Tradition, Karin Ohlsson-Leijon, Laila Reppen, 2001

A house adapted to an existing building culture, built 1990–1991, in Sydkoster in Bohuslän. The solar collectors on the roof testify to its modernity.

Source: Architect Hans Arén

where every cultivable spot is valuable and therefore the buildings are built on cliffs, with the fishing sheds out in the water along the shoreline; not to mention the villages in Dalarna, with their harmonious and old-world character, with an open square and a Midsummer pole (maypole) in the middle of the village. Land division reforms, especially the Enskifte reform (collected land for each farm in one area) and the Laga reform in the beginning of the 1800s, involved great changes, above all through the dissolution of villages. The places where villages were not subject to the land reforms are today very valuable environments.

Town Character

In the Nordic countries there are many splendid small towns with a pronounced small-town character with low buildings, gardens with railings, gates and fences, narrow streets, squares and alleys. Even larger towns may possess a particular culturally determined character, like Gothenburg with its 'county council' buildings. The ground floor of 'county council' buildings were built with stone and the two upper storeys with wood. This type of house was built from 1875 to the 1940s. Fire regulations determined their special characteristics.

Extending Old Structures

In many places people have extended existing development according to old patterns and with respect for the traces of history. In Sweden there are several examples of attempts to adapt to old village configurations. Examples also exist of additions to larger and smaller towns, made in keeping with the old town structures. In Gothenburg, modern 'county council' houses have been built. In Finland, where there are many splendid small towns built of wood, projects called 'the modern wooden town' are under way, where people are attempting to continue building according to the small town tradition.

An older county governor house in Gothenburg

Source: *Så byggdes husen 1880–1980- Arkitektur, konstruktion och material i våra flerbostadshus under 100 år*, (How Houses were built 1880–1980 – Architecture, Construction and Material in our Blocks of Flats During a 100-Year Period,) Cecilia Björk, Per Kallstenius, Laila Reppen, 1984

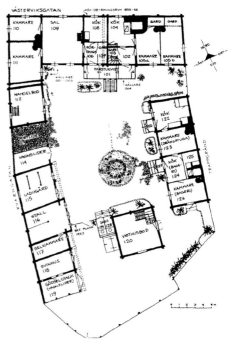

The old town settlement had a rural feel to it, with areas for animal care and all sorts of activities. The picture shows a reconstruction blueprint of the Grassa farmhouse in Strängnäs between 1846 and 1917.

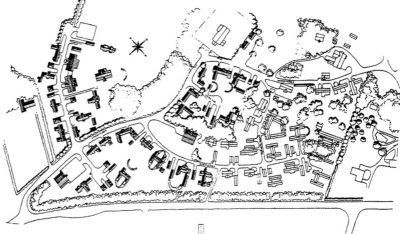

Håbo Tibble village in the Upplands-Bro municipality, a church village in Uppland that's been added to (from 1970 and onwards) with special contributions by the architect Sven Olof Nyberg. Here, additions to the village have been carried out according to the old tradition, creating a modern little community with roots in the old church-village.

Environments with an Interesting Cultural History

In Sweden, the county administrative boards take care of cultural environmental assets in the counties and the National Heritage Board (RAÄ) takes care of the country's historical monuments. These authorities have the right to determine any special considerations. In most of the counties, the county museums have done cultural inventories where culturally interesting buildings and environments have been described. Attempts have been made to classify the various objects according to their preservation value. Examples of important cultural-historical environments include the industrial communities in, for example, Bergslagen and Uppland, where the buildings lie symmetrically along the streets and the workplace hierarchy is reflected in the architecture; the church towns in Norrland that came into being because otherwise it was too far for people to go to church; as well as those communities where workers' dwellings and agricultural labourers' terraced housing grew up adjacent to country estates.

Cultural-Historical Buildings

There are a number of buildings in Sweden that are national monuments. Of the very old buildings scattered around the country the regional churches are the greatest treasure. There is an unusually large number of well-preserved manor houses and country seats which require large sums for maintenance and renovation. Adjacent to farms and estates, handsome farm buildings in the form of barns, storehouses, cowsheds, stables, smithies, etc., are often found. With their residences, they form beautiful complexes that should be preserved in order to provide a picture of life in former times. Watermills and windmills were built in suitable locations. Due to their locations, they're often exposed to water, weather and winds, and should be given extra care in order to preserve them. There are important secular buildings that are worth taking care of such as schools, shops and meeting places, courthouses, cinemas, people's amusement parks, inns, taverns, etc. Sweden also has many industrial buildings from different epochs. Tenement soldiers' cottages, officers' housing, regimental grounds and barracks are interesting examples of military buildings. Even outbuildings such as storerooms, root cellars, woodsheds, outside lavatories, saunas, boathouses and smaller barns can be worthy of care.

Older Infrastructure

Originally, old roads were laid in a sensitive way through the landscape, for example, high up on the ridges where there was a dry and firm foundation. When old roads passed over wetlands, streams and rivers, bridges and causeways were built, often in a way characteristic of the period. Many of the old roads are lined with milestones and mileposts which bear witness to a time when travel went at a speed totally different to today. It would be very worthwhile to integrate the old roads with new networks of cycle and pedestrian paths so that they're kept open. The canals and the narrow-gauge railways were prerequisites for the early industrialization.

Relics

There are many relics located in the fertile and long-inhabited parts of Sweden. They are protected by law and must, of course, be identified in an area that is being considered for development. If we follow ecological principals in attempting to find the best location with regard to water access, microclimate, cultivation and views, we often end up in places where people have lived before. The most common relics in Sweden are graveyards from the Iron and Bronze Ages. The county custodian of antiquities doesn't usually want building to take place close to graves, but the distance to development can be discussed with the authorities, especially if someone is willing to take responsibility for the care of the relics site. Old dwelling sites can often be identified by the colour of the soil, a high phosphorus content in the soil, by house foundations or cultivated plants found out in the country. In such cases, the county custodian of antiquities must be consulted to see if the relics must be preserved, excavated or are considered to lack cultural-historical value. This also applies to ruins.

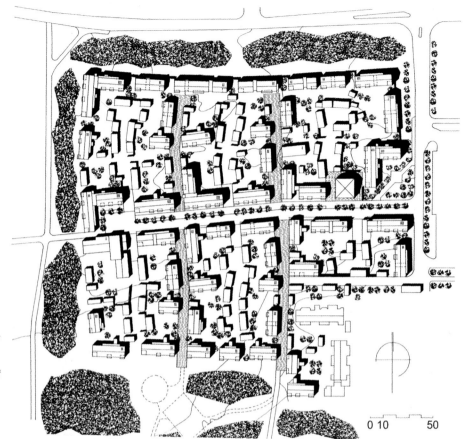

Syynimaa (Linnanmaa), Uleåborgs newly built town district with wooden buildings that continue the pattern of the old wooden town.

Source: Arkitektursektionen på Uleåborgs Universitet, Finland, Prof Jouni Koiso-Kanttila

Wooden buildings in the new town district in Uleåborg

In Gårdsten, Gothenburg, Sweden, an environmentally friendly rebuilding has been carried out. The photo shows a building that had extra insulation added to it and it was equipped with an air solar collector.

Source: Architect: Christer Nordström

4.3 Existing Buildings

With regard to the environment and buildings, existing buildings are most important since not so many new buildings are built each year. This is a matter of making existing buildings healthy, resource efficient and adapted to ecological cycles. Consideration should also be given to site-specific aspects. Poor quality material is replaced, and operation and running costs are made as low as possible. Following these measures, an investigation is carried out to determine if rebuilding is necessary for the sake of energy efficiency or other reasons.

4.3.0 The Use Phase

The environmental impact of building a house is one thing, but the fact is that during the lifetime of a building, 85 per cent of total energy consumed is during the use phase. So it is an important phase to focus on. The same principles apply to existing buildings as to new constructions.

Making an Inventory

The existing building stock is large. Management companies should have a strategy for improving the building stock. A careful inventory of every building would cost too much, so representative buildings are chosen. By looking at the overall status, a decision is made about which houses to

Energy in building and management.

Source: *Hus i Sverige – perspektiv på energianvändningen*, BFR T2:1996

Prior to rebuilding, an inventory of the existing building (area) is required. The Institute for Building Ecology (IBE), now in Tyréns in Stockholm, has developed a method for carrying out an inventory in several steps, which avoids unecessarily inventorying everything and allows effort to be concentrated on the most important areas.

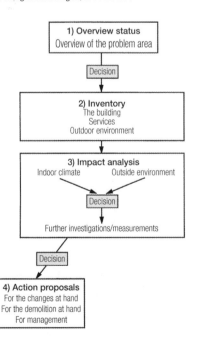

inventory. When a decision has been made to carry out an inventory, then the building as well as its services and outside environment are inspected. An impact analysis is made based on the results of the inventory of the indoor climate and outside environment. After that, a decision is made about whether further investigations and/or measurements are required. Once these have been carried out, new decisions are made about an action package. In this way, the worst buildings can be tended to first.

Business Ratios

More and more property managers are using so-called business ratios. Business ratios make it possible to compare different functions in different properties, e.g. resource consumption such as heating (kWh/m^2), electricity (kWh/m^2), and water ($litre/m^2$ or per flat). Many other factors can also be compared, such as costs (SEK/m^2), surface efficiency (people/m^2) and quality (e.g. hours per measure or air temperature deviation), or how many of the property management company's customers have received environmental training.

Business ratios can also serve as an alarm. A ratio that strongly deviates from the norm is a signal that something is wrong. Ratios have a diagnostic function as well. Properties can be compared to determine which ones should be improved. Furthermore, ratios have a prognostic function. For example, future goals can be set, such as a goal for a certain energy consumption level for all the properties.

Environmental Business Ratios

The Swedish Association of Municipal Housing Companies (SABO) has developed ratios for environmental work, which are organized into seven different categories:

1. **Environmental consciousness**
   How many employees have received environmental training? How many times have the residents received environmental information? How many flats have access to an environmental representative?

2. **Waste**
   How much unsorted household waste is generated in relation to the number of flats? How many sorting fractions is there access to? How many people have the opportunity to compost?

3. **Energy**
   How much electricity per m^2 of rented space is used? How much energy per m^2 is used for heating? What percentage of the flats have energy-efficient appliances?

4. **Transport and motorized equipment**
   How many flats can be managed per driven kilometre? How many litres of fuel per kilometre do the company's vehicles use? How many vehicles run on renewable fuels or electricity?

5. **Water and sewage**
   How many litres of water are used per m^2? How many of the flats have water-efficient equipment?

6. **The outside environment and gardening**
   How much artificial fertilizer is used per m^2 green area? How much chemical biocide is used per m^2 green area? How many of the flats have access to a garden plot?

7. **Healthy buildings and the indoor environment**
   What proportion of the flats consists of environmentally friendly material? What proportion of the flats have had their ventilation inspected? What proportion of the cleaning is carried out with environmentally friendly cleaning methods?

In addition, SABO has developed four ratios for social quality:

1 landlord with a good reputation;

2 contact and involvement;

3 order and security; and

4 status and stability.

Environmental Classification

For owners of either a single property or a large number of properties, environmental work is facilitated by carrying out an environmental inventory and environmental classification (environmental declaration) of the buildings concerned (see Section 1.4.1). These methods show what is good or bad from an environmental perspective and what should be taken care of first to reduce the negative environmental impacts of the buildings. There are several different environmental classification methods. All involve doing an inventory of the building with respect to a number of characteristics that affect the environment. To obtain a building's environmental status, every characteristic is rated on a scale from good to bad (the scale can be divided into 5 or 10 steps, where 0 is the worst and 5 or 10 the best). The result can be presented in an easy to understand diagram, where the more green, the better the building is from an environmental perspective. Such information also has an economic aspect, affecting the value of the property.

Characteristics that can be inventoried are for example:

- resource consumption such as use of heat, electricity and water;

- use of materials such as healthy materials, energy-intensive materials and materials that emit toxins;

- use of chemicals in the running, cleaning and management, and coolants in heat pumps;

- environmentally hazardous materials that should be removed, such as asbestos, lead, cadmium, mercury, PCBs, or others;

- waste management such as sorting, reusing and composting, and how hazardous waste is managed;
- comfort such as room temperature, surface temperatures, asymmetry of radiation (temperature), relative humidity and draughts;
- the indoor environment such as light, noise, vibrations, cooking odours, other odours and static electricity;
- radiation such as radon, electric and magnetic fields;
- aspects that affect health such as damp and mould, vermin, smoking and the risk of legionnaire's disease;
- the outdoor environment such as air pollutants, traffic noise and access to green areas;
- the environmental impact of transport, including transport of people.

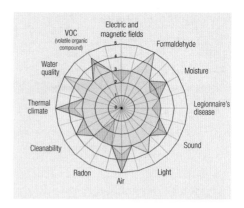

Example of a results compilation for the sub-area indoor environment, according to the evaluation method *Miljöstatus för byggnader 1999* (Environmental Status of Buildings 1999) by Jacobson & Widmark. Green area = estimated environmental status. Orange area = goal status.

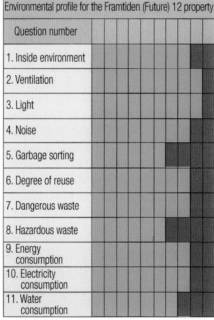

Environmental evaluation by the Building Ecology Institute (IBE), Tyréns, Stockholm.

Compilation of results. The EcoEffect method shows estimates of environmental impacts in different areas. The method was developed at the Royal Institute of Technology in Gävle, Sweden.

4.3.1 Operation and Management

The goal of property management is to maintain, and if possible improve, the technical, hygienic and environmental running and standard of buildings. Management costs are determined to a great degree when a building is constructed, as that is when the frame, technical systems, furnishings, surfaces and design are chosen. In some countries, those who build, own and manage a building bear the main responsibility for any environmental problems the building causes.

Management Organization

Many property owners make some form of long-term, rolling maintenance plans or area programme based on an analysis of technical needs and economic possibilities. Property management consists of three parts: administration, economic and practical. Administration includes rental matters, inspection, staff planning, management and operation planning, maintenance planning, procurement, purchasing, etc. Economic aspects include acquisition of capital, money management, rental fees, payments, budgeting, accounting and balancing the books. Practical work includes janitorial tasks, regular maintenance, operation supervision, heating, rubbish collection, cleaning and minor repairs.

Quality-Certified Management

How management is organized influences results and environmental work. Therefore, it is currently possible to quality-certify management with regard to environmental aspects. For a company to be certified, they must adopt an environmental policy, set environmental goals, and allocate adequate resources for achieving the goals within the planned timeframe. All staff are trained so that they understand what the environmental work is intended to achieve. The buildings are then inventoried from an environmental perspective to provide a basis for technical measures and maintenance planning. Plans are made for the measures required for inspection and maintenance. The concept also includes an internal audit so that the company gets feedback which helps them to improve the efficiency and reliability of the system. It is important for management to ensure that the environmental work routines are current, and it can help to have the audits checked by a third party.

Environmental Aspects of Management

How management is carried out can also have a direct impact on the environment. For example, it is possible to use electric vehicles, environmentally friendly fuel for both lawn mowers and leaf blowers, environmental cleaning chemicals and environmental weed control. It is of course also very important to choose environmentally friendly materials and surface treatments.

Maintenance Frequency

The purpose of maintenance is to keep a building and its components in good working order and repair. Maintenance is usually divided up into periodic maintenance, and corrective or acute maintenance (repairs). With every approach chosen, a maintenance requirement is also chosen. The better the quality and longer the lifetime of a building's different components, the less maintenance required. However, sooner or later structural parts must be maintained or replaced. Different products require maintenance at different intervals. The length of the interval of course depends a lot on the users and on how exposed the product is to the weather and wind.

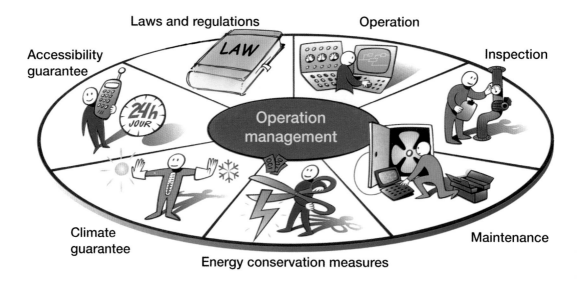

In property management, the lines between different improvement measures are not especially clear. There is often an overlap between operation, maintenance and rebuilding measures.

Source: ABB, 1997

Resource Consumption During Operation

Heat, electricity and water requirements are determined to a large degree by the choice of resource conservation methods for buildings and equipment. Whenever equipment must be replaced, the most resource-efficient replacement on the market with the least environmentally hazardous substances should be chosen. To quickly track malfunctions and set up long-term savings goals, management should have easy access to reliable and current operation statistics for heat, electricity, and hot and cold water consumption.

Operations Monitoring

Automatic operations monitoring can be used to control different systems and to quickly determine when something goes wrong. Computerized control and monitoring systems have always had the three following clearly distinguishable levels:

1 Sensors, controlling elements, etc. that are closest to the process itself.

2 Computer control centres that control, regulate and monitor.

3 Main computers (databases) that collect and present information from one or several computer control centres. The main computer can be placed in an operations centre or in the building.

Components that can be controlled include fans and pumps, heating and cooling equipment, temperature and ventilation, lighting, etc. It is possible to regulate for the summer and winter, weekdays and weekends, day and night, empty rooms and occupied rooms, etc.

Various types of sensors can sense the need for control. When a breakdown occurs, a repair person can use their laptop computer on site to determine what needs to be done. Today, computer systems are commonly used for control, monitoring and administration of buildings.

Maintenance Intervals

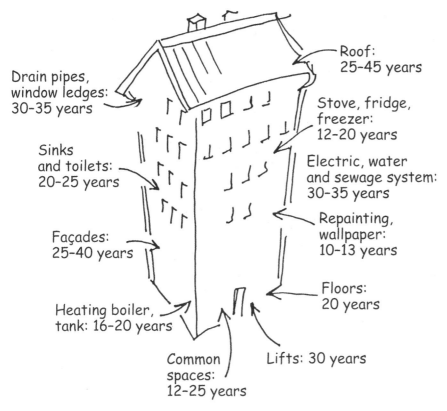

Approximate maintenance intervals for different parts of a building. It has been shown that preventative maintenance is profitable in the long term. Some measures are significantly cheaper if carried out in combination with others rather than done separately.

Source: Cover picture by Gunnel Eriksson Layout AB for the publication *Att planera sitt underhåll*, published by Sveriges Fastighetsägare, 1989

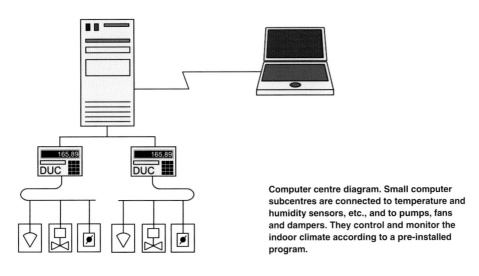

Computer centre diagram. Small computer subcentres are connected to temperature and humidity sensors, etc., and to pumps, fans and dampers. They control and monitor the indoor climate according to a pre-installed program.

4.3.2 Energy Conservation

Many old buildings have a high energy consumption, up to 300kWh/m²/year. Experience has shown that systematic implementation of energy conservation measures can reduce energy consumption to 100kWh/m²/year. After the 1970s oil crises, an entire profession started working in the area of energy conservation. At that time, many rather unsuccessful energy conservation measures were taken. Currently better knowledge of and experience with a variety of sensible energy conservation measures are available.

Energy Conservation Possibilities

Energy conservation measures are usually divided into four categories: user habits, operation and maintenance, adjustments and simple measures, as well as rebuilding and changing systems.

It is common to start by making adjustments, and adjusting and controlling heating systems and ventilation. Then, there is a search for weak points in seals, insulation and thermal bridges. Measures that are usually the most profitable are installation of a third pane on windows, supplementary insulation of loft floors, and for outer walls, adding extra insulation to windowless gables and basement roofs is simple and inexpensive. Some heating systems must be reinstalled in order to satisfactorily adjust the heat in different flats. In some cases, ventilation systems must be rebuilt in order to provide a good indoor climate and prevent energy over-consumption.

Inspection

Many existing buildings have weak points where there are draughts and large energy losses. An energy conservation adviser can go through a building to find weak points and propose simple and inexpensive measures for saving energy. Equipment used for such an inspection include a smoke puffer for finding draughts and a fast thermometer for determining cold bridges and other poorly insulated locations. One interesting method is infrared camera technology. A building is photographed from the outside on a cold winter day and infrared radiation gives a visual image with different colours for different surface temperatures. Infrared cameras are also used to check insulation in new buildings. Buildings are also inspected to check their ventilation systems.

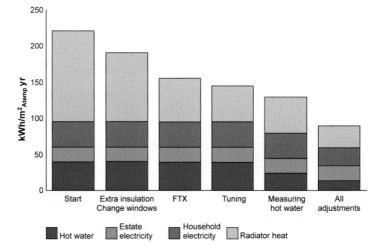

By realistic measures energy consumption from a typical building in the 1960s can be reduced from 220kWh/m²/year to 90kWh/m²/year.

Source: "Renovering pågår ...", VVS Företagen and Svensk Ventilation, 2008

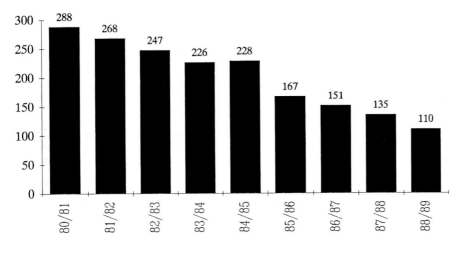

Falkenberg municipality leads the way in efficient use of energy. In 10 years the municipal housing company succeeded in reducing energy consumption by 62 per cent in a district consisting of 150 apartments. This was achieved by gradual technical improvements, better maintenance and giving information to tenants. The illustration shows purchased energy per m² flat floor area/year.

Source: *Nya grepp om ekonomi, energi och miljö på lokal nivå*, NUTEK B 1994:5, Naturvårdsverkets Rapport 4284, 1994

Saving Energy

Energy can be saved by changing to the latest energy-efficient lighting technology (for example LEDs), reducing the number of lights (if possible), using time-clocks for lighting and ventilation fans as well as heating and water system pumps. Washing machines and dryers can be replaced with new, energy-efficient models when appropriate.

Adjustment of Heating Systems

Careful adjustment is required to get the desired temperature and reduce excess temperatures in rooms and flats in a building. A heating system functions well when each radiator has the right amount of water in relation to its heat-emanating surface area. Fine-tuning a regulating system involves setting the different adjustment valves

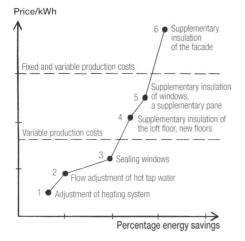

The relationship between energy savings and the cost of various conservation measures. Items 1–3 are economically justified for individuals. Items 4 and 5 are justified from a national economic perspective. Item 6 is only justifiable from a long-term environmental perspective.

Source: *Effektivare energianvändning*, SOU 1986:16

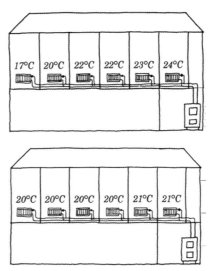

The top illustration shows the temperature conditions in a building prior to adjustment. The bottom illustration shows the same building and radiators after adjustment.

Three different techniques that can be used to determine a building's energy losses.

Source: adapted from *Trimma täta isolera – Åtgärder för energihushållning i hus,* en handbok utarbetad av Bygginfo på uppdrag av statens planverk, 1978

Smoke gun.

such as the bypass, main and radiator valves. The control system also needs to be adjusted and it is important to have thermostats and temperature sensors correctly located. Heating systems should be made so that the radiators are turned off when airing, so that extra heat isn't added when ventilating. The heating system should also be able to sense when there is free heat available from people, solar insolation and electrical equipment, and then turn off the heating in order to take advantage of the free heat. Some rooms have greatly varying heating requirements, e.g. assembly halls.

Hot water consumption influences energy requirements. Improvements can be made in the areas of individual monitoring, pipe insulation, reduction of water temperature, limiting of pressure and flow, as well as adjustment of the flow of hot water in the heating system.

Adjustment of Ventilation Systems

A careful adjustment of air flow to different rooms is required to get the desired air flow. Windows and doors must be sealed in order to be able to control ventilation. Assistance by experts is often needed to achieve good results. Ventilation should

Some energy efficiency measures that can be made in a school.
1. Windows: U value 1.0WM/m²K. Some windows can be opened at the top, which means less risk of draughts and better security.
2. Motion detectors that control lighting and ventilation.
3. Well-adjusted radiator system.
4. Motorized damper.
5. Intake air fixtures that allow intake of low temperature air without risk of draughts.
6. Adjustment of air flow to every room depending on occupancy.
7. Lighting: 10W/m², which is controlled by motion detectors.
8. Room sensors for controlling radiators, and blocking the supply to all radiators when the room temperature exceeds 19°C.

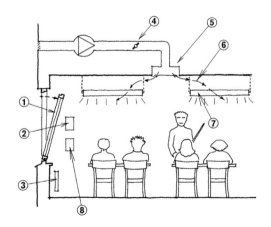

be adapted to needs. More ventilation is needed in the summer than in the winter, and less ventilation is needed when no one is home. Some rooms have a ventilation requirement that varies greatly, e.g. assembly halls.

Ventilation systems can be adjusted manually or automatically. Ventilation is influenced by the outdoor temperature, heat from solar insolation, and wind conditions, and should be adjustable according to weather conditions. Ventilation adjustment systems can be controlled by temperature, pressure drop or relative humidity. Control according to CO_2 levels is considered obsolete. Ventilation systems should be cleaned regularly. Ducts easily become clogged with dirt, dust and grease, which can significantly reduce air flow. Air intake and exhaust fixtures, fans and grates should also be cleaned regularly. Any filters in the ventilation system must be replaced regularly.

Tuning Combustion Systems

Most combustion systems must be tuned and adjusted. This is normally done during installation by the installer or a specialist. The soot must be removed from combustion areas and the areas cleaned regularly, and at the same time a check can also be made to see if anything needs to be adjusted. All this of course influences the level of efficiency and amount of emissions.

Four different types of measurements can be carried out to determine the extent of loss from flue gases, and thus the combustion efficiency of the boiler. It can also be determined how air adjustment and smoke dampers should be adjusted to improve heat transfer. Parameters measured are the amount of soot, level of CO_2, flue-gas temperature and draught. Fine-tuning should consist of a gradual adjustment of the air inflow and flue-gas speed. Combustion, heat absorption and flue-gas temperature are thus influenced.

Sealing

Older buildings are often draughty, which means that there are large, unintended energy losses. Air leakage is easily detected using a smoke puffer. Spaces between window casements and frames should be sealed. There is special rubber weather-stripping for this, e.g. made of EPDM rubber. The space between window frames and walls should also be sealed, which can be done with environmentally friendly packing material. The joins between window glass and casements also need to be sealed. Putty often needs to be reapplied to old windows. The space between doors and frames and between door frames and walls should also be sealed. The seal between outer walls and connecting floor structures and cross-walls should be checked. There can be leaks at electric boxes, entry points for electric and water services and at skirting boards. Points that should not be sealed include openings in foundation walls, ventilation openings

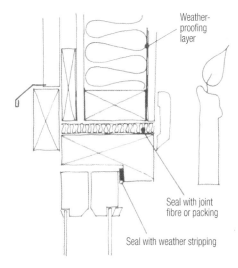

Example of a simple way to determine if the seals around windows and outer doors are good enough. Hold a candle close to the area and see how much the flame flickers.

Source: brochure 'Fakta om energibesparing', Energisparkommittén, 1978 to lofts, air ducts behind outer boards, and gaps at the lower edge of façade walls. These are needed to keep the construction dry and fresh. Air-intake fixtures that supply the building with fresh air should also not be sealed. In some buildings, fresh air is taken in through window gaps, and if these are sealed, it disrupts the ventilation.

Supplementary Insulation

Older buildings are often poorly insulated and can be improved with supplementary insulation. Buildings from the Million Programme have only about 10cm of insulation in the walls and 15cm in the roof. When the installation is replaced a layer twice as thick should be put in.

It is simplest and easiest to add insulation to loft floors, but it is important not to obstruct ventilation at the eaves. It is important to remember to insulate loft hatches. The lower part of balcony doors are often poorly insulated or not insulated at all. Outside doors are sometimes poorly insulated and supplementary insulation can be added to them or they can be replaced (modern outside doors have a U-value of 0.7 W/m²K). Supplementary insulation of basement ceilings may be a good measure to take if the ceiling height allows it. Old insulation may have become compressed, and in such cases it should be supplemented or removed and replaced. The edges of floor structures may need supplementary insulation. Spandrel beams in conjunction with slab foundations can be great thermal bridges and so should receive extra insulation. When outside basement walls are moisture damaged and require extra protection, it can be a good opportunity to add insulation to the outside of basement walls. Supplementary insulation of façades is generally not a cost-effective measure and often distorts the character of a building. In exposed areas, e.g. gables, insulating in a well-defined area can be considered. If insulation is added to outside walls, it must be done in a way that allows moisture to escape through the insulation.

Window Improvement Measures

To reduce energy loss, a third pane can be added to older windows. The extra pane can be placed on the inside so that the appearance of the window from the outside does not change, or it can be put on the outside to preserve the interior finish. Another possibility is to replace one of the casements with a new insulating pane, or to rebuild one of the casements so that one of the panes can be replaced with an insulating pane. There are also systems where an extra pane can be added to an existing casement and sealed, which in principle results in an insulating pane. It is also possible to replace the glass in double-glazed windows with energy-efficient glass, but energy savings are then smaller.

There has been much development in the area of energy-efficient windows. The best modern windows are three times as energy-efficient as the most common windows in multi-family dwellings. In some cases the best solution is to replace old windows with new ones.

Gables should be insulated on the outside and floors above crawl spaces from underneath.

Source: *Smalhus – Energisparande och fasadisolering*, Råd och riktlinjer Stockholms byggnadsnämnd, 1979

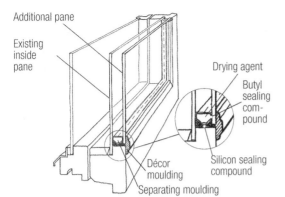

If a person wants to go further with energy conservation measures than adjustments and control, then building alterations are the next step. One of the simplest is to add a third window pane.

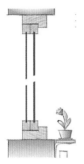

The inside glass is replaced with 4mm hard-coated energy-efficient glass. U-value: as low as 1.8W/m²K.

The inside glass is replaced with a double insulating argon-filled pane with soft-coated energy-efficient glass (if allowed by the construction of the casing). U-value: as low as 1.3W/m²K.

Supplementing with an inside 4mm coated energy glass with edge protection. U-value: as low as 1.4W/m²K.

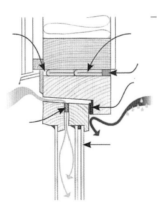

The glazier also checks that windows are properly sealed.

Source: "Fönsterrenovering med energiglas", Glasbranschföreningen.

Glassing-in

Glassing-in spaces can reduce energy losses and result in a certain amount of additional energy from solar insolation. Balconies or entire south-facing façades can be glassed-in. Glassed-in porches can be built at doors and entrances. Sun rooms can be built in attics or glassed-in verandas at ground level. In some cases, courtyards have been glassed-in to reduce heat loss. Energy is saved as long as a new glassed-in area is not heated, but glassing-in can be an expensive way to save energy and should actually only be done if the glassed-in areas also add other qualities.

Distortion of Buildings

In a few cases it is justified to add insulation to façades that have windows. In such instances, it is important to make sure the building is not distorted and made unattractive since the change to the exterior will be significant. Façade windows should be moved outwards so that they sit flush with the façade even after insulation is added. Materials and colours must be chosen carefully, and careful consideration must be given to how the wall connects to the foundation and roof projection, porches, outside stairwells and extensions. If windows are going to be changed, the new windows should have the same design, glazing bars, dimensions, details and proportions as the previous windows.

The greatest energy savings are from replacing the inside glass of two-glass windows with energy-saving glass, or insulated windows with energy-efficient glass. The building thus looks the same before and after replacement. If the windows don't need any outside maintenance it is easiest and most efficient to improve the windows with an energy-efficient glass construction on the inside. A 4mm-thick hard-coated energy-efficient glass can be mounted on the inside or outside of the window. Sound insulation is also improved.

On the left is an old window that originally sat flush with the façade. On the right is an unattractive example of how, seen from the outside, it ends up in a deep, dark niche if it isn't moved outwards.

Source: Olof Antell, Riksantikvarieämbetet

4.3.3 Decontamination

For a long time, especially during the 20th century, construction materials have been used that are now regarded as hazardous. Examples include radioactive lightweight concrete, fibres that can cause lung damage (such as asbestos), sealing compounds with PCBs, heavy metals and additives in materials. Current management and renovation practices involve an attempt to systematically remove these hazardous substances and materials from buildings. Some materials must be removed by law. There are special regulations about how this should be done and where the material should be taken.

Inventory of Materials

Carrying out an environmental inventory, e.g. when a building is acquired or prior to demolition, entails identification of all materials considered to be hazardous waste. Even other materials that could be considered hazardous as well as any leftover chemicals should also be identified. An appraisal should be made as to whether or not the activity in and around the building could have caused contamination of building material and/or the ground. When necessary, an analysis of potentially hazardous materials should be carried out.

Hazardous Waste

Hazardous waste is waste that can cause damage to the environment and health. Hazardous waste regulations define what waste is considered to be dangerous. Hazardous waste should not be mixed with other waste, but rather must be separated into categories. It is obligatory to ensure that hazardous waste is transported and dealt with by certified personnel in certified facilities. If it is not possible to declare that waste is free of hazardous substances, it should be considered as hazardous waste. Some examples of hazardous waste commonly found during building demolition follow:

Asbestos was mostly used in the 1950s and 1960s, and its use was not prohibited until 1976. Asbestos has been used as fire protection; heat insulation; reinforcement; sound insulation; condensation insulation; in pipes and ducts; as filler in paint, pasteboard and plastic; and in the form of asbestos cement in sheeting (eternite). Asbestos cement sheets have been used as floor, wall and roofing material. A layer of asbestos may also be found under plastic mats, in tile fixative, window putty, sprayed concrete, glue, sealing compounds and window sills. Internite and eternite are perhaps the most well-known brands. Asbestos removal should be carried out by certified decontamination companies. The material is transported in closed containers and is deposited at special sites.

Asphalt can be crushed and reused in new surface coatings on driving and warehouse surfaces. Reused asphalt is, however, not used on surfaces with a large amount of heavy traffic.

Batteries and accumulators contain heavy metals and environmentally hazardous electrolytes such as lead, nickel, cadmium and mercury.

Lead was used in large amounts before 1970 in the joints between cast iron sewage pipes (called caulking). Lead has been used as roof sheeting as well as under the joints between other metal materials (for example in roofs, balconies and bathrooms). Old electricity and telephone lines are often encased in lead. Lead was used to make gas and water pipes. Lead asphalt-impregnated felt was used as foundation insulation. Lead

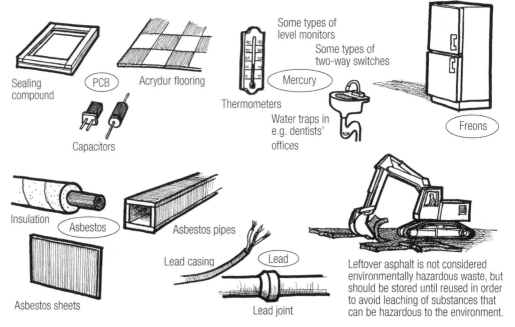

During demolition there can be many older products that contain environmentally hazardous substances such as materials and objects that contain lead compounds, mercury, asbestos, freon and PCBs.

Source: quoted freely from SIAB brochure

was also used in older window glass. Level transmitters in sump pumps also contain lead. Lead has been used as a stabilizer in plastic and as pigment for shades of white, yellow, green and red. Red lead oxide has been used as outdoor corrosion protection for metals. Lead should be separated out and deposited for metal recycling.

Fire protection equipment such as fire extinguishers may contain halogen. Smoke detectors and some fire alarms may contain radioactive material. There are special regulations for handling radioactive material.

Sealing compounds are used in many places in a building. Since the mid 1950s sealing compounds have been used to seal constructions and make attractive joints. A differentiation is usually made between elastic and plastic sealing compounds. Elastic compounds are able to absorb the greatest amount of movement. Binding agents used in sealing compounds include oil, acrylate, polyurethane, polysulphide, butyl, bitumen and silicone. Sealing compounds can contain one or more of the following environmentally hazardous substances: PCBs, phthalates, chlorinated paraffins, and biocides. It is often difficult to determine which type of sealing compound has been used. Polysulphide-based sealing compounds had their breakthrough in 1957. PCB (20 per cent) was used as a softener until it was prohibited in 1972. Polyurethane based (PUR) sealing compounds were introduced in the 1960s. PUR sealing compounds often contain phthalates as a softener (20–30 per cent). To set sealing compounds, thixotropic substances are added, often PVC (up to 20 per cent). Small quantities of organic tin compounds are often used as a catalyst.

Freons (CFCs = chlorofluorocarbons) can be found in refrigerators, freezers, heat pumps, and cooling and air conditioning plants. CFCs have been used to expand cellular plastic, e.g. extruded polystyrene (XPS) in insulating sheets; as hard polyurethane in insulating sheets, insulation for refrigerators and freezers and as packing around windows and doors; as well as for polystyrene foam for acoustic insulation. Cellular plastic that may contain CFCs is most often blue, red or yellow and never completely white. Regulations about CFCs and halogen, etc. regulate how to look after and destroy cooling equipment/appliances. Since 1995, refrigerators and freezers must be removed and destroyed in a controlled manner to avoid release of CFCs into the air.

Paint, varnish, solvents, sealants and glue often contain environmentally hazardous substances such as solvents, emulsifiers and softeners. Cans containing leftover paint and similar products are considered hazardous waste and should be sent for destruction. Wood products painted with environmentally hazardous paint should be combusted in special facilities. Metal products with a PVC (Plastisol) surface treatment cause a waste management problem because dioxins form during remelting. Leftover paint and varnish are separated into special containers. Solvents are separated out and dealt with on their own, as is leftover glue. They are all deposited at recycling stations to be sent for destruction.

Polluted land occurs mainly around and beneath industries, workshops, vehicle management facilities and heating plants. Polluted ground can contain oils, heavy metals, etc. When the levels are high, the ground is classified as hazardous waste. A pollution analysis should be carried out before measures are taken in order to determine the proper way to proceed.

Polluted land should be decontaminated on site using biological methods, soil washing or thermal treatment after receipt of a special permit, or taken away to a special landfill.

Impregnated timber is found in duckboards, piers, suspended foundations, roof trusses and joists, and other wooden constructions where there is a risk of decay. Impregnating agents used include CCA agents (chromated copper arsenate), which contain oxides of copper, chromium and arsenic. Organic tin compounds such as tributyltin oxide and tributyltin naphthenate were previously used as impregnating agents for window casements. Boron and phosphorus are also used in wood preservatives.

Spackle Paint waste Glue

Solvents Leftover oil, diesel oil Sealing compounds

At construction and demolition sites, hazardous waste should always be separated out and specially handled. It can be difficult to determine which waste is hazardous. At new construction sites, wall filler, glue, paint, solvents, leftover oil and sealing compounds are types of hazardous waste that may be present.

In the past, creosote oil was used, especially in railway ties and telephone poles. Between 1963 and 1978 much rough-sawn timber was dipped in fungicide made of pentachlorophenol. Impregnated timber is sent to special facilities for combustion. Timber damaged by dry rot or timber-destroying insects also needs to be incinerated.

Cadmium has been used as a stabilizer or pigment in plastic material (mostly during the 1960s and 1970s). Plastics in bright shades of yellow, orange and red were often based on cadmium pigment. Use of cadmium as a pigment is now prohibited, but materials containing cadmium may still be imported. Cadmium is present in nickel-cadmium batteries (NiCd).

Mercury is found in electronic components and measuring instruments. The sale of mercury has been prohibited for some time but older thermometers, thermostats, level monitors, pressure gauges, door bells, alarm equipment, hot water heaters, older freezers with lights that turn on automatically, and pressure controllers may still contain mercury. Mercury may also be found in electronic switches and contacts, e.g. in timer switches and relays, as well as in mercury lamps and fluorescent tubes. Devices containing mercury are considered hazardous waste and should be dealt with according to local by-laws. There are currently facilities that deal with fluorescent tubes and recover the mercury. If certain activities were located in a building, e.g. a dentist's practice, amalgam containing mercury may have accumulated in the sewage system.

Oil wastes come from oil cisterns, hydraulic fluid tanks, lift motors and other motors, lubrication oils, fuels, transformer oil and used oil. Older transformers may contain oil with PCBs. The oil should be analysed by certified decontamination companies before it is destroyed.

PCBs were first used in the building sector in the 1950s and peak use was reached in the 1960s. Use of PCBs were prohibited in 1972, but it may be present in buildings constructed before 1975. Buildings may contain polysulphide-based sealing compounds containing PCBs (up to 30 per cent) used to join wall panels and to seal insulating glass panes. PCBs were used as a softener in plastics, varnishes and paints and in slip-proof floors with quartz sand (called acrydur flooring). PCBs can also be found in condenser oil in washing machines, oil burners, and fluorescent tube fittings and transformers. According to the law, some building owners must make an inventory of their buildings with regard to the presence of PCBs and report on the inventory as well as measures planned. The Swedish Environmental Protection Agency and some municipalities have instructions on how this should be carried out.

PVC (polyvinyl chloride) plastic is the plastic that has been used the most in the building sector, and accounts for 60 per cent of the total plastic used. Rigid PVC is used for wall panels, boarding, ceilings, window casements, profiled mouldings, skirting boards, cover plates, cable housings, pipes and fittings. Softened PVC is used in carpets, cable insulation, insulating tape, sealing plugs, sheet metal coatings, barrier layers for insulating and wall panelling. The problems with PVC are the chlorine (PVC is the greatest source of chlorine in waste) and the additives. The environmentally hazardous additives most discussed are lead, brominated flame retardants, phthalates, chlorinated paraffins and organic tin compounds. Early imported plastics or PVC may also contain cadmium.

4.3.4 Rebuilding

For both rebuilding and new construction it is an advantage for those involved to be in agreement about their goals, the amount of consideration to give to environmental aspects and what they can afford. Will it be a cautious rebuilding with an attempt to keep costs down? Will residents stay or a workplace continue its operations during rebuilding, only moving out for a short time to a temporary dwelling or workplace? Or will there be total renovation to achieve an attractive living or work environment?

Cautious Rebuilding

Cautious should always apply when something is to be done to existing buildings. This means investigating existing conditions and qualities, and making use of them as much as possible to satisfy needs and goals (definition from Sonja Vidén, Ingela Blomberg and Eva Eisenhouer, BOOM group, KTH Architecture, Stockholm). From a resource conservation viewpoint it is natural to use existing buildings and not to rebuild more than necessary. Buildings can serve a new purpose or in any case be reused. Natural ventilation often results in a good indoor climate, but people have not always understood how old ventilation systems work, and if they haven't been maintained they may not work properly. Furthermore, most old buildings were built using healthy and proven building materials, but there are exceptions.

It is worthwhile preserving building qualities characteristic of when the building was constructed. The distinguishing features, qualities and technical condition of the building in question are identified. Deficiencies should be attended to and sometimes new goals need to be set. The social aspect is important: those who live or work at the location should be respected.

Requirements when Rebuilding

The regulations and legal requirements need to be followed. As well as meeting the technical performance standards, changes should be carried out thoughtfully so that distinctive features are taken into consideration and technical design, historical, cultural-historical and artistic merits are preserved. When rebuilding it is also important to face up to the new environmental requirements that communities are faced with. For example, it should be possible for people to separate rubbish and sewage, and the outside environment should respect ecocycles, with surface water, garden plots, greenhouses, composters, waste-sorting stations, etc.

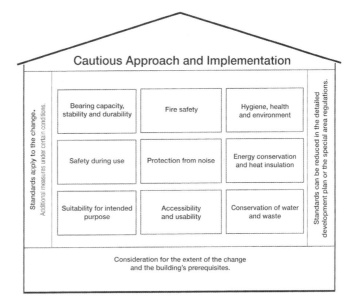

Schematic diagram of the entire regulatory system for changing buildings (PBL, Planning and Building Law). The nine boxes correspond to the nine principle technical performance standards in BVL (Construction Works Law) 2nd paragraph as developed in BVF (Construction Works Ordinance) paragraphs 3–14.

Source: *Allmänna råd om ändring av byggnad, BÄR*, 1996:4, changed 2006:1, Boverket (The Swedish Board of Housing, Building and Planning)

Environmental Rebuilding

Environmental rebuilding is more difficult than ordinary rebuilding. A lot of knowledge is required about former building techniques and the materials used. There are centres where it is possible to find old-fashioned components and building materials characteristic of different time periods. However, there is no guarantee that everything based on old knowledge is environmentally friendly.

Each country will have organizations to help with information for rebuilding correctly and sensitively for the period in question. In the UK there are the Georgian Society, Victorian Society, Society for the Preservation of Ancient Buildings, and so on.

In most buildings **water and sewage systems** have to be replaced after 30–60 years of use. It is uncertain when they will break, but when they start to leak costly damage can occur. There are methods to delay replacement of the water and sewage line system by a number of years. This can be accomplished by flushing, cleaning, inspecting and possibly lining the pipes. Lining of pipes is done with hardened plastics such as epoxy and glass reinforced polyester, and residents often do not need to move out while the renovation is taking place. Hard plastics are an occupational health problem, but using them results in avoiding creation of large amounts of construction waste, i.e. piping, for a time into the future.

An alternative is to plug old pipes and to place new ones in a new shaft where all pipes are accessible for inspection, replacement and extension. Prefabricated cassettes are available that contain sewage and water pipes, and a flushing cistern for the toilets connected to the cassette. In some cases a towel dryer is included. The cassette does not take up much more room than a toilet flushing cistern. There is another alternative where a small shaft and toilet seat are

Cautious rebuilding means that no more is done than necessary, that attractive façades, environments and parts of buildings are preserved. Properties usually end up being attractive and resources are saved as well.

attached to a projection that forms a bathroom shelf. These cassettes are more expensive than traditional fixtures but have the advantage of quick and easy installation as well as easy monitoring. Water-efficient toilets, showers and single-lever mixer taps are also installed at the same time.

Cold water pipes in older buildings can be made of galvanized pipe. They are more subject to corrosion than copper pipes. At the same time so-called 'thread joints' have been used for joining, which are also susceptible to corrosion. Such piping should be replaced. It is not a good idea to mix many different types of material since there is a risk that galvanic corrosion can occur. Water quality is a factor. Copper pipes can be damaged by acidic water and hard water is not good for stainless pipes. Alternatives are stainless pipes without nickel or cross-linked polyethylene.

Moisture damage in **bathrooms** is often due to poor or non-existent waterproofing and changed showering habits. The walls and floor should be renewed using an environmentally certified waterproof layer. Moisture-buffering wood-wool cement board is recommended for bathroom ceilings. Radiators can be replaced with a waterborne heated towel rail. If there are bathroom windows, the lower edge of the window sills should slope sharply downwards so that water runs off easily.

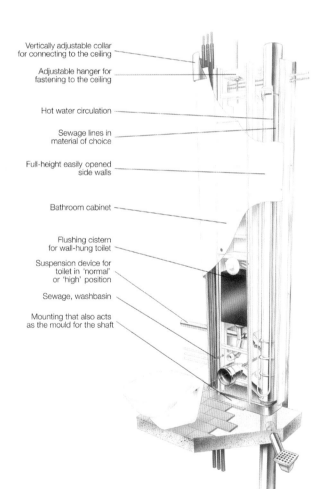

A wet area wall can be purchased incorporating a cassette with outlet pipes for cold water, hot water, and toilet and circulation.

Source: Columbi

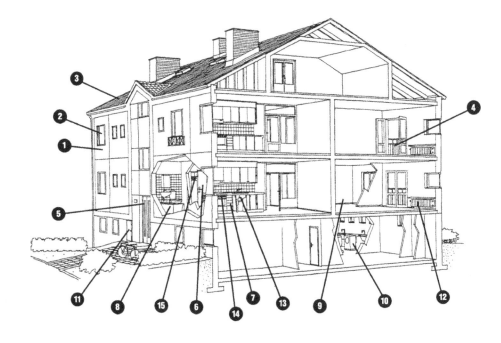

Environmental Rebuilding, A Design Tool (in Swedish) prepared by Birger Wärn for The Swedish Association of Municipal Housing Companies (SABO), is used to assess structural components from an environmental perspective.

Environmental profiles have been formulated for the following 15 structural components:

1. Outside walls
2. Windows
3. Roofs
4. Balconies
5. Entries and stairwells
6. Outside doors to flats
7. Kitchens
   i) woodwork
   ii) appliances
8. bathrooms/showers
   i) floor and wall waterproofing
   ii) floor surface layer
   iii) wall surface layer
9. Bedrooms and living rooms
   i) floor surface layer
   ii) wall surface layer
10. Laundry rooms – washing equipment
11. Garbage management
12. Heating systems
13. Water and sewage systems
14. Ventilation systems
15. Electric distribution boards and lines

Illustration: Leif Quist

The quality of **old windows** up until the 1960s is often excellent if they were made from close-grained and carefully dried timber. Before the 1960s, joinery had gentle profiles so that daylight came into a room gradually and the contrast between the outside and inside was moderated. They are often painted with linseed oil paint. Such windows should be renovated, reputtied, scraped, primed and painted with new linseed oil paint. Infrared heat and special tools may be used to remove old paint and putty. A third pane can be added to reduce energy losses. For windows in very exposed locations, the outside casement can be covered with aluminium sheeting, but it is important to provide the original timber with satisfactory ventilation.

Gates and outside doors should either be renovated or new ones ordered to resemble the old ones since they are often an important part of a building's character. The same goes for outside doors to blocks of flats and inside doors to individual flats. When replacing doors, however, energy losses, fire protection requirements, noise and burglary protection must be considered. Porches, especially in buildings from the 1960s and 1970s, may need to be redesigned.

Rubbish management should be carried out so that waste is separated into different fractions and each type of waste can be stored and picked up separately. In many cases a free-standing building is constructed in the yard. A rubbish building with a composting facility can produce a supply of humus for the vegetation in a nearby garden. The rubbish building can be placed in the yard so that it is experienced as a positive contribution to the area.

Using **natural ventilation** to ventilate sometimes does not work satisfactorily. There are new methods for reinforcing and adjusting existing natural ventilation systems. They reduce the need for large renovations to install space- and energy-demanding fan-controlled systems, and give users the option of adjusting the ventilation as required. Pressure and temperature controlled systems ensure that fans don't run unnecessarily. Some natural ventilation systems draw in air through narrow openings around windows. There are window mouldings (trickle vents) that diffuse the air flow and reduce the experience of draught.

Kitchens and joinery in older kitchens are often made from good materials, are attractive and have a well-defined purpose. These should be preserved. When appliances are replaced, energy-efficient ones should be chosen. New water-efficient single-lever mixer taps are installed. Electric cables are replaced with earthed wires and equipped with earth fault protection. An appropriately sized range hood can be installed over the cooker, which can function satisfactorily with natural ventilation. Larders should be kept or new, ventilated tall cupboards can be used to fulfil a similar function. Kitchen counters can be raised to 90cm, the current standard, either with a frame over the cupboard frame or a raised plinth.

Roofs and façades possess distinctive period specific characteristics that should be preserved during maintenance and repairs. Supplementary insulation of façades is generally not cost effective and often distorts the character of a building. In exposed areas, e.g. gables, insulating a limited area can be considered. Roof insulation is, however, often a cost-effective measure not visible from the outside. There are Danish solar collectors made to be installed under existing roof tiles. The energy recovery is not as great as those placed on top of roof tiles. Such solar collectors may be appropriate for protected buildings with an exterior that cannot be altered.

Balconies can be in poor condition due to corrosion of reinforcements or carbonation of the concrete. There are suppliers of period-specific balcony fronts (e.g. corrugated sheet metal). New balconies can be added if none exist, and existing balconies can be enlarged. This should, however, be done so that it preserves the character of the building. Glazing-in balconies provides a veranda-like space. Such balconies can reduce energy loss, but if heated, increase energy consumption. Balcony slabs can be a significant thermal bridge.

Lifts must be installed when an improvement is classified as extensive, in order to improve the building's accessibility. This often involves installing lifts that can be reached from the entrance lobby, either in an existing stairwell or using floor area previously taken up by flats. There are narrow lifts that can be built into stairwells by reducing the stair width. If there is enough land, the stairwell can be added on so that there is room for both the existing stair width and a lift. It is important to consider the noise made by a lift. There are hydraulic lifts, where the lift mechanism is located under the lift if there isn't enough space above it for the lift mechanism.

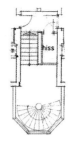

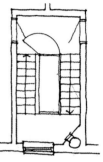

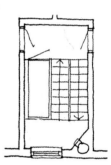

In some cases, to increase the accessibility of buildings being rebuilt, new lifts must be installed. There are several different methods of installing narrow lifts in existing stairwells.

Source: *Varsamt & Sparsamt – Förnyelse av 1950-talets bostäder*, Ingela Blomberg, 2003

Noise disturbance is one of the most common complaints in older buildings, where problems with noise from stairwells, neighbours, and services may occur. Installation of new soundproof security doors to stairwells and additional insulation in ceilings and walls between flats can often be carried out without distorting the character of a building too much.

Implementation of changes with a hurried contracting schedule should be avoided for several reasons. Small disruptions in deliveries or caused by the weather can cause delays that are impossible to recoup. This can result in additional costs and coordination difficulties, as well as problems for residents who plan their lives according to the schedule. The conduct of the building workers is very important for the residents.

Adaptation to Ecocycles

It is often possible to alter the outside environment so that it is more adapted to ecological cycles by planting gardens and kitchen gardens for flats, creating room to mature compost and the opportunity to provide nutrient-rich soil for potted plants. Rainwater can be drained into the ground or used for watering. Runoff can be emphasised with attractive dams and water features. Passive solar heat can be taken advantage of with glazed-in spaces, or solar heat can be actively utilized with roof solar collectors. Waste management can be made more practical with small enclosures for waste sorting and with space for people to exchange second-hand goods with each other. Such a facility can be integrated into local footpaths and bicycle routes. Use of bicycles can be facilitated through provision of covered bicycle stands and bicycle storage rooms.

A planned remodelling of the outside environment which includes kitchen gardens, composting, greenhouses or glassed-in outside spaces, bushes with edible berries and fruit trees, and gravelled footpaths to facilitate local infiltration of runoff.

Source: *Miljöarbete i bostadsförvaltning – från mirakeltrasa till miljöledning*, Ylva Björkholm och Örjan Svane, Byggforskningsrådet 1998

EXAMPLES OF ENVIRONMENTAL REBUILDING

Achieving a sustainable society involves not only building new environmental areas, but also transforming old buildings and quarters into healthy, resource-conserving localities adapted to ecocycles and sitting well in the surrounding area. In Sweden, there are about 800,000 energy-devouring apartments from the days of the Million Programme. If all these apartments were renovated to be highly energy efficient there would be a saving of 36TWh, which is equivalent to 8 per cent of Sweden's energy supply.

Gårdsten, Gothenburg
Source: Architect: Christer Nordström

Gårdsten was designed in the 1960s. Badly maintained deck-access blocks made of concrete gave a grey and dull impression. The buildings stand partly on pillars and the open walkways under the buildings were very unpleasant windy spots due to the west wind.

Former unpleasant windy places were made into pleasant conservatories with green plants where the building entrances are located. Well-lit, modern laundry rooms were placed between the entrances. The glazed-in conservatories under the buildings became a natural meeting place for the residents. A shared composter was also placed in each conservatory.

The low-rise buildings were made more energy efficient by adding supplementary insulation, new energy-efficient windows and heat exchangers in the ventilation system. The character of the surrounding yards was changed completely by painting the buildings different colours, making new porches with roofs and introducing new, appealing landscaping with trees and green plants.

Before renovation. The long side of the high deck-access blocks faces south, which was a good precondition for renovation. The rebuilding changed the area fundamentally with the addition of solar collectors on roofs, glazed-in balconies and glazing-in the ground floor, which transformed windy areas to conservatories.

Eriksgade, Copenhagen

Eriksgade, Vesterbro, Copenhagen. A dense urban area with small flats that lack modern conveniences. An area that was in great need of renovation and improvement. During rebuilding, there was an ambition to raise the environmental standards as well.

The flats were made a little larger by adding bay windows to the sides facing the yard. This made it possible to add bathrooms with a shower in the apartments. Solar cells were placed on the south-facing bay windows. Insulation was added to lofts, and windows were sealed and a third pane added.

A shelter was built in the yard where rubbish can be sorted in an environmental manner. Covered bicycle stands were also built where the residents can store and lock their bicycles. Trees and green plants were planted in the yard and the waste-sorting station and bicycle stand have grass roofs.

A large, top-quality laundry room was built in the basement, equipped with energy-efficient washing machines and dryers. The laundry room is glazed-in up to the ground level so that daylight comes into the basement and residents can sit in the yard and wait for their wash to be done. A social meeting point was thus created in the neighbourhood.

Navestad, Norrköping

Navestad in Norrköping is an area built in the 1960s that never succeeded in attracting residents. It consisted of two rings, an inner one of tall buildings and an outer one of lower buildings. The buildings were made of grey precast concrete and were poorly maintained. The area had a bad reputation and it was time for change, renovation and rebuilding.

Parts of the ring were demolished to open up the view and the tall buildings were reduced to three storeys to obtain a more human scale. Energy conservation measures were carried out – insulation was added, and windows were sealed and replaced. The buildings were renovated and freshened up with pleasant colours, and some of the buildings were made into workplaces.

Järnbrott, Gothenburg

Järnbrott in Gothenburg is made up of four-storey blocks of flats from the 1950s. Supplementary insulation was added and placed so that a small space was left between the old and new insulation to create an air gap. Air solar collectors were placed on the roof and fans are used to move the solar heat through the new air gap to heat the building and reduce the heating requirement from other sources.

The building was also equipped with a greenhouse where every flat has a space for cultivation, and the lawn was dug up to create kitchen gardens for the residents. There is also a common area in the greenhouse where residents can sit and drink coffee. The gardening opportunities are appreciated, people like it there and the residents have got to know each other.
Photo: Christer Nordström, architect.

The Brogården Housing Area, Alingsås

Brogården is a housing area in Alingsås with 300 apartments from the Million Programme in the 1970s that began a five-year renovation plan in 2008. The owner, Alingsåshem, made a thorough analysis of the merits and deficiencies in the area together with the residents. The work on improvements has taken place by partnering with, among others, the construction company Skanska and architects office Efem Arkitektkontor. All outside walls of have been torn down and replaced with new well-insulated ones. The façade consists of cement based sheets. New balconies are placed outside the weatherproofing shell to avoid thermal bridges in the homes. Heat exchangers, which are in a cabinet in each apartment, perform so well that the heating costs are marginal. About 10 days per year the temperature in the heat exchanger is supplemented with district heat. Energy consumption before renovation was $216 kWh/m^2/yr$, and after renovation about $90 kWh/m^2/yr$. Many of those involved with the project have received training to understand the whole picture.

Before renovation

After renovation

DESIGNING / DETAILING THE PUBLIC SPACES
A KEY WORD LIST

P R O T E C T I O N	**1. Protection against Traffic & Accidents** - traffic accidents - fear of traffic - other accidents	**2. Protection against crime & violence (feeling of safety)** - lived in / used - street life - street watchers - overlapping functions - in space & time	**3. Protection against unpleasant sense experiences** - wind / draught - rain / snow - cold / heat - pollution - dust, glare, noise
C O M F O R T	**4. Possibilities for WALKING** - room for walking - untiering layout of streets - interesting facades - no obstacles - good surfaces	**5. Possibilities for STANDING / STAYING** - attractive edges - defined spots for staying - supports for staying	**6. Possibilities for SITTING** - zones for sitting - maximizing advantages primary and secondary sitting possibilities - benches for resting
	7. Possibilities to SEE - seeing-distances - unhindered views - interesting views - lighting (when dark)	**8. Possibilities for HEARING / TALKING** - low noise level - bench arrangements	**9. Possibilities for PLAY / UNFOLDING / ACTIVITIES** - invitation to physical activities, play, unfolding & entertainment - day & night and summer & winter
E N J O Y M E N T	**10. Scale** - dimensioning of buildings & spaces in observance of the important human dimensions related to senses, movements, size & behaviour	**11. Possibilities for enjoying positive aspects of climate** - sun / shade - warmth / coolness - breeze / ventilation	**12. Aestetic quality / positive sense-experiences** - good design & good detailing - views / vistas - trees, plants, water

Research is being carried out at the Centre for Public Space Research in Copenhagen based on the work of Professor Jan Gehl, on quality criteria for shaping the pedestrian environment. The criteria are used in the analysis and design processes at the offices of Gehl Architects. The criteria make it possible to see local surroundings from a human perspective and the human senses. Other physical requirements are given lower priority.

Source: Lars Gemzøe, Architect M.A.A., Centre for Public Space Research/Realdania Research. Associate Partner Gehl Architects

The Apple Blossom day-care centre, in Norrköping, Sweden, by Asmussen's Architectural Office. The children's play area lies between the matron in the kitchen and the gardener in the garden. In the building, there are both rooms with high ceilings and small nooks. The glassed-in area provides a protected zone between the outside and inside.

4.4 People

Planning for a sustainable society involves a holistic perspective that takes into account ecological, economic and social aspects. An ecological society should not only reduce environmental problems but also provide conditions for a good life. Some important elements are private life, togetherness, participation in decision making, cooperation, comfort and beauty.

4.4.0 People's Needs

During the last century, Western lifestyle has changed rapidly from a peasant society to an industrialized society to an information technology (IT) society. People move relatively often and single-person households are much more common today than they were 25 years ago. More people live longer. We don't know which patterns will prevail in the future. Therefore, it is important for buildings and areas to have a built-in flexibility, so they can easily be adapted to new values and conditions.

Planning Based on Human Needs

Even in the context of planning, human needs are used as the point of departure. Ingrid Gehl (from Denmark), among others, has described three types of needs that must be considered:

1. Physical needs such as sleep, rest, food and drink, going to the toilet, keeping one's self and the immediate environment clean, sex, fresh air, sunshine and light.
2. Security needs, such as protection from wild animals, criminality, harmful assaults on our senses (noise, polluted air, humidity, extreme heat and cold), and accident prevention in the home and in traffic.
3. Psychological needs in the form of contact with others as well as self-chosen solitude, experiences, activity, play, beauty, as well as structure and identification, which involves understanding context and one's own role in it.

Planning thus deals with creating an environment where all these needs can be satisfied. This is achieved by working with sizes, components, placement and creating opportunities for stimulation.

A Threefold View

Within anthroposophy, human needs are seen as having a threefold dimension: there

Maslow's pyramid of human needs. People's first priority is to satisfy their basic physical needs. Once these are satisfied they can move on to seek social security and create relationships with other people. Intangible values dominate the higher levels of the pyramid, where our needs are increasingly often packaged in all sorts of experience services – of at least as great an economic importance as goods and basic services.

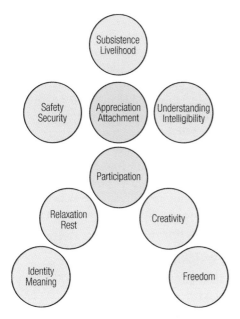

Fundamental basic human needs according to Manfred Max Neef.

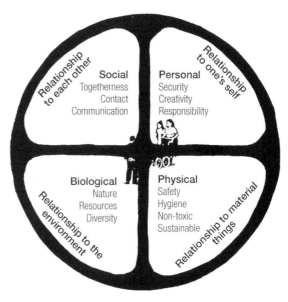

A sustainably planned society deals with peoples' relationships to themselves, to each other, to the environment and to material things.

is the need to move and do physical work, the need to think and reflect, and the relationship people have with themselves and others. These three, working life, intellectual life and social life, take place according to completely different sets of rules (impulses). Intellectual life deals with freedom. A person must be able to think freely and all limitations to intellectual freedom are bad. In social life, equality is what is important. All people are equal and should be treated equally by the legal system. Working life has to do with brotherhood and sharing the fruits of one's labour with others. This threefold division means that equality cannot apply within the intellectual sphere – differences inevitably develop. Freedom does not prevail in social life, where in the worst-case scenario people become involved in honour-related family feuds. Equality does not prevail in working life where a boss has more responsibility than an employee.

Spirituality

People's needs are not only physical and psychological in nature, but spiritual as well, where spirituality involves contact with higher values. In all cultures, both past and present, there are and have been places and buildings to fulfil these needs, including churches, temples, mosques and stupas. Their very special architecture (employing geometry, symmetry, etc.) is designed to facilitate contact with the spiritual. For many people, the spiritual dimension can be fulfilled through poetry, music, nature experiences, tranquility.

Territoriality and Sizes of Groups

Every person has a need for integrity and to be alone sometimes. There are many studies of human territoriality that illustrate this need and how it varies among cultures. Every actor in the building process must examine these conditions and plan for them.

In many cultures the family has traditionally been an important source of security. The need still exists even though an increasing number of people live by themselves. Therefore, it is perhaps extra-important to plan for small groups in the immediate neighbourhood.

Many eco-communities and ecologically oriented residential areas are built for between 5 and 60 households. This is regarded as a suitable size with regard to people getting to know and cooperate with each other.

Architect Anders Nyquist wrote the following about the recreational village Rumpan, near Sundsvall, Sweden, which he was involved in developing in the 1960s: 'The residents' group size of 25–30 families was very consciously chosen. A group of 75–90 people offers a diversity of individuals that makes it possible to both find sympathizers and to be with one's family or alone. Togetherness as well as isolation when desired are both available. If the size of the group increases, anonymity also increases. If the size of the group is smaller, conflicts increase. With 25 families, it is possible to keep in touch and create a feeling of togetherness. Twenty-five families is also a large enough group for carrying out large projects with small contributions per capita in the form of work, money, etc.'

The system biologist Folke Günther believes that planning sustainable communities must include parallel planning of ecocycles, social connections and technical systems. The strategy should be to conserve energy, exergy and materials, as well as the use of methods that benefit the whole system. He believes that a reasonable size for a settlement is 100–200 people.

Composition of Households

In highly developed countries at the beginning of the 21st century, single-person households are the most common household form. In about 30 per cent of households, there are two people, and families with children (mother, father and children) make up only 20 per cent of all households. About 5 per cent of households consist of a single parent with children. However, residential planning often follows the notion that homes should be designed for families with a mother, father and children.

Living Area Requirement

Conserving resources means striving for a reasonable amount of living space per person. Achieving a sustainable society means reducing the living area per capita rather than increasing it.

Flexibility

Since life continually changes, it is good to build flexibility into developments. Then people don't have to move just because their children leave home or they find a new partner. Flexibility can be achieved on different levels. Rooms can be designed so that the same room can be furnished for different functions. Homes can be designed so that the division into rooms can be done in different ways and then changed. Flats can be made so that they can be separated or joined together, equipped with patios or extended. This places demands on a building – that one area can have different floor plans, that walls can be moved, and that services are designed so that they don't impede changes. One flexible way of building a home is to make a flat within a flat, i.e. a home with an area that can be rented out, or perhaps inhabited by adult children or older family members. Flexibility, however, should not be exaggerated. It turns out that people are not so inclined to move walls. And construction needs to be carried out so that poor sound insulation is avoided.

An example of Mies van der Rohe's unconventional design for flats with flexible floor plans in a multi-unit dwelling.

Source: *Bau und Wohnung*, 1927

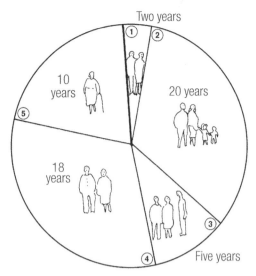

The composition of an average Swedish family during the different life phases A person's living situation changes a number of times during their life and homes should of course be able to function satisfactorily during the different phases.

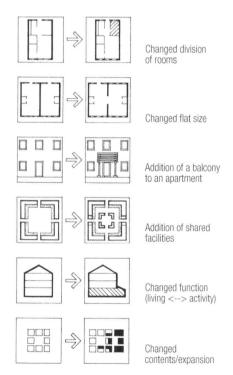

Various possibilities for change and additions to buildings.

Source: *Tid, människor och hus – Idéer kring ett experiment med föränderliga flerbostadshus i Malmö*, E. Nobis, H. Johannesson, J. Calsapeu-Layret, BFR, T47:1982

4.4.1 Comfort

Studies of what makes a residential area feel comfortable reveal some simple principles that are good for everyone involved in planning processes to be aware of. People need areas that are experienced as public, areas that can be semi-private, and totally private areas that are personal territory. An area should feel safe, it should have an identity, it should be cosy and should be easy to explain to visitors. It should also provide opportunities for spontaneous meetings between people.

Turnover Rate

One measure of satisfaction with an area is the turnover rate. Studies by Sjöfors and Orback led to some interesting observations about the appearance of areas where people would like to carry on living. The following characteristics were appreciated:

- A protective patio that cannot be observed from outside is appreciated very much.
- Closeness to semi-private parks and playgrounds.
- An area where it is easy look about and orient oneself.
- The area should have a character of its own that makes it possible for others to know which area it is and where it is located.

Cosiness

For most people, a home is their centre of existence, where they experience safety, trust, involvement, personal development and a sense of rootedness. These feelings

People like their homes to have an identity, to easily feel at home, and for it to feel pleasant and welcoming to come back to.

are often associated with conditions in the home environment: a big reading chair in a corner of the living room, a kitchen sofa at the kitchen table, and a veranda that looks out at a favourite view. Even simple things can be beautiful if the details have been carefully worked out. With an understanding of these factors, it is easy to see how dramatic moving can be for a person, especially if it is involuntary and forced. Necessary renovations should be carried out carefully and with great psychological insight. It is important for those involved to feel that the renovation is advantageous and will result in an improvement.

Security

There are several dimensions to security. One of them is to build an egalitarian society and towns without too much social segregation. Three different aspects of security can be examined:

1 Care and maintenance, e.g. the grass is cut, garbage is collected and playground equipment is in good repair. Wear and tear increase insecurity.

2 A structure and layout where the number of alleys and nooks where attackers can hide is minimized. Evening and night lighting provides greater security. Most attacks take place in poorly lit environments where it is easy to hide.

3 The layout of the estate or town, so that there aren't any 'no man's lands', desolate areas where it is easy to get lost.

Planning for security:

- Mix residences, shops and public offices.
- Avoid placing bus stops in out-of-the-way locations.
- Build small, round inner yards where any unauthorized entry can be seen immediately.
- Make sure that the rear of buildings and back streets are not experienced as unpleasant.
- The entrances, doorways and lifts of individual buildings should be easily and clearly seen.
- Clear away overgrown bushes.
- Improve lighting.
- Create 'safe routes' – lit and wide roads around parks and through residential areas where people can see each other.
- Avoid building dark cellars. Place storage areas directly adjacent to blocks of flats. Laundry rooms should be placed on a gable wall and be equipped with large panorama windows.
- Garages should be small and easy to see and observe.
- Avoid multiple copies of the same key. Instead, install an entry code system that can be changed often or use entry cards that can be cancelled.
- Think about security when planning underground passages and multi-storey car parks. They should be easy to see into and the whole area should be easy to scan.
- Particularly strategic areas could be monitored with cameras connected to a police operations centre.

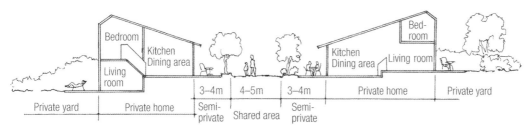

How the outdoor area in a residential area made up of single-family homes can be divided into private, semi-private and public zones.

Private – Semi-private – Semi-public – Public

A good way to plan a neighbourhood's immediate surroundings is to create semi-private zones that moderate the transition from totally private to public areas. Examples are shared open areas, front yards and communal premises. Older architectural designs often included these types of semi-private zones, but functionalist dreams of buildings in parks and large-scale areas of residential blocks lack these important zones. Sometimes the transition is divided up further into private, semi-private, semi-public and public zones.

Gardening Opportunities

Being able to garden close to one's home can be a great source of pleasure and peace of mind. The architects Charlotte Horgby and Lena Jarlöv have shown in their book about residential block gardens that the opportunity to garden near blocks of flats generates a number of positive effects, such as less vandalism and better social contacts.

Close Proximity to a Shared Park

Various studies have shown that living in close proximity to a shared park is one reason that people feel happy or choose not to move. Such parks should provide not only the opportunity for play and recreation but also gardening. Distinguishing characteristics of ecological parks are attempts to increase biological diversity with several different kinds of biotopes, e.g. a butterfly restaurant (plants that butterflies like), ponds and cultivatable areas.

Identification and Understanding

Identification with an area requires an understanding of how an area is demarcated and fits together, which is important to consider at the planning phase. An area should be graspable and identifiable. The technical systems of an area also need to be understood. We need to know how the technical systems work, who has responsibility for what, how decision-making processes work and how we can participate and influence the immediate surroundings. It is natural for a person to have a deeper understanding of a place if they know its history and how the place connects to its surroundings, both the natural and green areas and surrounding building development.

The Attractiveness of Different Types of Urban Developments

How can urban developments be designed so as to be as attractive to as many people as possible? Studies over several decades have shown that many of the most attractive urban developments are dominated by single-family homes, which doesn't prevent an area with both single-family homes and multiple-family dwellings from also being attractive. The attractive areas are primarily characterized by two qualities: (i) they are small-scale, and (ii) there are clear borders between public and private areas.

Types of urban developments regarded as very attractive include pre-industrial towns, areas with detached single-family homes, neighbourhood towns, and garden cities. Urban developments considered attractive include areas with terraced housing and single-family homes close together. Less attractive urban developments include three-storey slab blocks, tower blocks, two-storey slab blocks, and high-rises plus slab blocks. High-rise slab blocks on their own are considered the least attractive.

Spontaneous conversation between residents and a visitor in *Onkel Toms Hütte* in Berlin.

Source: Architect: Bruno Taut, 1920s and 1930s

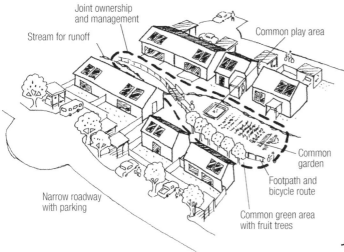

The residents of Davis, California are trying to make their town more environmentally friendly. They have established public pedestrian and bicycle routes as well as streams in the area for runoff. Common areas created for playing, trees and gardening, are managed by the people living in the eight single-family homes that surround the shared area. Ecological adaptation in the US means building in a more European way and not being totally dependent on cars and private areas.

A private patio with favourable climatic conditions, good growing conditions, and protected from view is something that is high on the wish list when people evaluate places to live.

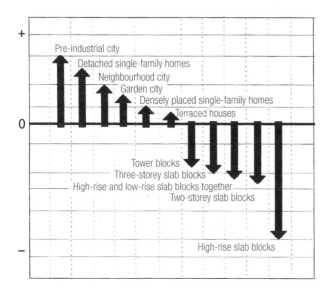

The attractiveness of different types of development. The length of the arrows indicate, respectively, the level of attractiveness and unattractiveness.

Source: *Stadstyp och kvalitet*, Johan Rådberg & Rolf Johansson

4.4.2 Room for Everyone

Sustainable development has not only to do with ecological sustainability but with social and economic sustainability as well. It is common in the context of international development to note that poverty is one of the greatest barriers to sustainable development. In industrialized countries, unemployment and segregation are factors that make it more difficult to achieve desired societal development.

Equality

Development has accentuated the gap between poor and rich countries, and the gap between the poor and rich within countries, and this must be actively counteracted. We cannot continue in a direction that will create a two-thirds society where almost one-third of the population do not feel that they belong or that they are needed. We have to ensure that employment opportunities exist to enable people to make a reasonable living. We need a society that can take care of people with difficulties. In addition, we cannot accept an economic situation where companies with the most inexpensive products (and the worst working conditions) drive socially responsible companies out of business.

Integration

There is an unfortunate division in the housing market, where some areas have gained a bad reputation. There should be active planning policies to counteract segregation and the creation of areas inhabited by only one ethnic or social group. It is necessary to oppose social segregation in many ways and problem areas must be improved. Problems should be solved place by place. Knowledge of different cultures is the key to understanding and cooperation.

Accessibility

What functions well for disabled people, children and the elderly is often good for others. Children should be able to walk to school safely, have after-school activities and play areas. A society where young and old, healthy and sick live side by side results in increased understanding.

Attainability

People should not feel cut off by large motorways or other acute limits. Such obstacles should be bridged with a focus on pedestrians and cyclists of all ages. Places that risk isolation include areas around cul-de-sacs, areas surrounded by major traffic routes, uncultivable land and/or natural areas. The goal should be to develop connections between the different parts of densely populated areas and towns.

Town Life

Many people appreciate the town, its diversity, cultural amenities, and social tolerance, where many different ways of life are accepted. Ecocities are an interesting challenge, and should of course be pleasant places to live in. Jan Gehl is a Danish architect with a special interest in town life and town spaces. He has worked for many years to make Copenhagen as pleasant a city as possible. He believes that pedestrian streets are the backbone of a city centre. Pedestrian streets link squares and open spaces, parks and water features. Pedestrian streets (which should be at ground level, not underground or raised) should be interrupted by squares and open areas. People should be able to sit down or go to an open-air café. The scale is important, small units and many doors as well as small streets with shops are favourable. Careful consideration of the climate is an important part of town planning: the

INDICATORS THAT SUPPORT THE AGENDA 21 PROCESS

Agenda 21 work in Stockholm has focused on outreach projects involving the active participation of and a broad dialogue with the public, the business sector, researchers, people active in clubs and associations, municipal employees and decision makers. One project developed, measured and reported on indicators that support and stimulate processes for sustainable development in Stockholm. Questions asked were: what can make Stockholm a better city to live in and what issues are important? Suggestions for indicators were developed through round-table discussions, reference groups, campaigns and seminars. After compilation and processing, the following 17 indicators were formulated:

Environment
- energy consumption per person
- amount of household waste per person
- amount of heavy metals entering the city
- emission of carbon dioxide per person
- number of days with good air quality
- proportion of passengers using public transport

Economics
- level of employment
- level of education
- sales volume of eco-labelled foods

Social development
- proportion of the population with asthma
- proportion of the population who feel financially secure
- proportion of the population afraid of encountering violence
- time children spend with adults while growing up

Democracy
- proportion of the population doing volunteer work
- election turnout among first-time voters
- proportion of the population who feel involved in society
- proportion of young people under the age of 25 who feel they can influence the development of society

Source: *Indicators for Sustainable Development,* City of Stockholm, 11 September 2003

---

sun's rays should be able to penetrate while the wind should be excluded. City life should be lively and varied, both during the day and at night, so it is good if people live in the city centre. What makes a town so attractive is all the activities found there. No single factor has greater significance for a city centre's vitality than the presence of educational facilities. Students use cities during different times of year and times of day. The town should act as an informal arena for both everyday activities and celebrations, and provide space for both cultural activities and other functions.

Town life is influenced by many factors. It is people and human activity that make a town attractive. In a 'good' town there should be places to sit down and relax and enjoy watching other people. 'People enjoy people' is an old saying from the Viking Age.

Town Space

Kevin Lynch, in his book *Site Planning*, writes that an open space size of about 25m is immediately perceived as pleasant from a social perspective. Spaces larger than 110m do not seem to exist in well-functioning cities. Practically all medieval squares were between 25 and 110m wide and long.

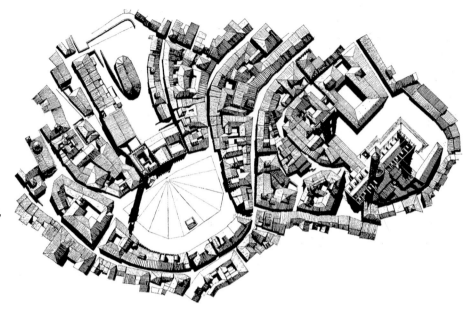

Architects have been inspired by Italian cities from the Middle Ages. Many people think that the irregular, semi-circular sloping square 'Piazza del Campo' in Siena, Italy is the most beautiful square in the world.

Source: *Livet mellem husene*, Jan Gehl, 1971

City life in Amsterdam.

4.4.3 Participation

We have responsibility for each other and the planet. No one is going to give it to us, rather we have to take responsibility ourselves, for other people and the environment. Morals deal with our attitude towards ethics and how this is transformed into action. We have a responsibility to become knowledgeable about environmental problems in the world and how they can be solved.
If one is unable to act personally, responsibility can be taken by joining or supporting an organization working in the area. It is important to influence the development of society, both politically and through day-to-day actions, e.g. as a consumer lobbyist.

Practical Democracy

Unfortunately, current methods of participation in society's development work poorly. Many people feel that they cannot be involved or influence development. This is a problem with democracy that must be taken seriously. New methods of participation that make it easier to be involved and do not require such a big time commitment need to be found.

Resident Involvement in Planning

Involving residents in planning is a difficult process. On the one hand, it is necessary to contact residents before planning the building, and on the other hand, the building process shouldn't take so long that residents decide to back out before the building is finished. Once a group is formed, it is common to begin the process by learning about basic architectural concepts. This can be done in study groups and with lectures. Once design work begins, work can continue in both the large group and smaller workgroups. Workgroups take on responsibility for various areas, e.g. the outdoor environment, shared areas, energy supply, etc. People's work on their own flats alternates between doing individual work (e.g. making a model of a dream flat) and meeting with the architect on a household by household basis. The drafts are influenced as the residents successively gain more knowledge.

Architect Frei Otto worked together with prospective residents on this residential building in Berlin to design the building and individual flats to suit each household.

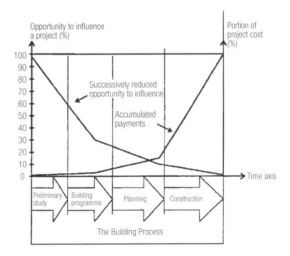

A building process is most easily influenced at the beginning. The opportunities for influence decrease the longer a project has been running and the more investments have been made.

The first level is the building and block (e.g. housing cooperative, eco-community association), where residents can be involved in planning and management. The next level is the neighbourhood or town district (e.g. town district committee or local association of householders). The third level is the city or region (e.g. the municipal and county council policies). The fourth level is the country or group of countries (e.g. Sweden, the Nordic countries, and the EU). The highest level is the global level.

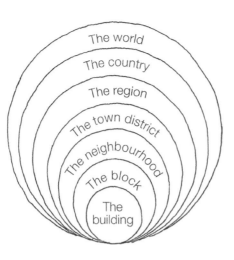

The Tuschteckningen 1 block of flats in Årsta, Stockholm (1990s), where involving residents in the planning was an important part of the planning process.

Source: Architect: Martin Wulff

The need to be able to influence one's surroundings applies to both the physical and the social dimensions. It is especially important to be able to influence activities and places that have a great emotional significance.

It is important to keep in mind four aspects when residents are involved in planning:

1 knowledge building among all those involved;

2 insight about realities;

3 the process takes a lot of time and is administratively cumbersome; and

4 the final product is better with than without residents' involvement in planning.

Participation in Implementation

Architect Anders Nyquist described the concept and implementation of planning principles at Rumpans holiday village outside Sundsvall as follows:

Primary jointly owned facilities, i.e. facilities that everyone uses and is responsible for, e.g. roads (through a road association), paths, green areas, pumphouse, summer waterpipes, jetties, and harbours (through a land-owner's association). Secondary jointly owned facilities, i.e. facilities that are voluntary, are established when enough residents are prepared to invest time, money and work. Opportunities are available for people who join at a later date to be involved in these facilities. Examples: saunas, winter waterpipes, trailer slipways, sports fields, etc. Establishment of primary and secondary jointly owned facilities, as well as their management, is based on a community working together. When like-minded people create something by working hard together, they feel more responsible than when someone 'supplies' the facility. At the same time, creativity is increased, and hidden talents are revealed in people who perhaps never before had a chance to flex their muscles and carry out 'great things'...

The final product is often of higher quality when residents are involved from the beginning, but the process takes more time. The project manager must find a balance between efficiency and sensitivity to the views of the residents. Purchasers and residents should be involved in project planning and have a relatively clear picture of what they want before involving other actors who cost money.

Charette

Charette is a method for creating participation in development planning where all stakeholders can participate. During a charette, which can last for two or three days, people draft and debate in order to arrive at proposals that have different emphases yet manage to reasonably balance existing interests. The method, which originated in France, is most used in the US, but has also been tried by municipalities in Sweden. The charette method requires a fast pace and great commitment. The intention is to avoid long, drawn-out discussions about small details and instead to focus on what is important for everyone involved. Earlier attempts at using this method have had good results. One example of a charette and the need to listen to residents occurred in Norrköping. There, a café, the most popular meeting place in town, was threatened with demolition, but because of its popularity was not demolished after all.

Public Consultation

The workbook method is a consultation process between government agencies/decision makers and people affected by a problem or a change. The method was developed so that things would not only be talked about but that there would also be an attempt to achieve change through a documented process. A project group with project managers has responsibility for the process and consults people from the area. Various workgroups are formed by the people affected. There are three steps to the process. The first step is to understand the problem and present it in words and pictures. The second step is to supplement

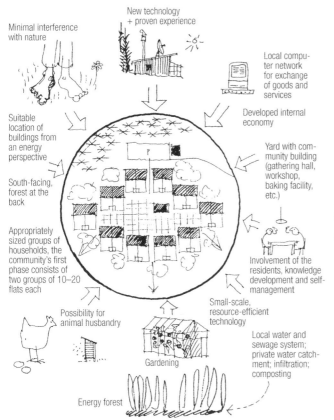

People's conception in an eco-community in Sundsvall, Sweden.

Source: Illustration: Anders Nyquist, architect

In simple terms, the workbook method can be described according to the illustration. Residents, planners, decision makers and others are involved, are influenced and can influence. When the method works optimally, a constructive communication arises between the different parties involved in planning and decision making.

Source: *Så här vill vi göra't – arbetsboksmetoden*, Bo Mårtensson och Lars Orrskog, Byggforskningsrådet 1986

the information and try to get an even more accurate view of what people want. Finally, in the third step, results are summarized and form the basis for the municipality's comments, views and actions.

Private Maintenance and Management

One way to keep living costs down is to take on maintenance and management yourself. This is a common strategy used in eco-communities, not only for financial reasons but because it is believed that it provides possibilities for greater participation and community spirit. A method often used is to form workgroups with special areas of responsibility, such as maintenance of a central boiler, organizing a car pool, grounds maintenance, etc. Division of responsibilities usually changes regularly, e.g. annually.

Different Kinds of Collectives

Collectively run blocks of flats 'get more for less' by sharing living areas and equipment, while at the same time creating a natural community of neighbours. Most collectively run blocks of flats in Sweden have between 10 and 50 flats. All the buildings have communal areas, but the use and size of the areas varies greatly.

In a collective house, a group of between 5 and 15 people share a home, usually with a bedroom for each person. This form of housing is common for disabled people and the elderly. All other areas are shared, e.g. the entrance, kitchen, living room, bathroom, etc. In the Danish '*bofælleskab*' people have their own residences, but communal areas play an important role. In a block of owner-occupied flats, people live on their own but share certain spaces for a variety of purposes. In the Understenshöjden eco-community in Stockholm, the living area per flat was reduced by about 10m^2 (about 10 per cent) in order to be able to build a community building containing a meeting room, kitchen, playroom, laundry, workshop and second-hand room. The meeting room is used for parties, meetings, play and recreation, as well as hobby activities. Other possibilities for communal areas include lounges and small film theatres.

The Private and the Shared

The idea of 'the good neighbourhood' encompasses community spirit, responsibility and cooperating to meet important needs, without loss of personal integrity. Parish councils of today and earlier times are examples of this.

It can be a matter of cooperatively meeting community needs such as childcare and care of the elderly, or transforming a shared courtyard into an oasis for people and animals to enjoy. Some neighbourhoods have established car pools in order to have access to a car that is less expensive than owning one personally. Other activities or facilities that can be shared are pubs, tennis courts, saunas, bakehouses, and root cellars for food storage.

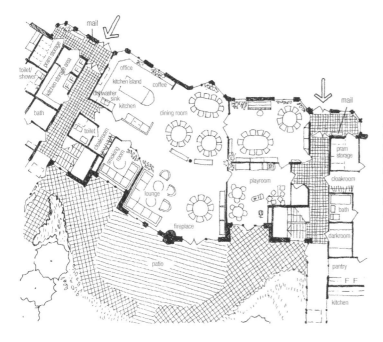

An example of well-planned communal areas which residents pass on the way to their flats and find out if there is something going on. The *Regnbågen* (Rainbow) area, Lund, Sweden, built 1988–1989, architects Rolf Lindström and Lotta Sundström.

Source: *Kök för gemenskap – i kollektivhus och kvarterslokaler*, Gunilla Lauthers, T1:1988

In the Tuberup Venge residential area outside Copenhagen, four flats share a large, glassed-in space. The area acts as an enclosed porch, conservatory and veranda where people can meet and have a cup of coffee. What is more the shared laundry facilities are also located in the same space.

Sharing a sauna is one way to make access to a sauna more economically reasonable. Common sauna evenings can be a pleasant social event in a residential area.

4.4.4 Beauty

There is a beauty deeper than surface aesthetics which designers and architects often bear in mind. Their points of departure may vary, e.g. designing according to the four or five basic elements: fire, water, air and earth. A fifth element that is sometimes included is a tree that symbolizes the life force. How our senses of taste, smell, vision, hearing and touch function can also be used as a point of departure. The past provides many examples of sources of inspiration.

An ecological building, at first glance, is not recognizable as such from its specific architectural design language. Ecological building goes deeper than that. Just as many environmental problems cannot be seen, it may not be possible to see that a building is environmentally adapted. However, there are many elements that characterize ecological architecture, such as vegetation, natural materials, solar collectors, orientation, etc.

The Senses

The word aesthetic originates from the Greek *aisthtikos*, which means 'of sense perception', i.e. what we experience with our senses.

> *Possibilities for variations in volume, scale, materials, etc., are required for an aesthetic and stimulating environment. An unambiguous and boring environment understimulates our brains, which is downright harmful for psychological health.* (Klas Tham, *Boendet och våra omätbara behov*)

Sight

Sight is our most developed sense, through which we register about 80 per cent of our total sensory impressions. Sight's field of use is great – faraway stars can be seen twinkling from light years away. To perceive other people we have to reduce the distance to 100m. At 70–100m it is possible to discern a person's age, gender, way of walking and approximately what they are doing. At a distance of about 30m facial features and hairstyle can be observed, and it is possible to recognize people we don't often meet. At 20–25m we perceive another person's mood and feelings. 'Within this distance it is possible to fall in love.' Normal conversation takes place at a distance of between 1 and 3m. At this distance, it is possible to perceive nuances and details that may be important.

Colour

Colour is usually perceived very quickly, often even before shape is registered. Colour is of course important symbolically and affects people psychologically and physiologically (e.g. a strong red light results in a faster pulse) as well. However, nuances can be more important than the actual colour choice. Pablo Picasso expressed it like this, 'If I don't have a red colour I use a green.' It is of course significant how the impression of colours is affected by their mutual relationship to each other with regard to shade, nuance, area, shine, shape, surface structure and light conditions. The impression of a colour shade can completely dominate in one context while in another it can appear minor.

Light, Light and Shade, and Darkness

Sunlight is one of the most important factors in the creation of a beautiful experience. Light determines a colour's lustre and provides light and shade that changes over time. The light may be direct, indirect or diffuse. Light also has a colour which influences the experience and shifts with place, season, weather and time of day.

Bruno Taut was a German architect who used colour in a very conscious way during the 1920s and 1930s. Pictured here is a multi-family dwelling at 'Onkel Toms Hütte' ('Parrot Town') in Berlin, painted with bright colours using silicone paint.

Windows play an important role as they allow both daylight into a room and a view of the surrounding area. The siting of windows and the design of a window recess are very important when it comes to creating beautiful, light rooms with interesting views. In well-insulated buildings with thick walls, angled window reveals should be used if possible.

It is also worth remembering that darkness has its qualities. Light coming into a building appears different, depending on the direction it comes from.

For some rooms, a south-facing location is not desirable, e.g. studios and kitchens, as the sunlight is too strong. Think about what it is like to sit in the twilight instead of turning on a light, and to slowly experience darkness creeping in and see the stars coming out; or to walk around a town at dusk and see the town's silhouettes darken. However, out of consideration for peoples' feeling of security, streets and footpaths should always be lit during the evening and night.

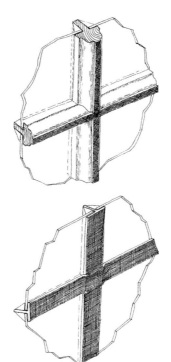

The design of window details is important. Glazing bar profiles create grey scales and soft transitions between light and dark. The risk of glare is reduced.

Sound

Human hearing is quite efficient within a distance of about 7m. Conversations can take place without difficulty up to this distance. Up to a distance of about 35m, it is possible to listen to lectures and have a question-and-answer session, but normal conversation is impossible. At a distance of more than 35m, it is possible to hear loud shouts, but difficult to distinguish words.

Sound affects us more than we think. It affects the heart, stomach, breathing, muscle reactions, blood vessels, hormones, sleeping rhythm and work pace. Silence is a luxury in today's society. Is there anything more irritating than humming ventilation, refrigerators and computers? Sound can also be regarded as a mood-creating factor. The most peaceful sounds are the crackle of a fire, water's murmuring, the whistling of the wind, the patter of rain on a window, and bird song from the garden. The best sound experiences can be composed of a mixture of sound impressions, e.g. listening to a string quartet in a beautiful building where a fire is crackling and rain is pattering on the windows while classical music floods into the room.

In Japanese temples there are wooden floors that sound like singing birds when someone walks on them. In the Japanese art of garden making, there are sometimes swinging bamboo containers called *sozu* that slowly fill with water, then tip over making a loud noise, and then slowly fill up again. In Thailand and China there are stairs made of bamboo or wood that are tuned so that when they are walked on, they play a certain melody or scale.

Smell

Smell is a powerful sense. Scents and smells are among the strongest memories people have. The sense of smell can instantly bring to mind long-forgotten experiences, and nauseating smells can quickly make a room

The scent of a newly tarred church (this one in Norway) is distinctive.

unpleasant. Positive experiences include the scent of flowers and vegetation, freshly baked bread and sensual perfumes. Inside buildings, scents or characteristic smells of wood, newly tarred roofs and façades, and newly soaped floors are discernable. A Finnish smoke sauna has a distinctive scent. At the same time, allergies and health aspects need to be considered.

Touch

The feel of a material is important, both surface texture and temperature. Think for example of the feeling when holding a wooden banister while going downstairs, or the feeling of cold when walking in stocking feet on a hard floor, or the feeling of luxury when your feet sink into a soft carpet under a meeting table.

The granary at Malingsbo in Västmanland is an example of the magnificent architecture that can still be seen in rural areas in Sweden. The holes in the walls are air intakes for the building's natural ventilation system for drying grain.

The Basic Elements

The four basic elements, air, water, fire and earth can prove very helpful when designing buildings. Many cultures include a fifth element that thrives among the four mentioned above. It is usually symbolized by a tree and refers to growth and life. This element is the only one that withstands the rules of entropy and the laws of thermodynamics. It builds up instead of breaking down, and concentrates instead of dispersing, creating more diversity as time goes by.

Air

Thinking about how air moves can result in better three-dimensional work in room design. How does choice of ventilation affect a building's design, and how does this manifest itself in the architecture? If natural ventilation is chosen, the architecture is affected by the shape of tall chimneys and special methods to increase or reduce natural ventilation, and inside, some rooms will have high ceilings, lofts and skylights.

Water

Water can be used outdoors or brought into buildings in many different ways, e.g. in fountains, ponds, waterfalls and streams. Water is beautiful as the sound of running water is soothing, water humidifies the air, and water can cool through evaporation. Artificial waterfalls are often found in entrances to public buildings, hotels, etc.

Fire

Nothing warms like a fire. The importance of a fireplace is deeply rooted: fire kept our ancestors warm, it was used to prepare meals, it brought light and security by scaring away wild animals and, it was thought, evil spirits. It is meditative to sit and gaze into a fire. Fireplaces are included in many ecological buildings. In order to achieve energy-efficient combustion, a fire should be enclosed. A compromise, allowing the fire to be seen, is to have stove doors that can be opened or that include panes of tempered glass.

Earth

For many people earth is connected to the dwelling place, to that piece of the Earth called home. In ecological building, this is the point of departure and there is an attempt to identify the soul of a place, which in Latin is called *Genius loci*. The earth symbolizes fertility, where life germinates; but it also symbolizes decay, where ecological cycles come to a close. Earth is one of the most common building materials in the world. It has many good characteristics for building. It is fireproof and noise-resistant, and it can store heat and buffer moisture and thus contribute to a pleasant indoor climate.

Water sculptures, so-called flow-forms, can be placed indoors to provide beauty, a pleasant sound, and increased air humidity.

Fire has played an important role in traditional buildings. A fire's radiant heat is experienced as pleasant warmth when winter cold holds outdoors in its grip.

4.4

The architect Bengt Warne tries to recreate paradise in his buildings. Paradise is interpreted here by the artist Barbro Hennius.

Simple designs can be beautiful if the details have been carefully carried out. The combination of a building with trees and vegetation can provide pleasant experiences.

Source: Illustration: Heinrich Tessenow

Architectural Quality

Architect Ola Nylander in his book, *Architecture of the Home* (2002) (*Bostaden som arkitektur*, 1999) emphasized questions such as: 'what are the unquantifiable architectural characteristics of a home?' and 'what is it that characterizes homes where people feel happy and where there are good conditions for creating a feeling of home?' His studies identified seven groups of attributes that he thinks are important for how a home is experienced. He calls these groups 'fields' of attributes, which he defines as a distinguishable complex of details, qualities and traits. He tries to describe them using case studies and his own method of identification.

Ola Nylander's seven identified fields of attributes:

1 Materials and details – design of materials and details has a central significance for residents' experience of a home.

2 Axiality – passages and movement can be used to connect rooms to each other. The experience of axiality implies a direct physical relation to the experience of a home's architecture.

3 Enclosure – whether a room feels closed or open is very significant for the way it is experienced. The feeling of cosiness is intensified when a person looks out from a home's enveloping seclusion.

4 Movement – the possibility of being able to move in different ways through the rooms of a house increases the richness of the experience. If the floor plan allows movement in a circle, then freedom of movement is not impeded. A room with two doors is more inviting than a dead end, except in the most private rooms.

Enclosed versus open areas.

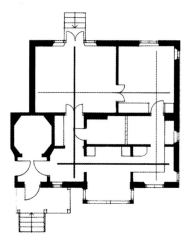
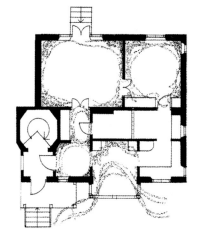
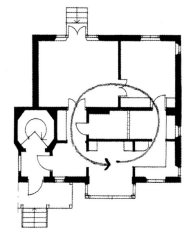

Single-family homes in Lindholmen, Gothenburg, Sweden, finished in 1992 (White Architects). The floor plans have several qualities, among others alignment (main axis and secondary axis), enclosure and circular movement.

Source: *Architecture of the Home*, Ola Nylander, 2002 (*Bostaden som arkitektur*, 1999)

5 Spatial figure – deals with the room's shape and proportions. In anthroposophic day-care centres there is always an attempt to include a large room with a high ceiling and stairs, and a little corner where children can play without being disturbed.

6 Daylight – light is an important factor. Daylight should be able to reach far into a room without dazzling. If daylight enters a room from two directions, the play of light and modulation on the interior furnishings is more complex.

7 Organization of rooms – it is important not to stumble right into the bedroom when coming in from outside. Work with an entrance, a hall, a communal space (kitchen and living room), and bedrooms in sheltered positions.

Sources of Inspiration

Ecological building has taken inspiration from many different sources including traditional building, building from other cultures, 'grass-roots architecture' (i.e. do-it-yourself building with imaginative architecture), the geodesic domes of the hippy era (efficient use of materials, large volume), passive solar homes in the US, underground buildings, buildings with a lot of vegetation (particularly in Denmark and Germany), buildings made from leftover materials, low-tech and high-tech buildings, anthroposophical architecture, organic architecture with shapes from the plant kingdom, nature (e.g. how different animals build their homes), and garden city architecture.

Traditional architecture has a lot to offer ecological building as it has developed from local conditions. It is shaped by the local climate, employs locally available building

Inside a Mongolian yurt.

Source: *Shelter*, Lloyd Kahn, 1973, book cover

self-build architecture can also be found in the district of Christiania in Copenhagen, Denmark. This way of building has been documented in several books, including *Handmade Houses* by Art Boericke and Barry Shapiro and *Woodstock Hand-made Houses* by Robert Haney and David Ballantine. Many people were inspired by the unconventional and artistic do-it-yourself constructions. The buildings were inexpensive since the simplest and most easily accessible materials were used to build them, such as wood, stone and glass.

The Mongolian yurt is an interesting dwelling that has been used for a very long time. It can be taken down since it is used in a nomadic culture. The wall elements are made up of collapsible trellises insulated with several layers of felt mats. The smoke hole in the roof acts as a sundial and calendar and gives the room a special character.

Source: *Shelter,* Lloyd Kahn, 1973

materials, and usually uses resources efficiently, e.g. by placing rooms that require heating around a chimney stack.

Building in other cultures, as documented in the book *Architecture Without Architects* by Bernard Rudofsky for example, demonstrates the huge variation and wealth found in world architecture. Within this great variety, it is of course possible to find exciting ideas from cultures with similar climatic conditions or access to similar local materials.

Earth is one of the most common building materials. Some of the oldest buildings in existence are made of it. Earth buildings of varied architectural styles have been built in many cultures, and in many parts of the world earth buildings are experiencing a revival.

Handmade houses or grass-roots architecture were especially popular in the US during the 1960s and 1970s. Such imaginative

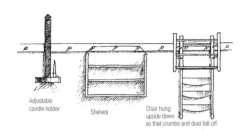

The Shakers were a religious group in the US during the 18th and 19th centuries. They strived for simplicity, beauty and functionality in the objects they created. In particular, their practical furnishings have served as a model for interior decorators and architects seeking environmentally friendly, sustainable and resource-conserving alternatives.

Source: *Commonsense Architecture*, John S.

The domes of the hippy era in the 1960s were inspired by the domes designed by US architect and engineer Buckminster Fuller. He discovered the geodesic dome and believed it to be the most material-efficient way to cover a large volume. The dome is a spherical construction made of triangle-shaped elements, which he patented in 1954. The first geodesic dome in the world was built in Jena, Germany in 1922.

Passive solar house architecture was most in vogue in the US in the 1960s and 1970s. These buildings were all oriented and glassed-in towards the south and in addition, built of heavy materials. At that time, a number of solar energy conferences were held in the US emphasizing this type of building.

Nibble school in Solvik outside Järna, near Stockholm, Sweden. The buildings are made using several different earth construction techniques. This building is made of wet loaves of loam stacked in a masonry pattern without mortar, and covered with clay plaster.

Residences in the shape of geodesic domes in the Torup eco-village, Denmark, built in the 1990s.

Greenhouses have inspired several architects working in the spirit of ecology. Constructing buildings beside greenhouses provides mutual advantages. The architect Bengt Warne has combined buildings with greenhouses in several of his projects. In the Fjällström home, there are three greenhouses, each with a different climate zone: Swedish summer where new potatoes and tomatoes are grown, Mediterranean climate where grapes are grown, and tropical climate where bananas are grown.

The Fredrikshøj area in Sydhavnen, Copenhagen, where residents were able to build themselves their own allotment-garden cottages with a maximum size of 100m², that could be lived in year-round.

Steve Baer's famous passive solar house in Arizona, US, where insulated shutters are opened and closed on the south-facing glassed-in façade. Water-filled gas barrels behind the windows store solar heat.

Hassan Fathy was an Egyptian architect who rediscovered the Nubian vault and was deeply involved in poor people's need for good architecture. He saw traditional building as a style that even the poor could afford.

Source: *Architecture for the Poor*, Hassan Fathy, 1973, book cover

The Fitzgerald underground house in Santa Fe, New Mexico.

Source: Architect: David Wright, US

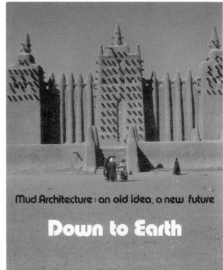

The interest in earth buildings blossomed after the 1981 exhibition, *Des Architecture de Terre*, at the Centre Pompidou in Paris.

Source: *Down to Earth. Mud Architecture: an old idea, a new future*, Centre George Pompidou, 1981

Underground houses became popular in the mid-1970s after the American architect Malcolm Wells declared that for ecological reasons, he would never again build on a square metre of the Earth's surface without replacing it on the building's roof. He published a book entitled *Underground Designs*, which started an underground building trend in the US.

Natural plant life and the homes animals make can serve as an interesting source of inspiration. Termite stacks are shaped so that they maintain a very stable temperature and relative humidity. They are cooled and heated with natural ventilation and underground channels. During the day, they can be heated from natural circulation between the channels in the south-facing wall and the interior of the stack, and during the night they are heated from natural circulation between the ground and the stack. The termite stack maintains a temperature of 28°C and a relative humidity of 90 per cent despite a highly variable outside climate.

Cross-section of a termite stack. The situation during the day is shown on the left and the situation during the night on the right.

Source: *Animal Architecture*, Museum of Finnish Architecture, 1995

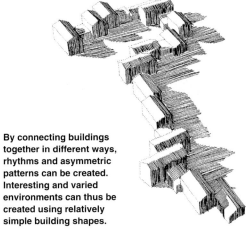

By connecting buildings together in different ways, rhythms and asymmetric patterns can be created. Interesting and varied environments can thus be created using relatively simple building shapes.

Proportions, Rhythm and Balance

No matter how buildings are designed, we want them to be attractive. Architecture is sometimes called frozen music. Structures, sequences, rhythms, patterns and variations are important features. This has meant that there has been a striving for harmony and balance in many cultures during many different periods of history. Some examples follow. One aid used in architecture to achieve harmonic proportions is the golden section. This is a popular term for the division of a line into two segments (a and b) so that the ratio of the length of the longer segment (a) to the length of the entire line is equal to the ratio of the length of the shorter segment (b) to the length of the longer segment (a), i.e. (a+b):a = a:b. That is not all. A rectangle with sides in proportion to each other according to the golden section can be divided into a square and a smaller rectangle, which in turn have the proportions of the golden section, and so on ad infinitum. This is also a way to draw a spiral as found in a snail shell.

A regular five-pointed star can be made by taking a strip of paper and making a common overhand knot with it. The result is a pentagon where the diagonal relates to the sides according to the golden section. This type of exercise is sometimes called golden or holy geometry. To the Pythagoreans, the five-pointed star was a symbol for good, and they liked to put one over the doorway into their homes.

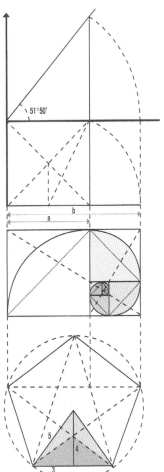

How to construct the golden section, as well as pentagons.

Source: *Duurzaam en Gezond Bouwen en Wonen – volgens de bio-ecologische principes*, Hugo Vanderstadt, 1996

Traditionally, the shape used to describe the heavens is a circle while a square is used for Earth. When these two shapes are joined by giving them the same surface area or perimeter it is referred to as 'the quadrature of the circle', which means that heaven and Earth, or the spiritual and material, are symbolically united.

Another example is that the movements of Venus as seen from Earth trace a very beautiful pattern. That is why the Goddess of Beauty is called Venus. This pattern is used, for example, in church windows.

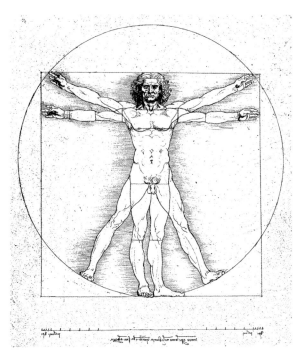

The quadrature of the circle, the unity of the heavens and Earth.

Source: Illustration: Leonardo da Vinci

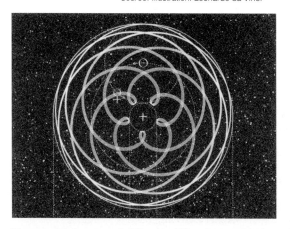
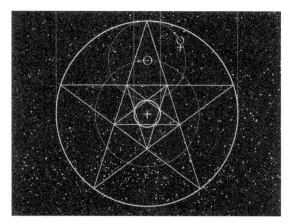

Diagram of the movement of the planet Venus across the sky as seen from Earth.

Source: *Solsystemet – Slump eller dolda strukturer*, John Martineau, 2001

Bibliography

Allard, Francis (ed) (2003) *Natural Ventilation in Buildings: A Design Handbook*, European Commission Directorate General for Energy Altener Program, Earthscan, London

Allen, Paul and Todd, Bob (1995) *Off the Grid: Managing Independent Renewable Energy Systems*, Centre for Alternative Technologies, Wales, CAT Publications

Allin, Steve (2005) *Building with Hemp*, Seed Press, Ireland

Anderson, Jane, Shiers, David and Sinclair, Mike (2002) *The Green Guide to Specification*, Blackwell Science, Oxford

Anink, David, Boonstra, Chiel and Mak, John (1996) *Handbook of Sustainable Building: An Environmental Preference Method for Selection of Materials for Use in Construction and Refurbishment*, James & James, London

Architects Council of Europe (1999) *A Green Vitruvius: Principles and Practice of Sustainable Architectural design*, James & James, London

Arendt, Randall et al (2000) *The Charter of the New Urbanism*, McGraw-Hill

Baggs, Sydney and Baggs, Joan (1996) *The Healthy House: Creating a Safe, Healthy and Environmentally Friendly Home*, Thames and Hudson

Baker, N., Fanchiotti, A. and Steemers, K. (1993) *Daylighting in Architecture: A European Reference Book*, James & James, London

Bear, Steve (1979) *Sunspots: An Exploration of Solar Energy through Fact and Fiction*, Cloudburst Press, Mayne Island and Seattle

Behling, Sophia and Behling, Stefan (1996) *Sol Power: The Evolution of Solar Architecture*, Prestel Verlag, Munich

Bell, Graham (1992) *The Permaculture Way: Practical Steps to Create a Self-Sufficient World*, Thorsons, London

Berg, Per G. (2009) *Timeless Cityland: Building the Sustainable Human Habitat*, Baltic University Press, Uppsala University

Berge, Bjørn (2009) *The Ecology of Building Materials*, 2nd edition, Architectural Press (Elsevier), Oxford

Betsky, Aaron (2002) *Landscrapers: Building with the Land*, Thames and Hudson, London

Boericke, Art and Shapiro, Barry (1973) *Handmade Houses: A Guide to the Woodbuchers Art*, Schrimshaw Press, San Francisco

Boyle, Godfrey and Harper, Peter (1976) *Radical Technology*, Pantheon Books, New York

Brodersen, Leif and Kraft, Per (2005) *Architectural Technology – A Survey of Contemporary Construction: Structures and Materials*, Royal Institute of Technology (KTH) School of Architecture, Stockholm

Brown, Lester R. (2001) *Eco-Economy: Building an Economy for the Earth*, Norton Books

Calthorpe, Peter (1993) *The Next American Metropolis: Ecology, Community and the American Dream*, Princeton Architectural Press

Compagno, Andrea (1999) *Intelligent Glass Facades: Material Practice Design*, Birkhäuser, Basel

Day, Christopher (1990) *Building with Heart: A Practical Approach to Self and Community Building*, Green Books, UK

De Asiain, Jaime Lopez (1992) *Open Spaces in the EXPO 92*, The Superior Technical School of Architecture of Seville

Dethier, Jean (1981) *Down to Earth: Mud Architecture – An Old Idea, a New Future*, Thames and Hudson, London

Douglas, Sholto J. and Hart, A. de J. (1972) *Forest Farming: Towards a Solution to Problems of World Hunger and Conservation*, Watkins, London

Dreiseitl, Herbert, Gran, Dieter and Ludwig, Karl H. C. (2001) *Waterscapes: Planning, Building and Designing with Water*, Birkhäuser Verlag, Basel

Edwards, Brian and Turrent, David (2000) *Sustainable Housing: Principles and Practice*, E. & F.N. Spon, New York

Ekman, Bo, Rockström, Johan and Wijkman, Anders (2009) *Grasping the Climate Crisis: A Provocation from the Tällberg Foundation*, Tällberg Foundation, Stockholm

Falkenberg, Haike (2008) *Eco Architecture – Natural Flair*, Evergreen, Cologne

Fathy, Hassan (1973) *Architecture for the Poor*, University of Chicago Press

France, Robert L. (2003) *Wetland Design: Principles and Practices for Landscape Architects and Land-Use Planners*, W. W. Norton & Company Ltd, London

Fromonot, Francoise (1997) *Glenn Murcott: Works and Projects*, Thames and Hudson, London

Gaffron, Philine, Huismans, Gé and Skala, Franz (eds) (2005) *Ecocity: Book I, A Better Place to Live*, Facultas Verlags- und Buchhandels AG, Vienna

Gaines, Jeremy and Jäger, Stefan (2009) *A Manifesto for Sustainable Cities: Think Local, Act Global*, Prestel Verlag, Munich

Gauzin-Müller, Dominique (2002) *Sustainable Architecture and

Urbanism: Concepts, Technologies, Examples, Birkhäuser, Basel

Gauzin-Müller, Dominique (2006) *Sustainable Living: 25 International Examples*, Birkhäuser, Basel

Gehl, Jan (2001) *Life Between Buildings: Using Public Space*, Arkitektens Forlag, Danish Architectural Press

Granstedt, Artur, Thomsson, Olof and Schneider, Thomas (2005) *Environmental Impact of Eco-Local Food Systems*, Final report from BERAS work package 2, Baltic Ecological Recycling Agriculture and Society (BERAS), Uppsala University

Grindheim, Barbro and Kennedy, Declan (1998) *Directory of Eco-villages in Europe*, Global Eco-village Network Europe, Steyrberg, Germany

Halliday, Sandy (2008) *Sustainable Construction*, Elsevier, Butterworth-Heinemann, Oxford

Hart, Robert (1988) *Forest Gardening*, Green Books, UK

Herzog, Thomas (1996) *Solar Energy in Architecture and Urban Planning*, Prestel Verlag, Munich

Herzog, Thomas (2000) *Architecture + Technology*, Prestel Verlag, Munich

Holdworth, Bill and Sealey, Antony F. (1992) *Healthy Buildings: A Design Primer for a Living Environment*, Longman Group, UK

Houben, Hugo and Guillard, Hubert (1994) *Earth Construction: A Comprehensive Guide*, CRAterre-EAG, Intermediate Technology Publications, London

Howard, Ebenezer (1902, reprint 1965) *Garden Cities of Tomorrow*, Faber and Faber, London

Ingram, Colin (1991) *The Drinking Water Book: A Complete Guide to Safe Drinking Water*, Ten Speed Press, Berkeley, CA

Jackson, Hildur and Svensson, Karen (2002) *Ecovillage Living: Restoring the Earth and Her People*, Green Books, UK

Japan Sustainable Building Consortium (2003) *CASBEE Manual, Comprehensive Assessment System for Building Environmental Efficiency: Design for Environment Tool*, Institute for Building and Energy Conservation, Tokyo

Johansson, Thomas B., Bodlund, Birgit and Williams, Robert H. (eds) (1989) *Electricity: Efficient End-Use and New Generation Technologies and Their Planning Implications*, Lund University Press

Johnston, Jacklyn and Newton, John (1991) *Building Green: A Guide to Using Plants on Roofs, Walls and Pavements*, London Ecology Unit, London

Kahn, Lloyd (1973) *Shelter*, Random House, London

Kennedy, Margrit and Kennedy, Declan (1997) *Designing Ecological Settlements*, European Academy of the Urban Environment, Dieter Reimer Verlag, Berlin

Kourik, Robert (1986) *Designing and Maintaining your Edible Landscape Naturally*, Metamorphic Press, Santa Rosa, CA

Krygiel, Eddy and Nies, Bradley (2008) *Green BIM: Successful Sustainable Design with Building Information Modeling*, Wiley Publishing, USA

Kvarnström, Elisabeth et al (2006) *Urine Diversion: One Step Towards Sustainable Sanitation*, EcoSanRes Publication Series, Stockholm Environment Institute (SEI), Stockholm

Lambertini, Anna and Leenhardt, Jacques (2007) *Vertical Gardens: Bringing the City to Life*, Thames and Hudson, London

Lloyd Jones, David (1998) *Architecture and Environment: Bioclimatic Building Design*, Laurence King Publishing, London

Lopez Barnett, Dianna and Browning, William D. (1995) *A Primer on Sustainable Building*, Rocky Mountain Institute. USA

Lyle, John Tillman (1994) *Regenerative Design for Sustainable Development*, John Wiley & Sons, New York

Lynch, Kevin (1984) *Site Planning*, 3rd edition, MIT Press

Malbert, Björn (1994) *Ecology-Based Planning and Construction in Sweden*, The Swedish Council for building Research

McNicholl, Ann and Lewis, Owen (1996) *Green Design: Sustainable Building in Ireland*, University Collage Dublin

Melet, Ed (1999) *Sustainable Architecture: Towards a Diverse Built Environment*, NAI Publishers, Rotterdam

Minke, Gernot (2000) *Earth Construction Handbook: The Building Material Earth in Modern Architecture*, WIT Press, Southampton, UK

Minke, Gernot and Mahlke, Friedemann (2005) *Building with Straw: Design and Technology of a Sustainable Architecture*, Birkhäuser, Basel

Mollison, Bill and Slay, Reny Mia (1991) *Introduction to Permaculture*, Tagari Publications, Tyalgum, Australia

Morrow, Rosemary (1993) *Earth User's Guide to Permaculture*, Kangaroo Press, Australia

Mostaedi, Arian, Broto, Carles and Minguet, Josep M. (2002) *Sustainable Architecture: Low Tech Houses*, Gingko Press, Barcelona

Ndubisi, Forster (2002) *Ecological Planning: A Historical and Comparative Synthesis*, Johns Hopkins University Press, Baltimore

Oesterle, Eberhard et al (2001) *Double-Skin Facades: Integrated Planning*, Prestel Verlag, Basel

Oijala, Matti (2001) *An ECO-Architectural Guide: A Summary of Ecological Building in Southern Finland*, EKO-SAFA RakAs 4/2001, Soumen Arkkitehtiliittory, SAFA, Finland

Olgyay, Victor (1963) *Design with Climate: Bioclimatic Approach to Architectural Regionalism*, Princeton University Press, Princeton

Olivier, Paul (1997) *Encyclopedia of Vernacular Architecture of the World*, Cambridge University Press, Cambridge, UK

Pallasma, Juhani (1995) *Eläinten Arkittehtuuri [Animal Architecture]*, Museum of Finnish Architecture, Helsinki

Papanek, Viktor (1995) *The Green Imperative: Ecology and Ethics in Design and Architecture*, Thames and Hudson

Pfaffenrott, Jens (2004) *Enhancing the Design and Operation of Passive Cooling Concepts*, Fraunhofer IRB Verlag, Stuttgart

Pratt, Simon (1996) *The Permaculture Plot: The Guide to Permaculture in Britain*, Permanent Publications, UK

Preisig, H. R. Dubach, W., Kasser, U. and Viriden, K. (2001) *Ecological Construction Practice: A–Z Manual for Cost-Conscious Clients*, Werd Verlag, Zürich

Reinberg, Georg W. (1998) *Architecture by Georg W. Reinberg*, Alinea, Firenze

Richards, J. M. Saregeldin, I. and Rasdorfer, D. (1985) *Hassan Fathy*, Concept Media, UK

Ridderstolpe, Peter (1999) *Wastewater Treatment in a Small Village: Options for Upgrading*, WRS (Water Revival Systems), Uppsala

Roaf, Sue, Fuentes, Manuel and Thomas, Stephanie (2003) *Ecohouse 2: A Design Guide*, Architectural Press (Elsevier), Oxford

Roodman, David Malin and Lenssen, Nicholas (1995) *A Building Revolution: How Ecology and Health Concerns Are Transforming Construction*, Worldwatch Paper No 124

Ruano, Miguel (1999) *Eco-Urbanism: Sustainable Human Settlements, 60 Case Studies*, Gustavo Gilli, SA, Barcelona

Rudofsky, Bernard (1964) *Architecture Without Architects*, Doubleday, New York

Rydén, Lars, Andersson, Magnus and Migula, Pawel (2003) *Understanding, Protecting, and Managing the Environment in the Baltic Sea*, Baltic University Press, Swedish University of Agricultural Sciences, Uppsala

Sartogo, Francesca (1999) *Saline Ostia Antica: Ecological Urban Plan with a 93% Integration of Renewable Energies – Ecology and Architecture*, Alinea, Firenze

Schiffer, Herbert (1979) *Shaker Architecture*, Schiffer Publishing Ltd

Schmitz-Günter, Thomas (2000) *Living Spaces: Ecological Building and Design*, Könemann, Cologne

Sick, Friedrich and Erge, Thomas (1996) *Photovoltaics in Buildings: A Design Handbook for Architects and Engineers*, Frauenhof Institute, Germany, and James & James, London

Smith, Peter F. (2001) *Architecture in a Climate of Change: A Guide to Sustainable Design*, Architectural Press (Elsevier), Oxford

Steel, James (1997) *Sustainable Architecture: Principles, Paradigms and Case Studies*, McGraw-Hill, New York

Steineck, S., Carlsson, G., Gustafson, A., Jakobsson, C. and Löfgren, S. (1997) 'Approaches to sustainable agriculture II: Good agricultural practice', in B. Bodin and S. Ebbersten (eds) *A Sustainable Baltic Region, Session 4, Food and Fibres. Sustainable Agriculture, Forestry and Fishery*, The Baltic University, Swedish University of Agricultural Sciences, Uppsala, pp37–40

Stiburek, Joseph and Carmody, John (1993) *Moisture Control Handbook: Principles and Practices for Residential and Small Commercial Buildings*, Van Nostrand Reinhold, New York

Strong, Steven J. and Scheller, William G. (1993) *The Solar Electric House: Energy for the Environmentally Responsive, Energy-Independent Home*, Sustainability Press, Massachusetts, USA

Taylor, John S. (1983) *Commonsense Architecture: A Cross-Cultural Survey of Practical Design Principles*, Norton

Theresia de Maddalena, Gudrun and Schuster, Matthias (2005). *Go South – Das Tübinger Modell*, Ernst Wasmuth Verlag, Tübingen and Berlin

Thoman, Randall (1996) *Environmental Design: An Introduction for Architects and Engineers*, Max Fordham & Partners, E. & F. N. Spon, London

Todd, Nancy Jack and Todd, John (1994) *From Eco-Cities to Living Machines: Principles of Ecological Design*, North Atlantic Books, Berkeley, CA

Todd, Nancy Jack and Todd, John (1979) *Tomorrow is Our Permanent Address*, Harper & Row, New York

Tong, Zhang (2009) *Green North: Sustainable Urbanism and Architecture in Scandinavia*, Southeast University Press, Nanjing

Turner, Bertha (1988) *Building Community: A Third World Case Book*, Building Community Books, London

Unwin, Raymond (1909, reprint 1996) *Town Planning in Practice*, Princeton Architectural Press, US

Vale, Brenda and Vale, Robert (1975) *The Autonomous House: Design and Planning for Self-sufficiency*, Thames and Hudson, London

Vale, Brenda and Vale, Robert (1991) *Green Architecture: Design for an Energy-Conscious Future*, Bulfinch Press, Thames and Hudson, London

Van der Ryn, Sim and Calthorpe, Peter (1991) *Sustainable Communities: A New Design Synthesis for Cities, Suburbs and Towns*, Sierra Club Books, San Francisco, CA

Van der Ryn, Sim and Cowan, Stuart (1996) *Ecological Design*, Island Press, Washington, DC

Van Hal, Anke (2000) *Beyond the Backyard: Sustainable Housing Experiences in their National Context*, Aeneas, The Netherlands

Von Weizsäcker, Ernst, Lovins, Amory and Lovins, Hunter (1997) *Factor Four: Doubling Wealth, Halving Resource Use – The New Report to the Club of Rome*, Earthscan, London

Walter, Bob, Arkin, Lois and Crenshaw, Richard (1992) *Sustainable Cities: Concepts and Strategies for Eco-City Development*, Eco-Home Media, Los Angeles, CA

Watkins, David (1993) *Urban Permaculture: A Practical Handbook for Sustainable Living*, Permanent Publications, UK

Watson, Donald (1983) *Climatic Design: Energy-Efficient Building Principles and Practices*, McGraw-Hill Book Company, New York

Weismann, Adam and Bryce, Katy (2006) *Building with Cob: A Step-By-Step Guide*, Green Books, UK

Wells, Malcolm (1981) *Underground Designs*, Brick House Publishing Company, Andover, MA

Whitefield, Patrick (1993) *Permaculture in a Nutshell*, Permanent Publications, England

Wilhide, Elizabeth (2002) *ECO: The Essential Sourcebook for Environmentally Friendly Design and Decoration*, Quedrille Publishing Ltd, London

Winblad, Uno and Simpson-Hébert, Mayling (2004) *Ecological Sanitation*, Stockholm Environment Institute, Stockholm

Wines, James (2000) *Green Architecture*, Taschen, Cologne

Wizelius, Tore (2006) *Developing Wind Power Projects, Theory and Practice*, Earthscan, London

Woolley, Tom (2006) *Natural Building: A Guide to Materials and Techniques*, The Crowood Press, Ramsbury, UK

Zeiher, Laura C. (1996) *The Ecology of Architecture: A Complete Guide to the Environmentally Conscious Building*, Whitney Library of Design, New York

Index

accumulation tanks 165, 377–382, 390, 398
acetic acid 96
acidification 340
acrylate compounds 80, 94, 98
acrylic paints 84
adaptation 482, 512–545, 594, 599, 633–634
additives 88–89
adhesives 19, 74
 see also glues
admixtures 18–19, 24
adobe houses 21–22
advanced daylight 309–313
advanced window openers 129
aerators 322
aeroplanes 569, 571–574
AGA stoves 392
Agenda 21 652–653
aggregates 17–18
agriculture 318, 478, 503–510
 see also farm level
air
 basic elements 663
 conditioning 412–413
 heat pumps 407–408
 movement in rooms 117–118
 open walls 268
 quality 112–113, 486
 solar energy 403, 611
airborne heating systems 168
aircraft 569, 571–574
air-jacketed stoves 393, 395
Airpod 567
Air Star 130
airtightness 56–58, 105, 176–177, 249–253
air-to-air heat exchangers 118–119, 129, 264–265
algal blooms 505
alkyd oil paint 75
allergies 488–489
allotments 500, 667
alternatives to electricity 309–316
aluminium 23, 51
aluminous cement-based liquid screed 73
ammonia 358
amorphous silicon 437

animal-based glues and pastes 97
animals 482, 501, 533, 668
anti-rust paints 84
appliances 223, 288–295, 306, 308, 326, 633
aquaculture 469
aquatrons 461–462
arable land 590
architecture 269–283, 312–313, 445–446, 490, 664–665
arsenic 340
artificial aspects
 fertilizers 504–505
 filters 464–465
 infiltration 332
 photosynthesis 443
 water surfaces 525
 wetlands 466–468, 470
asbestos 625–626
ashes 383, 387–388
asphalt
 binding agents 19
 decontamination 625
 glues 98
 ground cover 29
 infiltration 525, 527
 roofing 53–54
 weatherproofing 58–59, 61
attractiveness 650–651, 653–654
automatic regulation 113–114
axiality 664

baking ovens 392
balance 669–670
balconies 633
ballasts 303
Baltic Ecological Recycling Agriculture and Society (BERAS) 505–506
bamboo 68
bank loans 213–214
bark 30, 385
barrier effects 542
base loads 405
basements 107, 122, 175, 179, 622, 639
basic elements 662–663
basic needs 592–593

batch treatment 462
bathrooms 105–107, 136, 158, 190–191, 631
 see also wetrooms
batteries 419–421, 432–433, 443, 565–566, 625
beams 34, 102
BEAT 2000 13
beauty 660–670
bedrooms 91, 149
beekeeping 502
bentonite 58, 94
benzol 25, 39
BERAS *see* Baltic Ecological Recycling Agriculture and Society
bicycles 190, 558–561, 584
BIM *see* Building Information Modeling
binding agents 19–23, 36, 65, 77, 85, 88
biocides 85, 98
bio-diesel 568
biodynamic agriculture 507–509
bioenergy 383–397
biofuels 567–568
biogas 361, 475–478, 508, 567–568, 594
biological
 decontamination 517
 diversity 484, 532, 547
 nitrogen reduction 448
biological oxygen demand (BOD) 452
biomass 377–387, 419, 423–426
biotopes 530–531, 547
birch bark 59–60
bitumen *see* asphalt
Björn Berge's system 109–110
black liquor 383, 385, 425
black-water 361, 457–458, 460, 473–478
blasting 520
blood albumin glue 97
blue wedges 484
boarding 15, 45, 55
boats 571
BOD *see* biological oxygen demand

body heat   235
boilers   163–164, 167, 329–330, 388–390
boreholes   408–409
boron   340
brass   23
Brayton cycle   424
BREEAM *see* Building Research Establishment's Environmental Assessment Method
bricks   17–18, 29, 33, 44, 48, 71, 78, 222–223
briquettes   385, 387
Brogården Housing Area, Alingsås   641
brown soap   192
Building Information Modeling (BIM)   206–207
Building Research Establishment's Environmental Assessment Method (BREEAM)   203
Building Supply Assessment   7
buried pipes   158–161
buses   562–563
business ratios   612–613
butterfly gardens   501
butyl-based sealing compounds   94
butyl rubber (IIR)   60

cables   222
cadmium   628
calcium carbonate flooring   70
calcium silicate   37, 62
carbon   359
carbon dioxide   117, 294, 374, 383–384, 556–567, 569
  see also emissions; greenhouse gases
care and maintenance   172–173
carpentry oils   82–83
cars   374, 563–570, 577, 583, 598
CASBEE   204, 206
cascading energy   228
casein   75, 79–80, 97
caulking compounds   93
caustic liquor gasification   424
cautious rebuilding   629–641
ceilings   16
cell phones   569
cellular glass   32, 37–38
cellular plastic   38–39, 221
cellulose
  glues and pastes   100
  insulation   37–39, 44, 101
  joints   93

moisture   176
  wallpaper   67–68
cement
  adhesives   74
  binding agents   19–20
  cladding   47–48
  flooring   68–69
  glues   97
  roofing   48–49
  screed material   73
  sheets   61–62
  wood-wool   36, 44–45, 48, 65
centralized composters   361
centres for recycling   355–356
ceramic flooring   69, 192
certification   506–507, 510, 615
CFCs *see* chlorofluorocarbons
Charette   657
charging systems   163–165
chemicals
  cleaning   190–192
  construction materials   9–11
  decontamination   517
  sewage   449–450, 459, 465
chickens   480, 501
chimneys   136–137, 139–140, 397
chipboard   62, 64, 222
chips, fuel   30, 385–386
chlorine   25
chlorofluorocarbons (CFCs)   626–627
  see also freons
chloroprene   99
chromium   23
CIGS cells *see* copper-indium-gallium-selenide cells
circuit breakers   148–150
cities   484–489, 592
  see also district level; municipal level; towns; urban areas
cladding   45–48, 52, 78
classification, environmental   202–206, 613–614
classrooms   273
clay
  binding agents   20–23
  expanded   19, 34, 41–42, 69–70
  foundations   520
  moisture-buffering   105
  paints   80
  plaster and mortar   66–67
  reuse and recycling   221
  roofing   54–55

sheets   63
  structural materials   32, 34
  thermal insulation   41–42
  weatherproofing   58
cleaning   187–192, 216, 318–319, 382, 397
clean water   317–344
climate   482, 512, 518, 534–545, 594, 599
closed air solar heating systems   403
closing ecological cycles   372–375
clothes dryers   291
cloth facades   244
cob houses   21
coconut fibres   15, 31, 40, 93–94
Code for Sustainable Homes   204
cold air sinks   544
collectives   658
collectors, solar   398–399, 400–403, 408
colour   660–661
combined facilities   478
combined-function rotary terminals   137
combined heat and power technology   423–426
combined materials tests   104
combustion
  bioenergy   385–396
  hydrogen gas   421
  sewage   473
  tuning systems   621
  waste   347–349, 352–353, 385–386
comfort   112–113, 560–563, 648–651
communal areas   659
compact filters   464–465
compact fluorescent light bulbs   297–298, 304–305, 366
compact towns   550–551
companion planting   480
composting   352, 358–361, 454–455, 457, 473–474, 498
computer-controlled regulation   114–116
computers   153–154, 306–307, 366, 616
concentrating solar collectors   400
concrete
  description   18–19
  flooring   68, 192

ground cover   29
moisture   216
reuse and recycling   221–222
screed material   73
structural materials   32–34
surface treatments   78
thermal insulation   41, 43
condensation   176–177
condensation dryers   291, 315
congestion charges   556
conscious consumption   350
consciousness, environmental   613
conservation   2, 225–370, 531–532, 618–624
conservatories   635
constructed natural systems   466–470
construction
    costs   211
    environmental   196–198, 201
    green roofs   492
    materials and methods   2, 5–110, 180, 215–224, 625–628
    resource consumption   227
consultants   200–201
consumption
    conscious   350
    energy   226–229, 374–375, 618–619
    fuel-efficient cars   564–567
    goods transportation   572
    operational   616
    production proximity   589–590
    travel   560
contamination   334–335, 517
    see also decontamination; pollution
continuous flow mini purification   463–464
contractors   200–201, 208–209
control systems
    interior climate   113–116
    lighting   301–304
    shower stop buttons   323–324
    ventilation   621
    see also regulation
convection   249
cookers   136, 138, 292, 295
cooling, evaporative   135, 413
cooling systems
    conservation   226, 232–283
    flue-gas   382
    passive   122, 278–283
    recycling   366

renewable heat   372, 411–415
cooperation   201, 471–473
Copenhagen, Denmark   557, 558, 563, 637–638
copper   23, 52, 222
copper-indium-gallium-selenide cells (CIGS cells)   441
core board   63
cork   40, 70, 184
cork granule sealant   94
corridor effects   542
cosiness   648–649
costs   211–213, 438, 619
cotton   37
courtyards   275, 491, 538
cowls   397
cross-draughts   280
cross-flow turbines   329
cruise ships   572
cul-de-sacs   579–580
Culemburg, The Netherlands   582
cultivation   372, 471, 479–510, 531
cultural aspects   518, 603–610, 666
culverts   125, 160, 379
Curitiba, Brazil   562–563
current limiters   286–287
curtains   263
cycle paths   558–560

damp   174–177
    see also moisture
dampers   130
dangerous products   368–370
    see also hazardous waste
darkness   660–661
Davis, California   651
daylight   258, 262–263, 279, 296–297, 303, 309–313, 665
decentralized concentration   551, 589
decentralized ventilation systems   130
decontamination   517, 625–628
dehumidification   280–281
delivery services   573–574
democracy   653, 655–659
demolition   6, 219–220, 626
Denmark   126–127, 224, 557, 431, 478, 558, 563
desert climates   543
design
    architecture   269–271
    beauty   664

conservation and well-being   2
ecological   368–370
natural ventilation   123
participation   655
passive solar heat   274
permaculture   480–483
public spaces   654
radon protection   179
wind   541
windows   255
destruction facilities   365
detectors   303–304
developed countries   438–439
developed landscapes   531
development   516, 540–541, 595, 650–651
DGNB   204
diesel   567–568, 570
digested sludge   475–476
dimethyl ether (DME)   568
direct-driven ring generators   433
direct drive pumps   337–339
dirt traps   187
diseases   318
dishwashers   291–292, 326
distemper   81, 85–86
distillation   341, 344
distorted buildings   624
distribution   6, 146, 379
district level
    cooling   412
    heating   379–381, 383, 388–389, 423–425
    holistic towns   581–586
    new urbanism   579
    see also municipal level
ditches   527
diversity   480, 484, 530, 532–533, 547
DME *see* dimethyl ether
domes   667
domestic heating   377
doors   118, 221, 234, 252, 621, 632
double-glazing   127, 129, 255–256, 275–276
double walls   183–184
drainage
    basins   524
    floor   156–157
    hydrology   525
    liners   30–31
    permaculture   480
    pipes   89
    soil moisture   175–176

draught limiters   397
drilled wells   332–333, 336
driven-point wells   333–334
drum composters   358
dry-cleaning   294
drying cabinets   291, 313
dryness   542
drystone walls   18
dry toilets   454, 457, 460
ducts   122, 156, 185, 186
   see also pipes
dug-out buildings   269, 278, 280
dug wells   333–334
dyes   93
dynamic insulation   268

Earth   663, 670
earth
   green roofs   493
   houses   20–21, 280, 668
   leakage circuit breakers   148
   pipes   141–143
earthing   148–149, 152
earthquakes   522
eco-communities   657
ecocycles   359–361, 371–510, 590–591, 600, 633–634
ecoducts   485
EcoEffect method   206, 614
eco-houses   237
ecological aspects
   agriculture   503–510
   area principles   533
   city districts   581–586
   closing cycles   372–375
   design   368–370
   green areas   487
   sewage   448–450
   sustainability   548
Eco-Management and Audit Scheme (EMAS)   195–196
economics   109, 211–214, 549–550, 652
Eco-Saver   298
eco-villages   277
edge zones   480, 483–484
efficiency
   buildings   271
   construction costs   212–214
      electricity   226, 284–316
      glass   46
      heating   233–247
      insulation   248–253
      solar cells   437
      see also energy
egg oil tempera   75
EIA *see* environmental impact assessment

eight-sided houses   271
ejector tanks   473
electrical devices   306–308
electricity
   cleaning   190
   conservation   226, 230
   efficient use   284–316
   energy   235–236
   fuel cells   570
   heat pumps   405, 410
   renewable   372, 416–446
   services   144–154
   supply   2, 146, 148
   ventilation   120
electric vehicles   565–566, 569
electrochromic glass   258
electro-climate   112
electronic high-frequency (HF) ballasts   303
electronics   366, 428, 574
electrostatic charges   114
EMAS *see* Eco-Management and Audit Scheme
emissions   10, 348, 556, 561, 567, 569
   see also carbon dioxide; greenhouse gases
emulsion paint   75
enclosure   664
endangered species   532
energy
   bioenergy   383–397
   building   271, 612
   conservation   226–229, 233–247, 618–624
   crops   384–385, 600, 602
   ecocycles   373–375
   environmental business ratios   613
   glass   46
   hot water   327–328
   insulation   248–253
   labelling   288
   lighting   298–299
   location   536–537
   planning   594
   renewable heat   372
   sewage   448
   traffic   556, 560–561, 572
   windows   254–263
   see also efficiency
engines   567
enhanced filtration   464–465
entropy   228
environmental impact assessment (EIA)   516–517
environmental aspects
   adaptation   512

adhesive compounds   74
Agenda 21   652
business ratios   613–614
cladding   46, 48
classification   202–206, 613–614
construction   2–3, 11–13, 102–104
energy   234
fillers   73
flooring   72
fuel cells   569
glues and pastes   100
hazardous waste   354
heating   233
hydropower   427–429
implementation   2
inventories   625–628
joints   96
labelling   8, 322
management   194–198, 615
packaging   364
pipes   90, 630–631
planning   200
plaster and mortar   67
rebuilding   630–641
recycling   109, 364
roofing   54–55
sewage   448
sheet materials   64
structural materials   32
thermal insulation   37, 45
ventilation   120
wallcoverings   68
water quality   340
weatherproofing   58
EPDM *see* ethylene-propylene rubber
EPI glue *see* isocyanate glue
epoxy   25, 98
equality   652–653
Eriksgade, Copenhagen   637–638
erosion   522
erosion mats   31
e-TAP environmental label   322
ethanol   567–568
ethylene-propylene rubber (EPDM)   60, 97
ethylene vinyl acetate glue (EVA glue)   98
Europe   288, 352, 438, 505, 560, 581–586
eutrophication
evacuated solar collectors   400
EVA glue *see* ethylene vinyl acetate glue
evaporation   280

evaporative cooling   135, 413
exhaust air
   cooling   414
   heat pumps   118–119,
      265–266, 407
   heat recovery   264
   radon   181
   ventilation   123–124, 126,
      128–129, 136–143
Exhausto brand   132
existing buildings   512, 606,
   611–641
expanded clay   19, 34, 41–42,
   69–70
extending old structures   607
exterior treatments   87

facades   172, 244, 445, 633
Factor 4   226–227
faeces management   473
   see also sewage
fair trade   8
family   645, 647
fan-reinforced natural ventilation   118, 121, 123–128
fans   172, 186
farm level   471–473, 506–508,
   591, 604–605, 608
   see also agriculture
faxes   306
FCF *see* future contamination
   factor
feedback   209–210
fees   211
felt   53–54, 57, 59, 61
fertilizers   448–449, 451,
   472–473, 480, 504–505, 524
FFP panel   170
fibre   14–15
fibreboards   44, 48–49, 56–57,
   62–63
fibreglass   32–33
fibre optics   310–311
fibre paint   75–76
fillers   61, 73–74
filtration   336, 341–344,
   464–465
Finland   392–393, 484, 607
fire   626–663
fireproof paint   75
first generation biofuels   567
fish ladders   427–428
fittings and furnishings   91–93
five-pointed stars   669
five-wire systems
   (TN-S system)   151
flashing   104

flat-plate solar collectors
   400–401
flats
   collectives   658
   energy   238
   environmental rebuilding
      635–641
   gardens   498
   low-cost   214
   planning participation
      655–656
   sound   183–184
   see also multi-family dwellings
flax   15, 40–41, 92–93
flexibility   109, 646–647
Flipper   131
flooding   529
flood irrigation   469
flooring
   cleaning   191–192
   heating   167
   laying   217
   materials   68–72
   plumbing   156–158
   recycling   219
   sound   183–184
   surface treatments   78–79,
      87
   timber   15
floor plans   123, 272–273, 665
flora and fauna   516, 530–533
   see also vegetation
flow
   energy   234–236
   increasers   325–326
   restrictors   322
   sustainable societies
      594–595
   turbines   429–430
flower meadows   488
flue-gas cleaning   382, 397
flue-gas coolers   382
fluoride   340
flush toilets   324–326, 455,
   457, 460
foamed concrete   32, 43
foam glass   101, 109
foam sealants   96–97
fodder   503–504
food   314–316, 503, 508, 589,
   591–593
forests   384–385, 489, 499,
   503–510, 590–592
Forest Stewardship Council
   (FSC)   8, 510
formaldehyde   19, 36, 42,
   63–64, 72, 98

fossil-fuelled power stations   416
foundations
   damp   175–176
   energy   235
   insulation   251
   land   520
   laying   521
   plumbing   158
   suspended   90, 107–108,
      521
fountains   543
four-wire systems   151
fragmentation   553, 575, 577,
   579
Francis turbines   429
free cooling   411–412
freezers   289–290, 293
freon   367, 405, 626–627
fruit trees   491
FSC *see* Forest Stewardship
   Council
fuel
   cells   421, 566–570
   wood   30, 383, 386–387
fuel-efficient cars   564–567
fungicides   88–89
funnel effects   542
furnishings   91, 190
future aspects
   aircraft   573
   holistic towns   575–588
   plastics   25
   waste management   350–351
future contamination factor
   (FCF)   23–24

gallium arsenide   437
galvanized steel   23, 51
garden cities   587–588
gardens   490–491, 494,
   496–502, 543, 613, 650
Gårdsten, Gothenburg   635–636
gas   314, 424
gasification   424–425
gas-powered Stirling
   engines   410
gates   632
genetically modified organisms
   (GMOs)   506–507
geodesic domes   667
geographic information systems
   (GIS)   513
geology   516, 519–523
geomagnetic fields   153–154
geoproducts   30–31
geothermal energy   373, 418
geothermal heat   407, 409

Germany 127–128, 438, 554, 581, 583
GIS *see* geographic information systems
give-and-take relationships 483
glass
    cladding 45–47
    facades 172
    foam 101, 109
    materials 16–17, 32
    passive solar heat 274–276
    roofing 49
    thermal insulation 37–38
    wallcoverings 67
    windows 257–261
    wool 42
glassed-in spaces 127, 274, 536, 624
GlassXcrystal 259–260
glaze paint 76, 80
glazing 255–259
globalization 547
global level 375, 504, 535, 547
global warming potential (GWP) 406
glues 68, 81, 97–100, 627
    see also adhesives
glulam beams 34–35
GMOs *see* genetically modified organisms
golden section 669
goods traffic 570–574
Gothenburg, Sweden 607, 635–636, 640
Gotland, Sweden 605
granaries 663
grass 29, 172
grass roofs 50–51, 527
    see also green roofs
Grätzel cells 441–442
gravel 29–30, 527
graveyards 609
gravity systems 166, 522
great fen-sedge roofs 53
green
    areas 172, 486–487
    building concept 203
    cities 484–489
    roofs 49–51, 58–61, 281, 492–494
    towns 550–551
    vitriol 77, 80
    walls 281
    wedges 484
Green Building Rating System (LEED) 203–204
greenhouse gases 556

    see also carbon dioxide; emissions
greenhouses 469, 494–495, 640, 667
Green Star 204
grey water 458, 469–470, 472
grid connection 432–433, 439, 443
grocery stores 573
ground
    cover 28–31, 487, 525–527, 543, 545
    radon 180–181
    slabs 108
    waterproofing 58–61
grounding 150
ground-source heat pumps 407, 410
    see also heat pumps
groundwater 411
GRUDIS culvert trench 379
gutters 90
GWP *see* global warming potential
gypsum
    binding agents 20
    fillers 73
    materials 19
    plaster and mortar 66
    reuse and recycling 221
    sheet materials 62, 64
    surface treatments 78

habitat 591–592
Håbo Tibble, Sweden 608
halogen-free wiring 152
halogen light bulbs 297
Hamann turbines 330
handmade houses 666–667
handovers 209–210
hand pumps 337–338
hand-split roof shingles 54–55
hanging gardens 490–491
hard fibreboard 56–57
harmonic proportions 669–670
Haydite balls 19
hazardous waste 354, 362, 364, 365–366, 368–370, 625–628
HDF 63
HDPE *see* high-density polyethylene
health 2, 9–10, 27, 91–92, 178, 318, 567
healthy buildings 1–224, 613
heat /heating
    conservation 226, 230, 232–283

exchangers 141
islands 544–545
pipes 90
pumps 118–119, 265–266, 335, 404–410, 413
recovery 264–268
renewable 372, 376–415
solar 398–403
storage 101–102
surplus 117
systems 162–170, 619–620
water 327–330
heated glass 260
heated greenhouses 495
heavens 670
heavy buildings 101–102, 143, 234, 269, 272, 278
hedges 172
helium balloons 435
hemp 40–41, 63, 93
herb spirals 497
HEV *see* hybrid electric vehicles
HF *see* electronic high-frequency ballasts
high-density polyethylene (HDPE) 60
historical aspects 608–609
hobbit homes 520
holistic approaches 552, 575–588
holographic heliostats 313
home heat pumps 406–408
honeycomb brick 17, 44
hot-air stoves 395
hot compost 358–361
hot water 327–330
household level
    chemicals 450
    combustion devices 389–396
    energy consumption 286–287
    size 645–646
    waste 346–347, 352–354
    water use 318, 320
Houten, The Netherlands 560, 584
human
    needs 644–647
    waste 346–351
humidity 112, 132, 137
    see also moisture
hybrid electric vehicles (HEV) 565–566
hybrid ventilation systems 128
hydraulic lime plaster and mortar 66

hydraulic rams   339
Hydricity   569–570
hydrocarbons   390
hydroelectric power   374, 416–417, 419, 427–428
hydrogen   567
hydrogen gas   421–422, 442, 566, 568–570
hydrology   516, 524–529
hydropower   427–430
hypocaust systems   162, 170

icehouses   411
IIR *see* butyl rubber
impact assessment   202, 207–208
implementation   2, 193–224, 402, 633, 656
impregnated wood   76–77, 627–628
incandescent light bulbs   297, 300, 366
incineration of waste   347–349
indirect daylight   279
individual energy metering   228–229
indoor environments *see* interior environments
industrial level   381, 478, 503–505, 609
infiltration   332, 525, 527
infrastructure   514, 609
In Line Systems   328
inorganic materials   29
input materials   6
inserts, stove   396
in-situ water treatment   344
inspections   209, 618, 620
inspiration sources   665–670
instruction signs   158
insulation
    basements   107
    cellulose fibre   101
    cooking pans   295
    damp   175–177
    dynamic   268
    energy   248–253, 618
    existing buildings   611
    food storage   315
    glass   46, 257
    green roofs   50, 493
    hot water heaters   329
    lightweight studs   102
    moisture   216
    organic materials   56
    passive solar heat   378
    pipes   89

sound   183–185, 254
super-   277
supplementary   622
thermal   37–45
transparent   43, 259–260
walls   248, 251, 293
windows   254, 257, 262–263
intake air   122, 124, 126, 129–135, 141, 403, 414
integration   577, 579–580, 653
intellectual life   645
intensive gardening   496–502
intensive land usage   482
interior environments
    care and maintenance   172–173
    construction materials   3, 6–7
    environmental business ratios   613–614
    fittings and furnishings   91–93
    heating pipes   90
    services   112–143
    surface treatments   87
    vegetation   495
intermediate plugs   152
international level   285, 433
inventories   512–513, 518, 530, 612, 625–628
irrigation   469–470
ISO 14000 Series   194–196
isocyanate glue (EPI glue)   99
isocyanates   64
isolation transformers   148
Italy   654

Japan   438, 662
Järnbrott, Gothenburg   640
joints   93–97, 175–176
joists   34
junction boxes   151–153, 157–158
jute   67, 93

Kalle hood   138
kindergarten   245
Kiruna method   165
kitchen gardens   500
kitchens
    appliances   223, 288–295, 306, 308, 326, 633
    boilers   390–391
    environmental rebuilding   633
    larders   314

plumbing   157–158
    waste   354
KMP Neptuni stove   397
knowledge of materials   14–27
KRAV   506–507
KYRO system   218

labelling   8, 288–289, 322
lacto-fermentation   315
laminated flooring   72, 192
laminated sheets   63
lamp ballasts   303
land   517, 520, 590–592
landfills   349–350, 352, 354, 478
landscapes   513–514, 531, 594
lanterns   312–313
larch cladding   47
larders   289, 314–315, 633
large heat pumps   408–409
large-scale
    hydropower   427
    sewage separation   453
    solar energy   439–440
latex paints   84
laundry rooms   639
LCA *see* life-cycle analysis
LCC *see* life-cycle costs
LDF   63
LDPE *see* low-density polyethylene
leaching   349
lead   625–626
leakage
    airtightness   250
    nutrients   503, 505–506
    phosphorus   465
    radon   179–180
    sealing   621–622
    taps   322–323
    water   155–156, 340
leasing agreements   370
LED *see* light emitting diodes
lee   539–540
LEED *see* Green Building Rating System
LEGEP   207
Legionnaire's Disease   155, 327
leisure travel   576
life-cycle analysis (LCA)   5, 11–12, 102–103, 368
life-cycle costs (LCC)   212
LifeLine™   70
lifts   186, 633
light
    beauty   660–661
    fillers   73–74

quality 112
shelves 309–310
light emitting diodes
    (LED) 299–302, 305
lighting 153, 296–305, 366
lightweight structures 33–35,
    41, 78, 101–102, 222
lime 19–20, 65–66, 79
linear houses 371
lining pipes 630
linoleum 69–71, 192, 222
linseed oil paint 81
liquid screed 73
lithium ion batteries 420–421,
    565
living areas 646
load control 116
local level
    agriculture 503
    conditions 516–518
    electricity production
        418–419
    food production 508
    global perspective contrast
        547
    heating 378–379, 388–389,
        423
    runoff 525–527
location adaptation 512–514
lofts 176,–177, 493, 622
Lomma harbour 580
long-distance traffic 570–574
loose fill cellular glass 38
lorries 570
low-density polyethylene
    (LDPE) 61
low-energy buildings 238–245,
    277
low-energy houses 234
low-energy light bulbs 153, 298
lumber impregnation 77
lyes 81–83, 385

magnetic fields 144–154
magnetic refrigerators 293–294
main circuit breakers 148–150
maintenance 189, 212, 358,
    615–617, 649, 658
Malmö, Sweden 585
manholes 527
manor houses 609
manual regulation 113
maps 512–513, 530
Marby Area, Norrköping 518
Marlec 434
Maslow's pyramid of human
    needs 644

masonry 33, 394, 397
mass balancing 520
Material Intensity per Sequence
    model (MIPS model) 12
materials
    architectural quality 664
    construction 2, 5–110, 180,
        215–224, 625–628
    ecological design 369–370
MDF 63
mechanical ventilation
    118–121, 128–129
mercury-free compact fluorescent
    light bulbs 304–305
membrane filtering 343–344
membranes, geo- 31
mercury 366, 626, 628
Mesa Verde, US 535
metals 11, 23–24, 32–33, 51,
    83, 153, 397
metering energy 228–229
methane 504
methanol 568
methyl ketoxime 96
microclimates 516, 543
microfibre cleaning cloths 191
micro-hydropower 429
micro-organisms 466, 475
Miljöklassad Byggnad 206
mineral-based glues and
    pastes 97
minerals 16–19, 397
mineral wool 42, 93, 222
MINERGIE 203–204
mini-hydropower stations 428
mini purification 457, 460,
    462–465, 472
MIPS model see Material Intensity
    per Sequence model
mixer taps 321–322, 633
moderately sized towns 580
modified silicone (MS) polymer-
    based compounds 94–95,
    99
moisture
    climate adaptation 542
    construction 104, 216–217
    environmental rebuilding
        631
    insulation 250–251
    intake air 132
    properties 104–107
    ventilation 117
    wood 105, 176
    see also damp; humidity;
        relative humidity
Mölndal ventilation system 136

monastery gardens 500
Mongolian yurts 666
monitoring 178, 217, 616
monocrystalline solar cells 436,
    437
monoculture 505–506
mopping floors 192
moraine 520
Morsö heater 395
mortar 22, 65–67
motion detectors 303
motorized equipment 613
mould 174
mouldings 16
movable solar panel banks 446
MS see modified silicone
    polymer-based sealant
mud architecture 668
    see also earth
muffling sound 186
mull filters 470
multi-family dwellings 242, 498
    see also flats
multiple contractors 208
multipurpose products 370
multrums 454–455
municipal level
    district heating 379,
        380–381
    environmental planning 200
    passive houses 245
    purified water 331–332
    sewage 460
    sustainability 547–552,
        590–594, 596
    see also district level

nanofilters 341–342
nanomaterials 26–27
NAPS Magic Lantern 440
Närke houses 605
national monuments 609
natural aspects
    adaptation 512, 515–545
    circulation 167
    interior fibres 91–92
    landscapes 487–488
    paints 82
    rubber 60, 100
    sewage purification
        466–470
    succession 480
    ventilation 118–119,
        121–128, 133–134,
        141–143, 632–633
nature conservation 531–532
Nature Plus 8, 12

Navestad, Norrköping   639
needs, human   592–593, 644–647
neoprene   99
The Netherlands   560, 582, 584
network towns   551
new buildings   178–179
new old-style houses   604–610
new urbanism   577–580
nickel   23
nickel metal hydride batteries   420–421
night-soil composters   473
nitrogen
   agriculture   503, 505–506, 524
   composting   358
   leakage   340
   sewage   448, 451–452, 470
no-dig methods   159–161
noise   182–186, 633
nonylphenol ethoxylate   93
Nora®   69
Norway   289
notched boards   55
Nubian vaults   668
nuclear power stations   416
nutrients
   leakage   503, 505–506
   recovery   448
   recycling   372, 471–478

ocean thermal energy conversion (OTEC)   373, 418
offices
   cleaning   188
   comfort   113
   cooling   282–283, 411
   electrical devices   306
   electromagnetic fields   144
   lighting   298–299
   plus energy   247
   ventilation   126–128
   waste sorting   357
   windows   255
oils   75, 82–83, 86, 95–96, 383, 385, 527, 628
Öko Test label   8
OLED *see* organic light emitting diodes
OLF   118
Olgyay's bioclimate diagram   534
on-demand hot water   329–330
open air solar heating systems   403
open floor plans   272–273

operational level
   construction materials   7
   costs   211
   electricity   307–308
   existing buildings   615–617
   heating   164
   heat pumps   266
   instructions   209–210
organic farming   506
organic food   503
organic light emitting diodes (OLED)   305
organic materials
   biogas   475–476
   description   14–15
   ecocycles   600
   ground cover   30
   insulation   56
   pigments   83–84
   recirculating   372
   surface treatments   79
organic waste   478
oriented strand boards (OSB)   35, 63
OSB *see* oriented strand boards
osmotic energy   418
Ostia, Italy   599–600
OTEC *see* ocean thermal energy conversion
outdoor care and maintenance   172
outdoor construction materials   3
out-of-town shopping centres   577
outside doors   632
ozone   343, 405–406

packaging   362–364
paints   74–89, 145, 217–218, 627
Palma de Mallorca   601–602
panelling   15, 78
paper   14, 57, 59, 67–68, 78
paradise   664
Parans Solar Panels   310
parish councils   658–659
parks   650
parquet flooring   72
participation   655–659
partnerships   209
passive systems
   cooling   122, 278–283
   heat   272–278, 378, 536
   houses   239–245, 249, 277
   solar houses   667
pasteboard   65
pastes   97–100
   see also adhesives; glues

patios   651
pavements   525
paving materials   29–30
PCBs   626, 628
PE *see* polyethylene
peak loads   405
peat   43, 383–385
pedestrianized zones   557, 653
PEFC *see* Programme for the Endorsement of Forest Certification Schemes
pellets   238, 385–387, 389–391, 395
Pelton turbines   330
penetrability   335
pentagons   669
people   512, 643–670
pergolas   491–492
periscopes   312
perlators   322
perlite   42–43
permaculture   480–483
permanent fixtures   91
Personal Rapid Transit   562
PET *see* polyester
petrol   568
PF glue *see* phenol-formaldehyde glue
phenol-formaldehyde glue (PF glue)   98–99
phenol-resorcinol-formaldehyde (PRF) glue   99
PHEV *see* plug-in hybrid electric vehicles
phosphorus   451–452, 465, 467, 506
photocopiers   306
photography, thermal   253
photosynthesis   443
photovoltaics   247, 373, 436–446
   see also solar cells
physical needs   644
physical planning   511, 519
pigments   77, 83–85, 88
pigs   358
pillar systems   102
pipes
   care and maintenance   173
   construction   89–90
   environmental rebuilding   630–631
   hot water   327–328
   laying   522
   light   304
   plumbing   158–161
   radon   181

reuse and recycling 222
sound 185
transport 574
water leakage 155
place 511–670
planning
   demolition 219–220
   development 516, 540–541, 595
   environmental 197, 634
   floor 123, 272–273, 665
   gardens 496
   human needs 644–645
   hydrology 528
   implementation 198–210
   lighting 296
   maintenance 189
   participation 655–658
   permaculture 481
   physical 511, 519
   security 649
   sustainable municipalities 551–552, 594–595, 598–599
   town 589–602
   traffic 556
   vegetation 484–485
   waste 217
plant-based glues and pastes 100
plants 489, 496, 668
   see also trees; vegetation
plaster 22, 65–67, 78, 87, 176
plasterboard 56
plastic
   adhesive compounds 74
   construction materials 24–26
   flooring 71, 192
   paints 84
   reuse and recycling 221–222
   solar cells 444
   vapour-tight 176–177
   weatherproofing 56–58, 60–61
plastomer-asphalt board 61
plate heat exchanger hot water heaters 330
plinth foundations 108
plug-in hybrid electric vehicles (PHEV) 566
plugs 152
plumbing 155–161
   see also pipes
plus energy houses 246–247
plywood 62–63
pollution
   interior 117–118, 495
   land restoration 517

sewage 449–451
traffic water 527
polycrystalline solar cells 436–437
polyester (PET) 25
polyethylene (PE) 25, 31, 57, 60–61
polymer-based compounds 61, 94–95
polyolefin 60–61
polypropylene (PP) 60
polystyrene (PS) 25, 37–38
polyurethane (PUR) 25, 95–96, 99, 626
polyvinyl acetate glue (PVAc glue) 99
polyvinyl chloride (PVC) 25, 27, 628
ponds 466, 501, 543
porches 252, 624
Portland cement 19, 73
post mills 16
potassium 451–452
potato flour paste 100
potential equalization 150
poverty 652
power lines 145–147
power stations 416, 440, 443
PP *see* polypropylene
practical democracy 655–659
prams and buggies 190
precipitation 459, 525
prefabricated cassettes 630–631
prefabricated houses 238
preheated intake air method 129, 131, 134
preservatives 76–77
pressure
   insulation 250–251
   radon 180
   tanks 336
   tests 252–253
   ventilation 122
pressurized lines 522
pretreatments 461–462
PRF *see* phenol-resorcinol-formaldehyde glue
prices 211, 438, 619
primers 76
PRIO *see* Priority Setting Guide
Priority Setting Guide (PRIO) 10
prism glass 258–259
private 658–659
   electricity generation 418–419
   maintenance and management 658

water 335–336
water supply 342
zones 649–651
processing food 508
procurement 198–210
product information 7–8
production
   construction materials 6
   consumption proximity 589–590
   electricity 286, 417–420, 423–426
   textile fibres 92–93
   waste responsibility 350–351
Programme for the Endorsement of Forest Certification Schemes (PEFC) 510
project management 201
propeller turbines 430
property management 195
protected species 532
PS *see* polystyrene
psychological needs 644
public
   consultation 657–658
   spaces 653–654
   transport 554, 556–557, 560–563
   zones 649–650
pulp 14
pumps 118–119, 265–266, 333, 335, 337–339, 404–410, 413
PUR *see* polyurethane
purification
   interior vegetation 495
   land restoration 517
   mini 457, 460, 462–465, 472
   sewage 448, 452–454, 462–464, 466–470
   small–scale 461–463
   water 331–332, 341–344
PVAc glue *see* polyvinyl acetate glue
PVC *see* polyvinyl chloride

quadrature of the circle 670
quality
   air 112–113, 486
   architecture 664–665
   certified management 615
   construction site 215
   energy 227–228, 285
   light 297, 300
   management 194
   water 331–332, 340
quiet buildings 183, 185

radiant heat   162, 167, 169
radiation   281
radiators   115, 135, 137, 162, 166–167, 222
radioactivity   340
radon   178–181, 336, 343
railways   570
rain   175–176, 599
rammed earth houses   21
Rankine cycle   423
rapeseed methyl ester (RME)   567
rapid flow-slow flow   165–166
REACH *see* Registration, Evaluation, Authorisation and Restriction of Chemicals
rebuilding   629–641
reconditioning   369–370
reconstruction   180
recycling   109, 219–224, 347, 352–353, 355–356, 362–367
Redlist of Swedish Species   532
reed mats   65
reeds   53, 63
reflectors, sunlight   253, 261–262
refrigerants   405–406
refrigerators   289–290, 293–294
refuse rooms   173
regional agriculture   503–504
Registration, Evaluation, Authorisation and Restriction of Chemicals (REACH)   7, 9–10
regulation   113–116, 150, 629
   see also control systems
regulators   428–429
reinforced grass   29
relative humidity   112
   see also humidity; moisture
relics   609
renewable resources   227, 372, 376–446
see also organic materials
rental cars   565
repair   368, 370
residential areas   355, 484–485, 577
residual products   6
resorcinol   99
resorption   469
resource conservation   225–370
reusing materials   220–224, 369
   see also recycling
reverse osmosis   341, 343–344

rhombic grid town plans   539
rhythm   669–670
RME *see* rapeseed methyl ester
roads   335, 522, 609
rocks   17, 520
rock wool   42
roller blinds   263
roofing
   care and maintenance   172
   construction   48–56
   dynamic   268
   energy   234
   environmental rebuilding   633
   green   49–51, 58–61, 281, 492–494
   lightweight construction   102
   moisture   104
   passive solar heat   275
   ridge ventilators   137
   runoff   89–90
   rust protection   85
   solar   398, 445
   timber   15
   vegetation   490–492
   wooden   34
rooms   117–118, 665
root cellars   316
rope   95
rose figures   3–4
rot   174–175
rotating caps   137, 142
R-symbol   8
rubber   25–26, 31, 59–60, 69, 100
rubbish disposal grinders   458
Rumpans holiday village   656
runoff   89–90, 525–527
rural areas   356–357, 589–602, 609
rust protection   84–85
rye flour paste   100

SABO *see* Swedish Association of Municipal Housing Companies
safety glass   46
sailing ships   572
salsabil   280
salt water   340, 344
sand   520
sand fillers   73–74
sandlime bricks   33
sand ovens   393
sanitary systems   89, 222, 448
   see also sewage; toilets

saunas   16, 659
Savonius rotor   140
sawdust   385
SBR glue *see* styrene butadiene glue
SBS *see* sick building syndrome
SBTool *see* Sustainable Building Tool
scent   662
schools
   black-water   474
   comfort   113–114
   daylight   312
   energy efficiency   620
   natural ventilation   123–128, 142
   passive cooling   282
screed material   73
screening, solar   274
screw fittings   156
sculptures   446, 663
SEA *see* strategic environmental assessment
sea level   543
sealing compounds   93–97, 621–622, 626–627
seams   250, 252
seasonally adapted buildings   536
seasonal storing   399
seawater heat pumps   407
seaweed   52
second generation biofuels   567–568
second-hand centres   356
secular buildings   609
security   644, 649
sedum roofing   49–50, 493–494
selective dismantling   220
self-building   213
self-cleaning glass   260–261
self-installed wind power   433
self-sufficient small farms   507–508
semi-private/public zones   649–650
sense perceptions   660–662
sensors   115–116, 303–304
services   111–170, 185–186, 189–190
sewage
   ecological cycles   372, 447–478
   environmental business ratios   613
   environmental rebuilding   630–631

management   478
pipes   89, 522
planning   596
pollution   449–451
sustainable municipalities   594
waste-water heat exchangers   267
see also toilets
shading   279, 538, 660–661
shafts, service   158
Shakers   666
share clubs, cars   563–564
shared electricity use   307–308
shared spaces   658–659
sheep's wool   39, 93
sheet materials   61–65, 85, 104
shelterbelts   540
shielded electrical systems   151–152
shielded light fittings   153
shingles, roof   54–55
showers   106, 138, 157, 323–324
sick building syndrome (SBS)   3–4
sight   660
silence   662
silica aerogel   259
silica gel   105
silica plaster   67
silicate paints   84–85
silicon technology   95–97, 436–437
single-level taps   321–322
single-stage current limiters   287
sinks   221
sisal joints   93
site level construction   215–218
skifferit   52–53
skirting board heating   168, 170
slab foundations   107–108, 521
slate roofing   52–53
sling pumps   338–339
slope
  roofs   48–50, 52–54, 492
  runoff   525, 527
sludge
  ecological cycles   448, 449–450
  nutrient recycling   471–473, 475–476, 478
  purification   461, 469–470
small-scale aspects
  combined heat and power   425–426
  hydropower   427–429

permaculture   482
purification   461–463
self-sufficient farms   507–508
sewage separation   454–459
wind power   432–433
small towns   578
smart systems   114
smell   662
smoke guns   618, 620
Snorkel ventilation   134
snow   175, 529
snow-storage systems   411–412
soap   85–86
social level
  Agenda 21   652
  human needs   645
  structure   546–602
  sustainability   373, 512, 549
sockets   151–153
sodium silicate   97
sofas   369
soft distemper   81
soft parameters   209
soils   175–176
solar
  cells   436–446
  chimneys   137, 139–140
  collectors   310, 398–403, 408
  cooling   413
  energy   373, 375, 377–382
  heating   235, 330, 398–403
  hydrogen gas   421
  passive heat   273–277
  pumps   337–338
  tubes   311
  walls   131, 134
  water heaters   246
  see also photovoltaics
Solar water   318–319
solid wood elements   35–36
Solvatten   318–319
solvents   77, 82, 88, 98, 627
sorting
  sewage   452
  waste   217–219, 352–357
sound   112, 182–186, 254, 662
soya paste   100
spark-arresters   397
spatial figures   665
spirit stoves   426
spirituality   645
split-vision heat exchangers   267
sports halls   268
stack-induced ventilation   125
stainless steel   51

stairways   186, 299, 526–527, 573
standby power   306–307
starch-based compostable packaging   364
steam
  cleaners   192
  turbine cycle   423–424
  washing   294, 326
steel   23, 51
Stirling engines   410, 419, 425–426, 440
Stockholm, Denmark   523, 586
Stockholm ventilation   123
stone   29, 46, 71, 78, 192, 222
stop buttons, shower   323–324
storage
  cleaning equipment   190
  electricity   420–422
  food   314–316
  heat   101–102, 273, 399
  hot water   329–330
  tanks   162–164
  urine   473
stoves   377, 391–396, 397
strategic environmental assessment (SEA)   516–517
straw   14–15, 41–42, 53, 62
stray current   154
streetcars   549
street layout   587–588
street lighting   299
structural elements   31–36, 186, 484–489
studs   34, 102
styrene   25, 39
styrene butadiene glue (SBR glue)   99
submersible pumps   333
subsidies   214
sulphite lye glue   97
sunlight   261–262, 538, 660–661
sun protection glass   46
sunscreening   255, 261, 274, 276, 279, 445
Sunstrip absorbers   401
super-insulation   277
super-turbines   139
supplementary insulation   622
supply
  electricity   2, 146, 148
  energy   373
  water   331–340
surface finishes   74–89
surface temperatures   79, 543–544

suspended foundations 90, 107–108, 521
sustainability 373, 512, 547–552, 590–596
Sustainable Building Tool (SBTool) 203
Swan label 8
Sweden
   appliances 326
   biogas 477–478
   biomass 384, 386
   climate adaptation 536
   cooling systems 411–412
   cultural history 608–609
   district heating 380–381
   eco-communities 657
   ecological city districts 585
   electricity 285–287, 417–419
   endangered species 532
   energy 233, 373–374
   environmental classification 206
   family 647
   forests 510
   garden cities 587
   geology and topography 519, 523
   heat pumps 408–409
   hydroelectric power 427–428
   insulation 248
   low-energy buildings 238–245, 277
   natural ventilation 123–128, 141–143
   noise 182–183
   radon 178
   rebuilding 635–636, 640–641
   recycling 362–363
   sewage 453, 457, 468
   solar energy 377, 398, 400–401, 437–439
   technology cooperation 593
   traditional buildings 603–607
   travel 553, 575
   waste 346–349, 352, 360, 382
   water 319–320
   wave power 435
   wind power 422, 431–432, 434
Swedish Association of Municipal Housing Companies (SABO) 613–632

swimming pools 268, 398, 402
synthetic materials 24–27, 31, 59–60, 83, 92, 97–98
   see also plastic

Tainan, Taiwan 597
tall oil 383, 385
taps 321–322, 633
tars 86
taxes 211
tempera paint 75
temperature
   building design 269–270
   climate adaptation 543–544
   comfort 113–114
   control 115
   heat pumps 405–406
   hot water 327
   mixer taps 321
   sewage 471
temples 662
tendering 209–210
terazzo flooring 68–69, 192
termite stacks 668–669
terraced houses 239–240, 249
terracotta tiles 71
territoriality 645–646
textile fibre mates 43
textiles 31, 59, 61, 71–72, 91–93
thatched roofs 53
thermal aspects
   bridges 249–251
   comfort 112–113
   decontamination 517
   insulation 37–45
   mass 409
   photography 253
   ventilation 122
thermoaccumulators 414–415
thermochrome glass 258
thermoelectric generators 426
thermoplastics 24–25
thermostatic mixers 164–165, 323
Think City electric cars 565
thinners 82
three-storey height restrictions 587
tiles 54, 69, 71, 221, 392–393
timber
   construction 103
   decontamination 627–628
   ground cover 30
   moisture 216–217
   reuse and recycling 223
   sound insulation 184

surface treatments 78
   uses 14–16
   see also wood
titanium oxide 83
TN-S system *see* five-wire systems
TOD *see* transit-oriented development
toilets 190–191, 324–326, 454–460, 467
topography 516, 519–523, 601–602
Tornedal houses 606
touch 662
towns
   character 607–608
   green 550–551
   holistic 575–588
   life 653–654
   size 578, 580
   planning 539, 589–602
   small 578
traditional buildings 534, 603–607, 665–666
traffic 527, 553–574, 579, 594–595
   see also transport
trains 571
train-trams 555
trams 561
transformers 146, 148
transit-oriented development (TOD) 577
transparent insulation 43, 259–260
transparent solar panel banks 445
transport
   environmental business ratios 613
   public 554, 556–557, 560–563
   solar villages 601
   sustainable municipalities 591–593
   see also traffic
trees 487–489, 530
   see also vegetation
trellises 491–492
trenchless technology 159–161
triple glazing 236, 239, 254–255, 257, 277
Tubingen, Germany 583
tuning combustion systems 621
turbines 139, 329–330, 429–430
turnkey contracts 208–209
turnover rates 648–651

TWIN Model   13
two+two-glazed windows
  255–257
two-part fillers   74
tyres   367

UDT *see* unattended delivery
  units
UF *see* urea-formaldehyde
ultraviolet (UV) light   341–342
unattended delivery units
  (UDT)   574
underfloor heating   167
underground houses   668
underground pipes   133–135
Understenshöjden, Stockholm
  523
unheated greenhouses   494–495
upholstery   91
urban areas   484–489, 514,
  544–545, 550–551
    see also cities; district level;
      municipal level; towns
urbanism, new   577–580
urban–rural concept   589–590
urea-formaldehyde (UF)   96–97,
  99–100
urinals   455–456
urine
  management   473
  separation   324, 451–452,
    454–457, 460, 467
    see also sewage; toilets
user instructions   158
utility rooms   291
UV *see* ultraviolet light

vacuum solar collectors
  401–402
vacuum-fused silica insulation
  43–44
vacuum toilets   455–456, 458
vapour-tight plastic   176–177
varnish   80–81
Vauban area, Freiburg, Germany
  581
Vectrix Scooter   565
vegetation   372, 466, 479–510,
  540, 599, 601
    see also flora and fauna
Velco brand   132
venetian blinds   263
ventilation
  energy   234, 620–621
  environmental rebuilding
    632–633
  heat pumps   266

larders   314
moisture   216
passive cooling   279–280
quiet   185
radon   179–181
service systems   117–143
supply   2
Venturi effect   136
Venus   670
verandas   624
vermiculite   45
vertical gardens   490–491
vertical wind generators   434
village configurations   604–607
virtual water   319–320
viscose   92
Volvo's hybrid concept car   566
Vyredox   344

wallcoverings   67–68, 100
walled gardens   500
walls
  air open   268
  building shape   270
  energy   234–235
  green   281
  heating   162–163, 168, 170
  insulation   248, 251, 293
  lightweight construction   102
  moisture movement   104
  passive heat   272–273
  plumbing   157
  solar   131, 134
  sound   183–184
  vegetation   490–492
  wetrooms   157
Wasa stove   393
washing machines   289–291,
  293, 326
waste
  biogas   478
  conservation   226, 230,
    345–370
  construction   6, 217–219
  decontamination   625–628
  energy   235
  environmental business
    ratios   613
  grinders   361
  heat   381–382
  heat exchangers   267
  local food production   508
  planning   594
  rebuilding   632
  sustainable municipalities
    593–594
  water   476, 478

water
  basic elements   663
  care and maintenance   173
  clean   317–344
  cleaning   190
  conservation   226, 230
  energy   235–236
  environmental business
    ratios   613
  free cooling   411
  heat exchangers   267
  hydrology   516, 524–529
  leakage   155–156, 340
  passive cooling   280
  pipes   90
  planning   594–595
  radon   180–181
  rebuilding   630–631
  solar heaters   246
  supply   2
  sustainable municipalities
    591–592
  waste   476, 478
    see also taps
waterborne heat   162–169
water-damage-proof construction
  155–161
water-jacketed stoves   394, 396
watermills   609
water-proof barriers   56–61
waterproof layers   250
water-tables   524
water-vapour resistance   56
wattle-and-daub method   22
wave power   418, 435
waxes   87
weatherproofing   56–61, 176,
  249–253
weather rose for a sustainable
  municipality   548–552
weather stripping   97, 252
weeping willows   488
well-being   1–2
wells   332–336
wetlands   466–468, 470,
  528–529, 592
wet reactors   474–475
wetrooms   16, 58, 79, 105–107,
  157
    see also bathrooms
wet urine-separating systems   460
whiteners   81–82
willows   488
wind
  climate adaptation   539–541
  energy   373–374, 422
  environmental rebuilding   635

688   INDEX

power   418, 420, 422, 432–435
pumps   337
snow   529
vanes   136
ventilation   122
windmills   16, 609
windows
  care and maintenance   172
  cleaning   189, 191
  energy   234, 254–263, 621–624
  environmental rebuilding   632
  flashing   104
  light   661
  passive houses   244
  regulation   115–116, 129, 143
  reuse and recycling   221
  shading   279
  sound   184
  surface treatments   79
  weatherproofing   252
windproofing   176, 250, 252
  see also weatherproofing
wired glass   46
wiring   151–153
wood
  boilers   389–390
  burning appliances   166, 169–170, 391, 394–395, 397
  chips   30, 385–386
  circuit breakers   149
  cladding   45, 47–48
  construction   103
  fibreboard   64–65
  fibreboards   102
  flooring   69, 72, 192
  fuel   383, 386–387
  ground cover   30
  impregnation   76–77, 627–628
  moisture   105, 176
  organic materials   14
  passive houses   244
  roofing   54–55
  structural materials   34–35
  surface treatments   78–79, 82–83, 86–87
  thermal insulation   40, 44
  weatherproofing   59
  see also timber
wooden towns/buildings   610
wood–wool cement   36, 44–45, 48, 65
wool   39, 42, 57, 91
working environments   3, 645
worktops   16
woven wallcoverings   67–68

xylene   25, 39

yields   480
yurts   666

zeolites   293
Zeppelins   574
zinc   23, 51–52, 54, 223
zinc-air batteries   420–421
zoning   279, 481